Introduction to the Calculus of Variations

Introduction to the Calculus of Variations

Hans Sagan

DOVER PUBLICATIONS, INC., New York

Published in Canada by General Publishing Company, Ltd., 30 Lesmill Road, Don Mills, Toronto, Ontario.
Published in the United Kingdom by Constable and Company, Ltd., 3 The Lanchesters, 162–164 Fulham Palace Road, London W6 9ER.

This Dover edition, first published in 1992, is an unabridged, corrected republication of the work first published by the McGraw-Hill Book Company, New York, 1969, in its "International Series in Pure and Applied Mathematics."

Manufactured in the United States of America
Dover Publications, Inc., 31 East 2nd Street, Mineola, N.Y. 11501

Library of Congress Cataloging-in-Publication Data

Sagan, Hans.
Introduction to the calculus of variations / Hans Sagan.
p. cm.
Originally published: New York : McGraw-Hill, 1969, in series: International series in pure and applied mathematics.
Includes bibliographical references and index.
ISBN 0-486-67366-9 (pbk.)
1. Calculus of variations. I. Title.
QA315.S23 1992
515′.64—dc20 92-2356
CIP

PREFACE

The calculus of variations has attracted the attention of numerous eminent mathematicians, who have made significant and lasting contributions to its development since its inception 273 years ago. Early this century, when the problem of Bolza in its overwhelming generality was formulated and analyzed, most of the problems that were considered to lie within the province of the calculus of variations appeared to be solved, or at least solved in principle, and interest in the subject began to decline. Only in the last decade or so has a new upsurge of interest become noticeable. This is partly due to the novel demands of a new and sophisticated technology, in which space technology occupies a prominent position. Another factor in this revival of the calculus of variations is the availability of high-speed computing devices, which, together with newly developed mathematical algorithms, make it possible to solve many variational problems which heretofore could only be solved in principle.

The modern theory of optimal control, which has been developed in recent years and is still in a process of growth—in depth as well as in scope—has its roots in the calculus of variations, although many of its techniques are new and more easily adaptable to practical technological requirements than are the methods of the classical calculus of variations. It is the purpose of this text to lay a broad foundation for an understanding of the problems of the calculus of variations and of its many methods and techniques, with their many-faceted subtleties, and to prepare the reader for the study of modern optimal control theory. The treatment is limited to a thorough discussion of single-integral problems in one or more unknown functions, where the integral is employed in the riemannian sense. The functions that are admitted are at worst sectionally smooth. Such a setting seems general enough for most practical purposes and, at the same time, keeps the formulation and the proofs on a simple enough level. The problem in two or more independent variables is given peripheral treatment in the appendices to Chaps. 2 and 3 (Secs. A2.17 and A3.14).

The first three chapters deal with variational problems without constraints. The theory of the first variation and the theory of fields are developed as far as seems practical and to a point beyond which the arguments would become tiresome and repetitious.

Chapter 4 is a self-contained treatment of the homogeneous problem in the two-dimensional plane. The remaining material, with the exception of the appendix to Chap. 7, is entirely independent of the results developed in Chap. 4.

In Chap. 5, the minimum principle of Pontryagin as it applies to optimal control problems of nonpredetermined duration, where the state variables satisfy an autonomous system of first-order equations, is developed to the extent possible by classical means within the general framework of the Hamilton-Jacobi theory. A proof of Pontryagin's principle in a more general setting has been omitted. However, because of the importance of this principle, the appendix to Chap. 5 is devoted to a discussion of its application to time-optimal control problems and to control problems of fixed duration that are associated with a nonautonomous system of differential equations.

Chapter 6 is devoted to a derivation of the multiplier rule for the problem of Mayer with fixed and variable endpoints and its application to the problem of Lagrange and the isoperimetric problem.

In the last chapter, Legendre's necessary condition for a weak relative minimum and a sufficient condition for a weak relative minimum, which have already been established in Chap. 3, are derived within the framework of the theory of the second variation. A new result that is derived in this chapter is Jacobi's necessary condition for a weak relative minimum.

I have resisted the temptation to include the customary chapter on direct methods because to do this subject justice, one would have to devote an entire book to it. Not only is the method of attack quite distinct from the methods employed in this book, but the variational problems and the concept of the minimum itself change their character substantially owing to a different concept of the space of competing functions.

This book contains essentially the material I have developed during the past ten years in courses taught at the University of Idaho and North Carolina State University and in a series of lectures which I have delivered over the past five years to engineers and scientists at the Astromechanics Branch of the Space Mechanics Division of the National Aeronautics and Space Administration at Langley Field, Virginia.

I have used elementary functional analytical methods in the discussion of the foundations of the theory of the first and the second variation, not to appear modern at any price, but principally because the functional analytical setting provides a much better and deeper understanding of the fundamental concept of the space of admissible variations and the concepts of a weak and a strong relative minimum of an integral. I have tried to avoid repetitive arguments whenever feasible by using different approaches to the same concept at various stages of generality. Thus Bellman's fundamental partial differential equation is derived first by a method of Carathéodory and then, in a more general setting, by Bellman's method of needle-shaped variations. The theory of fields and the Weierstrass excess function are introduced on three separate occasions by three different methods. Some of the interrelations of the various approaches to the solution of variational problems are spelled out in detail, some are merely hinted at, and some are left for the reader to discover.

This book was written primarily for the student. To achieve greater clarity, care has been taken to spell out difficult steps in some detail, often at the expense of mathematical elegance. Every theorem is either preceded immediately by a derivation or succeeded by a proof, with one exception. The exception is the fundamental theorem on underdetermined systems on which the multiplier rule for the Mayer problem is based. To give the reader a chance to familiarize himself with the notation and with the content of this theorem, the proof is postponed until after the theorem has been applied to a number of special cases.

To illustrate a complicated situation with a complicated problem compounds the difficulties. Therefore, the examples discussed in the main body of the text (which are set aside from the main text by the use of smaller print) have been kept at a simple and technically uncomplicated level, uncluttered by extraneous matters and stripped of all unnecessary constants. There are 421 exercises interspersed throughout the text. Some of these exercises are intended to develop mastery of the formalism, some (which are marked by an asterisk) complement the text, and still others lead beyond the material presented in the text.

In order to read this book with profit or to take a course based on this book with a reasonable chance for success, the student should be able to use the implicit-function theorem in a variety of situations, he should be familiar with the Heine-Borel theorem, and he should know the standard existence and uniqueness theorems from the theory of ordinary differential equations. To put this in more conventional terms, the student should have had a course in advanced calculus and an intermediate course in ordinary differential equations. Some knowledge of the elements of matrix algebra would be helpful in Chaps. 6 and 7.

The material is arranged to give the user the greatest possible amount of flexibility. The material contained in the main portions of the chapters (excluding appendices) can be taught comfortably in a one-semester course, possibly by omitting Chaps. 4 and/or 7. The entire book (including appendices) can be taught in a two-semester course. Other combinations for a one-semester course are: Chaps. 1 (without Secs. 1.8 and 1.9) 2, 3, 5; 1, 2, 3, 4, 7; 1 (without Secs. 1.8 and 1.9) 2, 3, 6; or 1, 2, 3, 7 (without the Appendix). Other arrangements are possible if one notes that Chap. 4 is a prerequisite only for the appendix to Chap. 7 and that Chap. 6 (without Sec. 6.10) does not depend on Chaps. 3 and 5.

Hans Sagan

ACKNOWLEDGMENTS

In the preparation of this text I had help in many forms and from many quarters. It goes without saying that I am greatly indebted to the old masters—Oscar Bolza, Constantin Carathéodory, and Gilbert A. Bliss—who developed the subject extensively, and for their respective viewpoints in their well-known classical books. I am in an even greater measure indebted to the late Johann Radon, through whom I first became acquainted with the subject, and to Paul Funk, who, in his very own and unique style, imparted information and his interesting views on the subject.

I am grateful to Drs. Carl E. Langenhop and Nicholas J. Rose for their encouragement and constructive criticism at a time when the manuscript was in an intermediate state of development and to my colleagues Drs. Raimond A. Struble, David F. Ullrich, and Robert Silber, who helped me clarify my position in many private talks and discussions.

I thank my students Lawrence M. Hanafy and Stephen K. Park and my friend Dr. Karl Prachar for helping me with the burdensome task of proofreading.

Hans Sagan

CONTENTS

Introduction to the Calculus of Variations

CHAPTER 1

EXTREME VALUES OF FUNCTIONALS

APPENDIX

1.1 INTRODUCTION

The *calculus of variations* is a mathematical discipline that may best be described as a general theory of extreme values. The name of this discipline does not derive from the type of problems it is concerned with but rather from a specific technique—the technique of variation, which will be discussed in Chaps. 1 and 2—that is employed to obtain certain necessary conditions for the existence of extreme values.

To give the reader some indication of what "variational problems" are all about, we shall list and discuss a variety of such problems and then extract their essence to arrive at a fairly general formulation, which we shall then subject to mathematical analysis.

A. THE PROBLEM OF THE BRACHISTOCHRONE (βράχίστος = shortest, χρόνος = time)

There is hardly a book written on the subject of the calculus of variations that does not use this problem as a takeoff point, and we shall make no exception. The problem deserves merit not only because it was historically the first variational problem to be formulated mathematically but also because it serves well to illustrate the scope of applicability of this discipline to problems of an applied nature.

Let $P_a(a,0)$ and $P_b(b,y_b)$ denote two points in a vertical plane, with $a \neq b$, $y_b > 0$ (see Fig. 1.1). The problem is to find among all curves with a continuous derivative that join the point P_a to the point P_b the one along which a mass point under the influence of gravity will slide from P_a to P_b without friction in the shortest possible time.

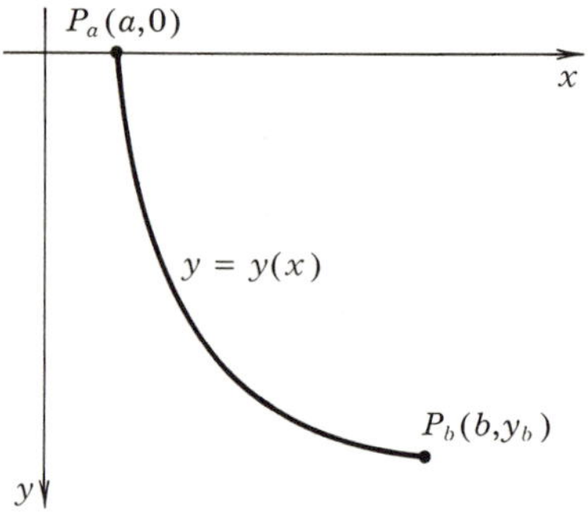

Figure 1.1

Before making an attempt to formulate this problem mathematically, let us emphasize a point that is of paramount importance: First, we consider all curves with a continuous derivative that join the given points P_a and P_b. Then we assign to each curve a number, namely, the amount of time it takes the mass point to slide from P_a to P_b. (All those curves along which the mass point will never get to P_b we either rule out from our consideration or else assign the "number" ∞.) Then, from among all these numbers (sliding times), we pick the smallest one—provided it exists—and call the curve that this smallest number is associated with the solution of our problem.

Now, let us move on to a mathematical formulation of the problem. Let $y = y(x)$ be a function with a continuous derivative that joins the point P_a to the point P_b:

$$y(a) = 0, \qquad y(b) = y_b.$$

If s represents the distance on $y = y(x)$ measured from P_a, we have

$$\frac{ds}{dt} = v,$$

where t represents the time and v the velocity of the mass point. Since the motion is, by assumption, frictionless with initial velocity 0 and is influenced only by the gravitational force, we have from the principle of the conservation of energy

$$\tfrac{1}{2}mv^2 = mgy,$$

where y is the vertical distance from the initial level. This yields

$$v = \sqrt{2gy}.$$

Hence

$$\frac{ds}{dt} = \sqrt{2gy}, \qquad \text{or} \qquad dt = \frac{ds}{\sqrt{2gy}} = \frac{\sqrt{1+y'^2}\,dx}{\sqrt{2gy}}.$$

So it would appear that the sliding time along $y = y(x)$ from P_a to P_b is given by

$$t = \int_a^b \frac{\sqrt{1 + y'^2(x)}\,dx}{\sqrt{2gy(x)}}.$$

This definite integral is to be rendered a minimum by a suitable choice of $y = y(x)$ within the class of functions with a continuous derivative which pass through the prescribed points.

As the solution of this problem, we obtain an arc of a *cycloid* (see Prob. 2.6.1 and Sec. 4.10). The problem was first proposed by Johannes Bernoulli (1667–1748) in 1696 and was solved by a method which, though ingenious, lacks mathematical rigor and applicability to more general problems of a similar type.†

B. THE PROBLEM OF MINIMAL SURFACES OF REVOLUTION

Given two points $P_a(a,y_a)$ and $P_b(b,y_b)$, $a \neq b$, in the plane. These two points are to be joined by a curve $y = y(x)$ with a continuous derivative in such a manner that the surface which is generated by rotation of this curve about the x axis has the smallest possible area (see Fig. 1.2).

Again, we realize that the formulation of this problem establishes a certain relationship between curves (functions) of a certain class and a subset of the set of real numbers, the smallest of which is to be found, provided that it exists.

The mathematical formulation of this problem poses no difficulties whatever. If S denotes the area of a surface generated by rotation of $y = y(x)$ about the x axis, where

$$y(a) = y_a, \qquad y(b) = y_b,$$

† A fairly detailed historical account of this problem can be found in P. Funk, "Variationsrechnung und ihre Anwendung in Physik und Technik," pp. 1ff and 614ff, Springer-Verlag OHG, Berlin, 1962.

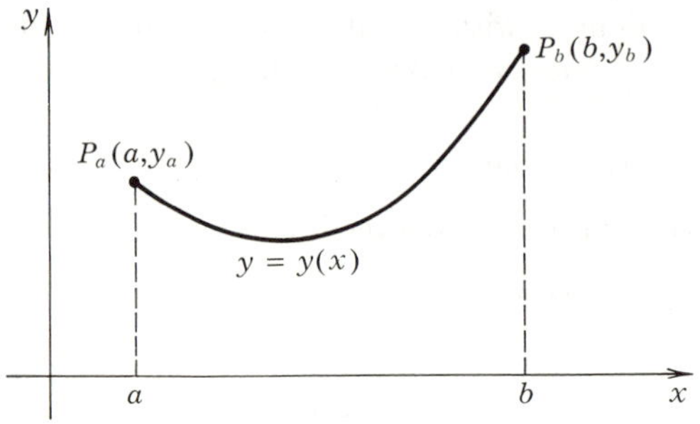

Figure 1.2

we have

$$S = 2\pi \int_a^b y(x)\sqrt{1 + y'^2(x)}\, dx,$$

which is to be rendered a minimum by an appropriate choice of $y = y(x)$.

The solution of this problem is—if the position of P_a relative to P_b satisfies certain additional conditions—a *catenoid*, that is, a surface of revolution that is generated by a *catenary* (see Sec. 2.6).†

An obvious generalization of this problem is the following: Given a closed *Jordan curve* in space. To be found is that surface which passes through the given curve and has the smallest possible area. This problem is known as the *problem of Plateau*.‡

The attentive reader cannot possibly fail to observe the striking similarity between problems A and B. In both cases, we look for a curve with a continuous derivative which satisfies the boundary conditions

$$y(a) = y_a, \qquad y(b) = y_b$$

and yields a minimum for an integral of the type

$$\int_a^b f(x,y(x),y'(x))\, dx,$$

where f is some given function of the three variables x, y, y'. Such a problem is often known as *the simplest problem of the calculus of variations*.

The problems which we shall discuss below will not fit into this narrow category, but they are nevertheless bona fide variational problems.

† For more details, see G. A. Bliss, "Calculus of Variations," The Open Court Publishing Company, La Salle, Ill., 1925.

‡ For a lucid discussion of this problem and many literature references, see R. Courant, "Dirichlet's Principle," pp. 95ff, Interscience Publishers, Inc., New York, 1950.

C. THE SIMPLEST ISOPERIMETRIC PROBLEM

This problem, too, is a classic. (Its formulation and solution—by pure intuition—are credited to Queen Dido of Carthage, about 850 B.C.) Among all curves with a continuous derivative that join two given points P_a and P_b and have the given length L, the one is to be found that encompasses the largest possible area, where the x axis and the vertical lines $x = a$ and $x = b$ serve as the supplementary boundary.

The mathematical formulation of this problem is fairly obvious: Among all curves $y = y(x)$ for which

$$y(a) = y_a, \qquad y(b) = y_b, \qquad \text{and} \qquad \int_a^b \sqrt{1 + y'^2(x)}\, dx = L,$$

find the one for which

$$\int_a^b y(x)\, dx$$

yields the largest possible value. A circular arc turns out to be the solution of this problem. (See Secs. 6.5 and 6.6.)

A more general formulation of this problem is the following: Among all possible simple closed curves of a given perimeter (hence the name "isoperimetric" problem), find the one that encompasses the largest possible area, or equivalently, among all simple closed curves that encompass an area of given magnitude, find the one of shortest perimeter. The circle is the solution in either case.

Problem C differs from problems A and B in that an additional condition in the form of an integral constraint is imposed on the class of competing functions. The nature of this additional constraint makes the problem less easily accessible to mathematical analysis than problems A and B.

D. A PROBLEM OF NAVIGATION

Given a river with parallel straight banks, b units apart. One of the banks coincides with the y axis. The stream velocity $\hat{v} = (u(x,y),v(x,y))$ at every point (x,y) is given by

$$u = 0, \qquad v = v(x).$$

A boat with constant speed $c\,(c^2 > v^2)$ in still water is to cross the river in the shortest possible time, using the point $(0,0)$ as point of departure. (See Fig. 1.3.)

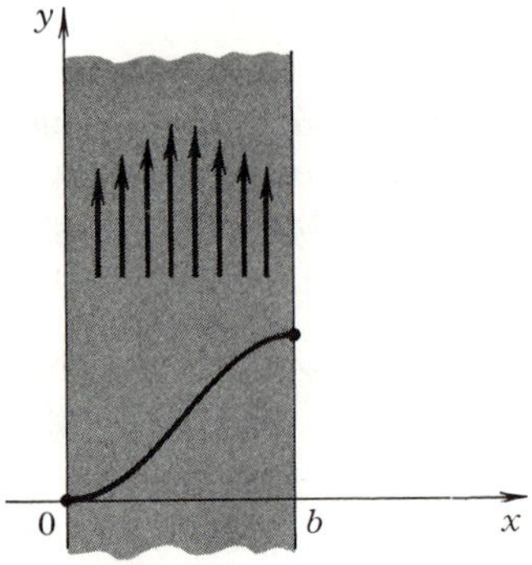

Figure 1.3

If α denotes an angle that depends on the course of the boat, then the actual velocity of the boat in the river is given by

$$\frac{dx}{dt} = c \cos \alpha, \qquad \frac{dy}{dt} = v(x) + c \sin \alpha.$$

For the path $y = y(x)$ on which the boat moves, we have

$$y' = \frac{dy/dt}{dx/dt} = \frac{v + c \sin \alpha}{c \cos \alpha}.$$

The time to cross is given by

$$t = \int_0^b \frac{dt}{dx}\, dx = \int_0^b \frac{dx}{c \cos \alpha}.$$

We have from the preceding equation

$$c \cos \alpha \cdot y' = v \pm c\sqrt{1 - \cos^2 \alpha}.$$

We solve this equation for $1/(c \cos \alpha)$ in terms of v, c, y' and obtain

$$t = \int_0^b \frac{\sqrt{c^2(1 + y'^2(x)) - v^2(x)} - v(x)y'(x)}{c^2 - v^2(x)}\, dx,$$

where $v = v(x)$ is a known function of x.

This integral, which is to be minimized, is of the same type as the integrals in the preceding problems; the class of competing functions, however, is less restrained than before. Only one boundary condition is imposed, namely, the one at the beginning point

$$y(0) = 0.$$

The endpoint is allowed to move freely along the other bank $x = b$. This is, of course, quite reasonable because different terminal points will, in general, yield different minimal crossing times—and we are, after all, interested in the shortest crossing time, no matter where we might land!

Because of the special structure of the given velocity field, the solution of the problem will be entirely independent of the choice of the beginning point. However, if we consider a more general velocity field, with u, v, as functions of y also, then different minimal crossing times can be obtained by varying the beginning point also. Further generalizations of this problem are easy to develop. Let the banks of the river be represented by some continuous curves (preferably with a continuous derivative) rather than by straight lines, assume a velocity field of the general type $u = u(x,y)$, $v = v(x,y)$, and leave the point of departure and point of arrival unspecified. Then you will have a fairly general problem that falls into the category of "free-endpoint problems." (See Sec. 4.6.)

E. A SIMPLE OPTIMAL CONTROL PROBLEM

As M. R. Hestenes so aptly remarked,† had Johannes Bernoulli lived in our time, he would probably have formulated his problem as follows: To be found is the path of minimal travel time of a rocket under the influence of gravity and with a thrust force of constant magnitude and variable direction. The rocket is to be fired from a given point with a given initial direction and is to arrive at another given point with a given terminal direction.

If T denotes the constant magnitude of the thrust force, $u(t)$ its direction as represented by the angle with the positive x axis, and t the time, then we obtain for the equations of motion

$$\frac{d^2x}{dt^2} = T \cos u(t)$$

$$\frac{d^2y}{dt^2} = T \sin u(t) - g.$$

If (a,y_a) and (b,y_b) represent the coordinates of the given initial and terminal points and if y'_a and y'_b represent the given initial and terminal slopes, then the boundary conditions can be expressed in terms of the unknown duration t_1 of the process as follows: $x(0) = a$, $y(0) = y_a$, $x'(0) = 1$, $y'(0) = y'_a$, $x(t_1) = b$, $y(t_1) = y_b$, $x'(t_1) = 1$, $y'(t_1) = y'_b$. In addition to these conditions, t_1 has to be minimized.

† M. R. Hestenes, Elements of Calculus of Variations and Optimum Control Theory, "Space Mathematics," part 2, pp. 212ff, American Mathematical Society, 1966.

A formulation of this problem that is more easily accessible to analysis may be obtained in terms of the functions

$$y_1(t) = x(t), \qquad y_2(t) = y(t), \qquad y_3(t) = \frac{dx(t)}{dt}, \qquad y_4(t) = \frac{dy(t)}{dt}.$$

Then we obtain, instead of the previously listed equations of motion,

$$\begin{aligned} y_1' &= y_3 \\ y_2' &= y_4 \\ y_3' &= T\cos u(t) \\ y_4' &= T\sin u(t) - g \end{aligned}$$

and the new boundary conditions

$$\begin{aligned} y_1(0) &= a, \qquad & y_2(0) &= y_a, \qquad & y_3(0) &= 1, \qquad & y_4(0) &= y_a' \\ y_1(t_1) &= b, & y_2(t_1) &= y_b, & y_3(t_1) &= 1, & y_4(t_1) &= y_b'. \end{aligned}$$

The problem consists in finding a suitable function $u = u(t)$, the range of which may or may not be subjected to constraints, so that

$$t_1 \to \text{minimum}.$$

We may view the above system of first-order differential equations as an *underdetermined* system of four first-order differential equations for the five unknown functions y_1, y_2, y_3, y_4, u. A choice of $u = u(t)$ will turn this into a determined system, the solutions of which will, in general, be determined by the initial conditions. If there are functions $u = u(t)$ for which the terminal conditions can also be satisfied, then the problem consists of finding among these functions the one for which t_1 assumes the smallest possible value. Since the direction of the thrust force at any time t controls the motion of the rocket, we call it a *control function*, or simply a *control*. The control that yields the minimal travel time is called the *optimal control*.

Again, we have a problem that fits our general framework: There is a relation between functions (controls) and numbers (duration of process), and that function that yields the shortest duration is to be found. (Although the quantity to be minimized in our formulation of the problem is not expressed as an integral, as on previous occasions, this can easily be remedied by some mathematical trickery—see Prob. 1.1.8.)

A problem such as this is variously called a *Mayer problem*, a *Lagrange problem*, or an *optimal control problem*. (See Chaps. 5 and 6.)

PROBLEMS 1.1

1. Give examples of curves with a continuous derivative that have to be ruled out from the competition in problem A, even though they satisfy the given boundary conditions.
2. Find the parametric representation of a *cycloid* that is generated by a point on the circumference of a circle of radius a that rolls along the x axis.
3. Formulate mathematically: Find a curve of length L joining the points P_a and P_b in the x,y plane of such shape that its center of gravity is as low as possible. Also, impose suitable differentiability conditions on the curve.
4. Give a mathematical formulation of the navigation problem (problem D) for the case where the velocity field of the river is given by $u = u(x,y)$, $v = v(x,y)$, $u^2 + v^2 < c^2$ for all x, y, and where the banks are represented by curves with a continuous derivative.
5. Formulate mathematically: Given is a closed curve Γ in space with a simple closed curve as projection into the x,y plane. To be found is the surface $u = u(x,y)$ of smallest possible area that possesses Γ as boundary.
6. Formulate mathematically: Two given points in the plane are to be joined by a curve of shortest possible length.
7. A surface in space is given by a parametric representation. Join two points on this surface by a curve of shortest possible length.
8. Consider problem E. Introduce $y_o = t$ with $y_o(0) = 0$, $y_o(t_1) = t_1$ as a new unknown function, and formulate the problem as a problem of an underdetermined system of five differential equations in six unknown functions with five initial conditions, four terminal conditions, and a minimum condition that is imposed on a definite integral with an unknown upper limit.

1.2 FUNCTIONALS

The discussion in the preceding section was, by necessity, rather sketchy. All the problems that were discussed will be dealt with in much greater detail at the appropriate time. For the time being, this superficial survey will have to suffice for the purpose of illustrating the type of problem we shall deal with. The discussion, nevertheless, brought out one main point of immediate interest:

No matter what the particular trimmings, in each one of these problems a certain class of functions defined by differentiability conditions, boundary conditions, constraints of various types, etc., is considered, and by the formulation of the problem, there is associated with each element of this class a real number. The solution of the problem will be that element of the class of functions which is associated with the smallest (largest) real number—provided such a number exists.

Now, this certainly sounds familiar. In calculus, we consider real-valued functions of a real variable which are defined on a certain subset of the real line (e.g., closed interval, open interval). With each element of this subset of the reals, the function associates a real number. The theory of extreme values in calculus is concerned with finding that element in the domain in which the function is defined with which the smallest (largest) value of the function is associated. This is actually the problem we are concerned with now, except that the class on which the relationship (function) is defined is not a subset of the set of real numbers but is a specific subset of the set of all functions.

In other words, instead of a mapping of a subset of the set R of real numbers (domain of the function) onto a subset of R (range of the function), we now consider a mapping of a subset of the set of all functions onto a subset of R. So to speak, we are dealing with functions of functions. Such things are called *functionals*.

We give the following definition:

Definition 1.2 *Let S be a set of well-defined elements. If F denotes a mapping of S into R such that to every element $f \in S$ there corresponds exactly one real number, then F is called a* functional on S.

Symbolically, we write

$$F = F[f], \qquad f \in S, \qquad F[f] \in R.$$

Examples of functionals abound in mathematical analysis. Let $C[0,1]$ denote the set of all real-valued functions that are defined and continuous on $[0,1]$. Then

$$I[y] = \int_0^1 y(x)\,dx, \qquad y \in C[0,1]$$

is a functional on $C[0,1]$.

Let $C^1[0,1]$ denote the set of all functions that are defined and differentiable and have a continuous derivative on $[0,1]$. Then

$$I[y] = \int_0^1 (y^2(x) + y(x)y'(x))\,dx, \qquad y \in C^1[0,1],$$

is a functional on $C^1[0,1]$. If $f = f(x,y,y')$ is continuous for all x, y, y', then

$$I[y] = \int_a^b f(x, y(x), y'(x))\,dx, \qquad y \in C^1[a,b]$$

is a functional on $C^1[a,b]$. This latter case clearly embraces all the integrals that have been considered in the preceding section wherever the integrands are continuous functions of x, y, y'.

Functionals do not necessarily have to be of the particular nature given in these preceding examples.

$$D[y] = \left(\frac{dy}{dx}\right)_{x=1} = y'(1), \qquad y \in C^1(1 - \varepsilon, 1 + \varepsilon)$$

is also a legitimate functional, and so is

$$K[y] = \int_0^1 k(x)y(x)\,dx, \qquad y \in C[0,1],$$

where $k(x)$ is a given continuous function.

Now that we have found such a simple and obvious generalization of the concept of a function, we shall try to pursue this line of inquiry further in our search for a solution of the general extreme-value problem. Again, we shall see that within reasonable limits, the classical arguments from the extreme-value theory of real-valued functions of a real variable will find their counterparts in the theory of extreme values of functionals.

The next section will be devoted to a re-examination of the extreme-value problem in calculus, and in the sections after that, we shall generalize these concepts and procedures so that they will sensefully apply to the theory of functionals.

PROBLEMS 1.2

1. What is the customary name for a functional that is defined as a mapping of the n-dimensional (cartesian) space into R?
2. An obvious generalization of the concept of a functional is a mapping that admits image sets other than subsets of R. Such a mapping is called an *operator*, or more precisely, a *functional operator*. Give examples of functional operators.
3. Try to define continuity of a functional by a straightforward generalization of the definition of continuity of a function. What generalized concept is missing and not immediately obvious?
4. A theorem of Weierstrass states: A function that is continuous on a closed interval $[a,b]$ will assume on $[a,b]$ its maximum value and its minimum value.† We call f lower [upper] semicontinuous at x_o if for any $\varepsilon > 0$ there exists a $\delta(\varepsilon)$ such that $f(x_o) - f(x) < \varepsilon$ $[f(x) - f(x_o) < \varepsilon]$ for all $|x - x_o| < \delta(\varepsilon)$. Prove that if f is lower (upper) semicontinuous in $[a,b]$, then f will assume its minimum (maximum) value in $[a,b]$, by adapting the proof of Weierstrass' theorem in a suitable manner.

1.3 NECESSARY CONDITIONS FOR RELATIVE EXTREME VALUES OF REAL-VALUED FUNCTIONS OF ONE REAL VARIABLE

We shall devote this section to a review of well-known necessary conditions for relative extreme values of functions of a real variable. The argument we shall supply, however, will differ from the customary reasoning that is to be found in treatments of this subject. The advantage of our argument is that it is amenable to immediate generalization.

† Angus E. Taylor, "Advanced Calculus," p. 496, Ginn and Company, Boston, 1955.

We know that, by a celebrated theorem of Weierstrass (see Prob. 1.2.4), a continuous function assumes its maximum value and its minimum value in a closed interval, and on the other hand, we know that direct calculus methods fail to locate all relative extreme values of functions with even a continuous derivative in a closed interval. (The relative extrema at the endpoints are the ones that present difficulties.) For this reason, we shall restrict our investigation to functions on open intervals.

To formulate the definition of a relative extreme value in practical terms, let us first introduce the concept of a neighborhood:

Definition 1.3.1 *A subset $N^\delta(x_o)$ of R is called a δ neighborhood of x_o if it contains all points x for which $x_o - \delta < x < x_o + \delta$—in other words, if it contains all points $x_o + h$ for which $|h| < \delta$.*

Now we can proceed to a definition of a relative extreme value (see also Fig. 1.4):

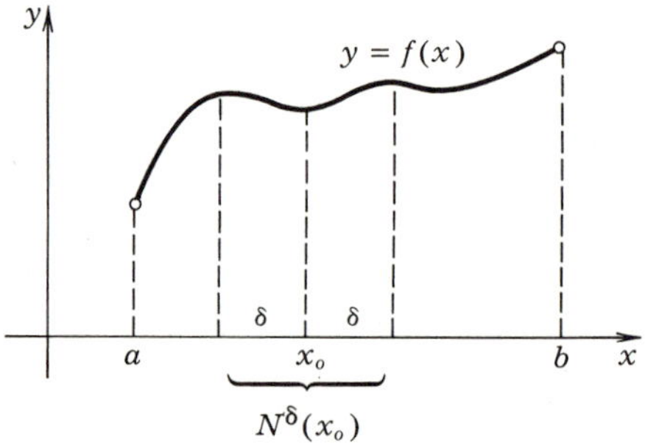

Figure 1.4

Definition 1.3.2 *Let $y = f(x)$ represent a real-valued function of a real variable which is defined on the open interval (a,b). $y = f(x)$ possesses a relative minimum (maximum) at $x_o \in (a,b)$ if there exists a δ neighborhood of x_o, $N^\delta(x_o) \subset (a,b)$, such that $f(x) - f(x_o) \geq 0$ (≤ 0) for all $x \in N^\delta(x_o)$, or in other words, if $f(x_o + h) - f(x_o) \geq 0$ (≤ 0) for all $|h| < \delta$.*

With a view toward later generalizations, let us now define the derivative of a function in the following, somewhat unusual manner, assuming that f is defined in $N^\delta(x_o)$:

Definition 1.3.3 *The number $f'(x_o)$ is the derivative of $y = f(x)$ at x_o if and only if there exists a $\delta > 0$ such that*

$$f(x_o + h) - f(x_o) = f'(x_o)h + \varepsilon(h) \qquad \text{for all } |h| < \delta \tag{1.3.1}$$

where $\lim_{h \to 0} [\varepsilon(h)/h] = 0$.

We contrast this definition to the classical one, which is to be found in most treatments of the differential calculus:

Definition 1.3.3a *The number* $f'(x_o)$ *is the derivative of* $y = f(x)$ *at* x_o *if and only if*

$$\lim_{h\to 0} \frac{f(x_o + h) - f(x_o)}{h} = f'(x_o). \tag{1.3.2}$$

These two definitions are tied together by the following theorem:

Theorem 1.3.1 *Definitions 1.3.3 and 1.3.3a are equivalent.*

Proof: (*a*) Definition 1.3.3 implies Definition 1.3.3*a*.
We obtain from (1.3.1) after division by h:

$$\frac{f(x_o + h) - f(x_o)}{h} = f'(x_o) + \frac{\varepsilon(h)}{h}, \qquad |h| < \delta,\ h \neq 0.$$

Consequently,

$$\lim_{h\to 0} \frac{f(x_o + h) - f(x_o)}{h} = f'(x_o) + \lim_{h\to 0} \frac{\varepsilon(h)}{h} = f'(x_o).$$

(*b*) Definition 1.3.3*a* implies Definition 1.3.3.
Let

$$\frac{f(x_o + h) - f(x_o)}{h} - f'(x_o) = \varepsilon_1(h), \qquad 0 < |h| < \delta.$$

By (1.3.2),

$$\lim_{h\to 0} \varepsilon_1(h) = 0,$$

and (1.3.1) follows immediately with $\varepsilon(h) = \varepsilon_1(h)h$.

Let $C'(a,b)$ denote the class of functions that possess a derivative at all points of the interval (a,b). We now assume that the function $f \in C'(a,b)$ has a relative minimum at $x_o \in (a,b)$. By Definition 1.3.2, there exists a $N^\delta(x_o) \subset (a,b)$ such that

$$f(x_o + h) - f(x_o) \geq 0 \qquad \text{for all } |h| < \delta.$$

(This δ is, in general, not the same as the one in Definition 1.3.3. We shall agree that whenever the two δ's are different, we shall work with the smaller one.)

Since $f \in C'(a,b)$, we have from (1.3.1),

$$f(x_o + h) - f(x_o) = f'(x_o)h + \varepsilon(h) \qquad \text{for all } |h| < \delta,$$

where $[\varepsilon(h)/h] \to 0$ as $h \to 0$.

Choose $h_o \neq 0$ such that $|h_o| < \delta$, and take $|\lambda| \leq 1$. Then

$$f(x_o + \lambda h_o) - f(x_o) = f'(x_o)\lambda h_o + \varepsilon(\lambda h_o) \geq 0 \qquad \text{for all } |\lambda| \leq 1.$$

In particular, we have, for all $0 < \lambda \leq 1$, that

$$\frac{f(x_o + \lambda h_o) - f(x_o)}{\lambda} = f'(x_o)h_o + \frac{\varepsilon(\lambda h_o)}{\lambda} \geq 0.$$

Since $\lim_{\lambda \to 0} [\varepsilon(\lambda h_o)/\lambda] = 0$, it follows that

$$f'(x_o)h_o \geq 0.$$

Similarly, for all $-1 \leq \lambda < 0$,

$$\frac{f(x_o + \lambda h_o) - f(x_o)}{\lambda} = f'(x_o)h_o + \frac{\varepsilon(\lambda h_o)}{\lambda} \leq 0$$

and by the same reasoning as before,

$$f'(x_o)h_o \leq 0.$$

Since h_o is any number for which $|h_o| < \delta$, we have the following necessary condition for a relative minimum:

$$f'(x_o)h = 0 \qquad \text{for all } |h| < \delta. \tag{1.3.3}$$

The familiar necessary condition $f'(x_o) = 0$ follows immediately after division by h.

Next, we define the second derivative of $y = f(x)$ in terms of the first derivative, assuming that the first derivative $f'(x_o + h)$ exists for all $|h| < \delta$.

Definiton 1.3.4 *The number $f''(x_o)$ is the second derivative of $y = f(x)$ at $x = x_o$ if there exists a δ such that*

$$f'(x_o + h) - f'(x_o) = f''(x_o)h + \varepsilon(h) \tag{1.3.4}$$

for all $|h| < \delta$, where $\lim_{h \to 0} [\varepsilon(h)/h] = 0$.

A representation of the total difference $f(x_o + h) - f(x_o)$ in terms of the first and second derivatives of $f(x)$ is given in the following theorem:

Theorem 1.3.2 *If $f''(x)$ exists in a neighborhood of x_o and is continuous at*

$x = x_o$, then

$$f(x_o + h) - f(x_o) = f'(x_o)h + \tfrac{1}{2}f''(x_o)h^2 + \varepsilon_1(h) \tag{1.3.5}$$

for all $|h| < \delta$, where $\lim_{h \to 0} [\varepsilon_1(h)/h^2] = 0$.

Proof: By Taylor's formula,

$$\begin{aligned} f(x_o + h) - f(x_o) &= f'(x_o)h + \tfrac{1}{2}f''(x_o + \Theta h)h^2 \\ &= f'(x_o)h + \tfrac{1}{2}f''(x_o)h^2 \\ &\quad + \tfrac{1}{2}h^2[f''(x_o + \Theta h) - f''(x_o)], \qquad |\Theta| \le 1. \end{aligned}$$

Since $f''(x)$ is assumed to be continuous at $x = x_o$, we have $[f''(x_o + \Theta h) - f''(x_o)] \to 0$ as $h \to 0$, and hence

$$\frac{\varepsilon_1(h)}{h^2} = \tfrac{1}{2}[f''(x_o + \Theta h) - f''(x_o)] \to 0 \qquad \text{as } h \to 0.$$

Suppose now that $f \in C^2(a,b)$, where $C^2(a,b)$ denotes the class of all functions with a continuous second-order derivative in (a,b). Suppose further that $y = f(x)$ posesses a relative minimum at $x = x_o$. Then, by (1.3.3), $f'(x_o) = 0$, and we have from (1.3.5) that

$$f(x_o + h) - f(x_o) = \tfrac{1}{2}f''(x_o)h^2 + \varepsilon_1(h) \ge 0$$

for all $|h| < \delta$, where $\lim_{h \to 0} [\varepsilon_1(h)/h^2] = 0$.

We pick an h_o so that $|h_o| < \delta$, and we obtain

$$f(x_o + \lambda h_o) - f(x_o) = \tfrac{1}{2}f''(x_o)\lambda^2 h_o^2 + \varepsilon_1(\lambda h_o) \ge 0$$

for all $|\lambda| \le 1$.
Since

$$\tfrac{1}{2}f''(x_o)\lambda^2 h_o^2 + \varepsilon_1(\lambda h_o) = \tfrac{1}{2}\lambda^2 h_o^2 \left(f''(x_o) + \frac{2\varepsilon_1(\lambda h_o)}{\lambda^2 h_o^2}\right)$$

and since $\lim_{\lambda \to 0} [\varepsilon_1(\lambda h_o)/\lambda^2 h_o^2] = 0$, we have by necessity that $f''(x_o) \ge 0$.

Collecting all the results that have been obtained thus far, we can state:

Theorem 1.3.3 *If $y = f(x)$ is differentiable in (a,b) and if $y = f(x)$ possesses a relative minimum (maximum) at $x = x_o \in (a,b)$, then, by* *necessity,* $f'(x_o) = 0$. *If* $f \in C^2(a,b)$, *then the additional condition* $f''(x_o) \geq 0$ (≤ 0) *has to hold.*

Finally, let us observe that if $f''(x_o) > 0$ while $f'(x_o) = 0$, then

$$f(x_o + h) - f(x_o) > 0$$

for all $0 < |h| < \delta$ for some $\delta > 0$, and we see that $f(x_o)$ is indeed a relative minimum of $y = f(x)$. Hence:

Theorem 1.3.4 *If* $f \in C^2(a,b)$ *and if* $f'(x_o) = 0$, $f''(x_o) > 0$ (< 0), *then* $y = f(x)$ *has a relative minimum (maximum) at* $x = x_o$.

PROBLEMS 1.3

1. Assume that f,g are differentiable functions of x. Show that $f'g + fg'$ is the derivative of fg by Definition 1.3.3.
2. Show that $f'(x_o)$ is uniquely defined by Definition 1.3.3 and that $f''(x_o)$ is uniquely defined by Definition 1.3.4.
3. Suppose that (1.3.5) holds. Show that

$$f''(x_o) = \lim_{h \to 0} \frac{f(x_o + h) - 2f(x_o) + f(x_o - h)}{h^2}.$$

4. Suppose that $f'(x_o)$ is defined by

$$f(x_o + h) - f(x_o) = f'(x_o)h + \alpha(h)h$$

where $\alpha(h) \to 0$ as $h \to 0$.

   (*a*) Show that if $f'(x_o)$ exists in this sense, then it also exists in the sense of Definition 1.3.3.

   (*b*) Show that $f'(x_o)$ as defined here is unique, provided it exists.

   (*c*) Prove Theorem 1.3.1 for the case where the concept of the derivative is based on the definition given here rather than the one given in the text.
5. Demonstrate with an example that $\varepsilon(h)$ and $\varepsilon_1(h)$ in Definitions 1.3.3 and 1.3.4 also depend on x_o.

1.4 NORMED LINEAR SPACES

We shall now start to lay the foundation for a generalization of the ideas developed in the preceding section.

We note that if a function f is defined on an interval (a,b) then if $x_o \in (a,b)$, $f(x_o)$ exists and so does $f(x_o + h)$ provided that h is sufficiently small. Is this also true if f is a functional and x is a function rather than a real number?

This section will be devoted to a discussion of this problem. Suppose that a functional F is defined for all $y \in S$, where S is some specified class of functions but does not necessarily exist for functions that are not in S.

So, unless $y + h \in S$ if $y \in S$ and $h \in S$, $F[y + h]$ may not even exist. This line of thought leads us directly to the concept of a linear space. Roughly speaking, a class of functions forms a linear space if, with any two elements, it also contains all linear combinations of these two elements with coefficients from a given (number) field.

We give the following precise definition:

Definition 1.4.1 *The collection S of elements $x, y, z, \ldots$ is called a linear space over the field $\mathfrak{F}$ with elements $\lambda, \mu, \nu, \ldots$ if the following conditions are satisfied:*

1. *If $x \in S$, $y \in S$, then the sum of x and y, written $x + y$, is defined and $x + y \in S$.*
2. *Addition is commutative: $x + y = y + x$.*
3. *Addition is associative: $(x + y) + z = x + (y + z)$.*
4. *There exists an additive identity $0 \in S$ such that $x + 0 = x$ for all $x \in S$.*
5. *For each $x \in S$, there exists an additive inverse $(-x) \in S$ such that $x + (-x) = 0$.*
6. *Scalar multiplication of elements of S with elements of $\mathfrak{F}$ is defined; that is, if $x \in S$, $\lambda \in \mathfrak{F}$, then $\lambda x \in S$.*
7. *The scalar multiplication is associative: If $x \in S$, $\lambda,\mu \in \mathfrak{F}$, then $\lambda(\mu x) = (\lambda\mu)x$.*
8. *Scalar multiplication is distributive: If $x \in S$, $\lambda,\mu \in \mathfrak{F}$, then $(\lambda + \mu)x = \lambda x + \mu x$,* and *if $x,y \in S$ and $\lambda \in \mathfrak{F}$, then $\lambda(x + y) = \lambda x + \lambda y$.*
9. *For the multiplicative identity $1 \in \mathfrak{F}$, $1 \cdot x = x$ for all $x \in S$.*

The first five postulates express the fact that S is an *Abelian group*, with addition as group operation, while the remaining four postulates regulate the multiplication of elements from S by the scalars from $\mathfrak{F}$ in the customary manner. (See also Prob. 1.4.1.)

Examples of linear spaces are easy to find:

A. $C[0,1]$, the space of all real-valued functions that are defined and continuous on the interval $[0,1]$. The following two results are established in elementary calculus:

If $f \in C[0,1]$ and $g \in C[0,1]$, then $f + g \in C[0,1]$, and if λ is a real number, then $\lambda f \in C[0,1]$. 0, in particular, is a continuous function. All the other properties are so obviously satisfied that it is superfluous to dwell on them.

The reader, of course, should realize that the space of all real-valued continuous functions on $[0,1]$—or on any other interval, for that matter—is a linear space over the field of reals, as over the field of rationals, or any other field that is contained in R. Whenever we use the symbol $C[0,1]$, however, we always mean the space of all real-valued continuous functions on $[0,1]$ over the field of real numbers R.

B. $C^1[0,1]$, the space of all real-valued functions with a continuous derivative on the interval $[0,1]$. The two theorems quoted under example *A*, if applied to f' rather than to f, yield the result that this space, too, is a linear space over any field $\mathfrak{F} \subset R$. Again, we shall specifically mean "over R" whenever we use the symbol $C^1[0,1]$.

C. $C_s[0,1]$, the space of all *sectionally continuous functions* in the interval $[0,1]$. [A function is called "sectionally continuous in $[0,1]$" if it has at most finitely many "jump" discontinuities in $[0,1]$. If x_o is such a point of discontinuity, then $\lim_{x \to x_0 - 0} f(x) = f(x_o - 0)$ and $\lim_{x \to x_0 + 0} f(x) = f(x_o + 0)$ have to exist.] For definiteness, we assign the value $f(x_o + 0)$ to $f(x)$ at every point x_o of discontinuity. If f has a discontinuity at $x_o = 1$, then we define $f(1) = f(1 - 0)$. This is obviously a linear space over the field of real numbers.

$C_{sP}[0,1]$, the space of all sectionally continuous functions that have jump discontinuities at most at the points of a given (fixed) point set $P = (x_1,x_2, \ldots, x_n)$, is also a linear space over R, as the reader can easily convince himself.

D. $C_s{}^1[0,1]$ and $C_{sP}{}^1[0,1]$, the space of all *sectionally smooth functions* (i.e., *continuous* functions with a *sectionally continuous* derivative) and the space of all sectionally smooth functions of which the derivative has jump discontinuities at most at $P = (x_1,x_2, \ldots, x_n)$, P fixed, are also linear spaces.

Many more examples of linear spaces are discussed in Prob. 1.4.3. Henceforth, we shall restrict our discussion to linear spaces over the field of reals R.

It will be of paramount importance to us to measure the discrepancy between functions, i.e., to find a measure that will enable us to tell how near two functions are to each other. Since the difference of two functions in a linear space is again a function in the same linear space, this will amount to establishing a measure for the "magnitude" of a function. How to accomplish this task is suggested by the definition of length in an n-dimensional euclidean space E_n.

The length of a vector $x = (\xi_1, \ldots, \xi_n)$ is defined as

$$||x|| = \sqrt{\xi_1^2 + \xi_2^2 + \cdots + \xi_n^2}. \tag{1.4.1}$$

Of course, it is not at all easy to see how this specific definition of a length can be generalized so as to be applicable to general linear spaces. However, if we abstract from this concept its basic properties, the generalization will be obvious. We shall therefore proceed to define "length," henceforth called *norm*, in terms of its fundamental properties:

Definition 1.4.2 *Let S denote a linear space over the field of reals R. A functional $||x||$ which is defined on S is called a norm of $x \in S$ if it has the following properties:*

1. $||x|| > 0$ *for all* $x \neq 0$, $x \in S$.
2. $||x|| = 0$ *if* $x = 0$.

3. $||\lambda x|| = |\lambda| \, ||x||$ *for all* $x \in S, \lambda \in R$.
4. $||x + y|| \le ||x|| + ||y||$ (*triangle inequality*).

(Note that the norm, being a functional, assigns to every element $x \in S$ a real number.)

The length of a vector in E_n, as defined by (1.4.1), satisfies these conditions, as one can see without much difficulty (see Prob. 1.4.11). It is possible, however, to introduce concepts of a norm in the space of all n-tuples R_n that have nothing to do with the euclidean length of a vector, for example,

$$||x|| = \max_{(i)} |\xi_i|. \tag{1.4.2}$$

That postulates 1 to 3 of Definition 1.4.2 are satisfied is obvious. Regarding postulate 4, we note that

$$|\xi_i + \eta_i| \le |\xi_i| + |\eta_i| \le \max_{(j)} |\xi_j| + \max_{(j)} |\eta_j| = ||x|| + ||y||$$

for all $i = 1, 2, 3, \ldots, n$.
Hence the inequality will hold, in particular, for that subscript i for which $|\xi_i + \eta_i|$ assumes its maximum value. Therefore,

$$||x + y|| = \max_{(i)} |\xi_i + \eta_i| \le ||x|| + ||y||.$$

Another possibility for defining a norm in the space R_n of all n-tuples is indicated in Prob. 1.4.4. Note that the space R_n of all n-tuples is only called a *euclidean space* E_n if the norm is defined as in (1.4.1).

Definition 1.4.3 $\mathcal{S}$ *is a* normed linear space *over* R *if:*

1. $\mathcal{S}$ *is a linear space over* R.
2. *A norm which satisfies postulates 1 to 4 of Definition 1.4.2 is defined for all* $x \in \mathcal{S}$.

The following normed linear spaces will be of great interest to us in the subsequent developments:

E. $\mathcal{C}[0,1]$ is the space $C[0,1]$ in which the norm is defined as follows:

$$||f|| = \max_{[0,1]} |f(x)|. \tag{1.4.3}$$

Before demonstrating that this is indeed a norm, let us note that $||f||$ as defined in (1.4.3) exists for all functions in $C[0,1]$ since the absolute value of a continuous function is again a continuous function and a continuous function assumes its maximum in a closed interval.

We shall now check properties 1 to 4 from Definition 1.4.2:

1. If $f \not\equiv 0$ on $[0,1]$, then there exists an $x_o \in [0,1]$ such that $f(x_o) \neq 0$ and hence $||f|| \geq |f(x_o)| > 0$. (We remind the reader that $f = 0$ is the additive unity in $C[0,1]$.)
2. If $f(x) \equiv 0$ on $[0,1]$, then $\max_{[0,1]} |f(x)| = ||f|| = 0$.
3. $||\lambda f|| = \max_{[0,1]} |\lambda f(x)| = |\lambda| \max_{[0,1]} |f(x)| = |\lambda| \, ||f||$.
4. For all $x \in [0,1]$,

   $$|f(x) + g(x)| \leq |f(x)| + |g(x)| \leq \max_{[0,1]} |f(t)| + \max_{[0,1]} |g(t)| = ||f|| + ||g||,$$

   and hence this inequality holds in particular for that x for which $|f(x) + g(x)|$ assumes its maximum in $[0,1]$. Thus,

   $$||f + g|| = \max_{[0,1]} |f(x) + g(x)| \leq ||f|| + ||g||.$$

F. $\mathcal{C}^1[0,1]$ is the space $C^1[0,1]$ in which the norm is defined as follows:

$$||f|| = \max_{[0,1]} |f(x)| + \max_{[0,1]} |f'(x)|. \tag{1.4.4}$$

The existence of $||f||$ follows from $f, f' \in C[0,1]$. That (1.4.4) satisfies the requirements of a norm follows from an argument similar to that used in example *E*. (See Prob. 1.4.9.)

A norm as defined in (1.4.2), (1.4.3), or (1.4.4) is usually referred to as a *maximum norm* or a *Chebychev norm.*

G. $\mathcal{C}_s^1[0,1]$ and $\mathcal{C}_{sp}^1[0,1]$ are the spaces $C_s^1[0,1]$ and $C_{sp}^1[0,1]$, respectively (see example *D*), in which the norm is defined as follows:

$$||f|| = \max_{[0,1]} |f(x)| + \sup_{[0,1]} |f'(x)|. \tag{1.4.5}$$

We observe that $||f||$ exists for all $f \in C_s^1[0,1]$ or $f \in C_{sp}^1[0,1]$ (see Prob. 1.4.12) and that it satisfies all the requirements of a norm. (See Prob. 1.4.13.)

Now that we are in possession of the concepts of *linear space* (as a generalization of the "space" of real numbers) and *norm* (as a generalization of the concept of distance), we are ready to talk about open subsets of linear spaces (generalized open intervals) and, as a special case thereof, the concept of a neighborhood.

Definition 1.4.4 *Let $Y \subset \mathcal{S}$. Y is called an* open subset *of the normed linear space $\mathcal{S}$ if for every $f \in Y$ there is a $\delta > 0$ so that $f + h \in Y$ for all $h \in \mathcal{S}$ for which $||h|| < \delta$.*

To illustrate the importance of this concept to our future investigations, let us consider the functional

$$I[y] = \int_a^b f(x,y(x),y'(x))\, dx$$

where $f \in C^1(\mathcal{R})$, with $\mathcal{R}$ denoting a *domain* (*open, connected point set*) in (x,y,y') space. [$C^1(\mathcal{R})$ denotes the class of all functions with continuous partial derivatives of first order for all $(x,y,y') \in \mathcal{R}$.]

To provide for a practical and simple formulation for what is to come, let us introduce the concept of *lineal element*: Suppose that $y \in \mathcal{C}^1[a,b]$. Then we call the triple of numbers $(x_o,y(x_o), y'(x_o))$, $x_o \in [a,b]$, a *lineal element* of $y = y(x)$. Hereby it is understood that $y'(a) = y'(a+0)$ and $y'(b) = y'(b-0)$.

We shall assume that $y_o \in \mathcal{C}^1[a,b]$ is such that all its lineal elements lie in $\mathcal{R}$. Then, with $h \in \mathcal{C}^1[a,b]$, all lineal elements of $y_o + h$ will also lie in $\mathcal{R}$ provided that $||h|| < \delta$ for some $\delta > 0$. (See Prob. 1.4.15.) Hence we can say that $I[y]$ is defined on an open subset Y of $\mathcal{C}^1[a,b]$ if we define Y such that y lies in Y if all its lineal elements lie in $\mathcal{R}$.

By the same token, if all lineal elements of $y = y_o(x) \in \mathcal{C}_{sp}{}^1[a,b]$, including all $(x_i, y_o(x_i), y_o'(x_i \pm 0))$, $x_i \in P$, lie in $\mathcal{R}$, then all lineal elements of $y = y_o(x) + h(x)$ will also lie in $\mathcal{R}$ provided that $||h|| < \delta$ for some $\delta > 0$ and that $h \in \mathcal{C}_{sp}{}^1[a,b]$.

Definition 1.4.5 *The δ neighborhood of $f_o \in \mathcal{S}$, $N^\delta(f_o)$ denotes the open subset in $\mathcal{S}$ that consists of all $f \in \mathcal{S}$ for which*

$$||f - f_o|| < \delta,$$

or as we may put it, consists of all elements $f_o + h$, $h \in \mathcal{S}$, for which $||h|| < \delta$.

Figure 1.5 depicts a typical neighborhood in $\mathcal{C}[0,1]$.

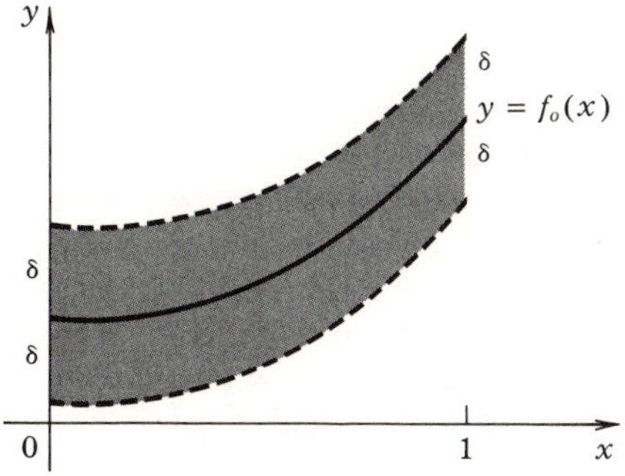

Figure 1.5

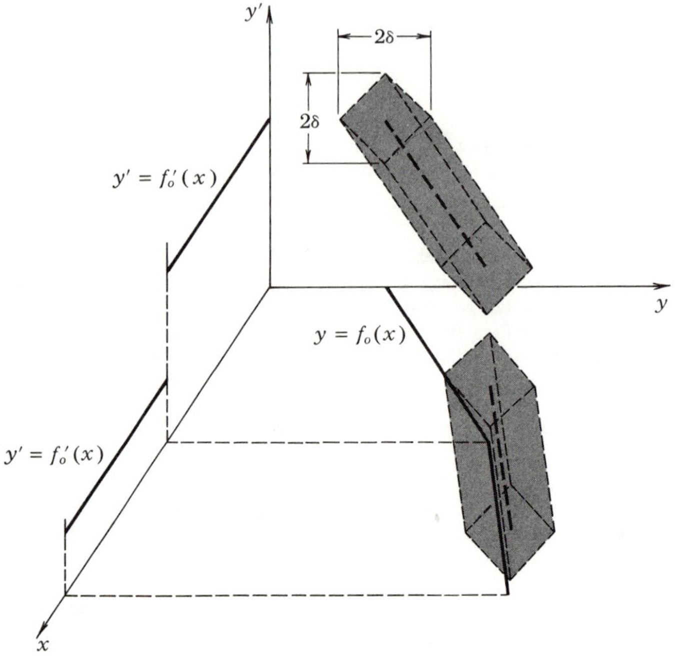

Figure 1.6

It is not possible to give such a simple geometric interpretation of a neighborhood in $\mathcal{C}^1[0,1]$ or $\mathcal{C}_s{}^1[0,1]$. However, since

$$|f(x) - f_o(x)| + |f'(x) - f_o'(x)| \leq \max_{[0,1]} |f(x) - f_o(x)| + \sup_{[0,1]} |f'(x) - f_o'(x)|,$$

we can say that if $f \in N^\delta(f_o)$ in $\mathcal{C}_s{}^1[0,1]$, then all lineal elements of f will lie in a (possibly broken) tube of diamond-shaped crossection as depicted in Fig. 1.6. ($|f| + |f'| < \delta$ represents a diamond with diagonals of length 2δ in f,f' plane.) However, not all functions f all lineal elements of which lie in this tube are also in a δ neighborhood of f_o. To make certain that $f \in N^\delta(f_o)$, we have to require that all lineal elements lie in a tube of this sort; but where the diagonals have length δ, since

$$|f(x) - f_o(x)| + |f'(x) - f_o'(x)| < \frac{\delta}{2}$$

implies

$$\max_{[0,1]} |f(x) - f_o(x)| < \frac{\delta}{2}, \qquad \sup_{[0,1]} |f'(x) - f_o'(x)| < \frac{\delta}{2};$$

and hence, $||f - f_o|| < \delta$. (See also Prob. 1.4.16.)

PROBLEMS 1.4

1. Show that condition 2 in the definition of linear space is redundant, i.e., that addition is commutative anyway on account of the other postulates.
2. Suppose, in the definition of a linear space, that the field $\mathfrak{F}$ of scalar multipliers is replaced by a ring with multiplicative identity of scalar multipliers. Are there any substantial losses?
3. Show that the following spaces are linear spaces:

   (*a*) The space of all continuous complex-valued functions of a real variable† $x \in [0,1]$ over the field of real numbers and over the field of complex numbers. (The latter space is denoted by $\bar{C}[0,1]$.)

   (*b*) The space of all complex-valued functions of a real variable $x \in [0,1]$ with a continuous derivative† over the field of complex numbers. (This space is denoted by $\bar{C}^1[0,1]$.)

   (*c*) The space of all real numbers over the field of rational numbers.

   (*d*) The space of all convergent sequences

   $$c = \{(\xi_1,\xi_2,\xi_3, \ldots) \mid \lim \xi_n \text{ exists}\}$$

   over the field of real numbers.‡

   (*e*) The space of all bounded sequences

   $$m = \{(\xi_1,\xi_2,\xi_3, \ldots) \mid |\xi_i| \leq M \text{ for some } M\}$$

   over the field of real numbers.‡

   (*f*) The space of all n-tuples of real numbers

   $$R_n = \{(\xi_1,\xi_2, \ldots, \xi_n) \mid \xi_i \text{ real}\}$$

   over the field of real numbers.‡
4. Given the space R_n of all n-tuples of real numbers (see problem 3*f*). Show that

   $$||x|| = |\xi_1| + |\xi_2| + \cdots + |\xi_n|$$

   satisfies the conditions for a norm.
5. Given the space of problem 4 with $n > 1$. Show that

   $$||x|| = |\xi_1|$$

   satisfies all the conditions for a norm except the first one.

† A complex-valued function of a real variable is continuous if and only if its real part is continuous and its imaginary part is continuous.

‡ Addition and scalar multiplication are defined as follows:

$$x + y = (\xi_1 + \eta_1, \xi_2 + \eta_2, \ldots, \xi_n + \eta_n, \ldots)$$
$$\lambda x = (\lambda\xi_1,\lambda\xi_2, \ldots, \lambda\xi_n, \ldots)$$

6. Given the linear space $C[0,1]$. Is

$$\|f\| = \sqrt{\int_0^1 f^2\,dx}, \qquad f \in C[0,1],$$

a legitimate norm?

7. Given the linear space $C^1[0,1]$. Is

$$\|f\| = \sqrt{\int_0^1 (f^2 + f'^2)\,dx}, \qquad f \in C^1[0,1],$$

a legitimate norm?

8. Given the space $\bar{C}[0,1]$ (see problem 3*a*). Is

$$\|f\| = \sqrt{\int_0^1 |f|^2\,dx}, \qquad f \in \bar{C}[0,1],$$

a legitimate norm?

*9. Show that $\mathcal{C}^1[0,1]$ is a normed linear space. [For definition of $\mathcal{C}^1[0,1]$, see (1.4.4).]

10. Do the functions $f_n(x) = (1/n)\sin n^2x$, $n = 2, 3, 4, \ldots$, lie in a $1/(n-1)$ neighborhood of $f(x) \equiv 0$ in:

(*a*) $\mathcal{C}[0,1]$? (*b*) $\mathcal{C}^1[0,1]$?

11. Show that (1.4.1) satisfies the conditions for a norm.

*12. Suppose that $f \in C_s[0,1]$ and that $f = f(x)$ has jump discontinuities at $x_1, x_2, \ldots, x_n \in (0,1)$. For $k = 0, 1, 2, \ldots, n$, with $x_0 = 0$, $x_{n+1} = 1$, let

$$f_k = \bar{f}_k(x) \equiv \begin{cases} f(x) & \text{for } x \in (x_k, x_{k+1}) \\ f(x_k + 0) & \text{for } x = x_k \\ f(x_{k+1} - 0) & \text{for } x = x_{k+1}. \end{cases}$$

Then $\bar{f}_k \in C[x_k, x_{k+1}]$. Let $\max\limits_{[x_k, x_{k+1}]} |\bar{f}_k(x)| = M_k$, $k = 0, 1, \ldots, n$, and let

$M = \max(M_0, M_1, \ldots, M_n)$. Show that

$$\sup_{[0,1]} |f(x)| = M.$$

*13. Show that (1.4.5) satisfies the conditions for a norm.

14. Show that E_n, the space R_n of problem 3*f* with

$$\|x\| = \sqrt{\xi_1^2 + \xi_2^2 + \cdots + \xi_n^2},$$

is a normed linear space.

*15. Given that all lineal elements of $y = y_0(x) \in \mathcal{C}^1[a,b]$ lie in $\mathcal{R}$ where $\mathcal{R}$ is a domain in (x,y,y')-space. Show that all lineal elements of $y = y_0(x) + h(x)$ lie also in $\mathcal{R}$, provided that $\|h\| < \delta$ and that $h \in \mathcal{C}^1[a,b]$.

16. Given the function

$$f = f(x) \equiv \begin{cases} \frac{2}{3}x & \text{for } 0 \le x \le \frac{1}{3} \\ \frac{2}{9} & \text{for } \frac{1}{3} < x \le \frac{2}{3} \\ \frac{1}{2}x - \frac{1}{9} & \text{for } \frac{2}{3} < x \le 1 \end{cases}$$

Show that $|f(x)| + |f'(x)| < 1$ for all $x \in [0,1]$ but

$$\|f\| = \max_{[0,1]} |f(x)| + \sup_{[0,1]} |f'(x)| > 1.$$

1.5 THE GÂTEAUX VARIATION OF A FUNCTIONAL

We are now ready to generalize the concept of the derivative to functionals that are defined on normed linear spaces over R, or at least on open subsets thereof. By doing this, we shall lay the foundation for a theory of extreme values of functionals.

The derivative, or rather the differential, of a functional can be defined in a number of ways. We shall mention here three commonly known concepts and then base our further investigations on the weakest one of the three.

We assume that $I[y]$ is defined in an open subset Y of a normed linear space $\mathcal{S}$. Then, if $y_o \in Y$ and $h \in \mathcal{S}$, we also have $y_o + h \in Y$ provided that $||h|| < \delta$ for some $\delta > 0$, and $I[y_o + h]$ is defined.

We call $L_f[h]$ the *Fréchet differential* of $I[y]$ at $y = y_o$ if there is a $\delta > 0$ such that for all $h \in \mathcal{S}$, $||h|| < \delta$,

$$I[y_o + h] - I[y_o] = L_f[h] + \varepsilon_1[h]$$

where $L_f[h]$ is a *linear functional* of h and where $\lim_{||h|| \to 0} (\varepsilon_1[h]/||h||) = 0.$†

We note that $L[h]$ is called a *linear functional* of h if it is additive, that is, if $L[h + k] = L[h] + L[k]$ for all $h,k \in \mathcal{S}$, and if it is continuous, that is, if $\lim L[h_n] = L[h]$ whenever $||h - h_n|| \to 0$, $h,h_n \in \mathcal{S}$. (See Prob. 1.5.2.)

A second, somewhat weaker concept of the differential of a functional is as follows: $L_g[h]$ is called the *Gâteaux differential* of $I[y]$ at $y = y_o$ if there is a $\delta > 0$ such that for all $h \in \mathcal{S}$, $||h|| < \delta$,

$$I[y_o + h] - I[y_o] = L_g[h] + \varepsilon_2[h]$$

where $L_g[h]$ is a linear functional of h and where $\lim_{t \to 0} (\varepsilon_2[th]/t) = 0$, t real.‡

The reader can see that the concept of the Gâteaux differential is somewhat weaker than the concept of the Fréchet differential, since in the case of the Fréchet differential, $\varepsilon_1[h]$ has to tend to zero uniformly in h, while in case of the Gâteaux differential, $\varepsilon_2[h]$ only has to tend to zero along each $h \in \mathcal{S}$. (See also Prob. 1.5.1.)

† This concept of a differential was chosen by I. M. Gelfand and S. V. Fomin as the basis for the theory of the first variation in their book "Calculus of Variations," Prentice-Hall, Inc., Englewood Cliffs, N.J., 1963.

‡ See M. Z. Nashed, Some Remarks on Variations and Differentials, *Amer. Math. Monthly*, vol. 73, No. 4, part II, pp. 63–76, 1966.

One can show easily that $L_g[h]$ is the Gâteaux differential of $I[y]$ at $y = y_o$ if and only if, for $t \in R$,

$$\frac{d}{dt} I[y_o + th]_{t=0} = L_g[h]$$

is a linear functional of h. (See Prob. 1.5.3.)

Our future discussions will be based on the still weaker concept of the *Gâteaux variation*:

Definition 1.5 *$\delta I[h]$ is called the* Gâteaux variation, *or the* first variation, *of $I[y]$ at $y = y_o$ if, for $t \in R$,*

$$\delta I[h] = \frac{d}{dt} I[y_o + th]_{t=0}$$

exists for all $h \in \mathcal{S}$.

In conjunction with this definition, we note that if $I[y]$ is defined in an open subset Y of $\mathcal{S}$ and if $y_o \in Y$, then $y_o + h \in Y$ for all $h \in \mathcal{S}$ as long as $\|h\| < \delta$ for some $\delta > 0$. Hence, for any given $h \in \mathcal{S}$, $I[y_o + th]$ is defined, provided that t is sufficiently small.

Note that the existence of the Gâteaux differential implies the existence of the Gâteaux variation, but not vice versa. (See Probs. 1.5.4 and 1.5.5.)

Lemma 1.5.1 *The* Gâteaux variation, *or* first variation, *$\delta I[h]$ of $I[y]$ at $y = y_o$ is homogeneous of the first degree; i.e., for all $h \in \mathcal{S}$ and all $\lambda \in R$,*

$$\delta I[\lambda h] = \lambda \delta I[h]. \tag{1.5.1}$$

Proof:

$$\delta I[\lambda h] = \frac{d}{dt} I[y_o + t\lambda h]_{t=0} = \lim_{t\to 0} \frac{I[y_o + t\lambda h] - I[y_o]}{t}$$

$$= \lim_{\tau\to 0} \lambda \frac{I[y_o + \tau h] - I[y_o]}{\tau} = \lambda \delta I[h].$$

The following representation of the *total variation* of $I[y]$ at $y = y_o$, namely, $I[y_o + h] - I[y_o]$, constitutes the key to all further applications:

Theorem 1.5 *$\delta I[h]$ is the* Gâteaux *variation, or first variation, of $I[y]$ at $y = y_o$ if and only if there is a $\delta > 0$ such that for all $h \in \mathcal{S}$, $\|h\| < \delta$,*

$$I[y_o + h] - I[y_o] = \delta I[h] + \varepsilon[h] \tag{1.5.2}$$

where $\delta I[h]$ is homogeneous of the first degree and where $\lim_{t\to 0} (\varepsilon[th]/t) = 0$.

Proof: (*a*) Suppose that $\delta I[h]$ exists. Then, by Lemma 1.5.1, $\delta I[h]$ is homogeneous of the first degree and

$$\delta I[h] = \lim_{t\to 0} \frac{I[y_\circ + th] - I[y_\circ]}{t}.$$

Hence,

$$\frac{I[y_\circ + th] - I[y_\circ]}{t} = \delta I[h] + \alpha[th]$$

where $\lim_{t\to 0} \alpha[th] = 0$.

We multiply both sides by t and obtain, in view of the homogeneity of $\delta I[h]$,

$$I[y_\circ + th] - I[y_\circ] = \delta I[th] + t\alpha[th]$$

as long as $||th|| = |t|\ ||h||$ is sufficiently small. Let $th = k$, and we have

$$I[y_\circ + k] - I[y_\circ] = \delta I[k] + \varepsilon[k]$$

for all $||k|| < \delta$ for some $\delta > 0$ and $\varepsilon[k] = t\alpha[k]$. Hence,

$$\frac{\varepsilon[tk]}{t} = \alpha[tk] \to 0 \text{ as } t \to 0.$$

(*b*) Suppose that the representation (1.5.2) is valid. Then,

$$I[y_\circ + th] - I[y_\circ] = t\delta I[h] + \varepsilon[th]$$

and hence,

$$\lim_{t\to 0} \frac{I[y_\circ + th] - I[y_\circ]}{t} = \delta I[h] + \lim_{t\to 0} \frac{\varepsilon[th]}{t} = \delta I[h].$$

As an example, we consider the functional

$$I[y] = \int_a^b f(x,y(x),y'(x))\,dx, \qquad y \in \mathcal{C}^1[a,b], \tag{1.5.3}$$

where we shall assume that $f \in \mathcal{C}^1(\mathcal{R})$ where $\mathcal{R}$ represents a domain in (x,y,y') space. (See also Sec. 1.4.)

If all lineal elements of $y = y_\circ(x) \in \mathcal{C}^1[a,b]$ lie in $\mathcal{R}$, then

$$I[y_\circ + th] = \int_a^b f(x,y_\circ(x) + th(x),y_\circ'(x) + th'(x))\,dx$$

is defined for all $h \in \mathcal{C}^1[a,b]$, provided that t is sufficiently small, and we obtain†

$$\delta I[h] = \frac{d}{dt} I[y_o + th]_{t=0}$$
$$= \int_a^b [f_y(x,y_o(x),y_o'(x))h(x) + f_{y'}(x,y_o(x),y_o'(x))h'(x)]\,dx, \qquad (1.5.4)$$

which we recognize easily as a homogeneous functional of h.

Before we close this section, we shall quickly establish the uniqueness of the *Gâteaux* variation (first variation):

Lemma 1.5.2 *The* Gâteaux *variation, or first variation, as defined in Definition 1.5, is* uniquely *determined, provided it exists.*

Proof: Let us assume to the contrary that $I[y]$ has two Gâteaux variations at $y = y_o$, namely, $\delta_1 I[h]$ and $\delta_2 I[h]$. Then, by Theorem 1.5,

$$I[y_o + h] - I[y_o] = \delta_1 I[h] + \varepsilon_1[h]$$
$$I[y_o + h] - I[y_o] = \delta_2 I[h] + \varepsilon_2[h].$$

Then
$$\delta_1 I[h] - \delta_2 I[h] = \varepsilon_2[h] - \varepsilon_1[h].$$

We note that $L[h] = \delta_1 I[h] - \delta_2 I[h]$ is homogeneous of the first degree. Hence,

$$\frac{\delta_1 I[th] - \delta_2 I[th]}{t} = \frac{L[th]}{t} = L[h] = \frac{\varepsilon_2[th] - \varepsilon_1[th]}{t}.$$

Suppose that there is an $h_o \in \mathcal{S}$ such that $L[h_o] = c \neq 0$. Then

$$L[h_o] = c = \frac{\varepsilon_2[th_o] - \varepsilon_1[th_o]}{t}$$

where
$$\lim_{t \to 0} \frac{\varepsilon_2[th_o] - \varepsilon_1[th_o]}{t} = 0.$$

Thus we arrive at a contradiction, unless $L[h] = 0$ for all $h \in \mathcal{S}$, that is, $\delta_1 I[h] = \delta_2 I[h]$.

PROBLEMS 1.5

1. Show: If $I[y]$ possesses a Fréchet differential at $y = y_o$, then $I[y]$ also possesses a Gâteaux differential at $y = y_o$.
2. Show: A linear functional $L[h]$, $h \in \mathcal{S}$, where $\mathcal{S}$ is a normed linear space over R, is homogeneous of the first degree in h, that is, $L[\lambda h] = \lambda L[h]$ for all $\lambda \in R$.

† See R. Courant, "Differential and Integral Calculus," vol. 2, p. 218, Interscience Publishers, Inc., New York, 1959.

3. Show: $L_g[h]$ is the Gâteaux differential of $I[y]$ at $y = y_o$ if and only if

$$\frac{d}{dt} I[y_o + th]_{t=0} = L_g[h]$$

is a linear functional.

4. Show: If $I[y]$ possesses a Gâteaux differential at $y = y_o$, then $I[y]$ also possesses a Gâteaux variation (first variation) at $y = y_o$.

5. Let $\mathcal{S} = R_2$. Then the elements of $\mathcal{S}$ are represented by $y = (x_1,x_2)$, where x_1, x_2 are real numbers. Consider the functional

$$I[y] = \begin{cases} \dfrac{x_1 x_2^2}{x_1^2 + x_2^2} & \text{if } (x_1,x_2) \neq (0,0) \\ 0 & \text{if } (x_1,x_2) = (0,0). \end{cases}$$

Show that this functional possesses a Gâteaux variation at $y = 0$ but does not possess a Gâteaux differential at $y = 0$.

6. Impose suitable conditions on $f(x,y)$ and $f(x,y')$ and find $\delta I[h]$ for:

(a) $I[y] = \int_a^b f(x,y(x))\, dx$ (b) $I[y] = \int_a^b f(x,y'(x))\, dx$

7. Assume that $\delta I[h]$ exists. Find $\delta E[h]$ where $E[y] = e^{I[y]}$.

8. Calculate $\delta I[h]$ for:

(a) $I[y] = \int_0^1 \sqrt{1 + y'^2(x)}\, dx$ at $y_o = ax + b$

(b) $I[y] = \int_0^{\pi/2} (y'^2(x) - y^2(x))\, dx$ at $y_o = a \cos x + b \sin x$

(c) $I[y] = \int_0^1 y(x) \sqrt{1 + y'^2(x)}\, dx$ at $y_o = b \cosh \dfrac{x - a}{b}$

9. Impose suitable conditions on $f = f(x,y_1, \ldots, y_n, y_1', \ldots, y_n')$ and find

$$\delta I[h_k] = \frac{d}{dt} I[y_1, \ldots, y_{k-1}, y_k + th_k, y_{k+1}, \ldots, y_n]_{t=0}$$

where $I[y] = \int_a^b f(x,y_1(x), \ldots, y_n(x), y_1'(x), \ldots, y_n'(x))\, dx.$

10. Given two functionals $I_1[y]$ and $I_2[y]$ which are both defined in the same open subset $Y \subset \mathcal{S}$ and where it is assumed that $I_2[y] \neq 0$ for all y for which $\|y - y_o\| < \delta$ for some $\delta > 0$, $y_o \in Y$. Assume that $\delta I_1[h]$ and $\delta I_2[h]$ exist at $y = y_o$ and find $\delta(I_1 + I_2)[h]$, $\delta(I_1 \cdot I_2)[h]$, and $\delta(I_1/I_2)[h]$ at $y = y_o$.

1.6 THE SPACE OF ADMISSIBLE VARIATIONS

Let us consider a typical variational problem: To be found is a function $y = y(x) \in \mathcal{C}^1[a,b]$ such that

$$I[y] = \int_a^b f(x,y(x),y'(x))\, dx \to \text{minimum}$$

whereby $y(a) = y_a$, $y(b) = y_b$ are given. (See examples A and B of Sec. 1.1.)

The notation $\int_a^b f(x,y(x),y'(x))\,dx \to$ minimum, which we shall use throughout this book, indicates that $\int_a^b f(x,y(x),y'(x))\,dx$ is to be made a minimum; it does *not* indicate a limit process.

In analyzing this problem, we are not interested in all functions $y \in \mathcal{C}^1[a,b]$ but only in the ones which satisfy the stated boundary conditions. The set of all those functions from $\mathcal{C}^1[a,b]$ which also satisfy the boundary conditions, we call for obvious reasons the *space of competing functions* for this particular problem, and we denote the set by Σ.

In our particular case, we have

$$\Sigma = \{y \mid y \in \mathcal{C}^1[a,b],\ y(a) = y_a,\ y(b) = y_b\}. \tag{1.6.1}$$

This space is *not* a linear space since the sum of two of its elements does not satisfy the boundary conditions—unless $y_a = y_b = 0$, of course.

Among the functions in Σ, we try to locate the one which renders $I[y]$ a relative minimum (maximum). We give the following definition:

Definition 1.6.1 *Let $\Sigma \subset \mathcal{S}$ represent a space of competing functions. $y_o \in \Sigma$ is said to yield a relative minimum (maximum) for $I[y]$ in Σ if*

$$I[y] - I[y_o] \geq 0 \qquad (\leq 0)$$

for all $y \in \Sigma$ for which $\|y - y_o\| < \delta$ for some $\delta > 0$.

In terms of the space of competing functions, we now introduce the so-called *space of admissible variations*. We shall first explain this concept in conjunction with some examples.

Take the space of competing functions as defined in (1.6.1). If $y_o \in \Sigma$, then $y_o + h \in \Sigma$ provided that $h \in \mathcal{C}^1[a,b]$ and that $h(a) = h(b) = 0$. We call the space

$$\mathcal{H} = \{h \mid h \in \mathcal{C}^1[a,b],\ h(a) = h(b) = 0\},$$

a space of admissible variations of Σ.

In this particular case, $\mathcal{H}$ is a normed linear space. $\mathcal{H}$ is normed since $\mathcal{H} \subset \mathcal{C}^1[a,b]$, and it is linear because with any two $h_1,h_2 \in \mathcal{H}$, $\lambda_1 h_1 + \lambda_2 h_2 \in \mathcal{H}$, $\lambda_1,\lambda_2 \in R$, as the reader can easily convince himself.

We note that

$$H_1 = \left\{h \mid h \in \mathcal{C}^1[a,b],\ h(a) = h(b) = 0,\ h\left(\frac{a+b}{2}\right) \neq 0\right\}$$

is also a space of admissible variations of Σ since $H_1 \subset \mathcal{H}$, but H_1 is *not* a linear space.

In general, the space of admissible variations of Σ is not uniquely determined because any subset of a space of admissible variations is again a space of admissible variations.

The next example will show that certain spaces of competing functions do not admit *linear* spaces of admissible variations.

Consider the isoperimetric problem (example C, Sec. 1.1). In that case, the space of competing functions is given by

$$\Sigma = \left\{ y \mid y \in \mathcal{C}^1[a,b],\ y(a) = y_a,\ y(b) = y_b,\ \int_a^b \sqrt{1 + y'^2(x)}\, dx = L \right\}. \qquad (1.6.2)$$

Suppose that Σ admits a linear space of admissible variations $\mathcal{H}$. Then, with $y_0 \in \Sigma$, $h \in \mathcal{H}$, $y_0 + h \in \Sigma$. If $\mathcal{H}$ is a linear space, then $(\lambda/||h||)h \in \mathcal{H}$ for all $\lambda \in R$. But, in general, $y_0 + (\lambda/||h||)h \notin \Sigma$, as we so drastically demonstrate in Fig. 1.7.

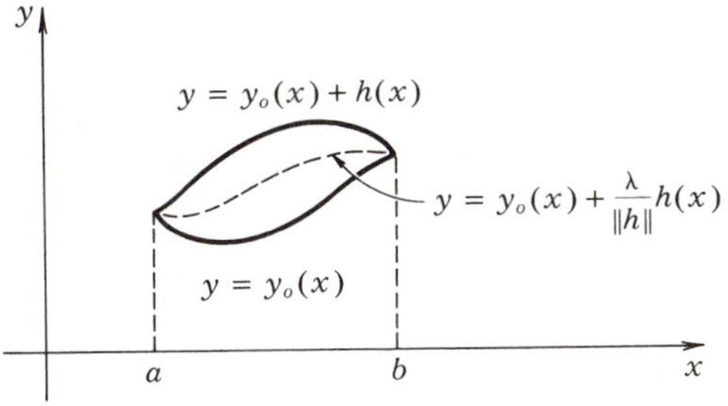

Figure 1.7

We give now the following definition:

Definition 1.6.2 *For a given space of competing functions* $\Sigma \subset \mathcal{S}$, $H \subset \mathcal{S}$ *is called a* space of admissible variations *of* Σ *if, for all* $y \in \Sigma$, $h \in H$, $y + h \in \Sigma$.

In these terms, we can state:

Lemma 1.6 *If* $I[y]$ *assumes a relative minimum (maximum) at* $y_o \in \Sigma$ *relative to elements* $y \in \Sigma$, *then it is necessary that*

$$I[y_o + h] - I[y_o] \geq 0 \qquad (\leq 0)$$

for all $h \in H$, *where* H *is a space of admissible variations of* Σ, *so long as* $||h|| < \delta$ *for some* $\delta > 0$.

The proof follows readily from Definitions 1.6.1 and 1.6.2.

By the methods which we have developed thus far, we shall only be able to deal with relative extreme values of functionals in those spaces of competing functions that admit a space of admissible variations H with the property that with each $h \in H$, $\lambda h \in H$ for all $\lambda \in R$. But then, Σ admits a linear space of admissible variations for the following reason: If $h,k \in H$, then $\lambda h, \mu k \in H$ for $\lambda,\mu \in R$, and hence, if $y \in \Sigma$, then $y + \lambda h = y_1 \in \Sigma$ and $y_1 + \mu k \in \Sigma$. Since $y_1 + \mu k = y + \lambda h + \mu k$, we have that $y + \lambda h + \mu k \in \Sigma$ for all $h,k \in H$, $\lambda,\mu \in R$.

Since H is a subset of a normed linear space, it follows that H is a normed linear space if it is a linear space. We give the following definition:

Definition 1.6.3 *$\mathcal{H} \subset \mathcal{S}$ is a (normed) linear space of admissible variations of $\Sigma \subset \mathcal{S}$ if*

1. *For all $y \in \Sigma$ and all $h \in \mathcal{H}$, $y + h \in \Sigma$.*
2. *For all $h,k \in \mathcal{H}$ and $\lambda,\mu \in R$, also $\lambda h + \mu k \in \mathcal{H}$.*

We note that the space of competing functions defined in (1.6.2) does not admit a linear space of admissible variations.

PROBLEMS 1.6

1. Prove: If the space of competing functions Σ is a linear space, then Σ admits a linear space of admissible variations.

*2. Define the space of competing functions for every one of the following variational problems and show that in every instance, the space of competing functions admits a linear space of admissible variations:

(a) $\displaystyle\int_a^b f(x,y(x),y'(x))\,dx \to \text{minimum},\ y(a) = y_a,\ y(b) = y_b.$

(b) $\displaystyle\int_a^b f(x,y(x),y'(x))\,dx \to \text{minimum},\ y(a) = y_a,\ y(b)$ not given.

(c) $\displaystyle\int_a^b f(x,y(x),y'(x))\,dx \to \text{minimum},\ y(a),\ y(b)$ not given.

*3. Show that

$$\Sigma = \{y \mid y \in \mathcal{C}_s^1[a,b],\ y(a) = y_a,\ y(b) = y_b\}$$

possesses

$$\mathcal{H} = \{h \mid h \in \mathcal{C}_s^1[a,b],\ h(a) = h(b) = 0\}$$

as well as

$$\mathcal{H}_P = \{h \mid h \in \mathcal{C}_{sP}^1[a,b],\ h(a) = h(b) = 0\}$$

as linear spaces of admissible variations, whereby $P = (x_1,x_2, \ldots,x_n)$ is a fixed pointset in (a,b).

1.7 FIRST NECESSARY CONDITION FOR A RELATIVE MINIMUM OF A FUNCTIONAL

Let the functional $I[y]$ be defined on a normed linear space $\mathcal{S}$, and let $\Sigma \subset \mathcal{S}$ denote the space of competing functions. We assume that Σ admits a linear space of admissible variations $\mathcal{H}$.

Suppose that $y_o \in \Sigma$ yields a relative minimum for $I[y]$ in Σ. Then, by Lemma 1.6,

$$I[y_o + h] - I[y_o] \geq 0 \qquad (1.7.1)$$

for all $h \in \mathcal{H}$ for which $||h|| < \delta$ for some $\delta > 0$.

If $I[y]$ possesses a Gâteaux variation $\delta I[h]$ at y_o, then by Theorem 1.5,

$$I[y_o + h] - I[y_o] = \delta I[h] + \varepsilon[h] \tag{1.7.2}$$

for all $h \in \mathcal{S}$ for which $||h|| < \delta$ for some $\delta > 0$. [This δ is not necessarily the same as the one above. Rather than distinguish between these δ's by using confusing subscripts, we shall agree to always use the smaller one whenever several δ's are floating around.]

Since (1.7.2) has to hold for all $h \in \mathcal{S}$ for which $||h|| < \delta$, it must hold, in particular, for all $h \in \mathcal{H}$ for which $||h|| < \delta$. From (1.7.1) and (1.7.2), we have

$$\delta I[h] + \varepsilon[h] \geq 0 \tag{1.7.3}$$

for all $h \in \mathcal{H}$ for which $||h|| < \delta$. We choose an arbitrary $h_o \in \mathcal{H}$ for which $||h_o|| < \delta$. Then, because $\mathcal{H}$ is a linear space, we have $th_o \in \mathcal{H}$ for all $t \in R$, and if $|t| \leq 1$, we have $||th_o|| < \delta$. From (1.7.3), we obtain, in view of the homogeneity of $\delta I[h]$,

$$t\delta I[h_o] + \varepsilon[th_o] \geq 0 \qquad \text{for all } |t| \leq 1.$$

If $0 < t \leq 1$,

$$\delta I[h_o] + \frac{\varepsilon[th_o]}{t} \geq 0$$

and if $-1 \leq t < 0$,

$$\delta I[h_o] + \frac{\varepsilon[th_o]}{t} \leq 0.$$

Since $\lim_{t \to 0} (\varepsilon[th_o]/t) = 0$ by Theorem 1.5, we obtain $\delta I[h_o] = 0$.

Since h_o was an arbitrary element of $\mathcal{H}$ with $||h_o|| < \delta$, we obtain immediately the necessary condition

$$\delta I[h] = 0 \tag{1.7.4}$$

for all $h \in \mathcal{H}$ for which $||h|| < \delta$ for some $\delta > 0$.

We dispose of the restriction $||h|| < \delta$ once and for all by means of the following:

Lemma 1.7 *If $L[h]$ is a homogeneous functional of the first degree which is defined on a normed linear space $\mathcal{H}$, then $L[h] = 0$ for all $||h|| < \delta$ if and only if $L[h] = 0$ for all $h \in \mathcal{H}$.*

Proof: If $L[h] = 0$ for all $h \in \mathcal{H}$, then this is trivially true for all $h \in \mathcal{H}$ for which $||h|| < \delta$.

Conversely, let us assume that $L[h] = 0$ for all $h \in \mathcal{H}$, $||h|| < \delta$. Let $h_o \in \mathcal{H}$, $h_o \neq 0$, represent an otherwise arbitrary element of $\mathcal{H}$, and take

$$h^* = \frac{\delta}{2}\frac{h_o}{||h_o||}, \qquad \text{that is, } h_o = \frac{2||h_o||}{\delta} h^*.$$

Then
$$L[h_o] = \frac{2||h_o||}{\delta} L[h^*] = 0$$

because $||h^*|| < \delta$, $h^* \in \mathcal{H}$.

From (1.7.4) and Lemma 1.7, we have:

Theorem 1.7 *If the functional $I[y]$, which is presumed to possess a Gâteaux variation at $y_o \in \Sigma \subset \mathcal{S}$, assumes a relative minimum (maximum) in Σ at $y = y_o$ and if Σ admits a linear space of admissible variations $\mathcal{H}$, then it is necessary that*

$$\delta I[h] = 0 \qquad \textit{for all } h \in \mathcal{H}.$$

Note that $I[y]$ need not be defined on the entire space Σ as long as it is defined in an open subset $Y \subset \mathcal{S}$ that contains y_o.

Theorem 1.7 is called the first necessary condition for a relative minimum (maximum) of a functional, and that part of the calculus of variations that deals with the practical interpretation of this condition is called the *theory of the first variation*.

As an application of this theorem, we consider the variational problem

$$I[y] = \int_a^b f(x,y(x),y'(x))\,dx \to \text{minimum}$$

with the boundary conditions $y(a) = y_a, y(b) = y_b$.

As in Sec. 1.5, we assume that $f \in C^1(\mathcal{R})$, where $\mathcal{R}$ denotes a domain in (x,y,y') space. Then $I[y]$ is certainly defined on an open subset $Y \subset \mathcal{C}^1[a,b]$ and the first variation exists. [See (1.5.4).] Let $y_0 \in Y$ yield a relative minimum for $I[y]$ in

$$\Sigma = \{y \mid y \in \mathcal{C}^1[a,b], y(a) = y_a, y(b) = y_b\}.$$

Then
$$\mathcal{H} = \{h \mid h \in \mathcal{C}^1[a,b], h(a) = h(b) = 0\}$$

is a normed linear space of admissible variations of Σ, and we obtain from Theorem 1.7 and in view of (1.5.4), the following necessary condition for a relative minimum (maximum):

$$\delta I[h] = \int_a^b [f_y(x,y_0(x),y_0'(x))h(x) + f_{y'}(x,y_0(x),y_0'(x))h'(x)]\,dx = 0 \qquad (1.7.5)$$

for all $h \in \mathcal{H}$.

PROBLEMS 1.7

1. For the following variational problems, define the space of competing functions and the linear space of admissible variations, and show that in every instance, $\delta I[h] = 0$ for all elements of the space of admissible variations, for $y = y_o(x)$ as specified. (It is assumed that $\mathcal{S} = \mathcal{C}^1[a,b]$.)

(a) $\displaystyle\int_0^1 (1 - y'^2(x))^2\,dx \to \text{minimum},\ y(0) = 0,\ y(1) = -1,\ y_o(x) \equiv -x$

(b) $\displaystyle\int_0^1 (1 + y'^2(x))\,dx \to \text{minimum},\ y(0) = 0,\ y(1) = 0,\ y_o(x) \equiv 0$

(c) $\displaystyle\int_0^{\pi/2} (y'^2(x) - y^2(x))\,dx \to \text{minimum},\ y(0) = 0,\ y(\pi/2) = 1,\ y_o(x) \equiv \sin x$

(d) $\displaystyle\int_0^{\log 2} y(x)\sqrt{1 + y'^2(x)}\,dx \to \text{minimum},\quad y(0) = 1,\quad y(\log 2) = \tfrac{5}{4},\quad y_o(x) \equiv$ $\cosh x$.

*2. Show that the variational problem

$$I[y] = \int_0^1 (y^2(x) + xy'(x))\,dx \to \text{minimum},\ y(0) = 0,\ y(1) = 1$$

has no solution in $C^1[0,1]$.

3. Suppose that $I[y]$ which is defined on an open subset $Y \subset \mathcal{S}$ possesses a Fréchet differential at $y = y_o \in Y$. Show that $L_f[h] = 0$ for all $h \in \mathcal{S}$ implies $L_g[h] = 0$ for all $h \in \mathcal{S}$, which in turn implies $\delta I[h] = 0$ for all $h \in \mathcal{S}$. (See also Probs. 1.5.1 and 1.5.4.)

1.8 THE SECOND GÂTEAUX VARIATION AND A SECOND NECESSARY CONDITION FOR A RELATIVE MINIMUM OF A FUNCTIONAL†

As in Sec. 1.5, we assume that the functional $I[y]$ is defined on an open subset Y of a normed linear space $\mathcal{S}$ and that $y_o \in Y$. Then $I[y_o + h]$ is defined for all $h \in \mathcal{S}$ for which $||h|| < \delta$ for some $\delta > 0$. Hence, given any $h \in \mathcal{S}$, $I[y_o + th]$ is defined for all $t \in R$, provided that t is sufficiently small.

Definition 1.8 *$\delta^2 I[h]$ is called the* second Gâteaux variation, *or simply, the* second variation, *of $I[y]$ at $y = y_o$ if, for $t \in R$,*

$$\delta^2 I[h] = \frac{d^2}{dt^2} I[y_o + th]_{t=0}$$

exists for all $h \in \mathcal{S}$.

† The reader may delay the study of the material of this section until he is about to take up Chap. 7.

The second derivative $(d^2/dt^2)\, I[y_\circ + th]_{t=0}$ in this definition is to be understood as an iterated derivative, i.e., the derivative of $(d/dt)\, I[y_\circ + th]$ at $t = 0$. Hence, the existence of $\delta^2 I[h]$ presupposes the existence of $(d/dt)\, I[y_\circ + th]$ for sufficiently small t.

We have seen in Lemma 1.5.1 that the first (Gâteaux) variation is homogeneous of the first degree in h. Similarly, the second (Gâteaux) variation is homogeneous of the second degree in h, as we shall now show:

Lemma 1.8 *The second (Gâteaux) variation $\delta^2 I[h]$ of $I[y]$ at $y = y_\circ$ is homogeneous of the second degree in h; that is, for all $h \in \mathcal{S}$ and all $\lambda \in R$,*

$$\delta^2 I[\lambda h] = \lambda^2 \delta^2 I[h].$$

Proof:

$$\delta^2 I[\lambda h] = \lim_{t\to 0} \frac{(d/dt)\, I[y_\circ + t\lambda h] - (d/dt)\, I[y_\circ + t\lambda h]_{t=0}}{t}$$

$$= \lim_{\tau\to 0} \lambda^2 \frac{[(d/d\tau)\, I[y_\circ + \tau h] - (d/d\tau)\, I[y_\circ + \tau h]_{\tau=0}]}{\tau} = \lambda^2 \delta^2 I[h].$$

The following theorem will enable us to represent the *total variation* $I[y_\circ + h] - I[y_\circ]$ in terms of the first and second variations of $I[y]$ at $y = y_\circ$.

Theorem 1.8.1 *If for all $h \in \mathcal{S}$, $(d^2/dt^2)\, I[y_\circ + th]$ exists in a neighborhood of $t = 0$ and is continuous at $t = 0$, then*

$$I[y_\circ + h] - I[y_\circ] = \delta I[h] + \tfrac{1}{2}\delta^2 I[h] + \alpha[h], \tag{1.8.1}$$

where $\lim_{t\to 0} (\alpha[th]/t^2) = 0$.

Proof: Since the existence of $(d^2/dt^2)\, I[y_\circ + th]$ near $t = 0$ implies the existence of $\delta I[h]$, we have from Taylor's formula,

$$I[y_\circ + th] - I[y_\circ] = \delta I[h]t + \frac{d^2}{ds^2} I[y_\circ + sh]_{s=\theta t} \cdot \frac{t^2}{2}, \qquad \text{where } |\theta| \le 1.$$

Since

$$\frac{d^2}{ds^2} I[y_\circ + sh]_{s=\theta t} = \delta^2 I[h] + \frac{d^2}{ds^2} I[y_\circ + sh]_{s=\theta t} - \frac{d^2}{dt^2} I[y_\circ + th]_{t=0}$$

and since $(d^2/dt^2)\, I[y_\circ + th]$ is continuous at $t = 0$, we have

$$\beta[th] = \frac{d^2}{ds^2} I[y_\circ + sh]_{s=\theta t} - \frac{d^2}{dt^2} I[y_\circ + th]_{t=0} \to 0 \text{ as } t \to 0$$

and consequently, and in view of the homogeneity of $\delta^2 I[h]$ of the second degree,

$$I[y_o + th] - I[y_o] = \delta I[th] + \frac{1}{2}\delta^2 I[th] + \beta[th]\frac{t^2}{2}.$$

For $k = th$, we have

$$I[y_o + k] - I[y_o] = \delta I[k] + \tfrac{1}{2}\delta^2 I[k] + \alpha[k],$$

where

$$\alpha[k] = \frac{t^2}{2}\beta[k],$$

and hence,

$$\lim_{t\to 0} \frac{\alpha[tk]}{t^2} = \frac{1}{2}\lim_{t\to 0} \beta[tk] = 0.$$

As in Sec. 1.5, we consider the functional

$$I[y] = \int_a^b f(x,y(x),y'(x))\,dx, \qquad y \in \mathcal{C}^1[a,b], \tag{1.8.2}$$

where we shall now assume that $f \in C^2(\mathcal{R})$, where $\mathcal{R}$ represents a domain in (x,y,y') space. If all lineal elements of $y = y_o(x)$ lie in $\mathcal{R}$, then $I[y_o + th]$ is defined for all $h \in \mathcal{C}^1[a,b]$, provided that t is sufficiently small.

We obtain for

$$\frac{d}{dt}I[y_o + th] = \int_a^b [f_y(x, y_o + th, y_o' + th')h + f_{y'}(x, y_o + th, y_o' + th')h']\,dx,$$

and hence,

$$\frac{d^2}{dt^2}I[y_o + th] = \int_a^b [f_{yy}(x, y_o + th, y_o' + th')h^2 + 2f_{yy'}(x, y_o + th, y_o' + th')hh'$$
$$+ f_{y'y'}(x, y_o + th, y_o' + th')h'^2]\,dx.$$

[To avoid cumbersome notation, we have written y_o, y_o', h, h' instead of $y_o(x)$, $y_o'(x)$, $h(x)$, $h'(x)$ in the above integrals.] We observe that this derivative exists for sufficiently small t and is continuous at $t = 0$. Hence, the representation of the total variation (1.8.1) is valid for the functional (1.8.2) under the stated conditions.

In particular, we have

$$\delta^2 I[h] = \int_a^b [f_{yy}(x,y_o(x),y_o'(x))h^2(x)$$
$$+ 2f_{yy'}(x,y_o(x),y_o'(x))h(x)h'(x) + f_{y'y'}(x,y_o(x),y_o'(x))h'^2(x)]\,dx. \tag{1.8.3}$$

We assume now that $I[y]$ possesses a relative minimum at y_o in Σ, where Σ admits a linear space of admissible variations, and that the conditions of Theorem 1.8.1 are met. Then, in view of Lemma 1.6 and Theorem

1.7, we have

$$I[y_o + h] - I[y_o] = \tfrac{1}{2}\delta^2 I[h] + \alpha[h] \geq 0 \tag{1.8.4}$$

for all $h \in \mathcal{H}$ for which $||h|| < \delta$ for some $\delta > 0$, where $\lim_{t\to 0} (\alpha[th]/t^2) = 0$.

Let $h_o \in \mathcal{H}$, $||h_o|| < \delta$. Then $th_o \in \mathcal{H}$ and $||th_o|| < \delta$ for all $|t| \leq 1$, and we have from (1.8.4),

$$I[y_o + th_o] - I[y_o] = \frac{t^2}{2}\delta^2 I[h_o] + \alpha[th_o] \geq 0$$

for all $|t| \leq 1$. Division by $t^2 \neq 0$ yields

$$\frac{1}{2}\delta^2 I[h_o] + \frac{\alpha[th_o]}{t^2} \geq 0$$

for all $|t| \leq 1$, $t \neq 0$. Since $(\alpha[th_o]/t^2) \to 0$ as $t \to 0$, we have to have $\delta^2 I[h_o] \geq 0$, and since h_o was an arbitrary element of $\mathcal{H}$, with $||h|| < \delta$, we have to have $\delta^2 I[h] \geq 0$ for all $h \in \mathcal{H}$, $||h|| < \delta$.

Because of the homogeneity of $\delta^2 I[h]$ of the second degree, one can show that $\delta^2 I[h] \geq 0$ for all $h \in \mathcal{H}$, $||h|| < \delta$ if and only if $\delta^2 I[h] \geq 0$ for all $h \in \mathcal{H}$. (See Prob. 1.8.1.) Hence we can state:

Theorem 1.8.2 *If $I[y]$, which is presumed to be defined on an open subset Y of a normed linear space $\mathcal{S}$, assumes a relative minimum (maximum) at $y = y_o$ in Σ and if Σ admits a linear space of admissible variations $\mathcal{H}$ and if $(d^2/dt^2)\, I[y_o + th]$ exists near $t = 0$ and is continuous at $t = 0$ for all $h \in \mathcal{S}$, then it is necessary that*

$$\delta I[h] = 0 \qquad \text{and} \qquad \delta^2 I[h] \geq 0 \qquad (\leq 0)$$

for all $h \in \mathcal{H}$.

[Observe that to prove this statement for a relative maximum, all one has to do is reverse the inequality sign in (1.8.4).]

If $y = y_o(x) \in \mathcal{C}^1[a,b]$ yields a relative minimum for the functional (1.8.2) in $\Sigma = \{y \mid y \in \mathcal{C}^1[a,b],\ y(a) = y_a,\ y(b) = y_b\}$, then it is necessary by Theorem 1.8.2 that

$$\delta^2 I[h] = \int_a^b [f_{yy}(x,y_o(x),y_o'(x))h^2(x) + 2f_{yy'}(x,y_o(x),y_o'(x))h(x)h'(x) + f_{y'y'}(x,y_o(x),y_o'(x))h'^2(x)]\,dx \geq 0$$

for all $h \in \mathcal{H} = \{h \mid h \in \mathcal{C}^1[a,b],\ h(a) = h(b) = 0\}$.

We have seen in Sec. 1.3 that $f''(x_o) > 0$ (<0) together with $f'(x_o) = 0$ is sufficient for f to possess a relative minimum (maximum) at $x = x_o$

(Theorem 1.3.4). One might naively assume that a similar condition would hold for a relative minimum (maximum) of a functional. However, a simple argument will reveal that this is not possible.

Suppose that $\delta I[h] = 0$ and $\delta^2 I[h] > 0$ for all $h \in \mathcal{H}$, $h \neq 0$. Suppose there is for any $\delta > 0$, no matter how small, a h_δ such that $0 < \delta^2 I[h_\delta] < |\alpha[h_\delta]|$. Then,

$$I[y_o + h_\delta] - I[y_o] = \tfrac{1}{2}\delta^2 I[h_\delta] + \alpha[h_\delta] < \tfrac{1}{2}|\alpha[h_\delta]| + \alpha[h_\delta] < 0$$

if $\alpha[h_\delta] < 0$. Neither $\delta^2 I[h_\delta] < |\alpha[h_\delta]|$ nor $\alpha[h_\delta] < 0$ can be ruled out without bringing in extraneous conditions.

In Probs. 1.8.5 to 1.8.7, we indicated how a sufficient condition may be obtained in terms of the so-called second Fréchet differential. However, this condition is not very practical, as we shall point out in Sec. 7.4. A more practical sufficient condition in terms of the second Gâteaux variation will be derived in Sec. 7.5.

PROBLEMS 1.8

*1. Prove: $\delta^2 I[h] \geq 0$ for all $h \in \mathcal{H}$, $||h|| < \delta$, if and only if $\delta^2 I[h] \geq 0$ for all $h \in \mathcal{H}$, where $\mathcal{H}$ is a normed linear space.

2. Find $\delta^2 I[h]$ for the functionals in Prob. 1.7.1*a* to *d* at $y = y_o(x)$, as specified.

3. Find $\delta^2 I[h]$ for $I[y] = \int_a^b (y'^2(x) - 4y(x)y'(x) + 2xy'^4(x))\, dx$, $y(a) = y_a$, $y(b) = y_b$, at $y_o \in \mathcal{C}^1[a,b]$.

4. Find $\delta^2 E[h]$ for $E[h] = e^{I[h]}$, where $(d/dt)\, I[y_o + th]$ and $\delta^2 I[h]$ are presumed to exist.

*5. $B[y,y]$ is a quadratic functional on $\mathcal{S}$ if $B[x,y]$, $x,y \in \mathcal{S}$, is a linear functional in x for fixed y and a linear functional in y for fixed x. (See also Sec. 1.5.) Show that a quadratic functional is homogeneous of the second degree, that is, that $B[\lambda y, \lambda y] = \lambda^2 B[y,y]$ for all $\lambda \in R$.

*6. $B_f[h,h]$ is called the second Fréchet differential of $I[y]$ at $y = y_o$ if for all $h \in \mathcal{S}$, $||h|| < \delta$,

$$I[y_o + h] - I[y_o] = L_f[h] + \tfrac{1}{2}B_f[h,h] + \beta[h],$$

where $\lim_{||h|| \to 0} (\beta[h]/||h||^2) = 0$. Prove: The second Fréchet differential is uniquely determined.

*7. Prove: If $L_f[h] = 0$ and $B_f[h,h] \geq \mu||h||^2$ for all $h \in \{h \mid h \in \mathcal{S},\ h = y - y_o,\ \forall y \in \Sigma\}$ and some $\mu > 0$, where $L_f[h]$ and $B_f[h,h]$ are the Fréchet differentials of $I[y]$ at $y = y_o$, then y_o yields a relative minimum for $I[y]$ in Σ.

8. Under what continuity and differentiability conditions on $f = f(x,y,y')$ can one guarantee the existence of the second Fréchet differential of $I[y] = \int_a^b f(x,y(x),y'(x))\, dx$ at some $y_o \in \mathcal{C}^1[a,b]$?

9. Impose suitable conditions on $f = f(x,y_1, \ldots, y_n, y_1', \ldots, y_n')$, and find $\delta^2 I[h_1, \ldots, h_n]$, where

$$I[y_1, \ldots, y_n] = \int_a^b f(x, y_1(x), \ldots, y_n(x), y_1'(x), \ldots, y_n'(x))\, dx.$$

BRIEF SUMMARY

A functional $I[y]$ which is defined on an open subset Y of a normed linear space $\mathcal{S}$ is said to have a Gâteaux variation $\delta I[h]$ at $y = y_o \in Y$ if

$$\delta I[h] = \frac{d}{dt} I[y_o + th]_{t=0}$$

exists for all $h \in \mathcal{S}$ (Sec. 1.5).

$I[y]$ has a second Gâteaux variation $\delta^2 I[h]$ at $y = y_o$ if

$$\delta^2 I[h] = \frac{d^2}{dt^2} I[y_o + th]_{t=0}$$

exists for all $h \in \mathcal{S}$ (Sec. 1.8).

$\delta I[h]$ and $\delta^2 I[h]$ are homogeneous of the first degree and the second degree, respectively (Secs. 1.5 and 1.8).

The total variation $I[y_o + h] - I[y_o]$ may be represented as follows:

$$I[y_o + h] - I[y_o] = \delta I[h] + \varepsilon[h],$$

where $\lim_{t\to 0} (\varepsilon[th]/t) = 0$, if $\delta I[h]$ exists; and

$$I[y_o + h] - I[y_o] = \delta I[h] + \tfrac{1}{2}\delta^2 I[h] + \alpha[h],$$

where $\lim_{t\to 0} (\alpha[th]/t^2) = 0$, if $(d^2/dt^2)\, I[y_o + th]$ exists near $t = 0$ and is continuous at $t = 0$ (Secs. 1.5 and 1.8).

If $y_o \in \Sigma \subset \mathcal{S}$ renders $I[y]$ a relative minimum in the space of competing functions Σ and if Σ admits a normed linear space of admissible variations $\mathcal{H}$, then it is necessary that $\delta I[h] = 0$ for all $h \in \mathcal{H}$, provided that $\delta I[h]$ exists (Sec. 1.7) and that $\delta^2 I[h] \geq 0$ under the conditions under which the representation of the total variation in terms of the second variation is valid (Sec. 1.8).

If $I[y] = \int_a^b f(x,y(x),y'(x))\, dx$ and if $f \in C^1(\mathcal{R})$, where $\mathcal{R}$ represents a domain in (x,y,y') space, then if all lineal elements of $y = y_o(x) \in \mathcal{C}^1[a,b]$ lie in $\mathcal{R}$, we have at $y = y_o$

$$\delta I[h] = \int_a^b [f_y(x,y_o(x),y_o'(x))h(x) + f_{y'}(x,y_o(x),y_o'(x))h'(x)]\, dx$$

(Sec. 1.5).

If $f \in C^2(\mathcal{R})$, then

$$\delta^2 I[h] = \int_a^b [(f_{yy})_o h^2(x) + 2(f_{yy'})_o h(x)h'(x) + (f_{y'y'})_o h'^2(x)]\, dx,$$

where $(\quad)_o$ indicates that the expression so designated is to be evaluated for $(x,y_o(x), y_o'(x))$ (Sec. 1.8).

APPENDIX

A1.9 RELATIVE EXTREME VALUES OF REAL-VALUED FUNCTIONS OF n REAL VARIABLES

Now that we have all this powerful machinery available, we might as well take time out and check briefly how the notions introduced in the preceding sections apply to the theory of extreme values of real-valued functions of n real variables.

We may view a real-valued function F of the n real variables x_1, $x_2, \ldots, x_n$ as a functional that is defined on R_n, the space of all n-tuples of real numbers. This space is, as we saw in Sec. 1.4, a linear space if we define

$$\hat{x} + \hat{y} = (x_1 + y_1, x_2 + y_2, \ldots, x_n + y_n)$$
$$\lambda\hat{x} = (\lambda x_1, \lambda x_2, \ldots, \lambda x_n),$$

where $\hat{x} = (x_1, x_2, \ldots, x_n)$ and $\hat{y} = (y_1, y_2, \ldots, y_n)$.

With

$$||\hat{x}|| = \sqrt{x_1^2 + x_2^2 + \cdots + x_n^2},$$

R_n becomes the *normed linear space* E_n.

If F is defined on an open subset Y of E_n and if $\hat{x}_o \in Y$, then $\hat{x}_o + \hat{h} \in Y$, provided that $||\hat{h}|| < \delta$ for some $\delta > 0$. Hence,

$$F[\hat{x}_o + t\hat{h}] = F(x_1^o + th_1, x_2^o + th_2, \ldots, x_n^o + th_n),$$

where $\hat{x}_o = (x_1^o, x_2^o, \ldots, x_n^o)$, is defined for all

$$||\hat{h}|| = \sqrt{h_1^2 + h_2^2 + \cdots + h_n^2} < \delta$$

and all $|t| \le 1$, and we obtain, provided that $\partial F/\partial x_i$, $i = 1, 2, \ldots, n$, exist in a neighborhood of $\hat{x}_o$ and are continuous at $\hat{x}_o$,

$$\frac{d}{dt} F[\hat{x}_o + t\hat{h}]_{t=0} = \left(\frac{\partial F}{\partial x_1}\right)_o h_1 + \cdots + \left(\frac{\partial F}{\partial x_n}\right)_o h_n,$$

where $(\quad)_o$ indicates that the function in parentheses is to be evaluated at $(x_1^o, \ldots, x_n^o)$. This is the Gâteaux variation of F at $(x_1^o, \ldots, x_n^o)$. (See Definition 1.5.)

If F has a relative minimum (maximum) at $x = x_o$, then we obtain from Theorem 1.7 the necessary condition

$$\left(\frac{\partial F}{\partial x_1}\right)_o h_1 + \cdots + \left(\frac{\partial F}{\partial x_2}\right)_o h_n = 0$$

for all $h \in E_n$, that is, for all n-tuples of real numbers $(h_1, \ldots, h_n)$. (Clearly, in this case $\mathcal{H} = E_n$.) Since this condition has to hold, in particular, for $h = (0, \ldots, 0, h_k, 0, \ldots, 0)$, $k = 1, 2, \ldots, n$, $h_k \neq 0$, we obtain the familiar necessary condition

$$\frac{\partial}{\partial x_k} F(x_1^o, \ldots, x_n^o) = 0, \qquad k = 1, 2, \ldots, n.$$

In order to find a second necessary condition, we have to assume the existence and continuity of $\partial^2 F / (\partial x_i \partial x_k)$, for all $i,k = 1, 2, \ldots, n$, in a neighborhood of the point $\hat{x}_o$. Then

$$\frac{d^2}{dt^2} F[\hat{x}_o + t\hat{h}]_{t=0} = \sum_{i=1}^{n} \sum_{k=1}^{n} \frac{\partial^2}{\partial x_i \, \partial x_k} F(x_1^o, \ldots, x_n^o) h_i h_k,$$

which is the second Gâteaux variation. Since all the hypotheses of Theorem 1.8.2 are satisfied, we obtain the second necessary condition for a relative minimum (maximum) at x_o:

$$\sum_{i=1}^{n} \sum_{k=1}^{n} \left(\frac{\partial^2 F}{\partial x_i \, \partial x_k}\right)_o h_i h_k \geq 0 \qquad (\leq 0)$$

for all $\hat{h} \in E_n$.

Finally, if

$$F[\hat{x}_o + \hat{h}] - F[\hat{x}_o] = \sum_{i=1}^{n} \left(\frac{\partial F}{\partial x_i}\right)_o h_i + \sum_{i=1}^{n} \sum_{k=1}^{n} \left(\frac{\partial^2 F}{\partial x_i \, \partial x_k}\right)_o h_i h_k + \beta[\hat{h}],$$

where $(\beta[\hat{h}]/||\hat{h}||^2) \to 0$ as $||\hat{h}|| \to 0$ (see Probs. 1.8.6 and 1.8.7), then $\sum_{i=1}^{n} \sum_{k=1}^{n} (\partial^2 F/\partial x_i \, \partial x_k)_o h_i h_k$ is the second Fréchet differential and

$$\sum_{i=1}^{n} \sum_{k=1}^{n} \left(\frac{\partial^2 F}{\partial x_i \, \partial x_k}\right)_o h_i h_k \geq \mu ||\hat{h}||^2$$

is, together with $(\partial F/\partial x_k)_o = 0$, a sufficient condition for a relative minimum at $\hat{x}_o$.

For $n = 2$, this condition assumes the familiar form

$$\left(\frac{\partial^2 F}{\partial x^2}\right)_o h^2 + 2\left(\frac{\partial^2 F}{\partial x \, \partial y}\right)_o hk + \left(\frac{\partial^2 F}{\partial y^2}\right)_o k^2 \geq \mu(h^2 + k^2),$$

where we have used the customary notation x for x_1, y for x_2, h for h_1, and k for h_2.

PROBLEMS A1.9

1. Under what conditions on $F[\hat{x}]$ does $(\beta[\hat{h}]/||\hat{h}||^2) \to 0$ as $||\hat{h}|| \to 0$ hold where $\beta[\hat{h}]$ is defined by

$$F[\hat{x}_0 + \hat{h}] - F[\hat{x}_0] = \sum_{i=1}^{n} \left(\frac{\partial F}{\partial x_i}\right)_o h_i + \sum_{i=1}^{n} \sum_{k=1}^{n} \left(\frac{\partial^2 F}{\partial x_i \, \partial x_k}\right)_o h_i h_k + \beta[\hat{h}]?$$

2. Show that

$$Q = \sum_{i=1}^{n} \sum_{k=1}^{n} a_{ik} h_i h_k \geq \mu ||\hat{h}||^2, \qquad \mu > 0$$

is equivalent to

$$Q = \sum_{i=1}^{n} \sum_{k=1}^{n} a_{ik} h_i^o h_k^o \geq \mu$$

for all $||\hat{h}^o|| = 1$.

3. A quadratic form is called *positive definite* if

$$Q = \sum_{i=1}^{n} \sum_{k=1}^{n} a_{ik} h_i h_k > 0$$

for all $(h_1, h_2, \ldots, h_n) \neq (0, 0, \ldots, 0)$. Show that this definition is equivalent to the condition in problem 2.

CHAPTER 2

THE THEORY OF THE FIRST VARIATION

APPENDIX

2.1 WEAK AND STRONG RELATIVE EXTREME VALUES

We assume that the functional $I[y]$ is defined on an open subset Y of a normed linear space $\mathcal{S}$. The functionals we are dealing with in the calculus of variations are typically defined on subsets of the function spaces $C^1[a,b]$ or $C_s^1[a,b]$ in which a suitable norm has been defined. Depending on the type of norm which is introduced, different concepts of relative extreme values will emerge.

$$||f||_w = \max_{[a,b]} |f(x)| \tag{2.1.1}$$

exists in $C^1[a,b]$ as well as in $C_s^1[a,b]$ and satisfies in both cases the conditions for a norm. We call the norm defined in (2.1.1) a *weak* norm, and we denote the *weakly normed* linear spaces $C^1[a,b]$ and $C_s^1[a,b]$ by $\bar{\mathcal{C}}^1[a,b]$ and $\bar{\mathcal{C}}_s^1[a,b]$, respectively.

Also

$$||f|| = \max_{[a,b]} |f(x)| + \sup_{[a,b]} |f'(x)| \tag{2.1.2}$$

exists in $C^1[a,b]$ as well as in $C_s^1[a,b]$ and satisfies the conditions for a norm. (In the case of $C^1[a,b]$, $\sup_{[a,b]} |f'(x)|$ may be replaced by $\max_{[a,b]} |f'(x)|$.) We call the norm defined in (2.1.2) a *strong* norm, and we denote the *strongly normed* linear spaces $C^1[a,b]$ and $C_s^1[a,b]$ by $\mathcal{C}^1[a,b]$ and $\mathcal{C}_s^1[a,b]$, respectively.

Suppose that $f = f(x,y,y')$ is continuous for all $(x,y) \in \mathfrak{D}$, where $\mathfrak{D}$ is a domain in the x,y plane and for all $-\infty < y' < \infty$. Then

$$I[y] = \int_a^b f(x,y(x),y'(x))\,dx$$

is obviously defined on an open subset $Y \subset \bar{\mathcal{C}}^1[a,b]$ or $Y \subset \bar{\mathcal{C}}_s^1[a,b]$ because, if $y = y_o(x) \in Y$, where $Y = \{y \mid (x,y(x)) \in \mathfrak{D},\ x \in [a,b]\}$, then $y_o + h \in Y$, provided that $||h||_w$ is sufficiently small.

On the other hand, if $f = f(x,y,y')$ is continuous for all $(x,y,y') \in \mathcal{R}$, where $\mathcal{R}$ is a domain in (x,y,y') space, then $I[y]$ is defined on an open subset $Y \subset \mathcal{C}^1[a,b]$ or $Y \subset \mathcal{C}_s^1[a,b]$, as we have already discussed in Sec. 1.4, page 21.

Depending on whether we have a weakly or a strongly normed linear space, we speak of weak and strong neighborhoods:

Definition 2.1.1 *A δ neighborhood $N_w^\delta(y_o)$ $[N^\delta(y_o)]$ of $y = y_o(x) \in \bar{\mathcal{C}}^1[a,b]$ or $\bar{\mathcal{C}}_s^1[a,b]$ $[\mathcal{C}^1[a,b]$ or $\mathcal{C}_s^1[a,b]]$, that is, the pointset $\{(x,y) \mid x \in [a,b], |y - y_o(x)| < \delta\}$ $[\{(x,y,y') \mid x \in [a,b], |y - y_o(x)| < \delta, |y' - y_o'(x)| < \delta\}]$ is called a* weak [strong] *δ neighborhood of $y = y_o(x)$.*

We say that $y = y(x)$, $x \in [a,b]$, lies in a weak [strong] δ neighborhood of $y = y_o(x)$, $x \in [a,b]$, if

$$(x,y(x)) \in N_w^\delta(y_o) \; [(x,y(x),y'(x)) \in N^\delta(y_o)]$$

for all $x \in [a,b]$, and we denote it by $y \in N_w^\delta(y_o)[y \in N^\delta(y_o)]$. Clearly, if $||y - y_o||_w < \delta$, then $y \in N_w^\delta(y_o)$, and if $||y - y_o|| < \delta$, then $y \in N^\delta(y_o)$. (See also Probs. 2.1.7 and 2.1.8.) It is quite obvious that if $y \in N^\delta(y_o)$, then $y \in N_w^\delta(y_o)$. (See Prob. 2.1.1.)

Suppose we consider the variational problem

$$I[y] = \int_a^b f(x,y(x),y'(x))\,dx \to \text{minimum}$$

$$y(a) = y_a, \qquad y(b) = y_b.$$

Then, depending on whether $Y \subset \bar{\mathcal{C}}^1[a,b]$ or $Y \subset \mathcal{C}^1[a,b]$, the class of competing functions that also lie in Y will look either like those in Fig. 2.1 or like those in Fig. 2.2. Accordingly, we shall talk about *strong variations* in weakly normed spaces and *weak variations* in strongly normed spaces.

In the same vein, we have:

Definition 2.1.2 *$y_o \in \Sigma$ yields a* strong [weak] *relative minimum for $I[y]$ in Σ, where Σ denotes the space of competing functions, if*

$$I[y] - I[y_o] \geq 0$$

for all those functions $y \in \Sigma$ for which $y \in N_w^\delta(y_o)$ $[y \in N^\delta(y_o)]$, for some $\delta > 0$. (See also Prob. 2.1.9.)

The following theorem is fairly obvious:

Theorem 2.1 *If y_o yields a strong relative minimum (maximum) for $I[y]$,*

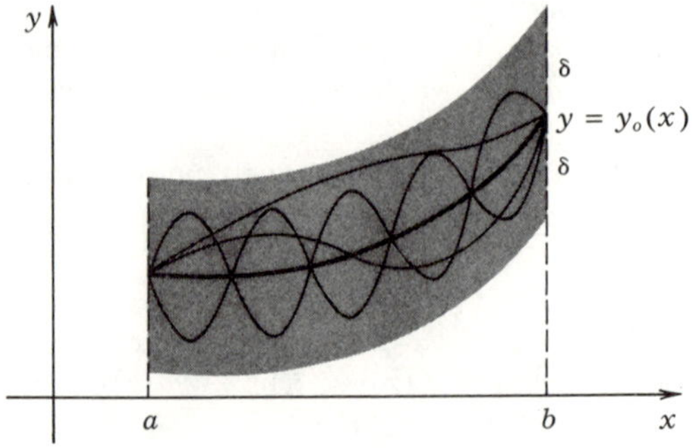

Figure 2.1

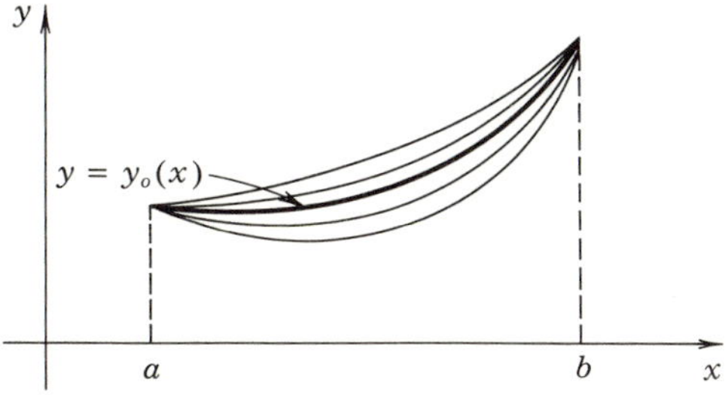

Figure 2.2

then it necessarily yields a weak relative minimum (maximum) for $I[y]$. (See Prob. 2.1.2.)

Hence, a necessary condition for a weak relative minimum (maximum) is also a necessary condition for a strong relative minimum (maximum).

In this chapter, we shall derive necessary conditions for weak relative minima (maxima), employing the theory that was developed in Chap. 1. In Chap. 3 (see Theorem 3.9.1), we shall derive a necessary condition for a strong relative minimum (maximum) directly, circumventing Theorem 1.7 and employing a technique that is not related to the theory of the first variation.

PROBLEMS 2.1

*1. Prove: If $y \in N^\delta(y_o) \subset \mathcal{C}^1[a,b]$, then $y \in N_w{}^\delta(y_o) \subset \bar{\mathcal{C}}^1[a,b]$.

*2. Prove Theorem 2.1. (Give proof by contradiction.)

3. Show that $y = y(x) \equiv 0$ cannot possibly yield a strong relative minimum for $\int_0^\pi y^2(x)(1 - y'^2(x))\,dx$, $y(0) = y(\pi) = 0$. (*Hint*: Consider $y = (1/\sqrt{n}) \sin nx$.)

4. In $y = y(x) \equiv \lambda x(1 - x)$, $x \in [0,1]$, find upper and lower bounds for λ so that:
   (*a*) $y \in N^\delta(0)$ for $\delta = \frac{1}{2}$ (*b*) $y \in N_w{}^\delta(0)$ for $\delta = \frac{1}{2}$

5. In $y = y(x) \equiv x + h(x)$, $x \in [0,1]$, choose a function $h(x)$ so that:
   (*a*) $y = x + h(x)$ is a weak variation of $y = x$.
   (*b*) $y = x + h(x)$ is a strong variation of $y = x$ but not a weak variation of x.

6. Consider the normed linear space of all $y \in C^2[a,b]$ with

$$||y|| = \max_{[a,b]} |y(x)| + \max_{[a,b]} |y'(x)| + \max_{[a,b]} |y''(x)|.$$

   Define open sets and neighborhoods in this space and interpret geometrically. Define also what you might call a "really weak minimum" of a functional that is defined on this space.

*7. Show: $y \in N_w{}^\delta(y_o)$, $y,y_o \in C[a,b]$, if and only if $||y - y_o||_w < \delta$.

*8. Show: If $||y - y_o|| < \delta$, $y,y_o \in C_s{}^1[a,b]$, then $y \in N^\delta(y_o)$, and if $y \in N^\delta(y_o)$, then $||y - y_o|| < 2\delta$.

*9. Show: If $I[y] - I[y_o] \geq 0$ for all $y \in \Sigma$ for which $y \in N^\delta(y_o)$, then $I[y_o + h] - I[y_o] \geq 0$ for all $h \in H$ for which $||h|| < \delta$, where H is the space of admissible variations of Σ.

2.2 FIRST NECESSARY CONDITION FOR THE SIMPLEST VARIATIONAL PROBLEM

We consider the variational problem

$$I[y] = \int_a^b f(x,y(x),y'(x))\, dx \to \text{minimum}, \tag{2.2.1}$$

where $y \in \Sigma$ or $y \in \Sigma_s$, whereby

$$\Sigma = \{y \mid y \in \mathcal{C}^1[a,b],\, y(a) = y_a,\, y(b) = y_b\} \tag{2.2.2}$$

and

$$\Sigma_s = \{y \mid y \in \mathcal{C}_s^1[a,b],\, y(a) = y_a,\, y(b) = y_b\}. \tag{2.2.3}$$

If we consider the problem in Σ, we choose as the space of admissible variations

$$\mathcal{H} = \{h \mid h \in \mathcal{C}^1[a,b],\, h(a) = 0,\, h(b) = 0\}, \tag{2.2.4}$$

which, with

$$||h|| = \max_{[a,b]} |h(x)| + \max_{[a,b]} |h'(x)|, \tag{2.2.4a}$$

is a normed linear space.

If we consider the problem in Σ_s and if the solution $y_o \in \Sigma_s$, $y = y_o(x)$, has a derivative that has jump discontinuities on the finite pointset $P = \{x_1,x_2, \ldots,x_n\}$, then we choose as space of admissible variations

$$\mathcal{H}_s = \{h \mid h \in \mathcal{C}_{s_P}{}^1[a,b],\, h(a) = 0,\, h(b) = 0\}, \tag{2.2.5}$$

which is really a subset of the "natural" space of admissible variations (see Prob. 1.6.3), and we introduce the norm

$$||h|| = \max_{[a,b]} |h(x)| + \sup_{[a,b]} |h'(x)| \tag{2.2.5a}$$

If we assume that $f \in C^1(\mathcal{R})$, where $\mathcal{R}$ is a domain in the (x,y,y') space that contains all lineal elements of $y = y_o(x)$, $x \in [a,b]$, including $(x_k, y_o(x_k), y_o'(x_k \pm 0))$ at all points $x_k \in [a,b]$ where y_o' has a jump discontinuity, then we obtain as a necessary condition for a weak relative minimum (maximum)—and also for a strong relative minimum (maximum), in view of Theorem 2.1—from Theorem 1.7, and in view of (1.7.5),

$$\delta I[h] = \int_a^b [f_y(x,y_o(x),y_o'(x))h(x) + f_{y'}(x,y_o(x),y_o'(x))h'(x)]\, dx = 0 \tag{2.2.6}$$

for all $h \in \mathcal{H}$ (or $h \in \mathcal{H}_s$).

This condition is anything but practical. We shall demonstrate in the next section how a more useful condition can be obtained from (2.2.6) by elimination of any reference to the space of admissible variations.

All the considerations of this and the following sections are, of course, based on the assumption that the functional (2.2.1) possesses a relative minimum (maximum) in Σ or in Σ_s. It is possible that a relative minimum (maximum) exists neither in Σ nor in Σ_s. It is also conceivable that a relative minimum (maximum) exists in Σ_s but not in Σ (see the example at the beginning of Sec. 2.9), that one exists in Σ but not in Σ_s, and that different relative minima (maxima) exist in Σ and Σ_s. More will be said about these various possibilities in Sec. 2.10.

PROBLEMS 2.2

1. Find f_y, $f_{y'}$, and $(d/dx)\, f_{y'}$ where:

   (a) $f = \dfrac{\sqrt{1 + y'^2}}{\sqrt{y}}$ (b) $f = n(x,y)\sqrt{1 + y'^2}$

   (c) $f = y\sqrt{1 + y'^2}$ (d) $f = y'^2 - y^2$

   (e) $f = (1 - y'^2)^2$
2. Let $f_{y'y'} = 0$ for all y'. Find the most general form of $f(x,y,y')$ under these circumstances, with the understanding that $f \in C^2(\mathfrak{R})$.
3. The variational problem

$$\int_0^1 \sqrt{1 + y'^2(x)}\, dx \to \text{minimum}, \qquad y(0) = 0, \qquad y(1) = 1$$

   has the solution $y = x$. Find a family of competing functions in Σ.
4. Show that $y = x$ satisfies condition (2.2.6) for the variational problem in problem 3.

2.3 THE EULER-LAGRANGE EQUATION

We shall now use the formulation of a necessary condition for a relative minimum (maximum) which we obtained in the preceding section, namely,

$$\int_a^b [f_y(x,y_o(x),y_o'(x))h(x) + f_{y'}(x,y_o(x),y_o'(x))h'(x)]\, dx = 0 \qquad (2.3.1)$$

for all $h \in \mathcal{H}$ ($h \in \mathcal{H}_s$), as a point of departure. It will suffice to consider the case $h \in \mathcal{H}_s$ since the case $h \in \mathcal{H}$ is obviously contained therein.

We denote

$$\int_a^x f_y(t,y_o(t),y_o'(t))\, dt = \phi(x) \qquad (2.3.2)$$

and note that

$$\phi \in C_{sp}{}^1[a,b] \qquad \text{if } y_o \in C_{sp}{}^1[a,b]. \qquad (2.3.3)$$

We apply integration by parts to the first term of (2.3.1) and obtain

$$\int_a^b f_y(x,y_o(x),y_o'(x))h(x)\, dx = h(x)\phi(x)\Big|_a^b - \int_a^b \phi(x)h'(x)\, dx.$$

Since $h(a) = h(b) = 0$ and since ϕ is continuous, we have

$$h(x)\phi(x)\Big|_a^b = 0,$$

and consequently,

$$\int_a^b [f_y(x,y_o(x),y_o'(x))h(x) + f_{y'}(x,y_o(x),y_o'(x))h'(x)]\,dx$$

$$= \int_a^b h'(x)[f_{y'}(x,y_o(x),y_o'(x)) - \phi(x)]\,dx.$$

In view of this, we can restate (2.3.1) as

$$\int_a^b h'(x)[f_{y'}(x,y_o(x),y_o'(x)) - \phi(x)]\,dx = 0 \qquad (2.3.4)$$

for all $h \in \mathcal{H}_s$.

The following lemma, which is due to *Dubois-Reymond*, will enable us to transform this condition still further.

Lemma 2.3: Lemma of Dubois-Reymond *If* $M \in C_{sP}[a,b]$ *and if* $\int_a^b M(x)h'(x)\,dx = 0$ *for all* $h \in \mathcal{H}_s$*, where* $\mathcal{H}_s$ *is defined in (2.2.5), then it is necessary that for some constant* C,

$$M(x) = C$$

for all $x \in [a,b]$ *except at* $x_1, x_2, \ldots, x_n$.

Proof: Let

$$C = \frac{1}{b-a}\int_a^b M(x)\,dx.$$

Then
$$\int_a^b [C - M(x)]\,dx = 0. \qquad (2.3.5)$$

If we choose

$$h = h(x) \equiv \int_a^x [C - M(t)]\,dt,$$

we see that $h \in C_{sP}{}^1[a,b]$ and, because of (2.3.5), $h(a) = h(b) = 0$. Hence, $h \in \mathcal{H}_s$

With this particular h and by the hypothesis of the lemma,

$$\int_a^b M(x)h'(x)\,dx = \int_a^b M(x)[C - M(x)]\,dx = 0. \qquad (2.3.6)$$

We multiply (2.3.5) by C and (2.3.6) by (-1) and add both equations:

$$C\int_a^b [C - M(x)]\, dx - \int_a^b M(x)[C - M(x)]\, dx = 0,$$

or as we may put it,

$$\int_a^b [C - M(x)]^2\, dx = 0.$$

Since $[C - M(x)]^2 \geq 0$, it follows readily that $M(x) = C$ for all $x \in [a,b]$, with the possible exception of $x_1, x_2, \ldots, x_n$. Hence, these points of discontinuity are, in fact, removable.

We shall now utilize this lemma to transform (2.3.4) into a more practical and more meaningful condition.

Since

$$f_{y'}(x,y_o(x),y_o'(x)) - \phi(x) = M(x) \in C_{sP}[a,b],$$

we obtain because of (2.3.4) from the lemma of Dubois-Reymond that

$$f_{y'}(x,y_o(x),y_o'(x)) - \phi(x) = C$$

for all $x \in [a,b]$ except $x_1, x_2, \ldots, x_n$. We may write this, in view of the definition of ϕ by (2.3.2), as

$$f_{y'}(x,y_o(x),y_o'(x)) = \int_a^x f_y(t,y_o(t),y_o'(t))\, dt + C.$$

Hence we can state:

Theorem 2.3: First necessary condition for a relative minimum *For $y_o \in C_{sP}{}^1[a,b]$ to yield a relative minimum (maximum) for the integral $I[y] = \int_a^b f(x,y(x),y'(x))\, dx$ in $\Sigma_s = \{y \mid y \in C_s{}^1[a,b],\ y(a) = y_a,\ y(b) = y_b\}$, where it is assumed that $f \in C^1(\mathfrak{R})$, $\mathfrak{R}$ representing a domain in the (x,y,y') space which contains all lineal elements of y_o, it is necessary that there be a constant C such that the integro-differential equation*

$$f_{y'}(x,y(x),y'(x)) = \int_a^x f_y(t,y(t),y'(t))\, dt + C \tag{2.3.7}$$

is satisfied by $y = y_o(x)$ for all $x \in [a,b]$ except for the points where y_o' has a jump discontinuity.

Equation (2.3.7) is called the *Euler-Lagrange equation in integrated form.*

Corollary 1 to Theorem 2.3 *If* $y_o \in C_{sp}^1[a,b]$ *minimizes the functional* $I[y]$ *under the conditions that are stated in Theorem 2.3, then it is necessary that every smooth portion of* y_o *satisfy the differential equation*

$$f_y(x,y,y') - \frac{d}{dx} f_{y'}(x,y,y') = 0. \tag{2.3.8}$$

To see this, we only have to note that

$$\phi(x) = \int_a^x f_y(t,y_o(t),y_o'(t))\,dt \in C^1(x_k,x_{k+1}), \qquad k = 0, 1, \ldots, n,$$

where $x_o = a$, $x_{n+1} = b$.

The differential equation (2.3.8) is called the *Euler-Lagrange equation* in honor of the Swiss mathematician *Leonhard Euler* (1707–1783) and the French mathematician *Joseph Louis de Lagrange* (1736–1813). Euler established this differential equation in 1744 as a necessary condition for a minimum of an integral of the type (2.2.1) by a heuristic method which, though ingenious, lacked mathematical rigor. He replaced the curve which is assumed to render the minimum by a polygon and then considered variations of the ordinates of the vertices one at a time.† He was justifiably criticized by Lagrange, who derived the equation essentially along the lines followed in this section, except that he applied integration by parts to the second term in the integral (2.3.1). His derivation was also incomplete, as we shall point out in the next section.

For the case where $y_o \in C^1[a,b]$, $h \in \mathcal{H}$, we obtain from Theorem 2.3:

Corollary 2 to Theorem 2.3 *For* $y_o \in C^1[a,b]$ *to yield a relative minimum (maximum) for* $I[y]$ *in* $\Sigma = \{y \mid y \in C^1[a,b],\ y(a) = y_a,\ y(b) = y_b\}$, *where it is assumed that* $f \in C^1(\mathcal{R})$, $\mathcal{R}$ *representing a domain of the* (x,y,y') *space which contains all lineal elements of* y_o, *it is necessary that* y_o *satisfy the differential equation*

$$f_y(x,y,y') - \frac{d}{dx} f_{y'}(x,y,y') = 0$$

for all $x \in [a,b]$.

If we carry out the total differentiation with respect to x in the second term of the Euler-Lagrange equation (2.3.8), we obtain, provided that y''

† A modernized version of Euler's method can be found in I. M. Gelfand and S. V. Fomin, "Calculus of Variations," pp. 27ff, Prentice-Hall, Inc., Englewood Cliffs, N.J., 1963.

exists and the second-order partial derivatives of $f(x,y,y')$ are continuous,

$$\frac{d}{dx} f_{y'} = f_{y'x} + f_{y'y}y' + f_{y'y'}y'',$$

and hence we obtain for the Euler-Lagrange equation

$$f_y - f_{y'x} - f_{y'y}y' - f_{y'y'}y'' = 0. \tag{2.3.9}$$

Since $f = f(x,y,y')$ is, in general, a nonlinear function of y and y', we see from (2.3.9) that the Euler-Lagrange equation is, in general, a nonlinear differential equation of second order. This equation and its solutions will be discussed in some detail in Sec. 2.5.

PROBLEMS 2.3

1. Give an alternate proof for Lemma 2.3 by choosing h' as indicated in Fig. 2.3 so that $h \in \mathcal{H}$.

Figure 2.3

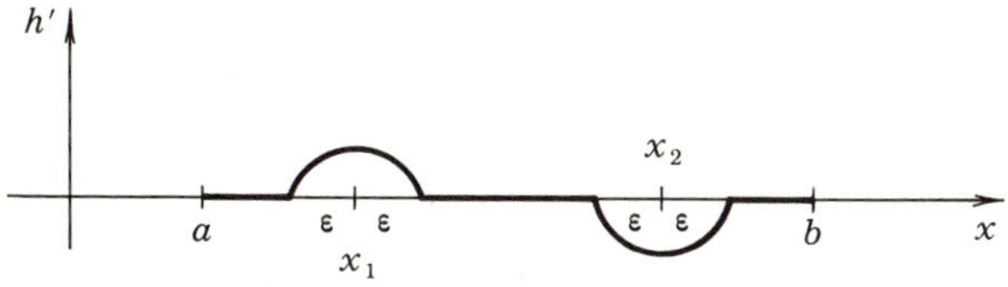

2. Show that the condition stated in Lemma 2.3 is also sufficient for $\int_a^b M(x)h'(x)\,dx = 0$.
3. $N \in C[a,b]$, $N^2(x) = 0$ for all $x \in [a,b]$. Show that $N(x) = 0$ for all $x \in [a,b]$.
4. Write the Euler-Lagrange equation for the following functions:

(a) $f(x,y,y') = \sqrt{1 + y'^2}$ (b) $f(x,y,y') = y\sqrt{1 + y'^2}$

(c) $f(x,y,y') = \dfrac{\sqrt{1 + y'^2}}{\sqrt{y}}$ (d) $f(x,y,y') = \sqrt{1 + y'^2} - \lambda y$

(e) $f(x,y,y') = y'^2 - y^2$ (f) $f(x,y,y') = \sin(xy')$

(g) $f(x,y,y') = \dfrac{x^2y'}{\sqrt{1 + y'^2}}$

5. Suppose that $f_{y'y'} = 0$, that is, $f(x,y,y') = \alpha(x,y) + \beta(x,y)y'$, where we shall assume that $\alpha(x,y),\beta(x,y) \in C^1\{(x,y) \mid a \le x \le b, -\infty < y < \infty\}$. Establish a necessary condition for $y = y_0(x)$ to yield a relative minimum for

$$I[y] = \int_a^b [\alpha(x,y(x)) + \beta(x,y(x))y'(x)]\,dx, \qquad y(a) = y_a, \qquad y(b) = y_b.$$

Interpret this condition.

2.4 LAGRANGE'S METHOD

In line with Lagrange's argument, we assume that $y_o \in C^1[a,b]$ yields a relative minimum for $I[y]$ in Σ, and we subject the second term of the integrand in (2.3.1) to integration by parts:

$$\int_a^b f_{y'}(x,y_o(x),y_o'(x))h'(x)\,dx = h(x)f_{y'}(x,y_o(x),y_o'(x))\Big|_a^b - \int_a^b h(x)\frac{d}{dx}f_{y'}(x,y_o(x),y_o'(x))\,dx.$$

In view of

$$h(a) = h(b) = 0 \qquad \text{and} \qquad f_{y'} \in C(\mathcal{R}),$$

we obtain, instead of (2.3.4), the following necessary condition:

$$\int_a^b h(x)\Big[f_y(x,y_o(x),y_o'(x)) - \frac{d}{dx}f_{y'}(x,y_o(x),y_o'(x))\Big]\,dx = 0 \quad (2.4.1)$$

for all $h \in \mathcal{H}$.

For the moment, we shall put aside the question about the circumstances under which this integration-by-parts process may be carried out and proceed with Lagrange's argument: Lagrange claimed without proof that from (2.4.1) it follows necessarily that the coefficient of $h(x)$ in the integrand has to vanish by itself. This is, of course, true, as we shall see in a moment, but as Euler pointed out in a communication to Lagrange, the statement is not evident and needs to be proved. Such a proof was supplied by Dubois-Reymond, who thus furnished the missing link in Lagrange's argument. Here it is:

Lemma 2.4: The fundamental lemma of the calculus of variations *If $M \in C[a,b]$ and*

$$\int_a^b M(x)h(x)\,dx = 0$$

for all $h \in \mathcal{H}$, then it is necessary that

$$M(x) = 0 \qquad \textit{for all } x \in [a,b].$$

(That the condition is also sufficient is obvious.)

Proof (by contradiction): Suppose that there exists an $x_o \in (a,b)$ such that $M(x_o) \neq 0$. Then we may assume without loss of generality that $M(x_o) > 0$. Since $M \in C[a,b]$, there exists an $N^\delta(x_o) \subset (a,b)$ such that

$$M(x) > 0 \qquad \text{for all } x \in N^\delta(x_o).$$

Let

$$h = h(x) \equiv \begin{cases} (x - x_o - \delta)^2(x - x_o + \delta)^2 & \text{for } x \in N^\delta(x_o) \\ 0 & \text{for } x \notin N^\delta(x_o). \end{cases}$$

Clearly, $h \in \mathcal{H}$.

With this choice of h,

$$\int_a^b M(x)h(x)\,dx = \int_{x_o-\delta}^{x_o+\delta} M(x)(x - x_o - \delta)^2(x - x_o + \delta)^2\,dx > 0,$$

which contradicts our hypothesis. Thus,

$$M(x) = 0 \qquad \text{for all } x \in (a,b).$$

That $M(x)$ also has to vanish at the endpoints of the interval follows from the continuity of M.

In order to apply this fundamental lemma of the calculus of variations to (2.4.1), we have to be sure that

$$M = f_y - \frac{d}{dx}f_{y'} = f_y - f_{y'x} - f_{y'y}y' - f_{y'y'}y'',$$

as a function of x, is continuous in $[a,b]$. Therefore, we shall have to make the additional assumption that $y'' \in C[a,b]$ and $f \in C^2(\mathcal{R})$. Now that we have made—have been forced into, as a matter of fact—this assumption we can see that the integration-by-parts process which we carried out in the beginning of this section is also justified.

Now, if $y'' \in C[a,b]$ and all the other continuity assumptions are satisfied, we can apply Lemma 2.4 to (2.4.1), and we arrive again at the Euler-Lagrange equation (2.3.8).

We can clearly see now that the derivation of the Euler-Lagrange equation which we presented in Sec. 2.3 is much superior to the derivation of this section, because the previous method did not require any assumption on y''—an assumption, incidentally, that is not at all motivated by the variational problem itself. We shall see in the next section under what additional conditions on f this assumption may be justified after the fact.

PROBLEMS 2.4

*1. Prove the fundamental lemma of the calculus of variations without the restriction $h(a) = h(b) = 0$.

2. Prove: If $M \in C[a,b]$ and $\int_a^b h''(x)M(x)\,dx = 0$ for all $h \in C^2[a,b]$ for which $h(a) = h(b) = h'(a) = h'(b) = 0$, then $M(x) = a_o + a_1x$ for some constants a_o, a_1.

3. Prove: If $M \in C[a,b]$ and $\int_a^b h^{(n)}(x)M(x)\,dx = 0$ for all $h \in C^{(n)}[a,b]$, where $h(a) = h(b) = h'(a) = h'(b) = \cdots = h^{(n-1)}(a) = h^{(n-1)}(b) = 0$, then $M(x) = a_o + a_1x + \cdots + a_{n-1}x^{n-1}$ for some constants $a_o, a_1, \ldots, a_{n-1}$.
4. Generalize the result of problem 3 to the case where $M \in C_s[a,b]$ and $h \in C_s^{(n)}[a,b]$, and show that the fundamental lemma of the calculus of variations and the lemma of Dubois-Reymond are special cases of this general result.
5. Make suitable assumptions and derive the Euler-Lagrange equation for the variational problem

$$\int_a^b f(x,y(x),y'(x),y''(x))\,dx \to \text{minimum},$$

$y(a) = y_a$, $y(b) = y_b$, $y'(a) = y'_a$, $y'(b) = y'_b$. (*Hint*: Use the result of problem 2.)
6. Same as in problem 5 for the variational problem

$$\int_a^b f(x,y(x),y'(x),y''(x),\ldots,y^{(n)}(x))\,dx \to \text{minimum},$$

$y(a) = y_a$, $y(b) = y_b$, $y'(a) = y'_a$, $y'(b) = y'_b$, $\ldots$, $y^{(n-1)}(a) = y_a^{(n-1)}$, $y^{(n-1)}(b) = y_b^{(n-1)}$. (*Hint*: Use the result of problem 3.)

2.5 DISCUSSION OF THE EULER-LAGRANGE EQUATION

As we noted at the end of Sec. 2.3, the Euler-Lagrange equation is a second-order differential equation which is, in general, nonlinear in y and y'. In (2.3.9) we wrote its explicit form:

$$f_y - f_{y'x} - f_{y'y}y' - f_{y'y'}y'' = 0,$$

where $f = f(x,y,y')$. We solve for y'' and obtain

$$y'' = \frac{1}{f_{y'y'}}(f_y - f_{y'x} - f_{y'y}y'). \tag{2.5.1}$$

We see that, so far as the solution of this equation is concerned, a great deal will depend on whether $f_{y'y'}$ vanishes or does not vanish.

In accordance with the terminology that is prevalent in the theory of ordinary differential equations, we introduce the following term:

Definition 2.5.1 (x_o,y_o,y'_o) *is called a* regular lineal element *if*

$$f_{y'y'}(x_o,y_o,y'_o) \neq 0,$$

and it is called a singular lineal element *if* $f_{y'y'}(x_o,y_o,y'_o) = 0$.

In order to streamline our statements concerning the solution of the Euler-Lagrange equation and to avoid awkward and complicated circumlocutions, we introduce the term *extremal*:

Definition 2.5.2 *A solution* $y = y_o(x) \in C^1[a,b]$ *of the Euler-Lagrange* *equation (2.3.8) is called an* extremal. *A* regular extremal *is an extremal that consists of regular lineal elements only.*

The choice of the term *extremal* is not a very fortunate one, but this term is so solidly entrenched in the literature on the calculus of variations that any attempt to eliminate it would be doomed to failure at the outset. The word "extremal" carries the connotation of something that renders an extremum, and this is not necessarily the case so far as the solutions of the Euler-Lagrange equations are concerned. All we have shown is that when a smooth function renders $I[y]$ a relative extremum, then it is necessarily an extremal. We shall see, however, that in many cases an extremal does not yield a relative extremum for $I[y]$; hence an extremal is not necessarily the solution of a variational problem. Considering that a point x_o for which $f'(x_o) = 0$ is called a *critical point* in the theory of extreme values of functions of a real variable, it may be more appropriate to call the smooth solutions of the Euler-Lagrange equation *critical functions.*

In this section we shall study the solutions of the Euler-Lagrange equation and their properties without regard to the boundary conditions that are associated with the variational problem. For this purpose, it is of advantage to transform the Euler-Lagrange equation into a system of two first-order equations so that the standard existence and uniqueness theorems from the theory of ordinary differential equations become applicable.

Although one may obtain such a system of first-order equations immediately from (2.5.1) by introduction of $z = y'$ as a new unknown function, we shall use a different transformation, which will have the advantage that the system thus obtained will satisfy the hypotheses of the existence and uniqueness theorems under weaker conditions on f than will the system that can be directly obtained from (2.5.1). (See Prob. 2.5.3.)

Instead of introducing $z = y'$ as a new unknown function, we introduce

$$p = f_{y'}(x,y,y'). \tag{2.5.2}$$

In order to study the possibility of solving this for y', we make the following hypotheses:

Let (x_o,y_o,y'_o) be a *regular lineal element*, and let $\mathcal{R}_o$ denote the three-dimensional domain (open, simply connected pointset)

$$\mathcal{R}_o = \{(x,y,y') \mid |x - x_o| < \alpha,\ |y - y_o| < \beta,\ |y' - y'_o| < \gamma\}. \tag{2.5.3}$$

Let $\mathcal{S}$ denote the four-dimensional domain

$$\mathcal{S} = \{(x,y,y',p) \mid |x - x_o| < \alpha,\ |y - y_o| < \beta,\ |y' - y'_o| < \gamma,\ |p - p_o| < \delta\},$$

where $p_o = f_{y'}(x_o,y_o,y'_o)$.

We shall assume that

$$f(x,y,y') \in C^2(\mathcal{R}_o).$$

Then
$$f_{y'}(x,y,y') - p \in C^1(\mathcal{S}),$$
$$f_{y'}(x_o,y_o,y'_o) - p_o = 0 \qquad \text{(by definition of } p_o\text{)},$$
$$f_{y'y'}(x_o,y_o,y'_o) \neq 0$$

[since (x_o,y_o,y'_o) is assumed to be a regular lineal element].

Hence, by the theorem on implicit functions,† there exists in $\mathcal{S}$ an open right parallelepiped

$$\mathcal{P} = \{(x,y,y',p) \mid |x - x_o| < \alpha_o, |y - y_o| < \beta_0, |y' - y'_o| < \gamma_o, |p - p_o| < \delta_o\}$$

such that in

$$\mathcal{P}_o = \{(x,y,p) \mid |x - x_o| < \alpha_o, |y - y_o| < \beta_o, |p - p_o| < \delta_o\}, \tag{2.5.4}$$

we may solve

$$f_{y'}(x,y,y') - p = 0$$

uniquely for y':

$$y' = \Phi(x,y,p) \tag{2.5.5}$$

where $|y' - y'_o| < \gamma_o$ for $(x,y,p) \in \mathcal{P}_o$ and where $\Phi \in C^1(\mathcal{P}_o)$.

Hence, if we define

$$\Psi(x,y,p) = f_y(x,y,\Phi(x,y,p)),$$

then
$$\Psi \in C^1(\mathcal{P}_o)$$

since $f \in C^2(\mathcal{R}_o)$ and $\Phi \in C^1(\mathcal{P}_o)$.

By the Euler-Lagrange equation (2.3.8),

$$\frac{dp}{dx} = \frac{d}{dx} f_{y'} \big|_{y'=\Phi(x,y,p)} = f_y(x,y,y') \big|_{y'=\Phi(x,y,p)},$$

and we obtain for the Euler-Lagrange equation the following system of first-order differential equations:

$$\frac{dy}{dx} = \Phi(x,y,p)$$

$$\frac{dp}{dx} = \Psi(x,y,p),$$

with $\Phi, \Psi \in C^1(\mathcal{P}_o)$.

† A. E. Taylor, "Advanced Calculus," p. 240, Ginn and Company, Boston, 1955.

We state this result as a lemma:

Lemma 2.5 *If* (x_o,y_o,y'_o) *is a regular lineal element and if* $f \in C^2(\mathcal{R}_o)$, *where* $\mathcal{R}_o$ *is defined in (2.5.3), then there exists an open parallelepiped* $\mathcal{P}_o$, *as defined in (2.5.4), such that the Euler-Lagrange equation (2.3.8) may be transformed by means of* $p = f_{y'}(x,y,y')$ *into the system of first-order differential equations*

$$\begin{aligned} y' &= \Phi(x,y,p) \\ p' &= \Psi(x,y,p) \end{aligned} \tag{2.5.6}$$

where $\Phi,\Psi \in C^1(\mathcal{P}_o)$.

From this lemma, the assertion of which is merely local in character, we obtain the following global theorem about the transformation of the Euler-Lagrange equation in a strong neighborhood of a regular extremal:

Theorem 2.5.1 *Let* $A < a$, $B > b$, *and assume that the extremal* $y = y_o(x) \in C^1(A,B)$ *is regular in* (A,B). *If* $f \in C^2(\mathcal{R})$, *where*

$$\mathcal{R} = \{(x,y,y') \mid A < x < B, |y - y_o(x)| < \alpha_1, |y' - y'_o(x)| < \beta_1\},$$

then there exists a domain

$$\mathfrak{D} = \{(x,y,p) \mid a_1 < x < b_1, |y - y_o(x)| < \delta, |p - p_o(x)| < \delta\}$$

for some $\delta > 0$, *where* $p_o(x) = f_{y'}(x,y_o(x),y'_o(x))$ *and* $A < a_1 < a$, $B > b_1 > b$, *such that the Euler-Lagrange equation (2.3.8) may be transformed by means of* $p = f_{y'}(x,y,y')$ *into the system of first-order differential equations*

$$\begin{aligned} y' &= \Phi(x,y,p) \\ p' &= \Psi(x,y,p) \end{aligned}$$

where $\Phi,\Psi \in C^1(\mathfrak{D})$.

Proof: By hypothesis, for any $x_o \in (A,B)$, $(x_o,y_o(x_o),y'_o(x_o))$ is a regular lineal element. Hence, by Lemma 2.5, there exists a right parallelepiped, as defined in (2.5.4), such that (2.5.6) holds and $\Phi,\Psi \in C^1(\mathcal{P}_o)$. We apply the same argument to every point of the interval $[a,b]$ and the corresponding lineal element of $y = y_o(x)$, and we obtain in this manner an open covering of the compact set $y = y_o(x)$, $p = p_o(x)$, $x \in [a,b]$, with right parallelepipeds. (See Fig. 2.4.) By the Heine-Borel theorem,† a finite number of these parallelepipeds, say $\mathcal{P}_1, \mathcal{P}_2, \ldots, \mathcal{P}_n$, will cover this compact set. Let $\mathcal{P} = \bigcup_{k=1}^{n} \mathcal{P}_k$, where we may assume without loss of generality that the $\mathcal{P}_k$ are enumerated in such a manner that consecutive

† *Ibid.*, p. 493.

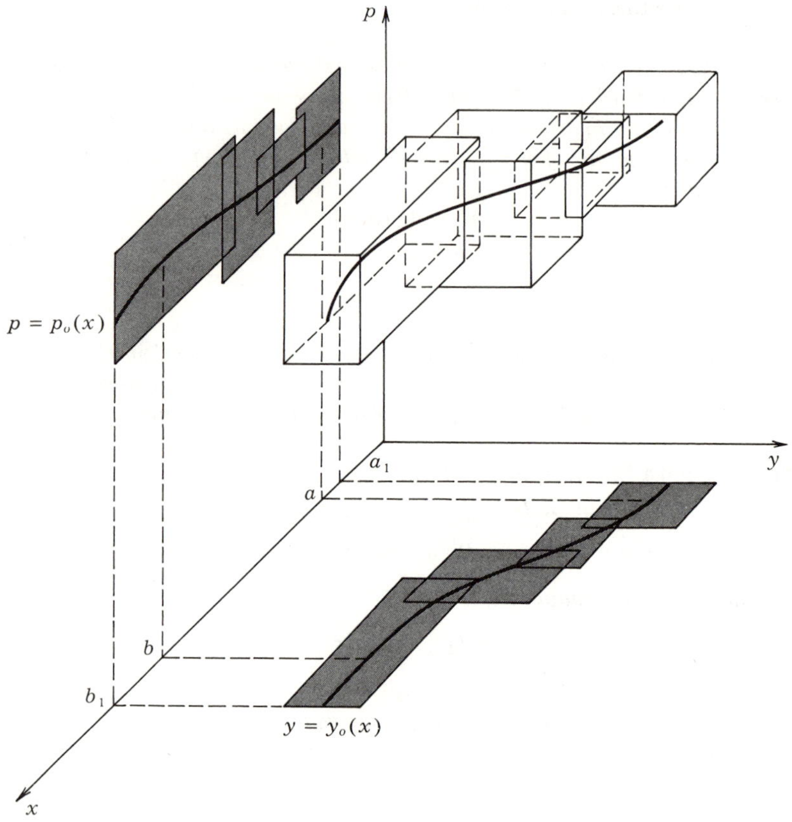

Figure 2.4

parallelepipeds overlap. Since there are only finitely many and since consecutive parallelepipeds overlap, there exists a domain $\mathfrak{D}$, as defined in the statement of the theorem, such that $\mathfrak{D} \subset \mathcal{P}$. Since Φ,Ψ are uniquely defined in each parallelepiped of the open covering, it follows that Φ and Ψ in $\mathcal{P}_{k-1}$ and Φ and Ψ in $\mathcal{P}_k (k = 2, 3, \ldots, n)$ have to coincide on the portion of overlap between $\mathcal{P}_{k-1}$ and $\mathcal{P}_k$. Hence, $\Phi,\Psi \in C^1(\mathfrak{D})$.

The application to (2.5.6) of the standard existence and uniqueness theorems on the solution of a system of first-order differential equations will lead to a statement about the qualitative behavior of regular extremals.

By Lemma 2.5, $\Phi,\Psi \in C^1(\mathcal{P}_o)$. Since $\mathcal{P}_o$ is convex, Φ,Ψ satisfy a Lipschitz condition in every closed and convex subset of $\mathcal{P}_o$.† Hence, we may conclude that through every point $(\xi,\eta,\pi) \in \mathcal{P}_o$, there passes exactly one solution $y = y(x)$, $p = p(x)$ of (2.5.6) which satisfies the initial condition $y(\xi) = \eta$, $p(\xi) = \pi$; and we have $y,p \in C^1(I_\xi)$, where I_ξ is some open interval that contains ξ.

As in the proof of Theorem 2.5.1, we conclude that there exists a domain $\mathfrak{D} = \{(x,y,p) \mid a_1 < x < b_1, |y - y_o(x)| < \delta, |p - p_o(x)| < \delta\}$ for some $\delta > 0$ such that through every point $(\xi,\eta,\pi) \in \mathfrak{D}$, there is a unique solution $y = y(x)$, $p = p(x)$ such that $y(\xi) = \eta$, $p(\xi) = \pi$ and such that $y,p \in C^1$ so long as y,p remain in $\mathfrak{D}$.

If $y = y_o(x) \in C^1(A,B)$ is a regular extremal in (A,B) and since $p_o(x) = f_{y'}(x,y_o(x),y_o'(x))$, we have for all $x \in (a_1,b_1)$ that $(x,y_o(x),y_o'(x)) \in \mathfrak{D}$, and we see that $y = y_o(x)$, $p = p_o(x)$ are uniquely determined in the entire interval $[a,b]$ and that $y_o,p_o \in C^1[a,b]$.

By (2.5.5), $y_o'(x) = \Phi(x,y_o(x),p_o(x))$. Hence

$$y_o''(x) = \Phi_x(x,y_o(x),p_o(x)) + \Phi_y(x,y_o(x),p_o(x))y_o'(x) + \Phi_p(x,y_o(x),p_o(x))p_o'(x) \in C[a,b]$$

since $\Phi \in C^1(\mathfrak{D})$, $(x,y_o(x),p_o(x)) \in \mathfrak{D}$, and $y_o,p_o \in C^1[a,b]$.

Hence we may state:

Theorem 2.5.2 *If $y = y_o(x) \in C^1(A,B)$, $A < a$, $B > b$, is a regular extremal in (A,B), then $y = y_o(x)$ is uniquely determined in $[a,b]$ by the initial conditions $y_o(a) = y_a$, $y_o'(a) = y_a'$, and $y_o \in C^2[a,b]$.*
[Note that y_a' determines $p_o(a)$ uniquely by $p_o(a) = f_{y'}(a,y_a,y_a')$.]

We can see now that Lagrange's argument as explained in Sec. 2.4 can be justified (after the fact) for variational problems that have regular extremals, i.e., variational problems with $f_{y'y'}(x,y,y') \neq 0$ in a domain that is sufficiently large to contain all the "possible" solutions. Such a variational problem is called a *regular variational problem.*

A very practical consequence of this theorem is that, when seeking the extremals in a certain domain $\mathfrak{a}$ of the x,y plane of a given variational problem which is regular in $\mathfrak{a}$, that is, for which $f_{y'y'}(x,y,y') \neq 0$ for all $(x,y) \in \mathfrak{a}$ and all $-\infty < y' < \infty$, one can consider the Euler-Lagrange equation immediately in differentiated form.

For example,

$$I[y] = \int_a^b y^n(x)\sqrt{1 + y'^2(x)}\,dx$$

is regular for $y > 0$ and $y < 0$ when $n \geq 1$ and for all x, y when $n = 0$.

† G. Birkhoff and G. C. Rota, "Ordinary Differential Equations," p. 20, Ginn and Company, Boston, 1962.

While Theorems 2.5.1 and 2.5.2 give us some insight into the nature of the solutions of the Euler-Lagrange equation, they are of no practical assistance in the solution of the boundary-value problem that is associated with the Euler-Lagrange equations, namely,

$$f_y - \frac{d}{dx} f_{y'} = 0, \qquad y(a) = y_a, \qquad y(b) = y_b.$$

We know from the theory of *linear* differential equations of second order that the boundary-value problem does not always have a solution and that a discussion of the various possibilities gets fairly complicated.† The situation is much more complicated in the case where the differential equation is nonlinear.

Although sufficient conditions have been established for the solvability of a boundary-value problem that is associated with a second-order nonlinear differential equation, none of these conditions is very satisfactory—or very practical, for that matter.‡ That the two integration constants which ordinarily appear in the solution of a second-order differential equation do not always suffice to satisfy the given boundary conditions will be demonstrated with an example in the next section.

PROBLEMS 2.5

1. Find the locus of all singular lineal elements of the extremals of the variational problem

$$\int_a^b y^2(x)(1 - y'(x))^2 \, dx \to \text{minimum}.$$

*2. Carry out the proof of Theorem 2.5.2 in detail.

*3. Let $y' = z$, and write down (2.5.1) as a system of two first-order differential equations in the unknown functions y, z. What conditions does one have to impose on f in order to apply the standard existence and uniqueness theorems for solutions of systems of ordinary first-order differential equations?

2.6 THE PROBLEM OF MINIMAL SURFACES OF REVOLUTION

We state the problem of minimal surfaces—see also example B, Sec. 1.1—as follows: To be found is a function $y \in C^1[a,b]$, $y(a) = y_a$, $y(b) = y_b$, where $y_a > 0$, $y_b > 0$, such that

$$\int_a^b y(x)\sqrt{1 + y'^2(x)} \, dx \to \text{minimum}.$$

† H. Sagan, "Boundary and Eigenvalue Problems in Mathematical Physics," 3rd printing, pp. 329ff, John Wiley & Sons, Inc., New York, 1966.

‡ See S. N. Bernstein, Sur les équations du calcul des variations, *Ann. Sci. École Norm. Sup.*, vol. 29, pp. 431–485, 1912.

We observe that $f_{y'y'} = y/(1 + y'^2)^{3/2} > 0$ for $y > 0$ and all y'. Since we are looking for an extremal in the positive half-plane, as dictated by the nature of our problem, we know from Theorem 2.5.2 that an extremal $y = y_o(x)$, if one exists, will be in $C^2[a,b]$, and we can then write the Euler-Lagrange equation in differentiated form with the total differentiation with respect to x carried out.

We observe that the integrand does not depend on x explicitly; that is, $f = f(y,y')$. Hence, $f_{y'x} = 0$, and the Euler-Lagrange equation reduces to

$$f_y - f_{y'y}y' - f_{y'y'}y'' = 0.$$

Since f does not explicitly depend on x,

$$\frac{d}{dx}(f - y'f_{y'}) = (f_y - y'f_{y'y} - y''f_{y'y'})y'.$$

Thus, for $y' \neq 0$, the Euler-Lagrange equation is equivalent to

$$\frac{d}{dx}(f - y'f_{y'}) = 0,$$

from which we obtain immediately

$$f - y'f_{y'} = \alpha \tag{2.6.1}$$

as a first integral. (For a first integral in the case that f does not explicitly depend on y, see Prob. 2.6.3. For a more general case, see Sec. A2.16.)

Hence we obtain the following first integral from the Euler-Lagrange equation for the minimal-surface problem:

$$y\sqrt{1 + y'^2} - \frac{y'^2 y}{\sqrt{1 + y'^2}} = \alpha,$$

from which we obtain, after a few obvious manipulations,

$$y' = \pm \frac{1}{\alpha}\sqrt{y^2 - \alpha^2}$$

for $\alpha \neq 0$. (For $\alpha = 0$, we simply obtain $y = 0$.)

Separation of variables and substitution of $y = \alpha \cosh t$ yield

$$\alpha t + \beta = x,$$

and hence,

$$y = \alpha \cosh \frac{x - \beta}{\alpha} \tag{2.6.2}$$

is a two-parameter family of extremals of the Euler-Lagrange equation for the minimal-surface problem.

These curves are called *catenaries*, and the surface of revolution which they generate is called a *catenoid of revolution.*

Without loss of generality, we choose for the coordinates of the beginning point P_a the values $a = 0$, $y_a = 1$. We shall demonstrate that there are points $P_b(b,y_b)$ which cannot be joined to P_a by an extremal (2.6.2), that is, for which the boundary-value problem has no solution.†

Let us first determine the relationship between α and β so that $y(0) = 1$:

$$1 = \alpha \cosh \frac{\beta}{\alpha}, \qquad \alpha = \frac{1}{\cosh (\beta/\alpha)}.$$

Denote $(\beta/\alpha) = \lambda$. Then (2.6.2) becomes

$$y = y(x, \lambda) \equiv \frac{\cosh (x \cosh \lambda - \lambda)}{\cosh \lambda}. \tag{2.6.3}$$

Since $y' = \sinh (x \cosh \lambda - \lambda)$, we have $y'(0) = -\sinh \lambda$, and we see that we obtain all possible initial slopes if we let λ run through all real numbers.

Let us now try to determine λ such that the curve (2.6.3) passes through the point $P_b(b,y_b)$. We obtain the equation

$$y(b,\lambda) \equiv \frac{\cosh (b \cosh \lambda - \lambda)}{\cosh \lambda} = y_b.$$

We shall demonstrate in the sequel that one can choose b and y_b in such a manner that this equation will not admit a solution.

We observe that

$$\cosh t = \tfrac{1}{2}(e^t + e^{-t}) \geq |t| \qquad \text{for all } t.$$

Hence

$$\frac{\cosh (b \cosh \lambda - \lambda)}{\cosh \lambda} \geq \left| \frac{b \cosh \lambda - \lambda}{\cosh \lambda} \right| = \left| b - \frac{\lambda}{\cosh \lambda} \right|$$

$$\geq |b| - \frac{|\lambda|}{\cosh \lambda}.$$

It is easy to see that

$$\max_{(\lambda)} \left| \frac{\lambda}{\cosh \lambda} \right| = \frac{|\lambda_o|}{\cosh \lambda_o},$$

† For a more general discussion of this problem, see G. A. Bliss, "Calculus of Variations," The Open Court Publishing Company, La Salle, Ill., 1925.

where λ_o is the positive root of the transcendental equation

$$\cosh \lambda_o - \lambda_o \sinh \lambda_o = 0.$$

(See Prob. 2.6.4.) Thus,

$$|b| - \frac{|\lambda|}{\cosh \lambda} \geq |b| - \frac{1}{\sinh \lambda_o},$$

and finally,

$$y(b,\lambda) \geq |b| - \frac{1}{\sinh \lambda_o} \qquad \text{for all } \lambda.$$

So, if we chose b such that $|b| - 1/(\sinh \lambda_o) = B > 0$ and subsequently $y_b < B$, then the boundary-value problem obviously has no solution.

One can show that the family (2.6.3) of catenaries possesses an envelope which lies in the positive half-plane and, consequently, that any point P_b which lies below that envelope cannot be joined to P_a by a catenary. A detailed discussion of the problem reveals† that any point on the envelope can be joined to the beginning point by exactly one catenary and that every point above the envelope can be joined to the beginning point by exactly two catenaries.

In the case that the endpoint lies on the envelope, the minimum is not rendered by the catenary but rather by the sectionally continuous function

$$y = y(x) \equiv \begin{cases} 1 & \text{if } x = 0 \\ 0 & \text{if } 0 < x < b \\ y_b & \text{if } x = b. \end{cases}$$

The "surface of revolution" in this case consists of the surfaces of the two boundary circles. This solution was discovered by *B. C. W. Goldschmidt* in 1831, but he failed to show that it does indeed yield a minimum. (In his honor, this particular solution is generally referred to as the *Goldschmidt solution.*)‡

If the endpoint lies above the envelope, then, among the two catenaries that are possible, the upper one—which does not touch the envelope—yields a strong relative minimum.

† See *ibid.* and also the following: I. Todhunter, "Researches in the Calculus of Variations," pp. 54ff, 1871, G. E. Stechert & Co., New York, (reprint 1924); L. A. Pars, "An Introduction to the Calculus of Variations," pp. 85ff, John Wiley & Sons, Inc., New York, 1962; L. A. Lyusternik, "Shortest Paths Variational Problems," pp. 90ff, The Macmillan Company, New York, 1964; C. Carathéodory, "Calculus of Variations and Partial Differential Equations of First Order," vol. 2, pp. 297ff, Holden-Day, Inc., San Francisco, Cal., 1967.

‡ The Goldschmidt solution renders a strong minimum. See N. I. Akhiezer, "The Calculus of Variations," pp. 173ff, Blaisdell Publishing Company, a division of Ginn and Company, Boston, 1962.

PROBLEMS 2.6

*1. Find all extremals of the Brachistochrone problem (problem A, Sec. 1.1).

2. Same as in problem 1 for $\int_a^b (y'^2(x) - y^2(x))\,dx \to$ minimum, $y(a) = y_a$, $y(b) = y_b$. What happens when $b - a = n\pi$?

*3. Show that $f_{y'} =$ constant is a first integral of the Euler-Lagrange equation if $f = f(x,y')$.

*4. Show that $\max\limits_{(\lambda)} |\lambda/\cosh\lambda| = \lambda_o/\cosh\lambda_o$, where λ_o is the positive root of the equation $\cosh\lambda - \lambda\sinh\lambda = 0$.

*5. Show that any point (b,y_b), where $b \neq 0$, $y_b \geq 0$, can be joined to the point $(0,0)$ by exactly one extremal of the Brachistochrone problem (cycloid) in problem 1.

2.7 NATURAL BOUNDARY CONDITIONS

As we pointed out in our discussion of the navigation problem—see problem D of Sec. 1.1—it is quite natural to ask for the curve of shortest crossing time of the river without specifying the point of departure and the point of arrival.

We formulate this problem as follows: Among all functions $y \in C_s^1[a,b]$, the one which minimizes

$$I[y] = \int_a^b f(x,y(x),y'(x))\,dx \tag{2.7.1}$$

is to be found. The x coordinates of beginning point and endpoint are given, namely, a and b, but the y coordinates remain unspecified.

We assume that $y_o \in C_s^1[a,b]$ renders the minimum, and we consider, in accordance with the formulation of the problem, the following space Σ of competing functions:

$$\Sigma_s = \{y \mid y \in \mathcal{C}_s^1[a,b]\}.$$

As space of admissible variations, we use

$$\mathcal{H}_s = \{h \mid h \in \mathcal{C}_{sP}^1[a,b]\},$$

where P is the set of points on which y_o' has jump discontinuities.

If $f \in C^1(\mathcal{R})$, where $\mathcal{R}$ is a domain of the (x,y,y') space that contains all lineal elements of $y = y_o(x)$, then the functional (2.7.1) has a Gâteaux variation at y_o, and we obtain the necessary condition

$$\delta I[h] = \int_a^b [f_y(x,y_o(x),y_o'(x))h(x) + f_{y'}(x,y_o(x),y_o'(x))h'(x)]\,dx = 0$$

for all $h \in \mathcal{H}_s$. With

$$\phi(x) = \int_a^x f_y(t,y_o(t),y_o'(t))\,dt,$$

we obtain by integration by parts, as in Sec. 2.3, that

$$\delta I[h] = h(x)\phi(x)\Big|_a^b + \int_a^b h'(x)[f_{y'}(x,y_o(x),y_o'(x)) - \phi(x)]\,dx = 0 \tag{2.7.2}$$

for all $h \in \mathcal{H}_s$. Since this relation has to hold for all $h \in \mathcal{H}_s$, it has to hold, in particular, for any subset of $\mathcal{H}_s$, such as

$$\mathcal{H}_s' = \{h \mid h \in \mathcal{C}_{sP}^1[a,b], h(a) = 0, h(b) = 0\} \subset \mathcal{H}_s,$$

and we obtain, as in Sec. 2.3, that

$$f_{y'}(x,y_o(x),y_o'(x)) - \phi(x) = C \tag{2.7.3}$$

for all $x \in [a,b]$ except where y_o' has a jump discontinuity.

Hence, (2.7.2) simplifies to

$$h(x)\phi(x)\Big|_a^b + Ch(x)\Big|_a^b = 0$$

for all $h \in \mathcal{H}_s$. From (2.7.3),

$$\phi(b) = f_{y'}(b,y_o(b),y_o'(b)) - C$$
$$\phi(a) = f_{y'}(a,y_o(a),y_o'(a)) - C,$$

and we obtain

$$h(b)[(f_{y'})_b - C] - h(a)[(f_{y'})_a - C] + C[h(b) - h(a)] = 0$$

for all $h \in \mathcal{H}_s$. In particular, this relationship has to hold for all $h \in \mathcal{H}_s$ for which $h(a) \neq 0$ but $h(b) = 0$ and for all $h \in \mathcal{H}_s$ for which $h(a) = 0$ but $h(b) \neq 0$. This will lead after cancellation by $h(a)$ or $h(b)$, respectively, to the *natural boundary conditions*:

$$\begin{aligned} f_{y'}(a,y(a),y'(a)) &= 0 \\ f_{y'}(b,y(b),y'(b)) &= 0. \end{aligned} \tag{2.7.4}$$

The name "natural boundary conditions" derives from the fact that these conditions arise "naturally" in a problem which, at the outset, is not equipped with boundary conditions.

We interpret these natural boundary conditions for the problem of a light ray emanating from a fixed point P_a, propagating in a plane through a medium with $n(x,y) > 0$ as index of refraction, and terminating at the vertical line $x = b$.

By *Fermat's principle*, the light ray will take that path $y = y(x)$ that minimizes the traveling time from P_a to $x = b$; that is,

$$\int_a^b n(x,y(x))\sqrt{1 + y'^2(x)}\,dx \to \text{minimum},$$

where $y(a) = y_a$ and where the endpoint moves freely on $x = b$.

As the second (natural) boundary condition, we obtain from (2.7.4) that

$$f_{y'}(b,y(b),y'(b)) = \frac{n(b,y(b))y'(b)}{\sqrt{1+y'^2(b)}} = 0$$

and hence

$$y'(b) = 0,$$

that is, the light ray will hit the boundary line at a right angle.

Observe that the Brachistochrone problem and the problem of minimal surfaces of revolution are special cases of the problem which we just discussed for $n(x,y) = 1/\sqrt{y}$ and $n(x,y) = y$, respectively.

PROBLEMS 2.7

1. Assume that $f \in C^2(\mathfrak{R})$ and that $y_o \in C^2[a,b]$. Derive the natural boundary conditions using Lagrange's integration-by-parts process. (See Sec. 2.4.)
2. Write down the natural boundary conditions for the problem of navigation, problem D in Sec. 1.1, and simplify.

2.8 TRANSVERSALITY CONDITIONS

We shall now consider a generalized version of the problem that was discussed in the preceding section. We shall let the endpoint move freely on a given curve while keeping the beginning point fixed for the time being.

Let us assume that $y = \varphi(x) \in C^1(-\infty,\infty)$. We seek a function $y = y(x) \in C_s^1[a,b]$ that emanates from the fixed beginning point $P_a(a,y_a)$ and terminates on the curve $y = \varphi(x)$ for some $x = b$ at the point $(b,\varphi(b))$, so that

$$I[y] = \int_a^b f(x,y(x),y'(x))\,dx \tag{2.8.1}$$

becomes a minimum. We assume that $f \in C^1(\mathfrak{R})$, where $\mathfrak{R}$ is some domain of the (x,y,y') space.

Suppose now that $y = y_o(x) \in C_s^1[a,b]$, with $y(a) = y_a$, $y(b) = y_b$, $y'(b) \neq \varphi'(b)$, all lineal elements of which lie in $\mathfrak{R}$, is the solution to our problem. We have as a space of competing functions

$\Sigma = \{y \mid y \in \mathfrak{C}_s^1[a,\beta], y(a) = y_a, y(\beta) = \varphi(\beta),$ where $\beta \in (a,\infty)$ is the smallest number for which the latter relationship is satisfied$\}$.

[This simply means that the competing functions have to be in C_s^1 between a and their first intersection point with $y = \varphi(x)$.] Since $y = y_o(x)$ is only defined in $[a,b]$ and since we need $y = y_o(x)$ beyond b in order to consider variations of y_o, we extend the definition of y_o continuously and linearly as follows:

$$y = y_o(x) \equiv \begin{cases} y_o(x) & \text{for } x \in [a,b] \\ y_o'(b-0)(x-b) + y_o(b) & \text{for } x \geq b. \end{cases}$$

We define in terms of

$$G = \{g \mid g \in C_{sp}{}^1[a,b], g(a) = 0\}$$

the following space of admissible variations of Σ:

$$\begin{aligned} \mathcal{H} = \{h \mid h(x) &= g(x) && \text{for } x \in [a,b], \\ h(x) &= g'(b-0)(x-b) + g(b) && \text{for } x \geq b, g \in G\}. \end{aligned}$$

$\mathcal{H}$ is obviously a linear space, and

$$||h|| = \max_{[a,b]} |h(x)| + \sup_{[a,b]} |h'(x)|$$

is a legitimate norm in $\mathcal{H}$. (See Probs. 2.8.2 and 2.8.3.) Note, in particular, that the maximum and the supremum in the definition of the norm of h are to be taken on the interval $[a,b]$ only and *not* on the entire interval on which h might be defined:

It is, of course, very unlikely that $y = y_o(x) + h(x)$ will actually penetrate the terminal curve $y = \varphi(x)$ for some x for all $h \in \mathcal{H}$. However, if we assume that $y_o'(b) \neq \varphi'(b)$, then we are assured that $y = y_o(x) + h(x)$ and $y = \varphi(x)$ intersect near $x = b$ so long as $||h||$ is sufficiently small. Beyond that, we simply adopt the position that the functional is not defined. (See also page 34.)

Now all we have to do is show that $I[y]$ possesses a Gâteaux variation $\delta I[h]$ at y_o and we shall then obtain

$$\delta I[h] = 0 \qquad \text{for all } h \in \mathcal{H}$$

as a necessary condition for a minimum. From Fig. 2.5 and in view of our hypotheses on φ, we have

$$\begin{aligned} I[y_o + th] &= \int_a^{b+\Delta(t)} f(x, y_o + th, y_o' + th')\, dx \\ &= \int_a^b f(x, y_o + th, y_o' + th')\, dx \\ &\qquad + \int_b^{b+\Delta(t)} f(x, y_o + th, y_o' + th')\, dx, \qquad (2.8.2) \end{aligned}$$

where $\Delta(0) = 0$. [y_o, y_o', h, h' in the above integral and on later occasions stand for $y_o(x)$, $y_o'(x)$, $h(x)$, $h'(x)$.]

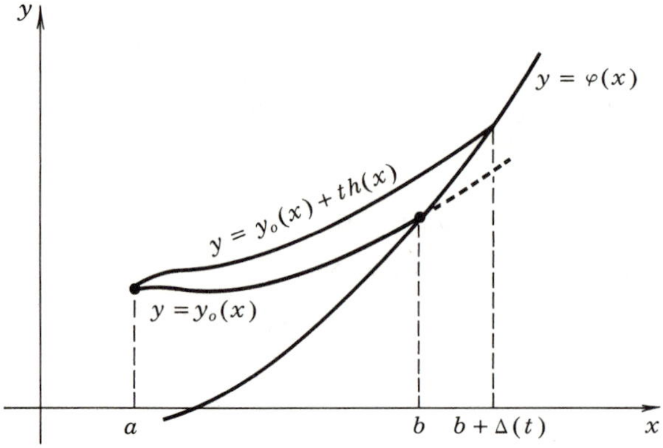

Figure 2.5

In order to find

$$\frac{dI\,[y_o + th]}{dt}$$

by the formula for the differentiation of an integral with respect to a parameter,† we have to convince ourselves that the integrands of the integrals in (2.8.2) are continuous in x in $[a,b]$ and $[b, b + \Delta(t)]$, respectively, and differentiable for $|t| \leq t_1$ for some t_1. We also have to assert the existence of $\Delta'(t)$ in some neighborhood of $t = 0$. The first question is easily settled in view of our hypotheses on f, h, and y_o. (In case y_o' and h' have jump discontinuities, we may break up the integration interval, carry out the differentiation process for the individual parts, and then put all the partial results together again.) The second question can be settled by resorting to the theorem on implicit functions.‡

We consider

$$F(t,\Delta) = \varphi(b + \Delta) - y_o(b + \Delta) - th(b + \Delta)$$

and observe that

$$F(0,0) = \varphi(b) - y_o(b) = 0$$

since $y = y_o(x)$ intersects $y = \varphi(x)$ at $x = b$. We note further that

$$F(t,\Delta)$$
$$F_t(t,\Delta) = -\,h(b + \Delta)$$
$$F_\Delta(t,\Delta) = \varphi'(b + \Delta) - y_o'(b + \Delta) - th'(b + \Delta)$$

† D. V. Widder, "Advanced Calculus," 2d ed., p. 353, Prentice-Hall, Inc., Englewood Cliffs, N.J., 1961.

‡ A. E. Taylor, "Advanced Calculus," p. 241, Ginn and Company, Boston, 1955.

are continuous in a neighborhood of $t = 0$, $\Delta = 0$. (Note that y_o' and h' may be considered to be continuous in a neighborhood of $x = b$.) Finally, since $y_o'(b) \neq \varphi'(b)$, we have that

$$F_\Delta(0,0) = \varphi'(b) - y_o'(b) \neq 0.$$

Therefore, near $t = 0$, we can solve for Δ in terms of t:

$$\Delta = \Delta(t)$$

such that

$$F(t,\Delta(t)) = \varphi(b + \Delta(t)) - y_o(b + \Delta(t)) - th(b + \Delta(t)) = 0 \quad (2.8.3)$$

and $\Delta(t)$, $\Delta'(t)$ are continuous near $t = 0$. (See also Prob. 2.8.6.)

$\Delta'(0)$ itself may be obtained from (2.8.3) by implicit differentiation:

$$\varphi'(b)\Delta'(0) = y_o'(b)\Delta'(0) + h(b),$$

and hence, if $\varphi'(b) \neq y_o'(b)$, as we had to assume when we constructed $\mathcal{H}$ and then again when we invoked the implicit-function theorem,

$$\Delta'(0) = \frac{h(b)}{\varphi'(b) - y_o'(b)} \quad (2.8.4)$$

Thus we obtain from (2.8.2), and in view of (2.8.4) and $\Delta(0) = 0$, that

$$\delta I[h] = \frac{d}{dt} I[y_o + th]_{t=0}$$

$$= \int_a^b [f_y(x,y_o(x),y_o'(x))h(x) + f_{y'}(x,y_o(x),y_o'(x))h'(x)\, dx]$$

$$+ f(b,y_o(b),y_o'(b)) \frac{h(b)}{\varphi'(b) - y_o'(b)}.$$

Hence we arrive at the necessary condition

$$\delta I[h] = \int_a^b [f_y(x,y_o,y_o')h + f_{y'}(x,y_o,y_o')h']\, dx$$

$$+ f(b,y_o(b),y_o'(b)) \frac{h(b)}{\varphi'(b) - y_o'(b)} = 0$$

for all $h \in \mathcal{H}$.

By the integration-by-parts process that was explained in Sec. 2.3, we obtain in terms of

$$\phi(x) = \int_a^x f_y(t,y_o(t),y_o'(t))\, dt$$

and in view of $h(a) = 0$,

$$\int_a^b h'(x)[f_{y'}(x,y_o(x),y_o'(x)) - \phi(x)]\,dx$$
$$+ h(b)\phi(b) + (f)_b \frac{h(b)}{\varphi'(b) - y_o'(b)} = 0$$

for all $h \in \mathfrak{H}$, where $(f)_b = f(b,y_o(b),y_o'(b))$.

Since $y = y_o(x)$ also has to yield a minimum, as compared with other curves joining the same fixed beginning point to the endpoint $P_b(b,y_o(b))$, it follows that $y = y_o(x)$ has to satisfy the Euler-Lagrange equation in integrated form, and we are left with

$$h(b)\left[C + \phi(b) + (f)_b \frac{1}{\varphi'(b) - y_o'(b)}\right] = 0$$

for all $h(b)$. [If $h(b) = 0$, then $\Delta(t) \equiv 0$, and we have a fixed-endpoint problem.] Here, C is the constant from the Euler-Lagrange equation in integrated form.

Since $\phi(b) = f_{y'}(b,y_o(b),y_o'(b)) - C$, we obtain after cancellation by $h(b)$, the condition

$$f_{y'}(b,y_o(b),y_o'(b))(\varphi'(b) - y_o'(b)) + f(b,y_o(b),y_o'(b)) = 0. \qquad (2.8.5)$$

This is the so-called *transversality condition*, which has to be satisfied at that point at which the extremal *transverses* the terminal curve $y = \varphi(x)$.

The same argument may now be applied to the case where the endpoint is fixed and the beginning point is allowed to move freely on a curve $y = \psi(x)$ and then applied to the case where both beginning point and endpoint move freely on curves $y = \psi(x)$ and $y = \varphi(x)$. (The variational problem where beginning point and endpoint are fixed is often referred to as the *point-point problem*, and the problems of this section as the *point-curve problem*, the *curve-point problem*, and the *curve-curve problem*.)

Equation (2.8.5) was derived under the condition that $y_o'(b) \neq \varphi'(b)$, that is, that the extremal does not terminate on $y = \varphi(x)$ tangentially. However, it is not difficult to see that the condition has to hold also for the case $y_o'(b) = \varphi'(b)$, where it simply reduces to $f(b,y_o(b),y_o'(b)) = 0$. (See Prob. 2.8.5.)

We shall formulate the results of this section in the following theorem:

Theorem 2.8 *If the function $y = y_o(x) \in C_s^1[a,b]$, which emanates at some $x = a$ from the curve $y = \psi(x) \in C^1(-\infty,\infty)$ and terminates for some $x = b$ on the curve $y = \varphi(x) \in C^1(-\infty,\infty)$, yields a relative minimum for $I[y] = \int_a^b f(x,y(x),y'(x))\,dx$, where $f \in C^1(\mathfrak{R})$, $\mathfrak{R}$ being a domain in the*

(x,y,y') space that contains all lineal elements of $y = y_o(x)$, then it is necessary that $y = y_o(x)$ satisfy the Euler-Lagrange equation in integrated form in the interval $[a,b]$ and that at the point of departure and the point of arrival, the transversality conditions

$$f_{y'}(a,y_o(a),y_o'(a))(\psi'(a) - y_o'(a)) + f(a,y_o(a),y_o'(a)) = 0$$
$$f_{y'}(b,y_o(b),y_o'(b))(\varphi'(b) - y_o'(b)) + f(b,y_o(b),y_o'(b)) = 0 \quad (2.8.6)$$

are satisfied. In the case that one of the points is fixed, then the transversality condition has to hold at the other point.

We see now that the natural boundary conditions (2.7.4) of the preceding section may be viewed as a special case of the transversality conditions for $\psi'(a) = \infty$ and $\varphi'(b) = \infty$.

If the initial curve or the terminal curve (or both) is a horizontal line, that is, if $\psi'(x) \equiv 0$ or $\varphi'(x) \equiv 0$ (or both), then we obtain from (2.8.6)

$$f_{y'}(a,y_o(a),y_o'(a))y_o'(a) - f(a,y_o(a),y_o'(a)) = 0$$

and an analogous condition for the endpoint.

Again we use *Fermat's principle* to illustrate condition (2.8.6). Since $I[y] = \int_a^b n(x,y(x))\sqrt{1 + y'^2(x)}\,dx$ and hence $f_{y'} = (n(x,y)y')/(\sqrt{1 + y'^2})$, we obtain after simple manipulations that

$$y'(a) = -\frac{1}{\psi'(a)}$$

and an analogous condition at the endpoint. Again, we see that in problems of this general type, transversality means orthogonality.

Before concluding this section, let us present a simple, geometric interpretation of the transversality condition.

We assume that the extremal $y = y_o(x)$ transverses the curve $y = \varphi(x)$ at the point $P_b(b,y_o(b))$. Now we plot $f(b,y_o(b),y')$ as a function of y'. The curve

$$\xi = y'$$
$$\eta = f(b,y_o(b),y')$$

thus obtained is called the *characteristic* of f at $(b,y_o(b))$. (See Fig. 2.6.)

The equation of the tangent line to the characteristic at the point $(\xi,\eta) = (y_o'(b),f(b,y_o(b),y_o'(b))$ is given by

$$\eta - f(b,y_o(b),y_o'(b)) = f_{y'}(b,y_o(b),y_o'(b))(\xi - y_o'(b)).$$

Hence we can see that the transversality condition (2.8.6) is satisfied if and only if the tangent line to the characteristic $\eta = f(b,y_o(b),\xi)$ at $\xi = y_o'(b)$, $\eta = f(b,y_o(b),y_o'(b))$ also passes through the point $\xi = \varphi'(b)$, $\eta = 0$.

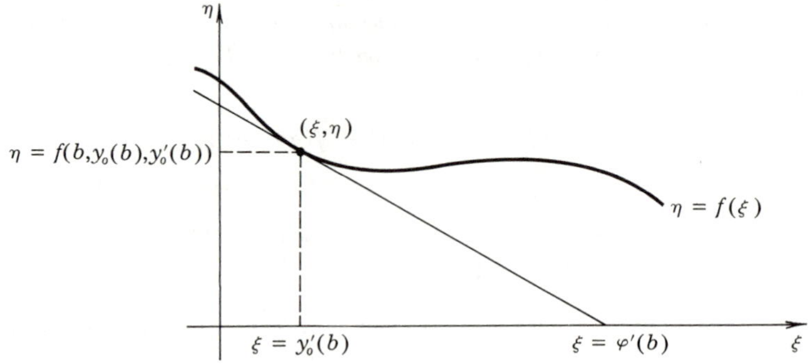

Figure 2.6

For the variational problem

$$\int_a^b \sqrt{1 - y'^2(x)}\, dx \to \text{minimum},$$

we obtain the characteristic

$$\eta = \sqrt{1 - \xi^2},$$

and we see that transversality may occur only if $|\varphi'(b)| > 1$. We leave it to the reader to explain this restriction. (See Prob. 2.8.1.)

PROBLEMS 2.8

1. Investigate all possibilities with regard to transversality for the problem $\int_a^b \sqrt{1 - y'^2(x)}\, dx \to$ minimum.
*2. Show that $\mathcal{H}$ as defined on page 69 is a linear space.
*3. Show that $\|h\|$ as defined on page 69 is a legitimate norm in $\mathcal{H}$.
4. Assume that $f \in C^2[\bar{\mathcal{R}}]$, where $\bar{\mathcal{R}}$ is a region of the (x,y,y') space which contains all lineal elements of $y = y_o(x)$ in its *interior*. Make suitable additional assumptions to ensure that $I[y]$ as defined in (2.8.1) has a Fréchet differential at $y = y_o$ (see Sec. 1.5), and derive the transversality conditions (2.8.6) on the basis of the Fréchet differential.
*5. Show that the transversality conditions also have to hold for the case where $y_o'(b) = \varphi'(b)$.
6. Show that $b + \Delta(t)$, where $\Delta = \Delta(t)$ was obtained on page 71 in the derivation of the transversality condition, yields for all $t \in N^\delta(0)$ for some $\delta > 0$ the x coordinate of the *first* intersection point of $y = y_o(x) + th(x)$ with $y = \varphi(x)$, provided that $x = b$ is the x coordinate of the first intersection point of $y = y_o(x)$ with $y = \varphi(x)$.

2.9 BROKEN EXTREMALS AND THE WEIERSTRASS-ERDMANN CORNER CONDITIONS

Let us consider the variational problem

$$I[y] = \int_{-1}^{1} y^2(x)(1 - y'(x))^2\,dx \to \text{minimum}$$

with the boundary conditions

$$y(-1) = 0, \qquad y(1) = 1.$$

Clearly, $I[y] \geq 0$, and we see that the lower bound 0 is actually assumed for $y = 0$, as well as for $y = x$ ($y' = 1$), and that these are the only smooth functions for which $I[y] = 0$. However, neither of these two functions satisfies the given boundary conditions. In order to obtain a function that satisfies the boundary conditions and still yields the value 0 for $I[y]$, we have to construct a sectionally smooth function that is suitably pieced together from portions of $y = 0$ and $y = x$.

Apparently,

$$y = y_o(x) \equiv \begin{cases} 0 & \text{for } -1 \leq x < 0 \\ x & \text{for } 0 \leq x \leq 1 \end{cases}$$

is such a function. Its derivative has a jump discontinuity at $x = 0$. Since $I[y_o] = 0$ and since 0 is certainly a relative minimum of $I[y]$—actually, 0 is even an absolute minimum in this case—$y = y_o(x)$ has to satisfy, by necessity, the Euler-Lagrange equation in integrated form, and any smooth portion thereof has to satisfy the Euler-Lagrange equation in differentiated form

$$y(1 - y')^2 + \frac{d}{dx}[y^2(1 - y')] = 0,$$

which, indeed, it does. We call an extremal such as $y = y_o(x)$ as defined above a *broken extremal.*

We shall see in the sequel that one can derive additional necessary conditions for a relative minimum (maximum) from the Euler-Lagrange equation in integrated form, which have to be satisfied by a broken extremal at the points at which the derivative has a jump discontinuity, and we shall then see that these conditions are indeed satisfied by the broken extremal of the above example.

Let us assume that $y = y_o(x)$ minimizes the functional

$$I[y] = \int_{a}^{b} f(x,y(x),y'(x))\,dx$$

and satisfies the boundary conditions $y(a) = y_a$, $y(b) = y_b$ or some other boundary conditions such as the natural boundary conditions of Sec. 2.7 or the transversality conditions of Sec. 2.8. We further assume that $y_o \in C_{sP}{}^1[a,b]$, where P is the set of points at which the derivative of $y = y_o(x)$ has a jump discontinuity. (See Fig. 2.7, where $P = \{c\}$.) Note that $y = y_o(x)$ itself remains continuous throughout the entire interval and that, in particular, $y_o(c - 0) = y_o(c + 0)$ for any $c \in P$.

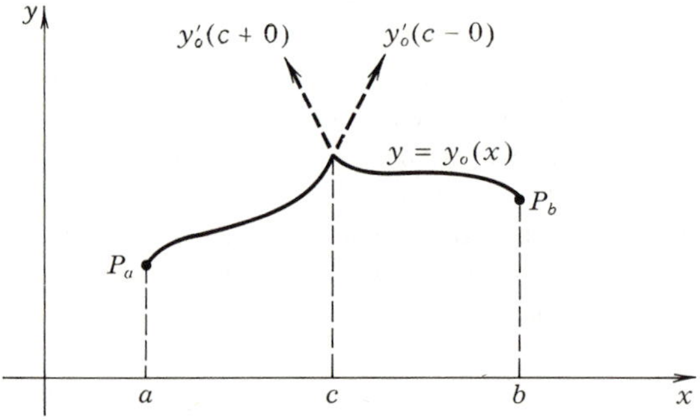

Figure 2.7

Regardless of whether we have a point-point problem, a point-curve problem, a curve-point problem, or a curve-curve problem, $y = y_o(x)$ has to satisfy, by necessity, the Euler-Lagrange equation in integrated form (2.3.7),

$$f_{y'}(x,y_o(x),y_o'(x)) = \int_a^x f_y(t,y_o(t),y_o'(t))\,dt + C,$$

everywhere in $[a,b]$ except where y_o' has a jump discontinuity, that is, for $x \in P$. Let $c \in P$. Then it follows from the continuity of $\int_a^x f_y(t,y_o(t),y_o'(t))\,dt$ in $[a,b]$ that

$$\int_a^{c-0} f_y(t,y_o(t),y_o'(t))\,dt = \int_a^{c+0} f_y(t,y_o(t),y_o'(t))\,dt,$$

and hence we obtain, in view of $y_o(c - 0) = y_o(c + 0)$, that

$$f_{y'}(c,y_o(c),y_o'(c - 0)) = f_{y'}(c,y_o(c),y_o'(c + 0)) \tag{2.9.1}$$

has to hold at such a corner.

A simple heuristic argument carried out with reference to the Euler-Lagrange equation in its differentiated form will lead us to another integrated form of the Euler-Lagrange equation. This, in turn, will lead us to still another condition that has to be satisfied at a corner of an extremal.

Since

$$\frac{d}{dx}(f - y'f_{y'}) = f_x + f_y y' + \cancel{f_{y'}y''} - \cancel{y''f_{y'}} - y'\frac{d}{dx}f_{y'}$$

$$= f_x + y'\left(f_y - \frac{d}{dx}f_{y'}\right)$$

it would appear that

$$\frac{d}{dx}(f - y'f_{y'}) = f_x,$$

and hence

$$f - y'f_{y'} = \int_a^x f_x\,dt + C$$

also has to be satisfied by an extremal. This is indeed so, as we shall prove in the following:

Lemma 2.9 *If $y = y_o(x) \in C_s^1[a,b]$ yields a relative minimum (maximum) for $I[y]$ under the conditions stated in Theorem 2.3, then it is necessary that $y = y_o(x)$ satisfy the integro-differential equation*

$$f(x,y_o(x),y_o'(x)) - y_o'(x)f_{y'}(x,y_o(x),y_o'(x)) = \int_a^x f_x(t,y_o(t),y_o'(t))\,dt + C \tag{2.9.2}$$

for all $x \in [a,b]$ except where y_o' has a jump discontinuity.

The lemma is still true if the point-point boundary conditions are replaced by natural boundary conditions or by transversality conditions.

Proof: In the proof, we shall restrict ourselves to the point-point problem because, in the final analysis of the point-curve and curve-curve problems, the extremal will have to yield a relative minimum (maximum) as compared with other curves that join the same beginning point to the same endpoint.

We subject the integral $I[y] = \int_a^b f(x,y(x),y'(x))\,dx$, together with its extremal $y = y_o(x)$, to the transformation

$$x = u + \alpha v$$
$$y = v$$

(see Fig. 2.8), where $|\alpha|$ is assumed to be small, so that if $y = y(x)$ is in a strong neighborhood of $y = y_o(x)$ and if $y = y(x)$ and $y = y_o(x)$ have the same initial point and the same endpoint, then $v = v(u)$ is in a strong

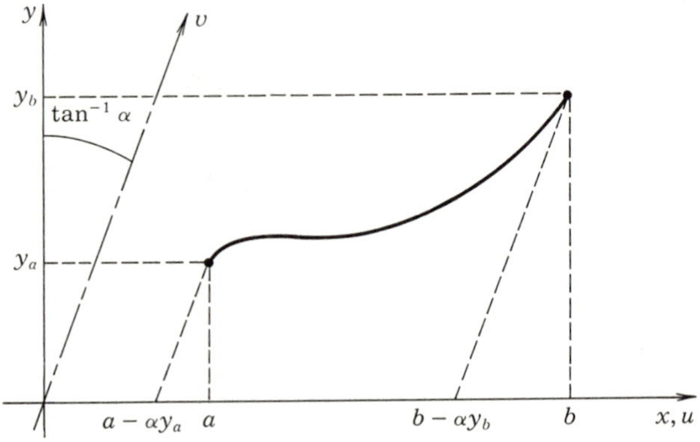

Figure 2.8

neighborhood of $v = v_o(u)$, where $v = v(u)$ is the image of $y = y(x)$ and $v = v_o(u)$ is the image of $y = y_o(x)$ in the u,v plane.

Since

$$dx = (1 + \alpha v')\, du$$

$$y' = \frac{v'}{1 + \alpha v'}, \qquad \text{where } v' = \frac{dv}{du},$$

we obtain

$$\begin{aligned} I[y] &= \int_a^b f(x,y(x),y'(x))\, dx \\ &= \int_{a-\alpha y_a}^{b-\alpha y_b} f\left(u + \alpha v(u), v(u), \frac{v'(u)}{1 + \alpha v'(u)}\right)(1 + \alpha v'(u))\, du \\ &= \bar{I}[v], \end{aligned}$$

while $y = y_o(x)$ is transformed into $v = v_o(u)$.

If $|\alpha|$ is sufficiently small, $v = v_o(u)$ will yield a relative minimum for $\bar{I}[v]$ with $I[y_o] = \bar{I}[v_o]$, and hence $v = v_o(u)$ will have to satisfy the Euler-Lagrange equation in integrated form (2.3.7); that is, if we denote for the moment

$$F(u,v,v') = f\left(u + \alpha v, v, \frac{v'}{1 + \alpha v'}\right)(1 + \alpha v'),$$

we have for $v = v_o(u)$,

$$F_{v'}(u,v_o(u),v_o'(u)) = \int_{a-\alpha y_a}^{u} F_v(t,v_o(t),v_o'(t))\, dt + C_1 \qquad (2.9.3)$$

wherever v_o' is continuous. Since

$$F_{v'} = \alpha f\left(u + \alpha v, v, \frac{v'}{1 + \alpha v'}\right) + (1 + \alpha v')f_{y'}\left(u + \alpha v, v, \frac{v'}{1 + \alpha v'}\right)\frac{1 + \cancel{\alpha v'} - \cancel{\alpha v'}}{(1 + \alpha v')^2}$$

$$= \alpha f + \frac{1}{1 + \alpha v'} f_{y'}$$

and $$F_v = (1 + \alpha v')\left[\alpha f_x\left(u + \alpha v, v, \frac{v'}{1 + \alpha v'}\right) + f_y\left(u + \alpha v, v, \frac{v'}{1 + \alpha v'}\right)\right],$$

we obtain for (2.9.3) in terms of $f, f_y, f_{y'}, f_x,$

$$\alpha f\left(u + \alpha v_o(u), v_o(u), \frac{v_o'(u)}{1 + \alpha v_o'(u)}\right) + \frac{1}{1 + \alpha v_o'(u)} f_{y'}\left(u + \alpha v_o(u), v_o(u), \frac{v_o'(u)}{1 + \alpha v_o'(u)}\right)$$

$$= \int_{a-\alpha y_a}^{u} \left[\alpha f_x\left(t + \alpha v_o(t), v_o(t), \frac{v_o'(t)}{1 + \alpha v_o'(t)}\right) + f_y\left(t + \alpha v_o(t), v_o(t), \frac{v_o'(t)}{1 + \alpha v_o'(t)}\right)\right](1 + \alpha v_o'(t))\, dt + C_1.$$

We note that

$$\frac{1}{1 + \alpha v'} = 1 - \alpha y' \qquad \text{since } \frac{dx}{du} = 1 + \alpha v' \qquad \text{and} \qquad \frac{du}{dx} = 1 - \alpha y'$$

and we rewrite the above relation in terms of x, y. This involves on the right side the following transformation of the integration variable:

$$\tau = t + \alpha v_o(t)$$
$$d\tau = (1 + \alpha v_o'(t))\, dt$$

where $t = a - \alpha y_a$ becomes $\tau = a - \alpha y_a + \alpha v_o(a - \alpha y_a) = a$, since $a = u + \alpha v$, $y_a = v$ corresponds to $u = a - \alpha v$, $v = y_a$, and where $t = u$ becomes $\tau = u + \alpha v_o(u) = x$. Therefore, we have

$$\alpha f(x,y_o(x),y_o'(x)) + f_{y'}(x,y_o(x),y_o'(x)) - \alpha y_o'(x) f_{y'}(x,y_o(x),y_o'(x)) = \int_a^x [\alpha f_x(\tau,y_o(\tau),y_o'(\tau)) + f_y(\tau,y_o(\tau),y_o'(\tau))]\, d\tau + C_1.$$

By Theorem 2.3,

$$f_{y'} = \int_a^x f_y \, d\tau + C_2,$$

and we obtain after cancellation by α for the extremal $y = y_o(x)$,

$$f - y'f_{y'} = \int_a^x f_x \, d\tau + C_1 - C_2 = \int_a^x f_x \, d\tau + C$$

for all $x \in [a,b]$ except where y_o' has jump discontinuities.

We can now utilize this form of the Euler-Lagrange equation to derive still another condition which has to hold at corners of the extremal.

Since $\int_a^x f_x(t,y_o(t),y_o'(t))\, dt$ is continuous for all $x \in [a,b]$, we conclude from (2.9.2) that $f - y'f_{y'}$ has to have the same value at $x = c - 0$ as at $x = c + 0$, that is,

$$\begin{aligned} &f_{y'}(c, y(c), y'(c-0))y'(c-0) - f(c, y(c), y'(c-0)) \\ &\quad = f_{y'}(c, y(c), y'(c+0))y'(c+0) - f(c, y(c), y'(c+0)). \qquad (2.9.4) \end{aligned}$$

This is the second corner condition. The two conditions (2.9.1) and (2.9.4) are generally referred to as the *Weierstrass†-Erdmann corner conditions*.

We summarize our results in the following theorem:

Theorem 2.9 *If $y = y_o(x) \in C_{sp}^1[a,b]$ yields a relative minimum (maximum) for $I[y] = \int_a^b f(x,y(x),y'(x))\, dx$ in Σ_s, then at all points $x = c \in P$ where y_o' has a jump discontinuity, the Weierstrass-Erdmann corner conditions*

$$[f_{y'}]_{c-0} = [f_{y'}]_{c+0}$$
$$[y'f_{y'} - f]_{c-0} = [y'f_{y'} - f]_{c+0}$$

have to be satisfied.

In the problem which we discussed in the beginning of this section, we have

$$f = y^2(1-y')^2, \qquad f_{y'} = -2y^2(1-y'), \qquad c = 0.$$

By the definition of y_o, we have $y_o(0) = 0$, $y_o'(0-0) = 0$, $y_o'(0+0) = 1$. Consequently,

$$[f_{y'}]_{0-0} = [f_{y'}]_{0+0} = 0, \qquad [y'f_{y'} - f]_{0-0} = [y'f_{y'} - f]_{0+0} = 0,$$

and we see that the corner of this extremal is indeed quite legitimate.

The corner conditions find a very simple and compelling geometric interpretation by means of the characteristic of f at $(c,y_o(c))$. (See Fig.

† Karl Weierstrass, 1815–1897.

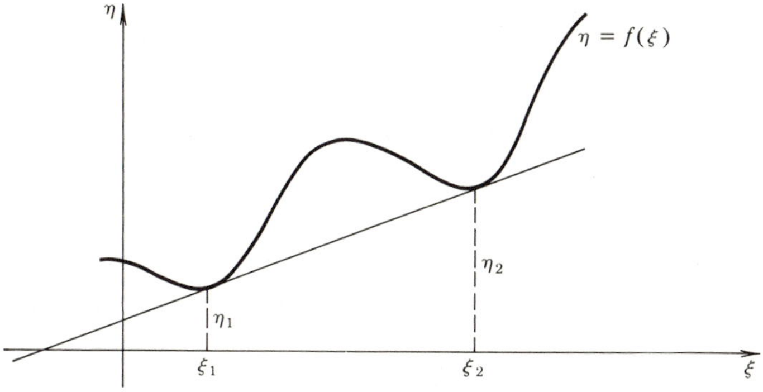

Figure 2.9

2.9.) The equations of the tangent lines to the characteristic of f at the points (ξ_1,η_1) and (ξ_2,η_2) are

$$\eta - \eta_1 = f_{y'}(c, y_o(c), \xi_1)(\xi - \xi_1)$$
$$\eta - \eta_2 = f_{y'}(c, y_o(c), \xi_2)(\xi - \xi_2)$$

If we write $y_o'(c-0)$, $y_o'(c+0)$ instead of ξ_1, ξ_2 and $f(c, y_o(c), y_o'(c-0))$, $f(c, y_o(c), y_o'(c+0))$ instead of η_1, η_2, then the equations of these tangent lines become

$$f(c,y_o(c),\xi) - f(c,y_o(c),y_o'(c-0)) = f_{y'}(c,y_o(c),y_o'(c-0))(\xi - y_o'(c-0))$$
$$f(c,y_o(c),\xi) - f(c,y_o(c),y_o'(c+0)) = f_{y'}(c,y_o(c),y_o'(c+0))(\xi - y_o'(c+0)).$$

We see that the first corner condition (2.9.1) states that the two tangent lines have to be parallel while the second one (2.9.4) states that they have to be actually identical. We also see that a corner at $(c,y_o(c))$ is thus possible only if the characteristic of f at that point permits a tangent line that touches in at least two points.

Corners may be ruled out immediately if the characteristic is convex ($f_{y'y'}(c,y_o(c),\xi) > 0$ for all ξ) or concave ($f_{y'y'}(c,y_o(c),\xi) < 0$ for all ξ).

PROBLEMS 2.9

1. Consider the transformation $x = u + \alpha v$, $y = v$.
   (*a*) Show that for suitably chosen α, the domain $\mathfrak{D} = \{(x,y,y') \mid |y'| < A\}$ is mapped one-to-one into a domain of the (u,v,v') space. Hereby, $v' = dv/du$.
   (*b*) Given $y = y_o(x) \in C^1[a,b]$ and $y = y(x) \in C^1[a,b]$, whereby $y_o(a) = y(a)$, $y_o(b) = y(b)$. Show: If $y = y(x)$ is in a strong δ neighborhood of $y = y_o(x)$, then it is possible to choose $|\alpha|$ sufficiently small so that its image $v = v(u)$ is in a strong δ^* neighborhood of the image $v = v_o(u)$ of $y = y_o(x)$. Also $\delta^* \to 0$ as $\delta \to 0$.

2. Consider the problem

$$I[y] = \int_0^1 (y'^2(x) - 1)^2\,dx \to \text{minimum}, \qquad y(0) = 0, \qquad y(1) = k.$$

Is there a solution with n corners for any given n? What about k?

3. Consider

$$I[y] = \int_a^b y^2(x)\,(1 - y'(x))^2\,dx \to \text{minimum}.$$

Are corners of extremals permissible anywhere except along $y = 0$?

4. Consider the functional $I[y]$ in problem 3. Is $I[y] = 0$ possible for a sectionally smooth function which satisfies the boundary conditions $y(a) = y_a$, $y(b) = y_b$ if:
 (*a*) $b - a \geq 2$, $0 \leq y_a \leq 1$, $-1 \leq y_b \leq 0$
 (*b*) $b - a < 2$, $y_a \geq 1$, $y_b \leq -1$
 (*c*) $y_a > 0$, $y_b > 0$
 (*d*) $y_a < 0$, $y_b < 0$

5. Give examples of functions $y = y(x)$ for which

$$I[y] = \int_a^b y^2(x)\,(1 - y'(x))^2\,dx = 0,$$

where:
 (*a*) $y = y(x)$ is continuous and has a sectionally continuous derivative.
 (*b*) y' has infinitely many discontinuities in $[a,b]$.

6. Consider the variational problem

$$I[y] = \int_0^1 y'^2(x)\,(1 + y'(x))^2\,dx \to \text{minimum},$$

$y(0) = 0$, $y(1) = m$, where $-1 < m < 0$, and find a broken extremal that yields the *absolute* minimum for $I[y]$. (See also Prob. 3.8.4.)

2.10 SMOOTHING OF CORNERS

We shall now prove and discuss a theorem that is likely to take some of the spotlight away from curves with corners as members of the space of competing functions. For lack of a better name, we shall refer to this theorem as the *fairing theorem*. What it is all about may be put very simply: What can be accomplished by a function from $C_s^1[a,b]$ can almost be accomplished by a function from $C^1[a,b]$, or to put it more precisely, if a functional assumes a certain value for a function with a sectionally continuous derivative, then one can always find in any weak neighborhood of this function and for any $\varepsilon > 0$ a function with a continuous derivative that joins the same two endpoints and is such that the value of the functional for this new function differs from the original value by less than ε.

Before we formalize this theorem, let us lay some groundwork: Suppose that $y = \eta(x) \in C_s^1[a,b]$. Then

$$\inf_{[a,b]} \eta'(x) = m, \qquad \sup_{[a,b]} \eta'(x) = M$$

exist. Let $\bar{\mathcal{R}}$ denote the compact region

$$\bar{\mathcal{R}} = \{ (x,y,y') \mid a \le x \le b, |y - \eta(x)| \le \mu, -M \le y' \le M \},$$

and let us assume that

$$f \in C[\bar{\mathcal{R}}].$$

Now we are ready to state the fairing theorem:

Theorem 2.10 *If $y = \eta(x) \in C_s^1[a,b]$ and if all lineal elements of $y = \eta(x)$ lie in $\bar{\mathcal{R}}$, then one can find for any $\varepsilon > 0$ and for any $\delta > 0$ a function $y = y(x) \in C^1[a,b]$ such that*

$$y(a) = \eta(a), \qquad y(b) = \eta(b), \qquad y \in N_w^\delta(\eta)$$

and

$$|I[y] - I[\eta]| < \varepsilon,$$

where $I[y] = \int_a^b f(x,y(x),y'(x))\,dx$ and where it is assumed that $f \in C[\bar{\mathcal{R}}]$.

Proof: For reasons of simplicity, we shall assume that η' has only one jump discontinuity at $x = c$, where $a < c < b$, and that it is continuous elsewhere. It will be quite obvious how one has to extend the proof to the case where η' has n jump discontinuities in $[a,b]$.

If it is possible to find for any $\delta > 0$ a $\Delta > 0$ such that the portion of $y = \eta(x)$ between $c - \Delta$ and $c + \Delta$ can be replaced by a smooth curve $y = \bar{\eta}(x)$ whose lineal elements lie in $\mathcal{R}$ and which connects with $y = \eta(x)$ at $x = c - \Delta$ and $x = c + \Delta$ with a continuous derivative and lies in $N_w^\delta(\eta)$, that is, $\max_{[c-\Delta,\, c+\Delta]} |\eta(x) - \bar{\eta}(x)| < \delta$, then we have for

$$y = y(x) \equiv \begin{cases} \eta(x) & \text{for } x \in [a, c - \Delta], x \in [c + \Delta, b] \\ \bar{\eta}(x) & \text{for } x \in (c - \Delta, c + \Delta) \end{cases}$$

that
$$|I[y] - I[\eta]| = \left| \int_a^b f(x,y(x),y'(x)) - f(x,\eta(x),\eta'(x))\,dx \right|$$

$$= \left| \int_{c-\Delta}^{c+\Delta} [f(x,\bar{\eta}(x),\bar{\eta}'(x)) - f(x,\eta(x),\eta'(x))]\,dx \right|$$

$$\le \int_{c-\Delta}^{c+\Delta} |f(x,\bar{\eta}(x),\bar{\eta}'(x)) - f(x,\eta(x),\eta'(x))|\,dx \le 4K\Delta$$

since $|f(x,y,y')| \le K$ for some K and all $(x,y,y') \in \bar{\mathcal{R}}$. Hence, if we choose $\Delta < \varepsilon/4K$, we have

$$|I[y] - I[\eta]| < \varepsilon.$$

Now all we have to do is show that such a function $y = \bar{\eta}(x)$ and such a $\Delta > 0$ can be found.

We proceed for this purpose as follows: By hypothesis,

$$\lim_{x \to c-0} \eta'(x) = \eta'(c-0), \qquad \lim_{x \to c+0} \eta'(x) = \eta'(c+0)$$

exist. Since $\eta'(x)$ is continuous to the left and to the right of $x = c$, we can find for any $\delta_1 > 0$ a $\Delta > 0$ such that

$$\begin{aligned} |\eta'(x) - \eta'(c-0)| &< \delta_1 \qquad \text{for } c - \Delta < x < c, \\ |\eta'(x) - \eta'(c+0)| &< \delta_1 \qquad \text{for } c < x < c + \Delta. \end{aligned} \tag{2.10.1}$$

The idea of the proof consists in replacing $\eta'(x)$ between two conveniently chosen points to the left and to the right of $x = c$ (but between $c - \Delta$ and $c + \Delta$) by a straight-line segment $\overline{A_t B_t}$ (see Fig. 2.10) in such a manner that the corresponding portion of $y = \eta(x)$ is replaced by a section of a parabola that joins the curve at both points continuously and with a continuous derivative. Of course, this would not work for just any straight-line segment. (Why?) However, by placing the line segment $\overline{A_t B_t}$ strategically, we shall see that our purpose can be accomplished.

We assume without loss of generality that $\eta'(c-0) > 0$, $\eta'(c+0) > 0$, and we choose a δ_1 such that

$$\delta_1 < \tfrac{1}{2}|\eta'(c-0) - \eta'(c+0)|.$$

Then there is a $\Delta > 0$ such that (2.10.1) holds for this Δ—and for any smaller one, for that matter. Let $\eta' = \bar{\eta}_o'(x)$ represent the line segment $\overline{A_o B_o}$, and let $\eta' = \bar{\eta}_1'(x)$ represent the line segment $\overline{A_1 B_1}$. From Fig.

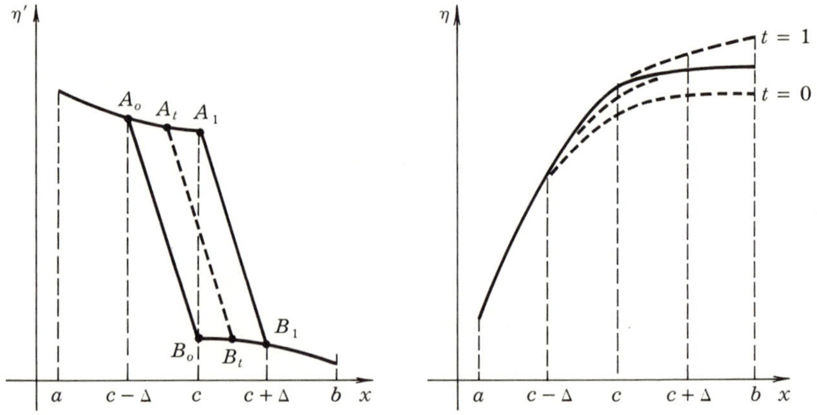

Figure 2.10

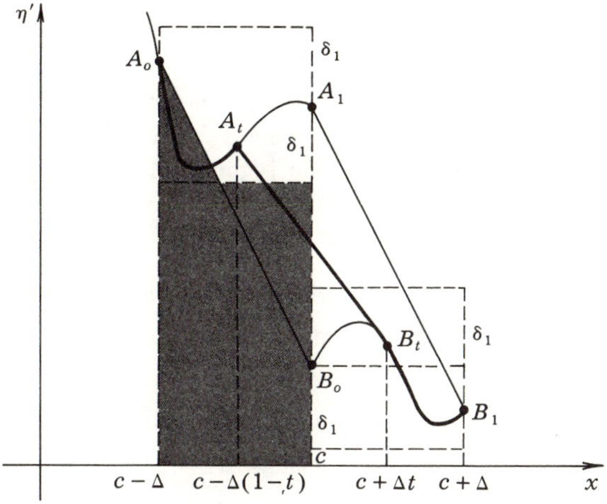

Figure 2.11

2.11, in view of the choice of δ_1 and because of (2.10.1), we have

$$\int_{c-\Delta}^{c} \bar{\eta}_o'(x)\,dx < \int_{c-\Delta}^{c} [\eta'(c-0) - \delta_1]\,dx < \int_{c-\Delta}^{c} \eta'(x)\,dx$$

$$\int_{c}^{c+\Delta} \eta'(x)\,dx < \int_{c}^{c+\Delta} [\eta'(c+0) + \delta_1]\,dx < \int_{c}^{c+\Delta} \bar{\eta}_1'(x)\,dx.$$

We define

$$\eta_o'(x) = \begin{cases} \eta'(x) & \text{for } a \le x \le c - \Delta \\ \bar{\eta}_o'(x) & \text{for } c - \Delta < x < c \\ \eta'(x) & \text{for } c \le x \le b \end{cases}$$

and

$$\eta_1'(x) = \begin{cases} \eta'(x) & \text{for } a \le x \le c \\ \bar{\eta}_1'(x) & \text{for } c < x < c + \Delta \\ \eta'(x) & \text{for } c + \Delta \le x \le b \end{cases}$$

and require that $\eta_o(a) = \eta(a)$, $\eta_1(a) = \eta(a)$. Then we obtain, in view of the above inequalities,

$$\eta_o(c+\Delta) = \eta_o(a) + \int_{a}^{c+\Delta} \eta_o'(x)\,dx$$

$$= \eta_o(a) + \int_{a}^{c-\Delta} \eta'(x)\,dx + \int_{c-\Delta}^{c} \bar{\eta}_o'(x)\,dx + \int_{c}^{c+\Delta} \eta'(x)\,dx$$

$$< \eta_o(a) + \eta(c-\Delta) - \eta(a) + \eta(c) - \eta(c-\Delta) + \eta(c+\Delta) - \eta(c)$$

$$= \eta(c+\Delta).$$

Similarly, we obtain

$$\eta_1(c + \Delta) > \eta(c + \Delta).$$

This means that the curve $y = \eta_o(x)$ undershoots the target point $(c + \Delta, \eta(c + \Delta))$ and that $y = \eta_1(x)$ overshoots this target point. We shall now try to piece together a function $y' = \eta_t'(x)$ from $y' = \eta'(x)$ and a strategically placed line segment $\overline{A_tB_t}$ so that the resulting function $y = \eta_t(x)$ will assume the value $\eta(c + \Delta)$ for $x = c + \Delta$.

Toward this end, we consider the one-parameter family of points

$$A_t(c - \Delta(1 - t), \eta'(c - \Delta(1 - t))), \qquad B_t(c + \Delta t, \eta'(c + \Delta t))$$

which, for $t = 0$, coincide with A_o, B_o and, for $t = 1$, coincide with A_1, B_1.

We define (see Fig. 2.11)

$$\eta_t'(x) = \begin{cases} \eta'(x) & \text{for } a \leq x < c - \Delta(1 - t) \\ \dfrac{\eta'(c + \Delta t) - \eta'(c - \Delta(1 - t))}{\Delta}(x - c - \Delta t) & \\ \quad + \eta'(c + \Delta t) & \text{for } c - \Delta(1 - t) \leq x \leq c + \Delta t \\ \eta'(x) & \text{for } c + \Delta t < x \leq b. \end{cases}$$

Then $\displaystyle \eta_t(c + \Delta) = \int_a^{c+\Delta} \eta_t'(x)\, dx$

is a continuous function of t in $0 \leq t \leq 1$, and we have $\eta_o(c + \Delta) < \eta(c + \Delta)$ and $\eta_1(c + \Delta) > \eta(c + \Delta)$. Hence there is a $t_o \in (0,1)$ such that $\eta_{t_o}(c + \Delta) = \eta(c + \Delta)$. (See Prob. 2.10.4.)

The function

$$y = y(x) \equiv \eta_{t_o}(x)$$

satisfies all the conditions that are attributed to it in the theorem:

$$y(a) = \eta(a), \qquad y(b) = \eta(b)$$

(see Prob. 2.10.1),

$$y \in C^1[a,b],$$

and

$$y \in N_w{}^\delta(\eta)$$

because $y(x) = \eta(x)$ for $a \leq x \leq c - \Delta$ and $c + \Delta \leq x \leq b$ and hence

$$|y(x) - \eta(x)| = \left| \int_{c-\Delta}^{x} (y'(x) - \eta'(x))\, dx \right| = \left| \int_{c-\Delta}^{x} (\eta_{t_o}'(x) - \eta'(x))\, dx \right|$$

$$\leq \int_{c-\Delta}^{c+\Delta} |\eta_{t_o}'(x) - \eta'(x)|\, dx \leq 4M\Delta$$

since $-M \leq \eta_{t_o}'(x) \leq M$ for all $x \in [c - \Delta, c + \Delta]$. If we choose $\Delta < \min(\delta/4M)$, then we have $|y(x) - \eta(x)| < \delta$ for all $x \in [a,b]$.

Finally, if we also choose $\Delta < \min(\delta/4M, \varepsilon/4K)$, then we have $|I[y] - I[\eta]| < \varepsilon$. (That all lineal elements of y lie in $\bar{\mathfrak{R}}$ is clear if we choose $\delta < \mu$; see Prob. 2.10.2.)

Corollary 1 to Theorem 2.10 *If, under the conditions of Theorem 2.10, $I[\eta] < \alpha$, where $\eta \in C_s^1[a,b]$, then it is possible to find a function $y \in C^1[a,b]$, with $y(a) = \eta(a)$, $y(b) = \eta(b)$, in any weak δ neighborhood of η such that $I[y] < \alpha$.*

Proof: Choose $\varepsilon = \alpha - I[\eta]$ and apply Theorem 2.10.

Corollary 2 to Theorem 2.10 *If $y_o \in C^1[a,b]$ yields a* weak *relative minimum for $I[y]$ in $\Sigma = \{y \mid y \in C^1[a,b], y(a) = y_a, y(b) = y_b\}$, then y_o also yields a* weak *relative minimum for $I[y]$ in $\Sigma_s = \{y \mid y \in C_s^1[a,b], y(a) = y_a, y(b) = y_b\}$.*

Proof: Assume to the contrary that there is a function $\eta \in N^\delta(y_o)$, $\eta \in \Sigma_s$, such that $I[\eta] < I[y_o]$ no matter how small $\delta > 0$ is chosen. Then we can round off the corners of $y = \eta(x)$ by the construction that was explained in the proof of Theorem 2.10 and obtain a function $y = y(x) \in \Sigma$ (see also Fig. 2.12) which, if Δ is chosen sufficiently small, is in a *strong* δ neighborhood of $y = y_o(x)$ and is such that $|I[y] - I[\eta]| < I[y_o] - I[\eta]$. The statement of the corollary now follows directly.

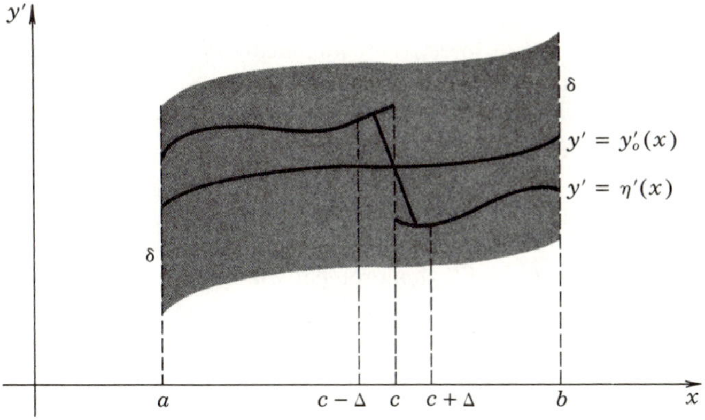

Figure 2.12

If we investigate a weak relative minimum in Σ_s that is rendered by a smooth function, then, in view of this last corollary, we see that the class of competing functions can be restricted to smooth functions without loss of generality. (It goes without saying that if $y_o \in C^1[a,b]$ yields a weak relative minimum for $I[y]$ in Σ_s, then it also yields a weak relative minimum for $I[y]$ in Σ because $\Sigma \subset \Sigma_s$.)

PROBLEMS 2.10

*1. Let y and η denote the functions in the proof of Theorem 2.10. Show that $y(a) = \eta(a)$, $y(b) = \eta(b)$.

*2. Show that $(x,y(x),y'(x)) \in \bar{\mathfrak{R}}$ for $x \in [a,b]$, where y is the function that has been constructed in the proof of Theorem 2.10 and where $\bar{\mathfrak{R}}$ is defined on page 83.

3. Show that Theorem 2.10 could not possibly hold if $N_w{}^\delta(\eta)$ in the statement of the theorem were replaced by $N^\delta(\eta)$.

*4. Show that $\eta_t(c+\Delta) = \int_a^{c+\Delta} \eta_t'(x)\,dx$ is a continuous function of t in $0 \le t \le 1$. Note that in the definition of η_t' on page 86,

$$\eta_0'(x) = \frac{\eta'(c+0) - \eta'(c-\Delta)}{\Delta}(x-c) + \eta'(c+0) \qquad \text{for } c-\Delta < x < c$$

$$\eta_1'(x) = \frac{\eta'(c+\Delta) - \eta'(c-0)}{\Delta}(x-c-\Delta) + \eta'(c+\Delta) \qquad \text{for } c < x < c+\Delta.$$

5. Give a detailed proof of Corollary 2 to Theorem 2.10.

2.11 GENERALIZATION TO MORE THAN ONE UNKNOWN FUNCTION

The theory developed in the preceding sections can easily be generalized to apply to variational problems that involve more than one unknown function.

A simple example of such a problem is given by the so-called *principle of least action* in mechanics (see also Sec. A2.15). A mass point of mass m, the position of which as a function of the time t is given by $x = x(t)$, $y = y(t)$, $z = z(t)$, will move between two instants t_o and t_1 under the influence of an external force, the components of which are given by the negative gradient of a potential function $U = U(x,y,z)$, along such a path for which

$$\int_{t_o}^{t_1} (T - U)\,dt \to \text{minimum},$$

where $T = (m/2)(\dot{x}^2 + \dot{y}^2 + \dot{z}^2)[\,\cdot = d/dt]$, represents the kinetic energy. When written out in detail, this problem will involve the minimization of the integral

$$\int_{t_o}^{t_1} \left[\frac{m}{2}(\dot{x}^2 + \dot{y}^2 + \dot{z}^2) - U(x,y,z)\right] dt.$$

As we can clearly see, the problem involves the determination of three unknown functions $x(t)$, $y(t)$, $z(t)$. (See Secs. A2.15 and A2.16.) Generally, we shall consider a problem of the type

$$I[y_1,\ldots,y_n] = \int_a^b f(x,y_1(x),\ldots,y_n(x),y_1'(x),\ldots,y_n'(x))\,dx \to \text{minimum}, \tag{2.11.1}$$

where the functions that are sought will have to satisfy certain boundary conditions at $x = a$ and $x = b$ and will have to belong to $C^1[a,b]$ or $C_s^1[a,b]$.

We introduce row-vector notation,

$$\hat{y} = (y_1, \ldots, y_n), \qquad \hat{y}(x) = (y_1(x), \ldots, y_n(x)),$$

and agree that

$$\hat{y} \in C^1[a,b] \times C^1[a,b] \times \cdots \times C^1[a,b] = C^1[a,b]^n$$

will mean that $y_1 \in C^1[a,b]$, $y_2 \in C^1[a,b]$, $\ldots, y_n \in C^1[a,b]$ (analogous for $C_s^1[a,b]^n$).

With the understanding that with $\hat{z} = (z_1, \ldots, z_n)$,

$$\hat{y} + \hat{z} = (y_1 + z_1, \ldots, y_n + z_n),$$

$$\lambda\hat{y} = (\lambda y_1, \ldots, \lambda y_n), \qquad \lambda \in R,$$

we can see that $C^1[a,b]^n$ (and $C_s^1[a,b]^n$) is a linear space. (See Prob. 2.11.1.) Both spaces $C^1[a,b]^n$ and $C_s^1[a,b]^n$ may be *weakly normed* by

$$||\hat{y}||_w = \sum_{i=1}^{n} \max_{[a,b]} |y_i(x)|. \tag{2.11.2}$$

We denote the normed linear spaces so obtained by $\bar{\mathcal{C}}^1[a,b]^n$ and $\bar{\mathcal{C}}_s^1[a,b]^n$. We may also introduce a *strong* norm by

$$||\hat{y}|| = \sum_{i=1}^{n} \left(\max_{[a,b]} |y_i(x)| + \sup_{[a,b]} |y_i'(x)| \right), \tag{2.11.3}$$

and we denote the normed linear spaces thus obtained by $\mathcal{C}^1[a,b]^n$ and $\mathcal{C}_s^1[a,b]^n$. (See Prob. 2.11.2.)

As in Sec. 2.1, we define a weak neighborhood $N_w^\delta(\hat{y}_o)$ of $\hat{y} = \hat{y}_o(x) \equiv (y_1^o(x), \ldots, y_n^o(x))$ as $\{(x,\hat{y}) \mid x \in [a,b], |y_i - y_i^o(x)| < \delta, i = 1, 2, \ldots, n\}$ and a strong neighborhood $N^\delta(\hat{y}_o)$ of $\hat{y} = \hat{y}_o(x)$ as $\{(x,\hat{y},\hat{y}') \mid x \in [a,b], |y_i - y_i^o(x)| < \delta,\ |y_i' - y_i'^o(x)| < \delta,\ i = 1, 2, \ldots, n\}$. (See also Prob. 2.11.3.)

We call $(x_o, y_1(x_o), \ldots, y_n(x_o), y_1'(x_o), \ldots, y_n'(x_o))$ a lineal element of $\hat{y} = \hat{y}(x)$ at x_o. Suppose $f \in C(\mathcal{R})$, where $\mathcal{R}$ is a domain in $(x, y_1, \ldots, y_n, y_1', \ldots, y_n')$ space. Then, if all lineal elements of $\hat{y} = \hat{y}(x)$ lie in $\mathcal{R}$ (including $(x_i, \hat{y}(x_i), \hat{y}'(x_i \pm 0))$ for all $x_i \in P$ for which $\hat{y}'$ has jump-discontinuities at $x_i \in P$—this is always to be understood in the future and will not be mentioned again), all lineal elements of $\hat{y} = \hat{y}(x) + \hat{h}(x)$ will also lie in $\mathcal{R}$ for all $\hat{h} \in \mathcal{C}^1[a,b]^n$ (or $\mathcal{C}_{sP}^1[a,b]^n$ if $\hat{y} \in \mathcal{C}_{sP}^1[a,b]^n$) for which $||\hat{h}|| < \delta$ for some $\delta > 0$. Hence $I[y_1, y_2, \ldots, y_n]$ as defined in (2.11.1) appears defined on an open subset of $\mathcal{C}^1[a,b]^n$ (or $\mathcal{C}_s^1[a,b]^n$).

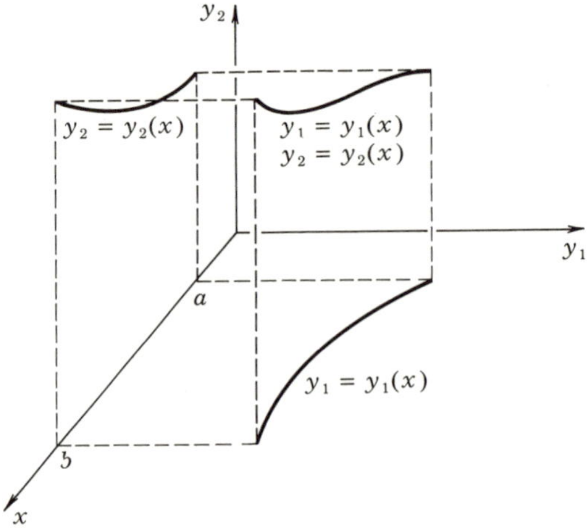

Figure 2.13

The concepts of space of competing functions and space of admissible variations have also an obvious generalization. For example, if the space of competing functions of $I[y]$ is defined by

$$\Sigma = \{\hat{y} \mid \hat{y} \in C_s^1[a,b]^n, \hat{y}(a) = \hat{y}_a, \hat{y}(b) = \hat{y}_b\},$$

where $\hat{y}_a = (y_1^a, \ldots, y_n^a)$ and $\hat{y}_b = (y_1^b, \ldots, y_n^b)$ are constant vectors, then

$$\mathfrak{H} = \{\hat{h} \mid \hat{h} \in \mathfrak{C}_s^1[a,b]^n, \hat{h}(a) = \hat{h}(b) = 0\}$$

is a normed linear space of admissible variations.

The concepts of strong and weak relative minimum (maximum) find simple and obvious generalizations. It is quite clear that a strong relative minimum (maximum) is necessarily a weak relative minimum (maximum), and this, in turn, implies that by necessity.

$$I[\hat{y}_o + \hat{h}] - I[\hat{y}_o] \geq 0 \qquad (\leq 0)$$

for all $\hat{h} \in \mathfrak{H}$ for which $||\hat{h}|| < \delta$ if $\hat{y}_o$ yields the weak minimum (maximum). (See also Lemma 1.6.)

As already indicated by the vector notation, we may view our problem geometrically as follows: To be found is a space curve $\hat{y} = \hat{y}(x)$ in $(x, y_1, \ldots, y_n)$ space that passes through the points with the coordinates $(a, y_1^a, \ldots, y_n^a)$ and $(b, y_1^b, \ldots, y_n^b)$ and is such that $I[\hat{y}] \rightarrow$ minimum. (See Fig. 2.13.) The components $y_i = y_i(x)$ of $\hat{y} = \hat{y}(x)$ appear as the projections of $\hat{y} = \hat{y}(x)$ into the x,y_i planes.

On the other hand, we may also view our problem as one of finding a curve $y_i = y_i(x)$ $(i = 1, 2, \ldots, n)$, in each one of the n x,y_i planes, that passes through the points (a, y_i^a) and (b, y_i^b) and is such that $I[\hat{y}] \rightarrow$ minimum.

We shall adopt the latter viewpoint in our subsequent derivation of a first necessary condition for a relative minimum (maximum). (For an alternate approach, see Probs. 2.11.4 and 2.11.5.)

We consider the problem

$$I[\hat{y}] = \int_a^b f(x,\hat{y}(x),\hat{y}'(x))\,dx \to \text{minimum},$$

with the space of competing functions

$$\Sigma = \{\hat{y} \mid \hat{y} \in C_s^1[a,b]^n,\ \hat{y}(a) = \hat{y}_a,\ \hat{y}(b) = \hat{y}_b\}.$$

We assume that $\hat{y} = \hat{y}_o(x) \equiv (y_1^o(x), \ldots, y_n^o(x)) \in C_s^1[a,b]^n$ is the solution of our problem, that all lineal elements of $\hat{y}_o$ lie in a domain $\mathcal{R}$ of the $(x,y_1, \ldots,y_n,y_1', \ldots,y_n')$ space, and that $f \in C^1(\mathcal{R})$.

Let k be any one of the integers 1, 2, ..., n. Suppose now that the components of $\hat{y}_o'$ have jump discontinuities at the points $P = \{x_1, \ldots,x_r\}$ and only there. We consider

$$\mathcal{H}_k = \{h_k \mid h_k \in C_{sP}^1[a,b],\ h_k(a) = h_k(b) = 0\}.$$

The functional

$$I[y_1^o, \ldots,\overset{o}{y}_{k-1},y_k,\overset{o}{y}_{k+1}, \ldots,y_n^o] = I_k[y_k]$$

may be viewed as a functional of y_k, which is defined on an open subset of $\mathcal{C}_s^1[a,b]$, and by hypothesis, it assumes its minimum for $y_k = y_k^o$. Hence we have to have

$$I_k[y_k^o + h_k] - I_k[y_k^o] \geq 0$$

for all $h_k \in \mathcal{H}_k$ for which $\|h_k\| < \delta$ for some $\delta > 0$.

As before, we find that by necessity

$$\left.\frac{dI_k[y_k^o + th_k]}{dt}\right|_{t=0} = \delta I_k[h_k] = 0 \qquad \text{for all } h_k \in \mathcal{H}_k,$$

provided that this derivative at $t = 0$ exists. (See Sec. 1.5.)

As in Sec. 1.5, we find that

$$\frac{d}{dt} I_k[y_k^o + th_k]_{t=0} = \int_a^b [f_{y_k}(x,\hat{y}_o(x),\hat{y}_o'(x))h_k(x)$$
$$+ f_{y'_k}(x,\hat{y}_o(x),\hat{y}_o'(x))h_k'(x)]\,dx,$$

and we obtain from $\delta I[h_k] = 0$ via the lemma of Dubois-Reymond (Lemma 2.3) and since k is any integer from 1, 2, ..., n, that

$$f_{y'_k}(x,\hat{y}_o(x),\hat{y}_o'(x)) = \int_a^x f_{y_k}(t,\hat{y}_o(t),\hat{y}_o'(t))\,dt + C_k, \qquad (2.11.4)$$

$k = 1, 2, \ldots, n$, has to hold for all $x \in [a,b]$ except at those points where at least one of the components of $\hat{y}_o'$ has a jump discontinuity. For any smooth portion of $\hat{y}_o$ (where none of the components of $\hat{y}_o'$ has a jump discontinuity), we obtain as before

$$f_{y_k}(x,\hat{y},\hat{y}') - \frac{d}{dx} f_{y'_k}(x,\hat{y},\hat{y}') = 0, \qquad k = 1, 2, \ldots, n. \tag{2.11.5}$$

We see that the Euler-Lagrange equation of the problem in one unknown function has now been replaced by n differential equations of second order, each of which has the same structure as the original Euler-Lagrange equation. We shall refer to (2.11.5) and (2.11.4) again as the *Euler-Lagrange equations* and the *Euler-Lagrange equations in integrated form,* respectively.

The *Weierstrass-Erdmann corner conditions* also find a simple and obvious generalization to this problem in n unknown functions. For the purpose of deriving one of them, we shall need a generalization of Lemma 2.9.

We subject the functional (2.11.1), together with the solution $\hat{y} = \hat{y}_o(x)$ of the associated variational problem, to the transformation

$$x = u + \alpha v_1$$
$$y_i = v_i, \qquad i = 1, 2, \ldots, n,$$

where $|\alpha|$ is presumed to be small, and we obtain

$$\bar{I}[\hat{v}] = \int_{a-\alpha y_1{}^a}^{b-\alpha y_1{}^b} f\left(u + \alpha v_1(u), \hat{v}(u), \frac{\hat{v}'(u)}{1 + \alpha v_1'(u)}\right)(1 + \alpha v_1'(u))\,du,$$

with $\hat{v} = \hat{v}_o(u)$ for $\hat{y} = \hat{y}_o(x)$.

As in Sec. 2.9, we argue that $\hat{v} = \hat{v}_o(u)$ has to satisfy the Euler-Lagrange equations in integrated form (2.11.4), of which we write the first:

$$\alpha f + f_{y'_1} - \frac{\alpha}{1 + \alpha v_1'} \sum_{i=1}^{n} v_i' f_{y'_i} = \int_{a-\alpha y_1{}^a}^{u} [\alpha f_x + f_{y_1}](1 + \alpha v_1')\,du + C_1{}^{(1)}.$$

(We have omitted all the arguments in this formula to avoid cumbersome notation.) This becomes, when written in terms of x and y:

$$\alpha f(x,\hat{y},\hat{y}') + f_{y'_1}(x,\hat{y},\hat{y}') - \alpha \sum_{i=1}^{n} y_i' f_{y'_i}(x,\hat{y},\hat{y}')$$
$$= \int_a^x (\alpha f_x(t,\hat{y}(t),\hat{y}'(t)) + f_{y_1}(t,\hat{y}(t),\hat{y}'(t))\,dt + C_1{}^{(1)}.$$

By (2.11.4),

$$f_{y'_1}(x,\hat{y},\hat{y}') = \int_a^x f_{y_1}(t,\hat{y}(t),\hat{y}'(t))\,dt + C_1^{(2)},$$

and we obtain as a result that

$$f(x,\hat{y},\hat{y}') - \sum_{i=1}^{n} y'_i f_{y'_i}(x,\hat{y},\hat{y}') = \int_a^x f_x(t,\hat{y}(t),\hat{y}'(t))\,dt + C \quad (2.11.6)$$

has to be satisfied by $\hat{y} = \hat{y}_o(x)$ for all $x \in [a,b]$ except where $\hat{y}'_o$ has a jump discontinuity, i.e., where at least one of its components has a jump discontinuity. [Note that the same result is obtained when one uses the transformation $x = u + \alpha v_\mu$, $y_i = v_i$, for fixed μ and then uses the μth equation from (2.11.4).]

Now we are ready to state the *Weierstrass-Erdmann corner conditions* for the problem in n unknown functions. Since the integrals on the right sides of (2.11.4) and (2.11.6) have to remain continuous at each point at which $\hat{y}'_o$ has a jump discontinuity, we obtain for such a point $x = c$ the $(n + 1)$ conditions

$$f_{y'_k}(c, \hat{y}(c), \hat{y}'(c-0)) = f_{y'_k}(c, \hat{y}(c), \hat{y}'(c+0)), \qquad k = 1, 2, \ldots, n$$

$$f(c, \hat{y}(c), \hat{y}'(c-0)) - \sum_{i=1}^{n} y'_i(c-0) f_{y'_i}(c, \hat{y}(c), \hat{y}'(c-0)) \qquad (2.11.7)$$

$$= f(c, \hat{y}(c), \hat{y}'(c+0)) - \sum_{i=1}^{n} y'_i(c+0) f_{y'_i}(c, \hat{y}(c), \hat{y}'(c+0)).$$

It is very easy to find a generalization of the natural boundary conditions that were derived in Sec. 2.7 for the problem in one unknown function. We consider the problem

$$I[\hat{y}] = \int_a^b f(x,\hat{y}(x),\hat{y}'(x))\,dx \to \text{minimum},$$

with the space of competing functions

$$\Sigma' = \{\hat{y} \mid \hat{y} \in C_s^1[a,b]^n\}.$$

This means geometrically that beginning point and endpoint are free to vary on the n-dimensional planes $x = a$ and $x = b$, respectively. We choose for every fixed k, $k = 1, 2, \ldots, n$, the space of admissible variations

$$\mathcal{H}'_k = \{h_k \mid h_k \in C_s^1[a,b]\}$$

and obtain as a necessary condition for a minimum

$$\delta I_k[h_k] = \int_a^b [f_{y_k}(x,\hat{y}_o(x),\hat{y}_o'(x))h_k(x) + f_{y'_k}(x,\hat{y}_o(x),\hat{y}_o'(x))h_k'(x)]\,dx$$

$$= \phi_k(x)h_k(x)\Big|_a^b + \int_a^b h_k'(x)[f_{y'_k}(x,\hat{y}_o(x),\hat{y}_o'(x)) - \phi_k(x)]\,dx = 0,$$

where $\phi_k(x) = \int_a^x f_{y_k}(t,\hat{y}_o(t),\hat{y}_o'(t))\,dt.$

Since $\hat{y} = \hat{y}_o(x)$ has to yield a relative minimum relative to all functions that join the same two endpoints, we obtain that $\hat{y}_o$ has to satisfy the Euler-Lagrange equations in integrated form (2.11.4), that is,

$$f_{y'_k}(x,\hat{y}_o(x),\hat{y}_o'(x)) - \phi_k(x) = C.$$

Hence, $$f_{y'_k}(x,\hat{y}_o(x),\hat{y}_o'(x))h_k(x)\Big|_a^b = 0, \qquad k = 1, 2, \ldots, n,$$

for all possible values of $h_k(a)$ and $h_k(b)$. As in Sec. 2.7, this leads immediately to the natural boundary conditions

$$f_{y'_k}(a,\hat{y}_o(a),\hat{y}_o'(a)) = 0 \tag{2.11.8}$$
$$f_{y'_k}(b,\hat{y}_o(b),\hat{y}_o'(b)) = 0, k = 1, 2, 3, \ldots, n.$$

We summarize the results of this section in the following:

Theorem 2.11 *If* $\hat{y} = \hat{y}_o(x) \in C_s^1[a,b]^n$ *yields a relative minimum for the functional*

$$I[\hat{y}] = \int_a^b f(x,\hat{y}(x),\hat{y}'(x))\,dx,$$

—where it is assumed that $f \in C^1(\mathfrak{R})$, $\mathfrak{R}$ *being a domain of the* $(x,\hat{y},\hat{y}')$ *space that contains all lineal elements of* $\hat{y} = \hat{y}_o(x)$*—in*

$$\Sigma = \{\hat{y} \mid \hat{y} \in C_s^1[a,b]^n, \hat{y}(a) = \hat{y}_a, \hat{y}(b) = \hat{y}_b\}$$

or in $$\Sigma' = \{\hat{y} \mid \hat{y} \in C_s^1[a,b]^n\},$$

then it is necessary that $\hat{y} = \hat{y}_o(x)$ *satisfy the Euler-Lagrange equations in integrated form (2.11.4), that every smooth portion of* $\hat{y} = \hat{y}_o(x)$ *satisfy the Euler-Lagrange equations (2.11.5), that the Weierstrass-Erdmann corner* *conditions (2.11.7) be satisfied at every point $x = c$ where $\hat{y}_o'$ has a jump discontinuity, and finally, in the case of Σ', that at the endpoints the natural boundary conditions (2.11.8) be satisfied.*

PROBLEMS 2.11

*1. Show that $C^1[a,b]^n$ and $C_s{}^1[a,b]^n$ are linear spaces.

*2. Show that both (2.11.2) and (2.11.3) are legitimate norms in $C^1[a,b]^n$ and $C_s{}^1[a,b]^n$.

*3. Show:

(a) If $||y - y_o||_w < \delta$, then $y \in N_w{}^\delta(y_o)$, and if $y \in N_w{}^\delta(y_o)$, then $||y - y_o||_w < n\delta$.

(b) If $||y - y_o|| < \delta$, then $y \in N^\delta(y_o)$, and if $y \in N^\delta(y_o)$, then $||y - y_o|| < 2n\delta$.

*4. Assume that $f \in C^1(\mathcal{R})$, and find

$$\frac{d}{dt} I[\hat{y}_o + t\hat{h}]_{t=0}, \qquad I[\hat{y}] = \int_a^b f(x,\hat{y}(x),\hat{y}'(x))\,dx,$$

where it is assumed that all lineal elements of $\hat{y}_o$ lie in $\mathcal{R}$.

*5. Derive the system (2.11.4) of Euler-Lagrange equations in integrated form from

$$\frac{d}{dt} I[\hat{y}_o + t\hat{h}]_{t=0} = 0 \qquad \text{for all } \hat{h} \in \mathcal{H}.$$

6. Show that $||\hat{h} - \hat{h}_k|| < \varepsilon$ for all $k > N(\varepsilon)$ is equivalent to $||h_i^{(k)} - h_i|| < \varepsilon^*$ for all $k > N(\varepsilon^*)$, where $\hat{h} = (h_1,h_2, \ldots ,h_n)$ and $\hat{h}_k = (h_1^{(k)},h_2^{(k)}, \ldots ,h_n^{(k)})$.

*7. Supply the details in the derivation of (2.11.6).

8. Find the Euler-Lagrange equations for variational problems with integrands of the type $f(x,y_1,y_2, \ldots ,y_n,y_1',y_2', \ldots ,y_n',y_1'',y_2'', \ldots ,y_n'')$. (See also Prob. 2.4.5.)

*9. $(x_o,y_1(x_o), \ldots ,y_n(x_o),y_1'(x_o), \ldots ,y_n'(x_o))$ is called a *regular lineal element* if

$$\det\,(f_{y'_i y'_k}(x_o,\hat{y}(x_o),\hat{y}'(x_o))) = \begin{vmatrix} f_{y'_1y'_1} \cdots f_{y'_1y'_n} \\ \vdots \\ f_{y'_ny'_1} \cdots f_{y'_ny'_n} \end{vmatrix} \neq 0.$$

Show that in a neighborhood of a regular lineal element, the Euler-Lagrange equations (2.11.5) may be solved explicitly for $y_1'', \ldots, y_n''$. (See also Sec. 2.12.)

2.12 THE EULER-LAGRANGE EQUATIONS IN CANONICAL FORM

As in the preceding section, we consider the variational problem in n unknown functions:

$$I[\hat{y}] = \int_a^b f(x,\hat{y}(x),\hat{y}'(x))\,dx \to \text{minimum}, \tag{2.12.1}$$

with fixed or variable endpoints.

We have seen in Sec. 2.11 that if $\hat{y} = \hat{y}_o(x) \in C_s{}^1[a,b]^n$ yields a relative minimum for $I[\hat{y}]$, then every smooth portion of $\hat{y}_o$ has to satisfy the n Euler-Lagrange equations

$$f_{y_k}(x,\hat{y},\hat{y}') - \frac{d}{dx} f_{y'_k}(x,\hat{y},\hat{y}') = 0, \qquad k = 1, 2, \ldots, n. \tag{2.12.2}$$

In Sec. 2.5, we demonstrated how the Euler-Lagrange equation of a variational problem in one unknown may be transformed into a system of two first-order differential equations. We shall now generalize these ideas to the problem in n unknown functions.

As in Sec. 2.5, we introduce new independent variables $p_i (i = 1, 2, \ldots, n)$ by means of

$$p_i = f_{y'_i}(x,\hat{y},\hat{y}'), \qquad i = 1, 2, \ldots, n, \tag{2.12.3}$$

and investigate under what conditions the y'_i may be uniquely expressed in terms of the p_i and all the other variables. From the theorem on implicit functions,† we have, with the notation

$$\hat{p} = (p_1, \ldots, p_n), \qquad f_{\hat{y}} = (f_{y_1}, \ldots, f_{y_n}), \qquad f_{\hat{y}'} = (f_{y'_1}, \ldots, f_{y'_n}),$$

that: If $\mathcal{S}$ is a $(2n + 1)$-dimensional domain in $(x,\hat{y},\hat{y}')$ space and if

$$f_{y'_i} \in C^1(\mathcal{S}),$$

$$\hat{p}_o \doteq f_{\hat{y}'}(x_o,\hat{y}_o,\hat{y}'_o) \qquad \text{for some point } (x_o,\hat{y}_o,\hat{y}'_o) \in \mathcal{S},$$

and

$$\left.\frac{\partial(f_{y'_1}, \ldots, f_{y'_n})}{\partial(y'_1, \ldots, y'_n)}\right|_{\substack{x=x_o\\ \hat{y}=\hat{y}_o\\ \hat{y}'=\hat{y}'_o}} = \left|\begin{matrix} f_{y'_1y'_1}\cdots f_{y'_1y'_n}\\ \vdots \\ f_{y'_ny'_1}\cdots f_{y'_ny'_n}\end{matrix}\right|_{\substack{x=x_o\\ \hat{y}=\hat{y}_o\\ \hat{y}'=\hat{y}'_o}} \neq 0,$$

then there exists a $(2n + 1)$-dimensional neighborhood $\mathfrak{I}$ of the point $(x_o,\hat{y}_o,\hat{p}_o)$ where (2.12.3) can be solved uniquely for $y'_1, \ldots, y'_n$:

$$y'_i = \varphi_i(x,\hat{y},\hat{p}), \tag{2.12.4}$$

where $\varphi_i \in C^1(\mathfrak{I})$, and

$$p_i = f_{y'_i}(x,\hat{y},\hat{\varphi}(x,\hat{y},\hat{p})), \qquad \hat{\varphi} = (\varphi_1, \ldots, \varphi_n),$$

for all $(x,\hat{y},\hat{p}) \in \mathfrak{I}$.

From the Euler-Lagrange equations (2.12.2), we have

$$\frac{d}{dx} f_{y'_k} = \frac{d}{dx} p_k = f_{y_k}.$$

Hence, we may replace the Euler-Lagrange equations (2.12.2), in view of (2.12.3) and (2.12.4), by

$$\begin{aligned} \hat{y}' &= \hat{\varphi}(x,\hat{y},\hat{p}) \\ \hat{p}' &= f_{\hat{y}}(x,\hat{y},\hat{\varphi}(x,\hat{y},\hat{p})). \end{aligned} \tag{2.12.5}$$

This system may be written in a more elegant and somewhat symmetric

† T. M. Apostol, "Mathematical Analysis," p. 147, Addison-Wesley Publishing Company, Inc., Reading, Mass., 1957.

fashion in terms of the *hamiltonian*

$$
\begin{aligned}
H(x,\hat{y},\hat{p}) &= \sum_{k=1}^{n} \varphi_k(x,\hat{y},\hat{p})p_k - f(x,\hat{y},\hat{\varphi}(x,\hat{y},\hat{p})) \\
&= \hat{\varphi}(x,\hat{y},\hat{p})\hat{p}^T - f(x,\hat{y},\hat{\varphi}(x,\hat{y},\hat{p})). \qquad (2.12.6)
\end{aligned}
$$

(The superscript T denotes the transpose, so that $\hat{p}^T$ is, in effect, a column vector.)

Then

$$dH = \hat{\varphi}(d\hat{p})^T + \hat{p}(d\hat{\varphi})^T - df.$$

Since

$$df = f_x\,dx + \sum_{i=1}^{n} f_{y_i}\,dy_i + \sum_{i=1}^{n} f_{y'_i}\,d\varphi_i,$$

$$\hat{\varphi}(d\hat{p})^T = \sum_{i=1}^{n} \varphi_i\,dp_i, \qquad \hat{p}(d\hat{\varphi})^T = \sum_{i=1}^{n} p_i\,d\varphi_i,$$

we have, using $p_i = f_{y'_i}$,

$$
\begin{aligned}
dH &= \sum_{i=1}^{n} \varphi_i\,dp_i + \sum_{i=1}^{n} p_i\,d\varphi_i - f_x\,dx - \sum_{i=1}^{n} f_{y_i}\,dy_i - \sum_{i=1}^{n} f_{y'_i}\,d\varphi_i \\
&= \sum_{i=1}^{n} \varphi_i\,dp_i + \cancel{\sum_{i=1}^{n} p_i\,d\varphi_i} - f_x\,dx - \sum_{i=1}^{n} f_{y_i}\,dy_i - \cancel{\sum_{i=1}^{n} p_i\,d\varphi_i} \\
&= \sum_{i=1}^{n} \varphi_i\,dp_i - f_x\,dx - \sum_{i=1}^{n} f_{y_i}\,dy_i.
\end{aligned}
$$

Hence,

$$H_x = -f_x, \qquad H_{y_k} = -f_{y_k}, \qquad H_{p_k} = \varphi_k. \qquad (2.12.7)$$

Thus (2.12.5) may now be written in the form

$$
\begin{aligned}
\hat{y}' &= H_{\hat{p}}(x,\hat{y},\hat{p}) \\
\hat{p}' &= -H_{\hat{y}}(x,\hat{y},\hat{p})
\end{aligned} \qquad (2.12.8)
$$

where

$$H_{\hat{p}} = (H_{p_1}, \ldots, H_{p_n}), \qquad H_{\hat{y}} = (H_{y_1}, \ldots, H_{y_n}).$$

Equations (2.12.8) are called the *canonical form* of the Euler-Lagrange equations (2.12.2).

We see from (2.12.7) that if $f \in C^2(\mathcal{S})$, then

$$H_{y_k} = -f_{y_k} \in C^1(\mathfrak{J}), \qquad H_{p_k} = \varphi_k \in C^1(\mathfrak{J}).$$

Hence, by the general existence and uniqueness theorems for systems of ordinary differential equations,† we have through every point of $\mathfrak{J}$ exactly

† G. Birkhoff and G. C. Rota, "Ordinary Differential Equations," pp. 99ff, Ginn and Company, Boston, 1962.

one solution $\hat{y} = \hat{y}(x)$, $\hat{p} = \hat{p}(x)$ of (2.12.8), where $\hat{y},\hat{p} \in C^1$ so long as these solutions stay in $\mathfrak{I}$. Since

$$y_k'' = H_{p_kx} + \sum_{i=1}^{n} H_{p_ky_i}y_i' + \sum_{i=1}^{n} H_{p_kp_i}p_i',$$

we have $y_k'' \in C$ so long as the solution stays in $\mathfrak{I}$. (See also Prob. 2.12.1.)

We summarize our result as a theorem:

Theorem 2.12 *If $(x_o,\hat{y}_o,\hat{y}_o')$ is a regular lineal element, i.e., if*

$$\left.\frac{\partial(f_{y'_1}, \ldots, f_{y'_n})}{\partial(y_1', \ldots, y_n')}\right|_{x_o,\hat{y}_o,\hat{y}'_o} \neq 0,$$

and if $f \in C^2(\mathcal{S})$, where $\mathcal{S}$ is a simply connected domain that contains the lineal element $(x_o,\hat{y}_o,\hat{y}_o')$, then the Euler-Lagrange equations (2.12.2) may be written in canonical form:

$$\hat{y}' = H_{\hat{p}}(x,\hat{y},\hat{p}), \qquad \hat{p}' = -H_{\hat{y}}(x,\hat{y},\hat{p}),$$

where the hamiltonian H is defined in (2.12.6), in a neighborhood $\mathfrak{I}$ of $(x_o,\hat{y}_o,\hat{p}_o)$, where $\hat{p}_o = f_{\hat{y}'}(x_o,\hat{y}_o,\hat{y}_o')$.

These equations—and hence the Euler-Lagrange equations—have a unique solution through every point of $\mathfrak{I}$ and $\hat{y},\hat{p} \in C^2$ so long as the solutions stay in $\mathfrak{I}$.

Corollary to Theorem 2.12 *If $\hat{y} = \hat{y}_o(x)$ is a regular extremal of $I[\hat{y}]$, that is, if*

$$\frac{\partial(f_{y'_1}(x,\hat{y}_o(x),\hat{y}_o'(x)), \ldots, f_{y'_n}(x,\hat{y}_o(x),\hat{y}_o'(x)))}{\partial(y_1', \ldots, y_n')} \neq 0$$

for all $x \in [a,b]$, then $\hat{y}_o \in C^2[a,b]$.

The transformation

$$\begin{aligned} \hat{p} &= f_{\hat{y}'}(x,\hat{y},\hat{y}') \\ H(x,\hat{y},\hat{p}) &= \hat{y}'f_{\hat{y}'}{}^T(x,\hat{y},\hat{y}') - f(x,\hat{y},\hat{y}') \end{aligned} \tag{2.12.9}$$

that leads from the Euler-Lagrange equations (2.12.2) to their canonical form (2.12.8) is called the *Legendre transformation*. By means of this transformation, $\hat{y}'$ and f are replaced by $\hat{p}$ and H. In Probs. 2.12.2 to 2.12.5 we indicate an alternate route that leads from the Euler-Lagrange equations to their canonical form. There, and also in Sec. 5.5, the *involutary character* of the Legendre transformation is utilized.

We have:

Lemma 2.12 *Suppose that* $(x,\hat{y},\hat{y}')$ *is a regular lineal element. Then the Legendre transformation (2.12.9) is an involution; that is,*

$$\hat{p} = f_{\hat{y}'}(x,\hat{y},\hat{y}'), \qquad H(x,\hat{y},\hat{p}) = \hat{y}'f_{\hat{y}'}^{T}(x,\hat{y},\hat{y}') - f(x,\hat{y},\hat{y}')$$

implies

$$\hat{y}' = H_{\hat{p}}(x,\hat{y},\hat{p}), \qquad f(x,\hat{y},\hat{y}') = \hat{p}H_{\hat{p}}^{T}(x,\hat{y},\hat{p}) - H(x,\hat{y},\hat{p}),$$

and vice versa. [This means that the Legendre transformation when viewed as a transformation from the $\hat{y}',f$ space into the $\hat{p},H$ space is its own inverse.]

Proof: The Legendre transformation has a unique inverse if

$$\frac{\partial(H,\hat{p})}{\partial(f,\hat{y}')} \neq 0.$$

Since

$$\frac{\partial(H,\hat{p})}{\partial(f,\hat{y}')} = -\frac{\partial(f_{y'_1}, \ldots, f_{y'_n})}{\partial(y'_1, \ldots y'_n)}$$

and the latter determinant is different from zero, the existence of the unique inverse follows. That the transformation listed in the lemma is indeed the inverse follows by direct computation.

PROBLEMS 2.12

*1. Show that under the conditions of Theorem 2.12, $\hat{p}'' \in C$ so long as $\hat{p} = \hat{p}(x)$ stays in $\mathfrak{I}$.

2. Find the Euler-Lagrange equations for the functional

$$I[\hat{y},\hat{p}] = \int_a^b [\hat{p}\hat{y}'^{T} - H(x,\hat{y},\hat{p})]\,dx.$$

3. Prove: $I[\hat{y}]$ is the minimum of $I[\hat{y},\hat{p}]$ if $\hat{y}$ is held fixed in the latter functional.

4. Prove: A minimum of $I[\hat{y},\hat{p}]$, where $\hat{y}$ and $\hat{p}$ are varied, is a minimum of $I[\hat{y}]$, and vice versa.

5. Establish Theorem 2.12 from the results of problems 2 to 4.

*6. Let $f = f(\hat{y},\hat{y}')$. Show that H = constant is a first integral of the Euler-Lagrange equations.

7. Given $\phi(\hat{y},\hat{p}) \in C^1$. Show that along each solution of the canonical system (2.12.8),

$$\frac{d\phi}{dx} = \sum_{i=1}^{n}\left(\frac{\partial\phi}{\partial y_i}\frac{\partial H}{\partial p_i} - \frac{\partial\phi}{\partial p_i}\frac{\partial H}{\partial y_i}\right).$$

8. Prove that $\phi(\hat{y},\hat{p})$ = constant is a first integral of the canonical system (2.12.8) if and only if

$$\sum_{i=1}^{n}\left(\frac{\partial\phi}{\partial y_i}\frac{\partial H}{\partial p_i} - \frac{\partial\phi}{\partial p_i}\frac{\partial H}{\partial y_i}\right) = 0.$$

(See also problem 7.)

9. Find the hamiltonian and write the Euler-Lagrange equation of the following variational problems in canonical form:

(a) $\int_0^1 \sqrt{1 + y'^2(x)}\, dx \to \text{minimum}$

(b) $\int_0^{\pi/2} (y'^2(x) - y^2(x))\, dx \to \text{minimum}$

(c) $\int_{-1}^1 y^2(x)(1 - y'^2(x))\, dx \to \text{minimum}$

(d) $\int_a^b n(x,y(x))\sqrt{1 + y'^2(x)}\, dx \to \text{minimum}$

BRIEF SUMMARY

Given the variational problem

$$I[y] = \int_a^b f(x,y(x),y'(x))\, dx \to \text{minimum},$$

where beginning point and endpoint of $y = y(x)$ are either fixed or vary on the vertical lines $x = a$ and $x = b$ or vary on the curves $y = \psi(x)$ and $y = \varphi(x)$, or any combination thereof.

If $y = y(x) \in C_s^1[a,b]$ is to yield a weak relative minimum (or a strong relative minimum, for that matter) for $I[y]$, then it is necessary that $y = y(x)$ satisfy the Euler-Lagrange equation in integrated form

$$f_{y'}(x,y(x),y'(x)) = \int_a^x f_y(t,y(t),y'(t))\, dt + C$$

for some constant C for all $x \in [a,b]$ except where y' has a jump discontinuity.

Every smooth portion of $y = y(x)$ has to satisfy the Euler-Lagrange equation

$$f_y(x,y,y') - \frac{d}{dx} f_{y'}(x,y,y') = 0$$

(Sec. 2.3).

At every point $x = c$ where y' has a jump discontinuity, the Weierstrass-Erdmann corner conditions

$$(f_{y'})_{c-0} = (f_{y'})_{c+0}, \qquad (y'f_{y'} - f)_{c-0} = (y'f_{y'} - f)_{c+0}$$

have to be satisfied (Sec. 2.9).

If beginning point and endpoint are fixed, then $y = y(x)$ has to satisfy the boundary conditions $y(a) = y_a$, $y(b) = y_b$. If beginning point and endpoint are allowed to vary on the vertical lines $x = a$ and $x = b$, then the natural boundary conditions

$$f_{y'}(a,y(a),y'(a)) = 0, \qquad f_{y'}(b,y(b),y'(b)) = 0$$

have to be satisfied (Sec. 2.7).

If beginning point and endpoint vary on the curves $y = \psi(x)$ and $y = \varphi(x)$, then the transversality conditions

$$f_{y'}(a,y(a),y'(a))(\psi'(a) - y'(a)) + f(a,y(a),y'(a)) = 0$$
$$f_{y'}(b,y(b),y'(b))(\varphi'(b) - y'(b)) + f(b,y(b),y'(b)) = 0$$

have to be satisfied at the point $(a,\psi(a))$ where the curve $y = y(x)$ leaves the initial curve $y = \psi(x)$ and at the point $(b,\varphi(b))$ where $y = y(x)$ arrives at the terminal curve $y = \varphi(x)$ (Sec. 2.8).

If $f \in C^2$ and $f_{y'y'} > 0$ (or <0), then the solutions $y = y(x)$ of the Euler-Lagrange equation are uniquely determined by any lineal element (x,y,y'), and $y'' \in C$ (Sec. 2.5).

If $y = \eta(x)$ is sectionally smooth, then one can find in any weak neighborhood of $y = \eta(x)$ a smooth function $y = y(x)$ satisfying the same boundary conditions so that $|I[y] - I[\eta]| < \varepsilon$ for any $\varepsilon > 0$.

If $y = y(x) \in C^1[a,b]$ renders $I[y]$ a weak relative minimum relative to smooth competing functions, then it also renders $I[y]$ a weak relative minimum relative to sectionally smooth competing functions (Sec. 2.10).

For the variational problem

$$I[\hat{y}] = \int_a^b f(x,y_1(x), \ldots,y_n(x),y_1'(x), \ldots,y_n'(x))\,dx \to \text{minimum}$$

with fixed or variable endpoints, one obtains the Euler-Lagrange equations in integrated form

$$f_{y'_k}(x,\hat{y}(x),\hat{y}'(x)) = \int_a^x f_{y_k}(t,\hat{y}(t),\hat{y}'(t))\,dt + C_k$$

$(k = 1, 2, \ldots, n)$ as a necessary condition for a weak relative minimum. The Weierstrass-Erdmann corner conditions and the natural boundary conditions find obvious generalizations to apply to this more general problem (Sec. 2.11).

If $f \in C^2$ and if $\det |f_{y'_i y'_k}| \neq 0$, then the Euler-Lagrange equations $f_{y_k} - (d/dx)\, f_{y'_k} = 0$ may be transformed by means of the Legendre transformation

$$\hat{p} = f_{\hat{y}'}(x,\hat{y},\hat{y}')$$
$$H(x,\hat{y},\hat{p}) = \hat{y}' f_{\hat{y}'}{}^T(x,\hat{y},\hat{y}') - f(x,\hat{y},\hat{y}')$$

into the canonical form

$$\hat{y}' = H_{\hat{p}}(x,\hat{y},\hat{p}), \qquad \hat{p}' = -H_{\hat{y}}(x,\hat{y},\hat{p})$$

to which the standard existence and uniqueness theorems of the theory of ordinary differential equations are applicable.

APPENDIX

A2.13 THE PROBLEM IN TWO UNKNOWN FUNCTIONS WITH VARIABLE ENDPOINTS

The problem in n unknown functions, where beginning point and endpoint are allowed to vary on surfaces (manifolds) of dimension $\leq n$, will be dealt with in a more general setting in Sec. 6.9. Here we shall consider only the case for $n = 2$, where the beginning point is assumed to be fixed and where the endpoint is allowed to vary on a two-dimensional surface. To wit, we seek a vector function $\hat{y} = \hat{y}(x) \equiv (y_1(x),y_2(x))$ such that

$$\int_a^b f(x,y_1(x),y_2(x),y_1'(x),y_2'(x))\,dx \to \text{minimum}, \qquad \text{(A2.13.1)}$$

whereby $y_1(a) = y_1^a$, $y_2(a) = y_2^a$ and where the endpoint lies on a surface that is given by

$$x = u(y_1,y_2).$$

We shall assume that both

$$\frac{\partial u}{\partial y_1} \neq 0, \qquad \frac{\partial u}{\partial y_2} \neq 0, \qquad \text{for all } (y_1,y_2)$$

—or at least in a neighborhood of the point of impact—and that

$$\frac{\partial u}{\partial y_1}, \frac{\partial u}{\partial y_2} \in C(-\infty < y_1 < \infty, -\infty < y_2 < \infty).$$

[For the problem in one unknown function with a variable endpoint, we had to assume that $\varphi' \in C(-\infty,\infty)$, where $y = \varphi(x)$ represented the terminal curve. This condition excludes tangent lines to the terminal curve that are parallel to the y axis. Likewise, $\partial u/\partial y_1 \neq 0$ and $\partial u/\partial y_2 \neq 0$ excludes tangent planes to the terminal surface that are parallel to the y_1 axis and/or the y_2 axis.]

We assume that $\hat{y} = \hat{y}_o(x) \in \mathfrak{C}_s^1[a,b]^2$ emanates from the fixed beginning point $P_a(a,y_1^a,y_2^a)$, terminates on $x = u(y_1,y_2)$ at some point

$(b,y_1^o(b),y_2^o(b))$, and renders

$$I[\hat{y}] = \int_a^b f(x,y_1(x),y_2(x),y_1'(x),y_2'(x))\,dx$$

a relative minimum.

As in the preceding section, we shall vary one component of $\hat{y} = \hat{y}_o(x)$ at a time. When varying $y_1 = y_1^o(x)$ while leaving $y_2 = y_2^o(x)$ unchanged, then the variation will take place within a cylindrical surface the equation of which is given by $y_2 = y_2^o(x)$. This cylindrical surface will intersect $x = u(y_1,y_2)$ in the space curve

$$\left.\begin{aligned} y_2 &= y_2^o(x) \\ x &= u(y_1,y_2) \end{aligned}\right\} \Gamma,$$

which has $y_2 = y_2^o(x)$ as projection into the x,y_2 plane and

$$x = u(y_1,y_2^o(x)) \tag{A2.13.2}$$

as projection into the x,y_1 plane. (See Fig. 2.14.) The variations of $\hat{y} = \hat{y}_o(x)$ within this cylindrical surface will terminate on Γ, and their projections into the x,y_1 plane will terminate on the curve that is given by (A2.13.2). With this, we have essentially reduced our problem to the one of Sec. 2.8. All we have to do now is investigate the curve that is given by (A2.13.2) and see that it satisfies the requirements which we have imposed on $y = \varphi(x)$ in Sec. 2.8.

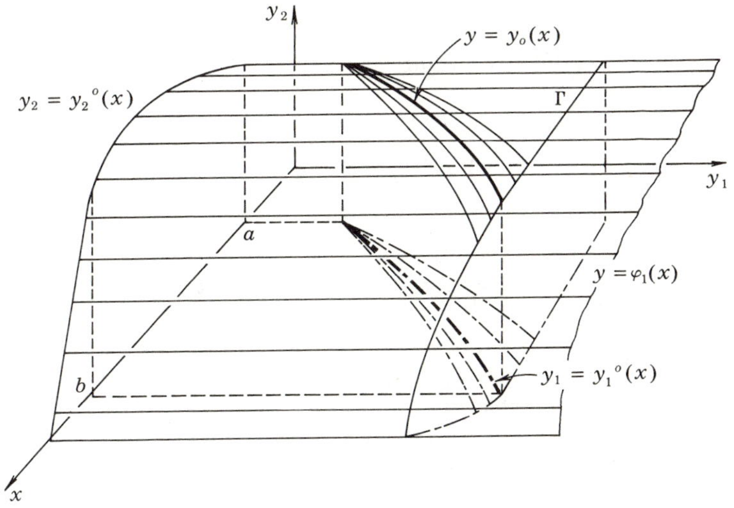

Figure 2.14

We extend $\hat{y} = \hat{y}_o(x)$ continuously and with a continuous derivative beyond $x = b$, for example, by defining $y_1^o(x) = y_1^{o\prime}(b-0)(x-b) + y_1^o(b)$, $y_2^o(x) = y_2^{o\prime}(b-0)(x-b) + y_2^o(b)$, for $x \geq b$.

Since, by hypothesis, $(\partial u/\partial y_1) \neq 0$ and since all the partial derivatives involved are assumed to be continuous near $x = b$, $\hat{y} = \hat{y}_o(b)$, we may solve (A2.13.2) on the strength of the implicit-function theorem for y_1, and we obtain

$$y_1 = \varphi_1(x) \in C^1(N^\delta(b)) \qquad \text{for some } \delta > 0.$$

Next we have to make sure that $y_1 = y_1^o(x)$ does not arrive at the terminal curve $y_1 = \varphi_1(x)$ tangentially, that is, $y_1^{o\prime}(b) \neq \varphi_1'(b)$. For this purpose, we first have to find $\varphi_1'(b)$. From (A2.13.2), we obtain by implicit differentiation,

$$1 = \frac{\partial u}{\partial y_1}\varphi_1' + \frac{\partial u}{\partial y_2} y_2^{o\prime}$$

and hence

$$\varphi_1'(b) = \frac{1 - (\partial u/\partial y_2)_b y_2^{o\prime}(b)}{(\partial u/\partial y_1)_b}. \tag{A2.13.3}$$

Now

$$\varphi_1'(b) - y_1^{o\prime}(b) = \frac{1}{(\partial u/\partial y_1)_b}\left[1 - \left(\frac{\partial u}{\partial y_1}\right)_b y_1^{o\prime}(b) - \left(\frac{\partial u}{\partial y_2}\right)_b y_2^{o\prime}(b)\right]$$

Since $\operatorname{grad}(x - u(y_1,y_2)) = (1, -(\partial u/\partial y_1), -(\partial u/\partial y_2))$ and since $T = (1, y_1^{o\prime}(b), y_2^{o\prime}(b))$ is a tangent vector to $\hat{y} = \hat{y}_o(x)$ at $x = b$, the point of impact,

$$1 - \left(\frac{\partial u}{\partial y_1}\right)_b y_1^{o\prime}(b) - \left(\frac{\partial u}{\partial y_2}\right)_b y_2^{o\prime}(b) = T \cdot \operatorname{grad}(x - u(y_1,y_2)) \neq 0 \tag{A2.13.4}$$

means that the curve $\hat{y} = \hat{y}_o(x)$ must not arrive at the terminal surface tangentially.

This is a straightforward generalization of the condition $y'(b) \neq \varphi'(b)$ which we had to impose when dealing with the problem in one unknown function and a variable endpoint. Incidentally, (A2.13.4) will also guarantee that $\varphi_2'(b) \neq y_2^{o\prime}(b)$, where $y = \varphi_2(x)$ is the terminal curve that arises when one varies $y_2 = y_2^o(x)$ and keeps $y_1 = y_1^o(x)$ fixed.

With all this out of the way, we may now continue as in Sec. 2.8.

We have

$$I_1[y_1^o + th_1] = \int_a^b f(x, y_1^o(x) + th_1(x), y_2^o(x), y_1^{o\prime}(x) + th_1'(x), y_2^{o\prime}(x))\,dx$$

$$+ \int_b^{b+\Delta(t)} f(x, y_1^o(x) + th_1(x), y_2^o(x), y_1^{o\prime}(x) + th_1'(x), y_2^{o\prime}(x))\,dx$$

and obtain, as in Sec. 2.8,

$$\frac{d}{dt} I_1[y_1^o + th_1]_{t=0} = \int_a^b [f_{y_1}(x,\hat{y}_o(x),\hat{y}_o'(x))h_1(x)$$

$$+ f_{y_1'}(x,\hat{y}_o(x),\hat{y}_o'(x))h_1'(x)]\,dx + f(b,\hat{y}_o(b),\hat{y}_o'(b))\Delta_1'(0),$$

where [see (2.8.4)]

$$\Delta_1'(0) = \frac{h_1(b)}{\varphi_1'(b) - y_1^{o\prime}(b)}.$$

Since $\hat{y} = \hat{y}_o(x)$ has to satisfy the Euler-Lagrange equations in integrated form (2.11.4), we obtain from $\delta I_1[h] = 0$ for all $h_1 \in \{h_1 \mid h_1 \in \mathcal{C}_{sp}{}^1[a,b],$ $h_1(a) = 0,\ h_1(x) = h_1'(b)(x - b) + h_1(b)$ for all $x \geq b\}$, in view of (A2.13.3),

$$\frac{\partial u}{\partial y_1} f + \left(1 - \frac{\partial u}{\partial y_1} y_1^{o\prime} - \frac{\partial u}{\partial y_2} y_2^{o\prime}\right) f_{y'_1}\bigg|_{x=b} = 0. \qquad \text{(A2.13.5}a\text{)}$$

If we vary $y_2 = y_2^o(x)$ and keep $y_1 = y_1^o(x)$ fixed, we obtain by an analogous argument (see Prob. A2.13.2),

$$\frac{\partial u}{\partial y_2} f + \left(1 - \frac{\partial u}{\partial y_1} y_1^{o\prime} - \frac{\partial u}{\partial y_2} y_2^{o\prime}\right) f_{y'_2}\bigg|_{x=b} = 0. \qquad \text{(A2.13.5}b\text{)}$$

These are the *transversality conditions* for the problem in two unknown functions with the endpoint varying on a two-dimensional surface under the restriction that the surface does not have a tangent plane parallel to the y_1 axis and/or y_2 axis at the point of impact.

As in Sec. 2.8, we can easily convince ourselves that the condition is still valid if the extremal $\hat{y} = \hat{y}_o(x)$ arrives at the terminal surface tangentially, that is, $1 - (\partial u/\partial y_1)y_1^{o\prime} - (\partial u/\partial y_2)y_2^{o\prime} = 0$. Then the condition simply reduces to $(f)_{x=b} = 0$. (See Prob. 2.8.5.)

Analogous conditions are obtained if the beginning point varies on a surface $x = v(y_1,y_2)$ which is subject to the same restrictions as $u(y_1,y_2)$.

The problem in n unknown functions, where the endpoint may vary on a surface of less than n dimensions, is vastly more complicated than the problem which we have just discussed because it is more difficult to hit such a lower-dimensional surface in an $(n + 1)$-dimensional space with variations of $\hat{y} = \hat{y}_o(x)$. For example, if the terminal manifold in (x,y_1,y_2) space were a curve rather than a two-dimensional surface, variations of the extremal, as we have considered them here, would ordinarily miss the terminal curve entirely. Considerable restrictions will have to be placed on the space of admissible variations to ensure that the varied curve reaches the terminal surface. Such spaces of admissible variations are, in general, not linear and do not possess linear subsets. More will be said about this case in Secs. 6.8 and 6.9, where we shall derive general transversality conditions, of which the ones just derived are merely a very special case.

Let us illustrate the transversality conditions (A2.13.5) by applying them to Fermat's principle of light propagation in a three-dimensional space.

We have

$$I[y] = \int_a^b n(x,y_1(x),y_2(x))\sqrt{1 + y_1'^2(x) + y_2'^2(x)}\,dx,$$

where $n(x,y_1,y_2) > 0$ denotes the index of refraction.

We assume a fixed beginning point $\hat{y}(a) = \hat{y}_a$ and a terminal surface $x = u(y_1,y_2)$. The transversality conditions on the terminal surface simplify to

$$y_1' + \frac{\partial u}{\partial y_1} + y_2'\left(\frac{\partial u}{\partial y_1}y_2' - \frac{\partial u}{\partial y_2}y_1'\right)\bigg|_b = 0$$

$$y_2' + \frac{\partial u}{\partial y_2} + y_1'\left(\frac{\partial u}{\partial y_2}y_1' - \frac{\partial u}{\partial y_1}y_2'\right)\bigg|_b = 0. \tag{A2.13.6}$$

Multiplication of the first equation by y_2', multiplication of the second equation by $(-y_1')$, and addition of the results yield

$$(1 + y_1'^2 + y_2'^2)\left(\frac{\partial u}{\partial y_1}y_2' - \frac{\partial u}{\partial y_2}y_1'\right)\bigg|_b = 0$$

and hence

$$\frac{\partial u}{\partial y_1}y_2' - \frac{\partial u}{\partial y_2}y_1'\bigg|_b = 0.$$

In view of this, the conditions (A2.13.6) will reduce to

$$y_1' + \frac{\partial u}{\partial y_1} = 0 \qquad \text{and} \qquad y_2' + \frac{\partial u}{\partial y_2} = 0.$$

Thus the vector cross product

$$(1,y_1',y_2') \times \left(-1, \frac{\partial u}{\partial y_1}, \frac{\partial u}{\partial y_2}\right) = \left(\frac{\partial u}{\partial y_2}y_1' - \frac{\partial u}{\partial y_1}y_2',\ -\frac{\partial u}{\partial y_2} - y_2',\ y_1' + \frac{\partial u}{\partial y_1}\right) = 0.$$

Since $(1,y_1',y_2')$ is a tangent vector to $\hat{y} = \hat{y}(x)$ at $x = b$ and since $(-1,\partial u/\partial y_1,\partial u/\partial y_2) = \text{grad}\,[u(y_1,y_2) - x]$, it follows that $\hat{y} = \hat{y}_o(x)$ penetrates the terminal surface $x = u(y_1,y_2)$ *orthogonally*.

PROBLEMS A2.13

*1. The transversality conditions for the problem in two unknown functions with the beginning point (or endpoint) varying on a two-dimensional surface are customarily† listed as

$$\left[f_{y'_1} + \frac{\partial u}{\partial y_1}\,(f - y'_1 f_{y'_1} - y'_2 f_{y'_2})\right]_a = 0$$

$$\left[f_{y'_2} + \frac{\partial u}{\partial y_2}\,(f - y'_1 f_{y'_1} - y'_2 f_{y'_2})\right]_a = 0.$$

Prove that these conditions are equivalent to the conditions (A2.13.5) if:

(a) $f|_a \neq 0$ and $\hat{y} = \hat{y}(x)$ is not tangent to $x = u(y_1,y_2)$ at $x = a$.

(b) $f|_a = 0$ and $\hat{y} = \hat{y}(x)$ is not tangent to $x = u(y_1,y_2)$ at $x = a$.

(c) $f|_a = 0$ and $\hat{y} = \hat{y}(x)$ is tangent to $x = u(y_1,y_2)$ at $x = a$.

(d) $f|_a \neq 0$ and $\hat{y} = \hat{y}(x)$ is tangent to $x = u(y_1,y_2)$ at $x = a$.

*2. Derive (A2.13.5b).

3. Find the transversality conditions for the variational problem

$$\int_a^b f(x,y_1,y_2,y_3,y'_1,y'_2,y'_3)\,dx \to \text{minimum}$$

with a fixed beginning point $\hat{y}(a) = \hat{y}_a$ and an endpoint that varies on the surface $x = v(y_1,y_2,y_3)$.

4. Try to generalize the result of problem 3 to n unknown functions and an n-dimensional terminal surface.

5. Find a curve which satisfies the Euler-Lagrange equations, the initial condition, and the transversality conditions for the variational problem

$$I[y] = \int_0^b (1 + y_1'^2(x) + y_2'^2(x))\,dx \to \text{minimum},$$

$$y_1(0) = y_2(0) = 0,$$

$$b + y_1(b) - y_2(b) = 1.$$

(The last condition means that the endpoint has to lie on the plane $x + y_1 - y_2 = 1$.)

6. Derive transversality conditions for a variational problem in two unknown functions for which the terminal surface is a cylindrical surface that is perpendicular to the x,y_1 plane, that is, $\partial u/\partial y_2 = 0$ everywhere.

A2.14 INVARIANCE OF THE EULER-LAGRANGE EQUATIONS

Occasionally, it is more convenient to formulate a variational problem in other than cartesian coordinates. Take, for example, the Fermat-type problem

$$\int_a^b \sqrt{x^2 + y^2(x)}\,\sqrt{1 + y'^2(x)}\,dx \to \text{minimum}$$

† I. M. Gelfand and S. V. Fomin, "Calculus of Variations," p. 61, Prentice-Hall, Inc., Englewood Cliffs, N.J., 1963; G. A. Bliss, "Lectures on the Calculus of Variations," p. 149, The University of Chicago Press, Chicago, 1946. (Bliss' condition yields the conditions of problem 1 if one takes first $dx = 0$ and then $dy = 0$.)

with suitable boundary conditions. Since the integrand contains x, y, y' explicitly, we cannot immediately write a first integral of the Euler-Lagrange equation. For the Euler-Lagrange equation itself [note that the problem is regular for $(x,y) \neq (0,0)$], we obtain after multiplication by $\sqrt{x^2 + y^2}\sqrt{1 + y'^2}$ and some obvious simplifications,

$$y - xy' - \frac{(x^2 + y^2)y''}{1 + y'^2} = 0,$$

which is quite a monstrous equation, by any standards. The particular appearance of the integrand, however, suggests that some simplifications might be obtained if polar coordinates were introduced.

Let

$$x = r \cos \theta, \qquad y = r \sin \theta.$$

Then $x^2 + y^2 = r^2$ and $\sqrt{1 + y'^2}\, dx = \sqrt{r^2 + r'^2}\, d\theta$, and our variational problem appears in the form

$$\int_\alpha^\beta r(\theta)\sqrt{r^2(\theta) + r'^2(\theta)}\, d\theta \rightarrow \text{minimum},$$

where α, β depend on a, b and the initial and terminal values that are prescribed for $y = y(x)$ by the boundary conditions in the original formulation of the problem.

Now, if it were true that the Euler-Lagrange equation of the original problem transforms into the Euler-Lagrange equation of the problem in polar coordinates, we would immediately obtain a first integral since the new integrand does not depend explicitly on θ.

By (2.6.1),

$$f - r'f_{r'} = C$$

is such a first integral. In our particular case, this yields

$$r^3 = C\sqrt{r^2 + r'^2},$$

which is much simpler than the Euler-Lagrange equation we obtained for the problem in the x,y formulation.

Since the derivation of the Euler-Lagrange equation (weak variations, etc.) was essentially based on the fact that the coordinates were cartesian coordinates, it is not immediately clear whether or not the formal appearance of the Euler-Lagrange equation remains unchanged under coordinate transformations. In particular, it is not at all obvious that the class of competing functions is the same in both formulations. Still, this is true under fairly liberal conditions on the integrand, the extremal, and the transformation itself, as we shall show in the sequel.

Let us consider the variational problem

$$I[\hat{y}] = \int_a^b f(x,\hat{y}(x),\hat{y}'(x))\,dx \to \text{minimum} \tag{A2.14.1}$$

$$\hat{y}(a) = \hat{y}_a, \qquad \hat{y}(b) = \hat{y}_b,$$

where we assume that $\hat{y} = \hat{y}_o(x) \in C^2[a,b]^n$ yields a weak relative minimum. We also assume that $f \in C^2(\mathcal{R})$, where the domain $\mathcal{R}$ shall contain all lineal elements of $\hat{y} = \hat{y}_o(x)$. Then, as we have seen in Sec. 2.11, $\hat{y} = \hat{y}_o(x)$ satisfies the n Euler-Lagrange equations

$$f_{y_k} - \frac{d}{dx} f_{y'_k} = 0, \qquad k = 1, 2, \ldots, n, \tag{A2.14.2}$$

and the boundary conditions in (A2.14.1).

We introduce $n + 1$ new variables $u, v_1, \ldots, v_n$ by means of the one-to-one transformation

$$\begin{aligned} u &= \varphi(x,y_1, \ldots,y_n) \\ v_1 &= \psi_1(x,y_1, \ldots,y_n) \\ &\vdots \\ v_n &= \psi_n(x,y_1, \ldots,y_n), \end{aligned} \tag{A2.14.3}$$

where we assume that

$$\varphi,\psi_i \in C^3, \qquad \frac{\partial(u,v_1, \ldots,v_n)}{\partial(x,y_1, \ldots,y_n)} \neq 0 \tag{A2.14.4}$$

for all $(x,y_1, \ldots,y_n)$ that lie in the projection $\mathcal{P}$ of $\mathcal{R}$ into the $(x,y_1, \ldots,y_n)$ space. Under these conditions, the transformation (A2.14.3) has a unique inverse:

$$\begin{aligned} x &= \bar{\varphi}(u,v_1, \ldots,v_n) \\ y_1 &= \bar{\psi}_1(u,v_1, \ldots,v_n) \\ &\vdots \\ y_n &= \bar{\psi}_n(u,v_1, \ldots,v_n), \end{aligned} \tag{A2.14.5}$$

where

$$\bar{\varphi},\bar{\psi}_i \in C^3, \qquad \frac{\partial(x,y_1, \ldots,y_n)}{\partial(u,v_1, \ldots,v_n)} = \frac{1}{\dfrac{\partial(u,v_1, \ldots,v_n)}{\partial(x,y_1, \ldots,y_n)}}\Bigg|_{\substack{x=\bar{\varphi}(u,\hat{v}) \\ y_i=\bar{\psi}_i(u,\hat{v})}} \neq 0 \tag{A2.14.6}$$

for all $(u,v_1, \ldots,v_n) \in \mathcal{P}'$, where $\mathcal{P}'$ is the image of $\mathcal{P}$ under the transformation (A2.14.3).

Suppose now that along $\hat{y} = \hat{y}_o(x)$ for some $x_o \in [a,b]$,

$$\frac{d\varphi\ (x_o,\hat{y}_o(x_o))}{dx} = \frac{d\psi_1\ (x_o,\hat{y}_o(x_o))}{dx} = \cdots = \frac{d\psi_n\ (x_o,\hat{y}_o(x_o))}{dx} = 0.$$

Then, because the linear homogeneous system

$$\frac{\partial \varphi}{\partial x} + \frac{\partial \varphi}{\partial y_1} y_1{}^{o\prime} + \cdots + \frac{\partial \varphi}{\partial y_n} y_n{}^{o\prime} = 0$$

$$\frac{\partial \psi_1}{\partial x} + \frac{\partial \psi_1}{\partial y_1} y_1{}^{o\prime} + \cdots + \frac{\partial \psi_1}{\partial y_n} y_n{}^{o\prime} = 0$$

$$\vdots$$

$$\frac{\partial \psi_n}{\partial x} + \frac{\partial \psi_n}{\partial y_1} y_1{}^{o\prime} + \cdots + \frac{\partial \psi_n}{\partial y_n} y_n{}^{o\prime} = 0$$

has the nontrivial solution $(1, y_1{}^{o\prime}(x_o), \ldots, y_n{}^{o\prime}(x_o))$, we must have

$$\begin{vmatrix} \frac{\partial \varphi}{\partial x} & \frac{\partial \varphi}{\partial y_1} & \cdots & \frac{\partial \varphi}{\partial y_n} \\ \frac{\partial \psi_1}{\partial x} & \frac{\partial \psi_1}{\partial y_n} & \cdots & \frac{\partial \psi_1}{\partial y_n} \\ \vdots & & & \\ \frac{\partial \psi_n}{\partial x} & \frac{\partial \psi_n}{\partial y_1} & \cdots & \frac{\partial \psi_n}{\partial y_n} \end{vmatrix} = \left. \frac{\partial(u, v_1, \ldots, v_n)}{\partial(x, y_1, \ldots, y_n)} \right|_{\substack{y = \hat{y}_o(x) \\ x = x_o}} = 0,$$

contrary to our hypothesis (A2.14.4). Hence we may assume without loss of generality that $d\varphi\,(x_o, \hat{y}_o(x_o))/dx \neq 0$ for some $x_o \in [a,b]$. (Otherwise, we simply renumber the new variables $u, v_1, \ldots, v_n$ and introduce one of the v_i as new independent variable.)

Then, by the theorem on implicit functions, we may solve

$$u = \varphi(x, \hat{y}_o(x))$$

in a neighborhood of $x = x_o$ for x, and we obtain

$$x = x(u) \in C^3 \qquad \text{for all } u \in N^\delta(u_o),$$

where $u_o = \varphi(x_o, y_o(x_o))$. If we now substitute this result into the remaining n equations in (A2.14.3), we obtain

$$v_1 = \psi_1(x(u), y_1{}^o(x(u)), \ldots, y_n{}^o(x(u))) \equiv v_1{}^o(u)$$
$$\vdots$$
$$v_n = \psi_n(x(u), y_1{}^o(x(u)), \ldots, y_n{}^o(x(u))) \equiv v_n{}^o(u),$$

or as we may write it,

$$\hat{v} = \hat{v}_o(u).$$

We see that $\hat{v} = \hat{v}_o(u) \in C^2(N^\delta(u_o))$.

Proof: Let $h_k(u) \in C^2[\alpha,\beta]$, $h_k(\alpha) = h_k(\beta) = 0$, where $1 \leq k \leq n$ and k is fixed. We denote

$$\hat{h}(u) = (0,0, \ldots,0,h_k(u),0, \ldots,0)$$

and consider the weak variation

$$\hat{v} = \hat{v}_o(u) + t\hat{h}(u)$$

of the image $\hat{v} = \hat{v}_o(u)$ of $\hat{y} = \hat{y}_o(x)$ under the transformation (A2.14.3).

The preimage in the $(x,y_1, \ldots,y_n)$ space of this weak variation is given by

$$\begin{aligned} x &= \bar{\varphi}(u, \hat{v}_o(u) + t\hat{h}(u)) \\ y_i &= \bar{\psi}_i(u, \hat{v}_o(u) + t\hat{h}(u)), \qquad i = 1, 2, \ldots, n. \end{aligned} \tag{A2.14.9}$$

For $t = 0$, we obtain

$$\begin{aligned} x &= \bar{\varphi}(u,\hat{v}_o(u)) \\ y_i &= \bar{\psi}_i(u,\hat{v}_o(u)), \qquad i = 1, 2, \ldots, n, \end{aligned}$$

which is equivalent to $\hat{y} = \hat{y}_o(x)$ by the definition of $\hat{v} = \hat{v}_o(u)$. Hence we have

$$\frac{d\bar{\psi}_i\,(u,\hat{v}_o(u))}{du} = \frac{dy_i^o\,(x)}{dx}\,\frac{d\bar{\varphi}\,(u,\hat{v}_o(u))}{du}. \tag{A2.14.10}$$

We wish to express the preimage of the variation of $\hat{v} = \hat{v}_o(u)$ as a function $\hat{y}$ of x and t. For this purpose, we have to solve the first of the $n + 1$ equations in (A2.14.9) for u in terms of x and t and then substitute the result into the remaining n equations. This is possible in a neighborhood of $t = 0$ if

$$\left.\frac{d\bar{\varphi}\,(u, \hat{v}_o(u) + t\hat{h}(u))}{du}\right|_{t=0} = \frac{d\bar{\varphi}\,(u, \hat{v}_o(u))}{du} \neq 0.$$

Suppose for a moment that this is not the case:

$$\frac{d\bar{\varphi}\,(u, \hat{v}_o(u))}{du} = 0,$$

that is,

$$\bar{\varphi}_u + \bar{\varphi}_{v_1}v_1^{o\prime} + \cdots + \bar{\varphi}_{v_n}v_n^{o\prime} = 0,$$

and because of (A2.14.10), also

$$\hat{\bar{\psi}}_u + \hat{\bar{\psi}}_{v_1}v_1^{o\prime} + \cdots + \hat{\bar{\psi}}_{v_n}v_n^{o\prime} = 0.$$

These $n + 1$ equations form a linear homogeneous system which has the nontrivial solution $(1, v_1^{o\prime}, \ldots, v_n^{o\prime})$, and hence

$$\begin{vmatrix} \dfrac{\partial\bar\varphi}{\partial u} & \dfrac{\partial\bar\varphi}{\partial v_1} & \cdots & \dfrac{\partial\bar\varphi}{\partial v_n} \\ \dfrac{\partial\bar\psi_1}{\partial u} & \dfrac{\partial\bar\psi_1}{\partial v_1} & \cdots & \dfrac{\partial\bar\psi_1}{\partial v_n} \\ \vdots & & & \\ \dfrac{\partial\bar\psi_n}{\partial u} & \dfrac{\partial\bar\psi_n}{\partial v_1} & \cdots & \dfrac{\partial\bar\psi_n}{\partial v_n} \end{vmatrix} = \frac{\partial(x, y_1, \ldots, y_n)}{\partial(u, v_1, \ldots, v_n)} = 0,$$

contrary to (A2.14.6).

Hence $(d\bar\varphi/du)|_{t=0} \neq 0$, and we obtain by the theorem on implicit functions from the first equation in (A2.14.9),

$$u = u(x,t),$$

where $u(x,t) \in C^2$ for all $x \in [a,b]$ and for t sufficiently small.

Then we obtain from the remaining n equations in (A2.14.9),

$$y_i = \bar\psi_i(u(x,t), \hat v_o(u(x,t)) + t\hat h(u(x,t))) = \eta_i(x,t),$$

where we note that $\hat y = \hat\eta(x,t) \in C^2([a,b]^n \times (-\varepsilon,\varepsilon))$ for ε sufficiently small. It also follows from the theorem on implicit functions that $\hat y = \hat\eta(x,t)$ lies in a strong neighborhood of $\hat y = \hat y_o(x)$ if t is sufficiently small. (See Prob. A2.14.2.) Hence its lineal elements will all lie in $\mathcal{R}$.

Then

$$\int_\alpha^\beta g(u, \hat v_o(u) + t\hat h(u), \hat v_o'(u) + t\hat h'(u))\,du = \int_a^b f(x, \hat\eta(x,t), \hat\eta'(x,t))\,dx$$

is true for all t that are sufficiently small. Hereby, $\hat\eta' = d\hat\eta/dx$. Differentiation of this identity in t with respect to t at $t = 0$ yields

$$\int_\alpha^\beta [g_{v_k}(u, \hat v_o(u), \hat v_o'(u))h_k(u) + g_{v'_k}(u, \hat v_o(u), \hat v_o'(u))h_k'(u)]\,du$$

$$= \int_a^b \left[\sum_{i=1}^n f_{y_i}(x, \hat y_o(x), \hat y_o'(x)) \frac{\partial\eta_i(x,0)}{\partial t} + \sum_{i=1}^n f_{y'_i}(x, \hat y_o(x), \hat y_o'(x)) \frac{\partial\eta_i'(x,0)}{\partial t} \right] dx.$$

Since we have $\hat y_o \in C^2[a,b]^n$ and $f \in C^2(\mathcal{R})$ anyway, we can apply integration by parts to the last n terms in the integral on the right, as in Sec.

2.4, and obtain, after collecting all results for the right side of the equation,

$$\sum_{i=1}^{n} f_{y'_i} \frac{\partial \eta_i(x,0)}{\partial t}\bigg|_a^b + \int_a^b \sum_{i=1}^{n} \frac{\partial \eta_i(x,0)}{\partial t}\left(f_{y_i} - \frac{d}{dx} f_{y'_i}\right) dx$$

if we also note that

$$\left(\frac{\partial \eta_i(x,0)}{\partial t}\right)' = \frac{\partial \eta'_i(x,0)}{\partial t}$$

in view of $\hat{y} = \hat{\eta}(x,t) \in C^2([a,b]^n \times (-\varepsilon,\varepsilon))$.

We have $\eta_i(a,t) = y_a$, $\eta_i(b,t) = y_b$ for all t since $(a,\hat{y}_a)$, $(b,\hat{y}_b)$ are mapped into $(\alpha,\hat{v}_\alpha)$, $(\beta,\hat{v}_\beta)$, and vice versa, and $h_k(\alpha) = h_k(\beta) = 0$. Hence

$$\frac{\partial \eta_i(a,0)}{\partial t} = \frac{\partial \eta_i(b,0)}{\partial t} = 0,$$

and the integrated term vanishes. By hypothesis, $\hat{y} = \hat{y}_o(x)$ is an extremal, i.e., satisfies the Euler-Lagrange equations (A2.14.2). Hence the integral also vanishes, and we obtain

$$\int_\alpha^\beta [g_{v_k}(u,\hat{v}_o(u),\hat{v}'_o(u))h_k(u) + g_{v'_k}(u,\hat{v}_o(u),\hat{v}'_o(u))h'_k(u)]\, du = 0$$

for all $h_k(u) \in C^2[\alpha,\beta]$ for which $h_k(\alpha) = h_k(\beta) = 0$.

We can now carry out integration by parts with respect to the first term, as in Sec. 2.3, and apply the lemma of Dubois-Reymond, or we can carry out integration by parts with respect to the second term and proceed as in Sec. 2.4. In either case, we can conclude that, by necessity,

$$g_{v_k} - \frac{d}{du} g_{v'_k} = 0,$$

which was to be proved.

Although we formulated Theorem A2.14 for a fixed-beginning-point, fixed-endpoint problem, it holds as well for the various cases and combinations of variable-endpoint problems because, in the final analysis, $\hat{y} = \hat{y}_o(x)$ always has to yield a relative minimum for a fixed-endpoint problem.

PROBLEMS A2.14

*1. Show that $g(u,\bar{v},\bar{v}') \in C^2(\mathcal{R}')$, where g is defined on page 112 and where $\mathcal{R}'$ is the image of $\mathcal{R}$ under the transformation (A2.14.3).

*2. Show that $\hat{\eta} = \hat{\eta}(x,t)$, which appears in the proof of Theorem A2.14, remains in a strong neighborhood of $\hat{y} = \hat{y}_0(x)$ if t is sufficiently small.

3. Find the extremals of $\int_a^b \sqrt{x^2 + y^2(x)}\sqrt{1 + y'^2(x)}\, dx$ by using polar coordinates.

4. Find the extremals of $\int_a^b \sqrt{y^2(x) + y'^2(x)}\, dx$ by introducing new coordinates $u = y \cos x$, $v = y \sin x$.
5. Obtain a first integral of the Euler-Lagrange equation of

$$\int_a^b \sqrt{2\left(E + \frac{\mu}{\sqrt{x^2 + y^2(x)}}\right)}\, \sqrt{1 + y'^2(x)}\, dx \to \text{minimum},$$

where E,μ are constants, in polar coordinates.

6. Show that $r = A/[\cos(\theta + B)]$ are the extremals of

$$\int_\alpha^\beta \sqrt{r^2(\theta) + r'^2(\theta)}\, d\theta \to \text{minimum}.$$

7. Show that $\theta = \cos^{-1}(A/r) + B$ are extremals of

$$\int_\alpha^\beta \sqrt{1 + r^2\theta'^2(r)}\, dr \to \text{minimum}.$$

A2.15 HAMILTON'S PRINCIPLE OF STATIONARY ACTION

We assume that n particles of mass $m_k (k = 1, 2, 3, \ldots, n)$ are located at time t at the points $(x_k(t), y_k(t), z_k(t))$ in (x,y,z) space, and we assume further that each particle is subjected to an external force, the components (u_k, v_k, w_k) of which are derived from a *potential function* (*potential energy*)

$$U = U(t, x_1, y_1, z_1, \ldots, x_n, y_n, z_n) \tag{A2.15.1}$$

by taking the negative gradient:

$$u_k = -U_{x_k}, \qquad v_k = -U_{y_k}, \qquad w_k = -U_{z_k}.$$

The *kinetic energy* T of the entire system is given by

$$T = \frac{1}{2} \sum_{k=1}^{n} m_k (\dot{x}_k^2 + \dot{y}_k^2 + \dot{z}_k^2), \tag{A2.15.2}$$

where a dot denotes the derivative with respect to the time t.

Definition A2.15 *$\hat{y} = \hat{y}(x)$ is said to yield a stationary value for $I[\hat{y}]$ if the first variation of $I[\hat{y}]$ vanishes for $\hat{y} = \hat{y}(x)$:*

$$\delta I[\hat{y}] = 0.$$

PRINCIPLE OF STATIONARY ACTION

The system of particles of mass m_k with locations at $(x_k(t), y_k(t), z_k(t))$ moves between any two instants t_o and t_1 along such curves

$$\left.\begin{aligned} x &= x_k(t) \\ y &= y_k(t) \\ z &= z_k(t) \end{aligned}\right\} \qquad t \in [t_o, t_1]$$

for which
$$\delta \int_{t_0}^{t_1} (T - U)\, dt = 0.$$

The integrand $T - U = L$ is called the *lagrangian*, and the integral $\int_{t_0}^{t_1} L\, dt$ is called the *action*. Hence the name "principle of stationary action."

To find the differential equations of stationary action for the mechanical system described here, we observe first that we are faced with a variational problem in $3n$ unknown functions in which the notation corresponds to the standard notation of previous sections and chapters, as follows:

$$t \Leftrightarrow x,\ x_1 \Leftrightarrow y_1,\ y_1 \Leftrightarrow y_2,\ z_1 \Leftrightarrow y_3,\ \ldots,\ x_n \Leftrightarrow y_{3n-2},\ y_n \Leftrightarrow y_{3n-1},\ z_n \Leftrightarrow y_{3n}.$$

With this in mind, we see that the Euler-Lagrange equations will appear as

$$L_{x_k} - \frac{d}{dt} L_{\dot{x}_k} = 0$$

$$L_{y_k} - \frac{d}{dt} L_{\dot{y}_k} = 0$$

$$L_{z_k} - \frac{d}{dt} L_{\dot{z}_k} = 0, \qquad k = 1, 2, 3, \ldots, n.$$

Since

$$L_{x_k} = -U_{x_k}, \qquad L_{y_k} = -U_{y_k}, \qquad L_{z_k} = -U_{z_k}$$

and
$$L_{\dot{x}_k} = m_k\dot{x}_k, \qquad L_{\dot{y}_k} = m_k\dot{y}_k, \qquad L_{\dot{z}_k} = m_k\dot{z}_k,$$

we obtain as Euler-Lagrange equations,

$$m_k\ddot{x}_k = -U_{x_k}, \qquad m_k\ddot{y}_k = -U_{y_k}, \qquad m_k\ddot{z}_k = -U_{z_k}, \qquad k = 1, 2, \ldots, n \tag{A2.15.3}$$

if we assume a nonrelativistic situation where the masses m_k are time independent. We recognize these equations as *Newton's equations of motion.*

We introduce canonical variables

$$p_k = L_{\dot{x}_k} = m_k\dot{x}_k$$

$$q_k = L_{\dot{y}_k} = m_k\dot{y}_k$$

$$r_k = L_{\dot{z}_k} = m_k\dot{z}_k$$

and

$$
\begin{aligned}
H(t,x_1, \ldots,z_n,p_1, \ldots,r_n) &= \left[\sum_{k=1}^{n} (p_k\dot{x}_k + q_k\dot{y}_k + r_k\dot{z}_k) - T + U \right]_{p,q,r} \\
&= \sum_{k=1}^{n} \frac{1}{m_k} (p_k^2 + q_k^2 + r_k^2) \\
&\quad - \sum_{k=1}^{n} \frac{1}{2m_k} (p_k^2 + q_k^2 + r_k^2) + U(t,x_1, \ldots,z_n) \\
&= \frac{1}{2} \sum_{k=1}^{n} \frac{1}{m_k} (p_k^2 + q_k^2 + r_k^2) + U(t,x_1, \ldots,z_n) \\
&= (T + U)_{p,q,r}.
\end{aligned}
$$

So we see that the hamiltonian of such a mechanical system represents the total energy $E = T + U$ of the system:

$$H = E. \tag{A2.15.4}$$

Since

$$H_{x_k} = U_{x_k}, \qquad H_{y_k} = U_{y_k}, \qquad H_{z_k} = U_{z_k}$$

and

$$H_{p_k} = \frac{p_k}{m_k}, \qquad H_{q_k} = \frac{q_k}{m_k}, \qquad H_{r_k} = \frac{r_k}{m_k},$$

we have for the canonical system,

$$
\begin{aligned}
\dot{x}_k &= \frac{p_k}{m_k}, & \dot{y}_k &= \frac{q_k}{m_k}, & \dot{z}_k &= \frac{r_k}{m_k} \\
\dot{p}_k &= -U_{x_k}, & \dot{q}_k &= -U_{y_k}, & \dot{r}_k &= -U_{z_k}.
\end{aligned}
\tag{A2.15.5}
$$

Integrals of these equations of motion for a number of special cases will be found in the next section.

PROBLEM A2.15

Show that if $U = U(x_1, \ldots,z_n)$ does not explicitly depend on the time t, then $H(x_1, \ldots,z_n,p_1, \ldots,r_n) = \text{constant}$ is a first integral of the canonical system (A2.15.5).

A2.16 NOETHER'S INTEGRATION OF THE EULER-LAGRANGE EQUATIONS—CONSERVATION LAWS IN MECHANICS

As a generalization of the observations made in Sec. 2.6, we shall now derive integrals of the Euler-Lagrange equation under certain invariance properties of $I[\hat{y}]$.

We consider a one-parameter family of transformations

$$\begin{aligned} \bar{x} &= \varphi(x,\hat{y},\varepsilon) \\ \hat{\bar{y}} &= \hat{\psi}(x,\hat{y},\varepsilon), \end{aligned} \tag{A2.16.1}$$

where we assume that $\varphi,\hat{\psi} \in C^2$ in an $(n+2)$-dimensional domain that will contain the interval $[a,b]$ of values x, some interval $(\varepsilon_1,\varepsilon_2)$ of values ε, and the entire $\hat{y}$ space. We further assume that the transformations (A2.16.1) contain the identity transformation for some ε_o in $(\varepsilon_1,\varepsilon_2)$:

$$\begin{aligned} \bar{x}\Big|_{\epsilon_o} &= \varphi(x,\hat{y},\varepsilon_o) = x \\ \hat{\bar{y}}\Big|_{\epsilon_o} &= \hat{\psi}(x,\hat{y},\varepsilon_o) = \hat{y}. \end{aligned} \tag{A2.16.2}$$

Then

$$\begin{aligned} \bar{x} = \varphi(x,\hat{y},\varepsilon) &= x + (\varepsilon - \varepsilon_o)[\varphi_\epsilon(x,\hat{y},\varepsilon_o) + \cdots] \\ \hat{\bar{y}} = \hat{\psi}(x,\hat{y},\varepsilon) &= \hat{y} + (\varepsilon - \varepsilon_o)[\hat{\psi}_\epsilon(x,y,\varepsilon_o) + \cdots], \end{aligned} \tag{A2.16.3}$$

and we see that

$$(\varphi_x)_{\epsilon_o} = 1, \qquad (\varphi_{\hat{y}})_{\epsilon_o} = 0, \qquad (\hat{\psi}_x)_{\epsilon_o} = 0, \qquad (\psi_{iy_j})_{\epsilon_o} = \delta_{ij}. \tag{A2.16.4}$$

We shall now assume that $I[\hat{y}]$ is invariant under the transformation (A2.16.1). To make this statement more precise, we consider any function $\hat{y} = \hat{y}(x) \in C^1[a,b]$ and subject it to the transformation (A2.16.1):

$$\begin{aligned} \bar{x} &= \varphi(x,\hat{y}(x),\varepsilon) \\ \hat{\bar{y}} &= \hat{\psi}(x,\hat{y}(x),\varepsilon). \end{aligned}$$

We solve the first equation for x,

$$x = \chi(\bar{x},\varepsilon),$$

(see Prob. A2.16.4) and substitute the result in the second equation,

$$\hat{\bar{y}} = \hat{\psi}(\chi(\bar{x},\varepsilon), \hat{\bar{y}}(\chi(\bar{x},\varepsilon)),\varepsilon) = \hat{\bar{y}}(\bar{x},\varepsilon). \tag{A2.16.5}$$

Thus $\hat{\bar{y}} = \hat{\bar{y}}(\bar{x},\varepsilon)$ is the image of $\hat{y} = \hat{y}(x)$ under the transformation (A2.16.1). We note that $\hat{\bar{y}} = \hat{\bar{y}}(x,\varepsilon)$ represents a one-parameter family of curves in $x,\hat{y}$ space.

We also note that the x coordinates of any two points $(x_1,\hat{y}(x_1))$, $(x_2,\hat{y}(x_2))$ on $\hat{y} = \hat{y}(x)$ are transformed into

$$\begin{aligned} \bar{x}_1 &= \varphi(x_1,\hat{y}(x_1),\varepsilon) = \bar{x}_1(\varepsilon) \\ \bar{x}_2 &= \varphi(x_2,\hat{y}(x_2),\varepsilon) = \bar{x}_2(\varepsilon). \end{aligned}$$

Now we are ready for:

Definition A2.16 *The functional* $I[\hat{y}]$ *is invariant on* $[a,b]$ *under the transformation* (A2.16.1) *if*

$$\int_{x_1}^{x_2} f(x,\hat{y}(x),\hat{y}'(x))\,dx = \int_{\bar{x}_1(\epsilon)}^{\bar{x}_2(\epsilon)} f(\bar{x},\hat{\bar{y}}(\bar{x},\varepsilon),\hat{\bar{y}}'(\bar{x},\varepsilon))\,d\bar{x}$$

for any curve $\hat{y} = \hat{y}(x) \in C^1[a,b]$, *for any* $a \leq x_1 < x_2 \leq b$, *and for all* $\varepsilon \in (\varepsilon_1,\varepsilon_2)$.

We note that if the integrand $f(x,\hat{y},\hat{y}')$ remains invariant under a transformation of the type (A2.16.1) and if $d\bar{x} = dx$, then, obviously, $I[\hat{y}]$ is also invariant under that transformation.

Let us assume now that $I[\hat{y}]$ is invariant under the transformation (A2.16.1) in the sense of Definition A2.16. Then

$$I(\varepsilon) = \int_{\bar{x}_1(\epsilon)}^{\bar{x}_2(\epsilon)} f(\bar{x},\hat{\bar{y}}(\bar{x},\varepsilon),\hat{\bar{y}}'(\bar{x},\varepsilon))\,d\bar{x}$$

is independent of ε; that is,

$$\frac{dI(\varepsilon)}{d\varepsilon} = 0$$

for all ε in $(\varepsilon_1,\varepsilon_2)$.

Let us now compute $I'(\varepsilon)$: By the formula for the differentiation of an integral with respect to a parameter,† we have

$$\frac{dI(\varepsilon)}{d\varepsilon} = \int_{\bar{x}_1(\epsilon)}^{\bar{x}_2(\epsilon)} \left[f_{\hat{y}} \left(\frac{\partial \hat{\bar{y}}}{\partial \varepsilon}\right)^T + f_{\hat{y}'} \left(\frac{\partial \hat{\bar{y}}'}{\partial \varepsilon}\right)^T \right] d\bar{x}$$
$$+ f(\bar{x}_2(\varepsilon),\hat{\bar{y}}(\bar{x}_2(\varepsilon),\varepsilon),\hat{\bar{y}}'(\bar{x}_2(\varepsilon),\varepsilon)) \cdot \frac{\partial \bar{x}_2}{\partial \varepsilon}$$
$$- f(\bar{x}_1(\varepsilon),\hat{\bar{y}}(\bar{x}_1(\varepsilon),\varepsilon),\hat{\bar{y}}_1'(\bar{x}_1(\varepsilon),\varepsilon)) \frac{\partial \bar{x}_1}{\partial \varepsilon}.$$

Integration by parts (see Sec. 2.3) yields, in terms of

$$\hat{\phi}(\bar{x}) = \int_{\bar{x}_1(\epsilon)}^{\bar{x}} f_{\hat{y}}(s,\hat{\bar{y}}(s,\varepsilon),\hat{\bar{y}}'(s,\varepsilon))\,ds,$$

$$\frac{dI(\varepsilon)}{d\varepsilon} = \int_{\bar{x}_1(\epsilon)}^{\bar{x}_2(\epsilon)} \left(\frac{\partial \hat{\bar{y}}'}{\partial \varepsilon}\right) (f_{\hat{y}'} - \hat{\phi})^T\,d\bar{x} + \hat{\phi}(\bar{x}) \left(\frac{\partial \hat{\bar{y}}}{\partial \varepsilon}\right)^T \Bigg|_{\bar{x}_1(\epsilon)}^{\bar{x}_2(\epsilon)}$$
$$+ (f)_{\bar{x}_2(\epsilon)} \frac{\partial \bar{x}_2}{\partial \varepsilon} - (f)_{\bar{x}_1(\epsilon)} \frac{\partial \bar{x}_1}{\partial \varepsilon}$$

† D. V. Widder, "Advanced Calculus," 2d ed., p. 353, Prentice-Hall, Inc., Englewood Cliffs, N.J., 1961.

Let us now assume that $\hat{y} = \hat{y}(x)$ is an extremal of $I[\hat{y}]$, that is, that it satisfies the Euler-Lagrange equations (in integrated form)

$$f_{\hat{y}'} = \hat{\phi}(x) + \hat{C} \tag{A2.16.6}$$

[see (A2.11.4)], where $\hat{C}$ is some constant vector.

Since $I'(\varepsilon) = 0$ for all $\varepsilon \in (\varepsilon_1,\varepsilon_2)$, we have, in particular, that $I'(\varepsilon_o) = 0$. But then we have, in view of (A2.16.2) and (A2.16.6),

$$\left.\frac{dI(\varepsilon)}{d\varepsilon}\right|_{\epsilon=\epsilon_o} = f_{\hat{y}'}\left(\frac{\partial\hat{\bar{y}}}{\partial\varepsilon}\right)^T_{\epsilon_o}\Bigg|_{x_1}^{x_2} + (f)_{x_2}\left(\frac{\partial\bar{x}_2}{\partial\varepsilon}\right)_{\epsilon_o} - (f)_{x_1}\left(\frac{\partial\bar{x}_1}{\partial\varepsilon}\right)_{\epsilon_o} = 0. \tag{A2.16.7}$$

We see from (A2.16.5) that

$$\frac{\partial\hat{\bar{y}}(\bar{x},\varepsilon)}{\partial\varepsilon} = \hat{\psi}_x\chi_\epsilon + \sum_{i=1}^{n}\hat{\psi}_{y_i}y'_i\chi_\epsilon + \hat{\psi}_\epsilon,$$

and in view of (A2.16.4),

$$\left.\frac{\partial\hat{\bar{y}}(\bar{x},\varepsilon)}{\partial\varepsilon}\right|_{\epsilon_o} = \hat{y}'(\chi_\epsilon)_{\epsilon_o} + (\hat{\psi}_\epsilon)_{\epsilon_o}.$$

Since $x = \chi(\bar{x},\varepsilon)$ is the solution of $\bar{x} = \varphi(x,\hat{y}(x),\varepsilon)$, we obtain from differentiation of the identity

$$\bar{x} = \varphi(\chi(\bar{x},\varepsilon),\hat{y}(\chi(\bar{x},\varepsilon)),\varepsilon)$$

with respect to ε,

$$0 = \varphi_x\chi_\epsilon + \varphi_{\hat{y}}(\hat{y}')^T\chi_\epsilon + \varphi_\epsilon,$$

and in view of (A2.16.4),

$$(\chi_\epsilon)_{\epsilon_o} = -(\varphi_\epsilon)_{\epsilon_o}.$$

Thus

$$\left.\frac{\partial\hat{\bar{y}}(\bar{x},\varepsilon)}{\partial\varepsilon}\right|_{\epsilon_o} = -\hat{y}'(\varphi_\epsilon)_{\epsilon_o} + (\hat{\psi}_\epsilon)_{\epsilon_o},$$

and we have fiom (A2.16.7) if we also note that $(\bar{x}_\epsilon)_{\epsilon_o} = (\varphi_\epsilon)_{\epsilon_o}$:

$$f_{\hat{y}'}(\hat{\psi}_\epsilon{}^T)_{\epsilon_o}\Big|_{x_1}^{x_2} + (f - \hat{y}'f_{\hat{y}'}{}^T)(\varphi_\epsilon)_{\epsilon_o}\Big|_{x_1}^{x_2} = 0$$

for all x_1,x_2 in $a \leq x_1 < x_2 \leq b$. Hence,

$$f_{\hat{y}'}(\hat{\psi}_\epsilon{}^T)_{\epsilon_o} + (f - \hat{y}'f_{\hat{y}'}{}^T)(\varphi_\epsilon)_{\epsilon_o} = \text{constant}$$

for $\hat{y} = \hat{y}(x)$ in $[a,b]$.

It is customary to denote

$$\varphi_\epsilon(x,\hat{y},\varepsilon)|_{\epsilon_o} = \xi(x,\hat{y})$$
$$\hat{\psi}_\epsilon(x,\hat{y},\varepsilon)|_{\epsilon_o} = \hat{\eta}(x,\hat{y}),$$

where $\xi(x,\hat{y})$ and $\hat{\eta}(x,\hat{y})$ are called the *infinitesimal transformations.* [$\xi(x,\hat{y})$ and $\eta(x,\hat{y})$ represent the "initial velocity" of the transformation (A2.16.1).] With this notation we have

$$f_{\hat{y}'}\hat{\eta}^T(x,\hat{y}) + (f - \hat{y}'f_{\hat{y}'}{}^T)\xi(x,\hat{y}) = \text{constant} \qquad \text{(A2.16.8)}$$

on each extremal of $I[\hat{y}]$.

In components, (A2.16.8) reads

$$\sum_{k=1}^{n} f_{y'_k}\eta_k + [f - \sum_{k=1}^{n} y'_k f_{y'_k}]\xi = \text{constant},$$

where η_k is the kth component of $\hat{\eta}(x,\hat{y})$.

This result, which is due to *E. Noether*, will now be summarized as a theorem:

Theorem A2.16 *If $I[\hat{y}]$ is invariant under the transformations (A2.16.1) in the sense of Definition A2.16, then*

$$f_{\hat{y}'}\hat{\eta}^T(x,\hat{y}) + (f - \hat{y}'f_{\hat{y}'}{}^T)\xi(x,\hat{y}) = \text{constant}$$

along the extremals of $I[\hat{y}]$.

Before we apply Noether's theorem to mechanical systems, as discussed in the preceding section, we shall introduce canonical coordinates:

$$\hat{p} = f_{\hat{y}'}$$
$$H(x,\hat{y},\hat{p}) = \hat{y}'f_{\hat{y}'}{}^T - f(x,\hat{y},\hat{y}').$$

Then (A2.16.8) becomes

$$\hat{p}\hat{\eta}^T(x,\hat{y}) - H(x,\hat{y},\hat{p})\xi(x,y) = \text{constant}. \qquad \text{(A2.16.9)}$$

Conservation of energy: Suppose that the potential energy does not explicitly depend on the time t:

$$U = U(x_1, \ldots, z_n).$$

Then $L = T - U$ is obviously invariant under the time translation

$$\bar{t} = t + \varepsilon$$

$$\bar{x}_k = x_k, \qquad \bar{y}_k = y_k, \qquad \bar{z}_k = z_k,$$

which contains the identity transformation for $\varepsilon_o = 0$ and $dt = d\bar{t}$.

Denoting the infinitesimal transformation by the corresponding Greek letters, we have

$$\tau(t,x_1, \ldots, z_n) = \left.\frac{\partial(t+\varepsilon)}{\partial\varepsilon}\right|_{\epsilon=0} = 1$$

$$\xi_k = \eta_k = \zeta_k = 0, \qquad k = 1, 2, \ldots, n.$$

Hence, by (A2.16.9) and in view of (A2.15.4),

$$E = \text{constant},$$

that is, the total energy of a mechanical system in which the forces are derived from a potential function which does not explicitly depend on the time (*conservative system*) remains unchanged. This is the principle of the conservation of energy.

Conservation of momentum: The vector with the components

$$P = \sum_{k=1}^{n} p_k, \qquad Q = \sum_{k=1}^{n} q_k, \qquad R = \sum_{k=1}^{n} r_k$$

is called the *total momentum vector* of the mechanical system.

Suppose now that the potential energy $U = U(t,x_1, \ldots, z_n)$ is *translation invariant* in the directions of the coordinate axes, i.e., invariant under the transformations

$$\left.\begin{aligned} \bar{t} &= t \\ \bar{x}_k &= x_k + \varepsilon \\ \bar{y}_k &= y_k \\ \bar{z}_k &= z_k \end{aligned}\right\}, \qquad \left.\begin{aligned} \bar{t} &= t \\ \bar{x}_k &= x_k \\ \bar{y}_k &= y_k + \varepsilon \\ \bar{z}_k &= z_k \end{aligned}\right\}, \qquad \left.\begin{aligned} \bar{t} &= t \\ \bar{x}_k &= x_k \\ \bar{y}_k &= y_k \\ \bar{z}_k &= z_k + \varepsilon \end{aligned}\right\}, \qquad \varepsilon_o = 0$$

We have, in the first case,

$$\tau = 0, \qquad \xi_k = 1, \qquad \eta_k = \zeta_k = 0, \qquad k = 1, 2, \ldots, n,$$

and hence we obtain, in view of (A2.16.9),

$$P = \sum_{k=1}^{n} p_k = \text{constant}.$$

In the other two cases,

$$Q = \sum_{k=1}^{n} q_k = \text{constant} \qquad \text{and} \qquad R = \sum_{k=1}^{n} r_k = \text{constant},$$

that is, the momentum remains unchanged if the potential energy is invariant under translations in the direction of the coordinate axes.

Conservation of angular momentum: If L is invariant under a rotation about one of the coordinate axes, say, about the z axis,

$$\begin{aligned} \bar{t} &= t \\ \bar{x}_k &= x_k \cos\varepsilon + y_k \sin\varepsilon \\ \bar{y}_k &= -x_k \sin\varepsilon + y_k \cos\varepsilon \\ \bar{z}_k &= z_k \end{aligned}$$

($\varepsilon_o = 0$). Then

$$\tau = 0, \qquad \xi_k = y_k, \qquad \eta_k = -x_k, \qquad \zeta_k = 0,$$

and we obtain from (A2.16.9),

$$\sum_{k=1}^{n} (p_k y_k - q_k x_k) = \text{constant},$$

that is, the z component of the total angular momentum is constant.

PROBLEMS A2.16

1. Show that total energy and angular momentum are preserved in a newtonian gravitational field where

$$U = \frac{1}{r} \qquad r = \sqrt{x^2 + y^2 + z^2}.$$

2. Extend the result of problem 1 to any field where particles are attracted by some time-independent force that is exerted by a fixed point uniformly in all directions.
3. Which of the three conservation laws holds in a field where particles are attracted by the z axis?

*4. Why is it possible to solve $\bar{x} = \varphi(x,\hat{y}(x),\varepsilon)$ for x?

5. Show that formula (2.6.1) and the formula of Prob. 2.6.3 are special cases of (A2.16.8).

A2.17 GENERALIZATION TO MORE THAN ONE INDEPENDENT VARIABLE

We shall restrict our discussion to variational problems in two independent variables and one unknown function and then summarize the more obvious generalizations to more than two independent variables and more than one unknown function. We shall also make fairly restrictive assumptions in order to ensure a path unencumbered by somewhat extraneous complications.

Let J: $x = x(t)$, $y = y(t)$, $a \leq t \leq b$ represent a *simple closed smooth* curve in the x,y plane; that is,

$$\begin{aligned} &x(t),y(t) \in C^1[a,b] \\ &\{x(t_1),y(t_1)\} \neq \{x(t_2),y(t_2)\} \qquad \text{for all } t_1 \neq t_2,\ a \leq t_i < b \\ &\{x(a),y(a)\} = \{x(b),y(b)\} \\ &\{x'(t),y'(t)\} \neq \{0,0\} \qquad \text{for all } t \in [a,b]. \end{aligned}$$

By a celebrated theorem of *Jordan*, such a curve divides the x,y plane into two disjoint point sets, namely, an "interior" and an "exterior." Let $\mathcal{a}$ denote the region that consists of the interior of J together with J. ($\mathcal{a}$ is simply connected, closed, and bounded in view of the definition of J.)

We now pose the problem of finding a function $u = u(x,y) \in C^1[\mathcal{a}]$ which satisfies the boundary condition

$$u(x(t),y(t)) = z(t), \qquad z(t) \in C^1[a,b], \qquad z(a) = z(b)$$

on J and is such that the functional

$$I[u] = \iint\limits_{\mathcal{a}} f(x,y,u(x,y),u_x(x,y),u_y(x,y))\,dxdy \qquad \text{(A2.17.1)}$$

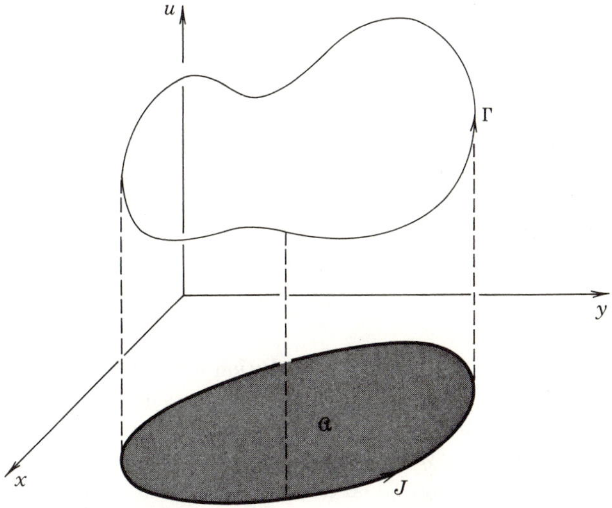

Figure 2.15

assumes its smallest value. Since $(x,y,z) = (x(t),y(t),z(t))$ represents a curve Γ in E_3, the boundary condition means geometrically that the surface $z = u(x,y)$ has to pass through the curve Γ. (See Fig. 2.15.)

We assume in the time-honored manner that $z = u_o(x,y) \in C^1[\mathcal{A}]$ is the solution to our problem, and we define the space of admissible variations as

$$\mathcal{H} = \{h \mid h \in C^1[\mathcal{A}], h(x(t), y(t)) = 0, t \in [a,b]\}.$$

If we define the norm in $\mathcal{H}$ as

$$||h|| = \max_{[\mathcal{A}]} |h(x,y)| + \max_{[\mathcal{A}]} |h_x(x,y)| + \max_{[\mathcal{A}]} |h_y(x,y)|, \quad \text{(A2.17.2)}$$

(see Prob. A2.17.1), then we see that it is necessary, if $u = u_o(x,y)$ is to yield a minimum, that

$$I[u_o + h] - I[u_o] \geq 0$$

for all $h \in \mathcal{H}$, $||h|| < \delta$ for some $\delta > 0$.

If we assume that $f \in C^1(\mathcal{R})$, where $\mathcal{R}$ is a domain in (x,y,u,u_x,u_y) space, and if we also assume that all "planar" elements

$$(x_o,y_o,u_o(x_o,y_o),u_{ox}(x_o,y_o),u_{oy}(x_o,y_o))$$

for all $(x_o,y_o) \in \mathcal{A}$ lie in $\mathcal{R}$, then $I[u]$ possesses a Gâteaux variation at

$u = u_o$ in the sense of Definition 1.5 because

$$\frac{d}{dt} I[u_o + th]_{t=0} = \iint_{\mathcal{A}} (f_u h + f_{u_x} h_x + f_{u_y} h_y)\, dx\, dy \quad \text{(A2.17.3)}$$

[where f_u, etc., and h, etc., stand for $f_u(x,y,u_o(x,y),u_{ox}(x,y),u_{oy}(x,y))$, etc., and $h(x,y)$, etc.] exists for all $h \in \mathcal{H}$. Hence we obtain

$$\delta \iint_{\mathcal{A}} f(x,y,u(x,y),u_x(x,y),u_y(x,y))\, dx\, dy$$
$$= \iint_{\mathcal{A}} (f_u h + f_{u_x} h_x + f_{u_y} h_y)\, dx\, dy = 0 \quad \text{(A2.17.4)}$$

for all $h \in \mathcal{H}$ as a necessary condition for a relative minimum.

It is our next task to transform this necessary condition into a form that is amenable to analysis. The role of Lagrange's integration by parts (Sec. 2.4) will now be assumed by *Green's theorem*, which states that under our assumptions on J and $\mathcal{A}$ (or less severe assumptions, for that matter),

$$\iint_{\mathcal{A}} (Q_x - P_y)\, dx\, dy = \int_J P\, dx + Q\, dy,\dagger$$

where $P = P(x,y)$, $Q = Q(x,y) \in C^1[\mathcal{A}]$.

If we assume that $f \in C^2(\mathcal{R})$ and $u = u_o(x,y) \in C^2[\mathcal{A}]$, then

$$P = -hf_{u_y}, \qquad Q = hf_{u_x}$$

satisfy the requirements of Green's theorem. Since we obtain with this choice of P and Q

$$Q_x - P_y = h_x f_{u_x} + h_y f_{u_y} + h\left(\frac{\partial}{\partial x} f_{u_x} + \frac{\partial}{\partial y} f_{u_y}\right),$$

we have from Green's theorem,

$$\iint_{\mathcal{A}} \left[h_x f_{u_x} + h_y f_{u_y} + h\left(\frac{\partial}{\partial x} f_{u_x} + \frac{\partial}{\partial y} f_{u_y}\right)\right] dx\, dy$$
$$= -\int_J h(f_{u_y}\, dx - f_{u_x}\, dy) = 0$$

since $h(x,y)|_{x,y \in J} = h(x(t),y(t)) = 0$. Thus

$$\delta I[h] = \iint_{\mathcal{A}} h\left[f_u - \frac{\partial}{\partial x} f_{u_x} - \frac{\partial}{\partial y} f_{u_y}\right] dx\, dy.$$

† A. E. Taylor, "Advanced Calculus," p. 420, Ginn and Company, Boston, 1955.

Hence the necessary condition for a minimum amounts to

$$\iint_{\mathcal{A}} h\left[f_u - \frac{\partial}{\partial x} f_{u_x} - \frac{\partial}{\partial y} f_{u_y}\right] dx\, dy = 0$$

for all $h \in \mathcal{K}$.

The next step requires a generalization of the fundamental lemma of the calculus of variations (Sec. 2.4):

Lemma A2.17 *If* $M \in C[\mathcal{A}]$ *and if* $\iint_{\mathcal{A}} h(x,y)M(x,y)\, dx\, dy = 0$ *for all* $h \in \mathcal{K}$, *then* $M(x,y) = 0$ *for all* $(x,y) \in \mathcal{A}$. (For proof, see Prob. A2.17.3.)

Since

$$M \equiv [f_u - f_{u_x x} - f_{u_x u}u_x - f_{u_x u_x}u_{xx} - f_{u_x u_y}u_{yx} - f_{u_y y} - f_{u_y u}u_y - f_{u_y u_x}u_{xy} - f_{u_y u_y}u_{yy}]_{u=u_o(x,y)} \in C[\mathcal{A}]$$

in view of our assumptions on f and u_o, then, by virtue of Lemma A2.17, we obtain the Euler-Lagrange equation for the two-dimensional problem as

$$f_u - \frac{\partial}{\partial x} f_{u_x} - \frac{\partial}{\partial y} f_{u_y} = 0. \qquad \text{(A2.17.5)}$$

This is a partial differential equation of second order which is, in general, nonlinear.

The question arises as to the possibility of deriving the Euler-Lagrange equation from $\delta I[h] = 0$ under less severe conditions on f and on u_o by using a procedure similar to the one we followed in Sec. 2.3, where we used the lemma of Dubois-Reymond and where we only assumed that $f \in C^1(\mathcal{R})$ and $y_o \in C_s^1[a,b]$. Based on a result by *A. Haar* one can show that this is indeed possible.† According to Haar, if the sectionally smooth surface $u = u_o(x,y)$ is to yield a relative minimum (maximum) for the functional $I[u] = \iint_{\mathcal{A}} f(x,y,u(x,y)u_x(x,y),u_y(x,y))\, dx\, dy$, $u(x(t),y(t)) = z(t)$, where it is assumed that $f \in C^1(\mathcal{R})$, it is necessary that

$$\iint_{\mathcal{A}} f_u(x,y,u_o(x,y)u_{ox}(x,y),\, u_{oy}(x,y))\, dx\, dy$$

$$= \int_j [f_{u_x}(x,y,u_o(x,y)u_{ox}(x,y),u_{oy}(x,y))\, dy - f_{u_y}(x,y,u_o(x,y)u_{ox}(x,y),u_{oy}(x,y))\, dx] \qquad \text{(A2.17.6)}$$

† See N. I. Akhiezer, "The Calculus of Variations," pp. 208 ff, Blaisdell Publishing Company, a division of Ginn and Company, Boston, 1962.

for any region $\mathfrak{a} \subset \mathfrak{A}$ that is encompassed by a simple, closed, and sectionally smooth boundary curve j. [$u = u_o(x,y)$ is a sectionally smooth surface on $\mathfrak{A}$ if $u_o \in C[\mathfrak{A}]$ and if $u_x, u_y \in C[\mathfrak{A}]$ except at finitely many isolated points in $\mathfrak{A}$ and on finitely many arcs that have a continuous tangent. A simple, closed, sectionally smooth curve j consists of finitely many simple smooth curves that are put together continuously so that no multiple points occur.]

Equation (A2.17.6) is the generalization of the Euler-Lagrange equation in integrated form for the problem in one independent variable. Equation (A2.17.5) follows from (A2.17.6) for any smooth portion of $u = u_o(x,y)$ if also $(\partial/\partial x)f_{u_x}$, $(\partial/\partial y)f_{u_y}$ are continuous.

We now illustrate the Euler-Lagrange equation (A2.17.5) with a discussion of two notable examples.

A. Dirichlet's problem:

$$D[u] = \iint_{\mathfrak{A}} (u_x^2(x,y) + u_y^2(x,y))\,dx\,dy \to \text{minimum},$$

$$u(x(t),y(t)) = z(t), \qquad t \in [a,b].$$

We see without difficulty that the *Laplace equation*

$$u_{xx} + u_{yy} = 0$$

is the Euler-Lagrange equation for Dirichlet's problem.

For a further study of the conditions on Γ, $\mathfrak{A}$, and the class of competing functions for the Dirichlet problem to admit a unique solution, we refer the reader to the literature.†

B. Plateau's problem (minimal surfaces): To be found is a surface of smallest area that is bounded by a given space curve Γ with projection $\mathfrak{J}$ into the x,y plane:

$$P[u] = \iint_{\mathfrak{A}} \sqrt{1 + u_x^2(x,y) + u_y^2(x,y)}\,dx\,dy, \qquad u(x(t),y(t)) = z(t), \qquad t \in [a,b],$$

where $\mathfrak{A}$ is the interior of J.

After some cumbersome manipulations that are not fit to print, we obtain as the Euler-Lagrange equation,

$$u_{xx}(1 + u_y^2) + u_{yy}(1 + u_x^2) - 2u_x u_y u_{xy} = 0.$$

This equation expresses that the mean curvature has to vanish at every point of the solution; i.e., the principal curvatures have to be equal in magnitude but opposite in sign. This means, in turn, that each point of a minimal surface is a saddle point.‡

† An elementary treatment can be found in L. A. Pars, "An Introduction to the Calculus of Variations," pp. 305ff, John Wiley & Sons, Inc., New York, 1962. For an extensive study of the problem, R. Courant, "Dirichlet's Problem," Interscience Publishers, Inc., New York, 1950, is recommended.

‡ For further study see Courant, *op. cit.*

So far as generalizations to higher dimensions and more unknown functions are concerned, suffice it to say that the Euler-Lagrange equations for a problem in n independent variables $x_1, x_2, \ldots, x_n$ with m unknown functions $u_1, u_2, \ldots, u_m$ have the form

$$f_{u_\mu} - \sum_{\nu=1}^{n} \frac{\partial}{\partial x_\nu} f_{u_\mu x_\nu} = 0, \qquad \mu = 1, 2, 3, \ldots, m.$$

(See also Prob. A2.17.5.)

PROBLEMS A2.17

*1. Show that $||h||$ as defined in (A2.17.2) has the properties of a norm.

*2. Show that the derivative in (A2.17.3) exists.

*3. Prove Lemma A2.17.

4. Solve Dirichlet's problem so that $|u(0,0)| < \infty$ for $\mathfrak{J}$: $x = \cos t$, $y = \sin t$, and $z = z(t)$, $t \in [0,2\pi]$. Assume that $z(t) \in C^1[0,2\pi]$ and $z(0) = z(2\pi)$.

*5. Find the Euler-Lagrange equations for

$$I[u,v] = \iiint_{\mathcal{V}} f\left(x, y, z, u, v, \frac{\partial u}{\partial x}, \frac{\partial u}{\partial y}, \frac{\partial u}{\partial z}, \frac{\partial v}{\partial x}, \frac{\partial v}{\partial y}, \frac{\partial v}{\partial z}\right) dx\, dy\, dz,$$

where $\mathcal{V}$ is a three-dimensional region encompassed by a closed rectifiable surface and where u,v are to assume given values on that surface.

6. Find the Euler-Lagrange equation of the variational problem

$$\int_{t_1}^{t_2} \int_0^1 \left[\rho\left(\frac{\partial u}{\partial t}\right)^2 - \tau\left(\frac{\partial u}{\partial x}\right)^2\right] dt\, dx, \qquad \rho, \tau \text{ constants},$$

$u(0,t) = u(1,t) = 0$, $u(x,t_1) = \phi(x)$, $u(x,t_2) = \psi(x)$, $\phi,\psi \in C^1[0,1]$, $\phi(0) = \psi(0) = \phi(1) = \psi(1) = 0$ (vibrating string).†

7. Find the Euler-Lagrange equation for the variational problem

$$\int_{t_1}^{t_2} \iint_{\mathfrak{A}} \left[\rho\left(\frac{\partial u}{\partial t}\right)^2 - \tau\left(\frac{\partial u}{\partial x}\right)^2 - \tau\left(\frac{\partial u}{\partial y}\right)^2 + 2f(t,x,y)u\right] dx\, dy\, dt, \qquad \rho, \tau \text{ constants},$$

where f is given. $u(x,y,t)|_{(x\,y)\in\mathfrak{J}} = 0$ ($\mathfrak{J}$ is the boundary of $\mathfrak{A}$), $u(x,y,t_1) = \phi(x,y)$, $u(x,y,t_2) = \psi(x,y)$, $\phi,\psi \in C^1[\mathfrak{A}]$ and vanish on $\mathfrak{J}$ (vibrating membrane under external force).†

8. Derive a natural boundary condition for the problem

$$I[u] = \iint_{\mathfrak{A}} f(x,y,u,u_x,u_y)\, dx\, dy \rightarrow \text{minimum}.$$

Geometrically, this problem may be viewed as the problem of finding a minimizing surface, the boundary of which is free to slide on the wall of the right cylinder

† H. Sagan, "Boundary and Eigenvalue Problems in Mathematical Physics," 3d printing, pp. 15 and 53, John Wiley & Sons, Inc., New York, 1966.

with trace $\mathfrak{J}$ in the x,y plane, where $\mathfrak{J}$ is the boundary of $\mathfrak{A}$. [*Hint*: Take $h \in H = \{h(x,y) \mid h(x,y) \in C^1[\mathfrak{A}]\}$. Then

$$\delta I[h] = \iint\limits_{\mathfrak{A}} h\left(f_u - \frac{\partial}{\partial x} f_{u_x} - \frac{\partial}{\partial y} f_{u_y}\right) dx\, dy + \int_{\mathfrak{J}} h(f_{u_x}\, dy - f_{u_y}\, dx).$$

Show that the Euler-Lagrange equation has to be satisfied, and derive a necessary condition for

$$\int_{\mathfrak{J}} h(f_{u_x}\, dy - f_{u_y}\, dx) = 0 \qquad \text{for all } h \in H.]$$

9. Derive the natural boundary condition (see problem 8) for

$$\iint\limits_{\mathfrak{A}} (u_x^2 + u_y^2)\, dx\, dy \to \text{minimum.}$$

(*Answer*: $du/dn = 0$ on $\mathfrak{J}$, where du/dn is the derivative in the direction of the outward normal to $\mathfrak{J}$.)

CHAPTER 3

THEORY OF FIELDS AND SUFFICIENT CONDITIONS FOR A STRONG RELATIVE EXTREMUM

APPENDIX

3.1 FIELDS

In Chap. 2 we derived a necessary condition for a weak and, by implication, a strong relative extremum of certain functionals. We obtained as such a necessary condition the Euler-Lagrange equation or Euler-Lagrange equations, depending on the specific nature of the functional. We saw that the Euler-Lagrange equation(s) together with the associated boundary conditions may or may not have a solution. Even in cases where a (unique) solution of the Euler-Lagrange equation(s) together with the boundary conditions exists, we still have no guarantee whatever that the function(s) so obtained will indeed yield a weak or strong relative extremum for the given functional. (An example of a variational problem where the uniquely determined extremal does not yield a strong relative minimum—or a weak relative minimum, for that matter—will be discussed in some detail in Sec. 3.4.)

When compared with the theory of extreme values of real-valued functions of a real variable, it appears that what we have done thus far is establish a procedure for finding critical points. Whether or not such critical points yield an extremum is, as we well know, quite another matter.

In the case of our variational problems, we shall have to investigate under what additional conditions it can be said that an extremal will indeed yield a weak or strong relative extremum. The investigations of this chapter will be concerned primarily with the establishment of a sufficient condition for a *strong relative minimum (maximum)*. That this investigation will also lead us to a sufficient condition for a *weak relative minimum (maximum)* and a further necessary condition for a *strong relative extremum* is an added fringe benefit.

The theory that will be developed here has no parallel in the extreme-value theory of real-valued functions of a real variable because in the latter, a distinction between weak and strong relative minima is inconsistent with the nature of the problem.

Initially, let us investigate the simplest variational problem:

$$I[y] = \int_a^b f(x,y(x),y'(x))\,dx \to \text{minimum}, \tag{3.1.1}$$

where $\quad y \in \Sigma = \{y \mid y \in C^1[a,b],\, y(a) = y_a,\, y(b) = y_b\}.$

If $y = y_o(x) \in C^1[a,b]$ is supposed to yield a *strong relative minimum* for $I[y]$, then it will be our task to establish a sufficient condition for

$$\Delta I = I[\bar{y}] - I[y_o] \geq 0 \tag{3.1.2}$$

to hold for all $\bar{y} \in \Sigma$ for which $\|\bar{y} - y_o\|_w < \delta$, or as we may put it, $\bar{y} \in N_w^\delta(y_o)$ for some $\delta > 0$. As in Chap. 1, we call ΔI the total variation of $I[y]$ at y_o.

Since the smooth function $y = y_o(x)$ is supposed to yield a relative minimum, y_o, by necessity, has to be a solution of the Euler-Lagrange equation (2.3.8)

$$f_y(x,y,y') - \frac{d}{dx} f_{y'}(x,y,y') = 0,$$

that is, it has to be an *extremal*. (See Definition 2.5.2.)

It is possible to transform the total variation (3.1.2) in such a manner that the end result will be more easily accessible to an examination. It is necessary for such a purpose to embed the extremal $y = y_o(x)$ in a so-called *field*. The originator of this idea was *K. Weierstrass*. In our treatment, however, we shall follow essentially the approach proposed by *D. Hilbert*.

Let us first define the concept of a field. Intuitively, we may arrive at this concept as follows: We consider a one-parameter family of extremals, i.e., smooth solutions of the Euler-Lagrange equation. We require that no two curves that correspond to different parameter values pass through the same point in a certain simply connected domain $\mathcal{A}$ in the x,y plane and that there are no points in $\mathcal{A}$ which are left out, i.e., through which there is no curve from the family. Then these curves will associate with each point (x,y) of $\mathcal{A}$ a vector $(1,y')$, where y' is the slope of the extremal passing through (x,y). The collection of all these directions that are associated with the points of $\mathcal{A}$ will be called a *field in* $\mathcal{A}$.

More precisely, we give the following definition:

Definition 3.1.1 *Let $\mathcal{A}$ denote a bounded and simply connected domain in the x,y plane. The vector function $(1,\phi(x,y))$ is said to define a field $\mathfrak{F} = \{(1,\phi(x,y)) \mid (x,y) \in \mathcal{A}\}$ in $\mathcal{A}$ for $I[y]$ if $\phi_x, \phi_y \in C(\mathcal{A})$ and if every solution of $y' = \phi(x,y)$ in $\mathcal{A}$ is an extremal of $I[y]$.*

In effect, a field determines a first-order differential equation, namely, $y' = \phi(x,y)$, which is satisfied by a subfamily of extremals, which in turn are solutions of the second-order Euler-Lagrange equation.

Definition 3.1.2 *The extremal $y = y_o(x) \in C^1[a,b]$, $y_o(a) = y_a$, $y_o(b) = y_b$, of $I[y]$ is* embeddable *in the field $\mathfrak{F}$ of $I[y]$ if:*

1. *$\mathfrak{F}$ is defined on a simply connected domain $\mathcal{A}$ that contains $N_w^\delta(y_o)$ and*
2. *$y = y_o(x)$ is a solution in $\mathcal{A}$ of $y' = \phi(x,y)$.*

[See Fig. 3.1, where the extremal $y = y_o(x)$ is labeled by E].

The following lemma will furnish a sufficient condition for the embeddability of a given extremal in a field:

Lemma 3.1 *If* $y = Y(x,c)$ *represents a one-parameter family of solutions of the Euler-Lagrange equation in* $\mathcal{S} = \{(x,c) \mid A < x < B, c_1 < c < c_2\}$ *such that* $Y(x,c_o) = y_o(x)$ *for some* $c_o \in (c_1,c_2)$ *and if* $Y, Y'', Y_c, Y'_c \in C(\mathcal{S})$ *and* $Y_c(x,c_o) \neq 0$ *for all* $x \in [a,b]$, *where* $[a,b] \subset (A,B)$, *then* $y = y_o(x)$, $x \in [a,b]$, *is embeddable in a field.*

(Note that $'$ denotes a partial derivative with respect to x and the subscript c denotes a partial derivative with respect to c.)

Proof: Let $\eta_o = y_o(x_o), x_o \in [a,b]$, and let $\mathcal{S}' = \{(x,y,c) \mid A < x < B, c_1 < c < c_2, -\infty < y < \infty\}$. Then

$$y - Y(x,c) \in C^1(\mathcal{S}')$$
$$\eta_o - Y(x_o,c_o) = 0$$
$$Y_c(x_o,c_o) \neq 0.$$

Hence, by the theorem on implicit functions, there exists in $\mathcal{S}'$ an open right parallelepiped $\mathcal{P} = \{(x,y,c) \mid |x - x_o| < \alpha_o, |y - \eta_o| < \beta_o, |c - c_o| < \gamma_o\}$ such that in $\mathcal{P}_o = \{(x,y) \mid |x - x_o| < \alpha_o, |y - \eta_o| < \beta_o\}$, we may solve

$$y - Y(x,c) = 0$$

uniquely for c,

$$c = C(x,y), \qquad (x,y) \in \mathcal{P}_o,$$

and $|c - c_o| < \gamma_o$ for $(x,y) \in \mathcal{P}_o, C(x,y) \in C^1(\mathcal{P}_o)$. Then $C(x,Y(x,c)) \equiv c$.

We repeat the same argument for each $x_o \in [a,b]$ and obtain thus an open covering of $y = y_o(x)$ on $[a,b]$ by rectangles. By the Heine-Borel theorem,† a finite number of these rectangles, say $\mathcal{P}_1, \mathcal{P}_2, \ldots, \mathcal{P}_n$, will cover

† A. E. Taylor, "Advanced Calculus," p. 493, Ginn and Company, Boston, 1955.

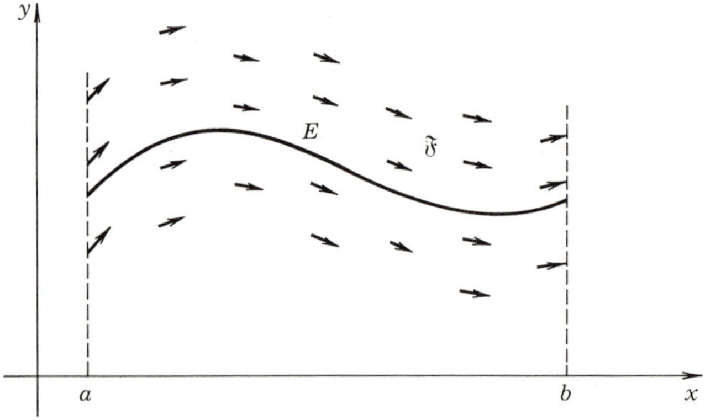

Figure 3.1

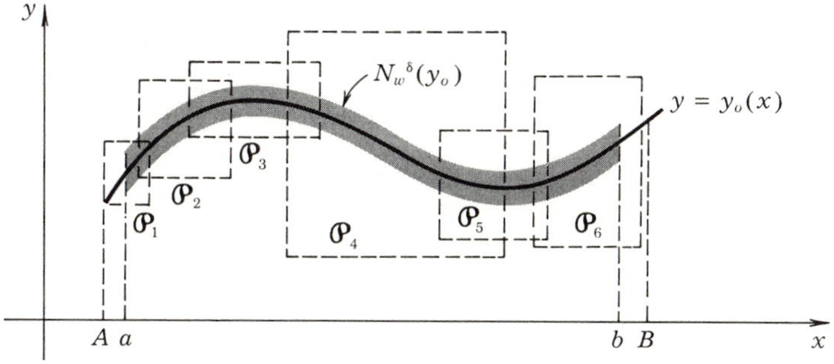

Figure 3.2

$y = y_o(x)$ on $[a,b]$. (See Fig. 3.2.) Let $\mathcal{P} = \bigcup_{k=1}^{n} \mathcal{P}_k$. Then there exists a $N_w^\delta(y_o) \subset \mathcal{P}$. We choose the enumeration of the rectangles $\mathcal{P}_k$ such that rectangles with consecutive subscripts overlap. Since $c = C(x,y)$ is uniquely determined in each $\mathcal{P}_k$, the values of $C(x,y)$ have to coincide in the portions of overlap of consecutive rectangles, and it follows that

$$c = C(x,y) \in C^1(N_w^\delta(y_o)).$$

Now we can show that $(1,\phi(x,y))$ defines a field on $N_w^\delta(y_o)$ where

$$\phi(x,y) = Y'(x,C(x,y)).$$

First:

$$\phi(x,y) = Y'(x,C(x,y)) \in C(N_w^\delta(y_o)),$$
$$\phi_x(x,y) = Y''(x,C(x,y)) + Y_c'(x,C(x,y))C_x(x,y) \in C(N_w^\delta(y_o)),$$
$$\phi_y(x,y) = Y_c'(x,C(x,y))C_y(x,y) \in C(N_w^\delta(y_o)).$$

Second:

$$\phi(x,Y(x,c)) = Y'(x,C(x,Y(x,c))) = Y'(x,c)$$

since $C(x,Y(x,c)) \equiv c$. Hence, the one-parameter family of extremals $y = Y(x,c)$ is a one-parameter family of solutions of $y' = \phi(x,y)$. Finally, $y_o(x) = Y(x,c_o)$, by hypothesis.

Thus, $\phi(x,y)$ satisfies all the conditions set forth in Definitions 3.1.1 and 3.1.2.

The following example may serve as a simple illustration of Lemma 3.1.

We consider

$$I[y] = \int_0^1 \sqrt{1 + y'^2(x)}\, dx \to \text{minimum}, \qquad y(0) = 0, \qquad y(1) = 1.$$

$y = y_o(x) \equiv x$ is the (uniquely determined) extremal of this problem. $y = Y(x,c) \equiv x + c$ is a one-parameter family of solutions of the Euler-Lagrange equation $y'' = 0$, and we have $Y(x,0) = x$ ($c_o = 0$ in this case). Further, $Y(x,c) = x + c$, $Y''(x,c) = 0$, $Y_c(x,c) = 1$, and $Y_c'(x,c) = 0$ are continuous everywhere, and $Y_c(x,c) = 1 \neq 0$. Hence, $y = x$, $x \in [0,1]$ is embeddable in a field $\mathfrak{F} = \{(1,1)\}$, since $\phi(x,y) = 1$.

$y = Y(x,c) \equiv cx + c - 1$ is another one-parameter family of solutions of the Euler-Lagrange equation and $Y(x,1) = x$ ($c_o = 1$ in this case). Since $Y(x,c) = cx + c - 1$, $Y''(x,c) = 0$, $Y_c(x,c) = x + 1$, and $Y_c'(x,c) = 1$ are continuous everywhere, and since $Y_c(x,c) = x + 1 \neq 0$ for all $x > -1$, we see that $y = x$, $x \in [0,1]$, is also embeddable in a field that is defined by $(1,(y+1)/(x+1))$, $x > -1$. [Since $C(x,y) = (y+1)/(x+1)$, we have $\phi(x,y) = Y'(x,C(x,y)) = (y+1)/(x+1)$ in this case.]

This example shows clearly that, at times, it is possible to embed a given extremal in a field in more than one way. This is really to be expected since the Euler-Lagrange equation has, in general, a two-parameter family of solutions, and if the circumstances are right, a suitable one-parameter family may be picked out in a number of ways. That a field is not necessarily uniquely defined is of no consequence for the subsequent investigation, so long as there is at least one field in which a given extremal may be embedded. That there is not even one field at times will be discussed in greater detail in Sec. 3.4.

Let $y = \bar{y}(x) \in C^1[a,b]$, $\bar{y}(a) = y_a$, $\bar{y}(b) = y_b$, and assume that

$$\|y_o - \bar{y}\|_w < \delta, \qquad \text{that is, } \bar{y} \in N_w^{\delta}(y_o).$$

We consider the total variation

$$\Delta I = I[\bar{y}] - I[y_o] = \int_a^b f(x,\bar{y}(x),\bar{y}'(x))\,dx - \int_a^b f(x,y_o(x),y_o'(x))\,dx$$

$$= \int_a^b [f(x,\bar{y}(x),\bar{y}'(x)) - f(x,y_o(x),y_o'(x))]\,dx.$$

We shall see that if $y = y_o(x)$ is embeddable in a field $\mathfrak{F}$, then we shall be able to transform this expression in such a manner that its sign can be checked with relative ease. The result of the next section will be the key to this transformation.

We wish to point out that we cannot simply expand the integrand by Taylor's formula and neglect terms of higher than first order because we are now working with a weak norm, as dictated by the nature of our problem, and hence $\max_{[a,b]} |\bar{y}'(x) - y_o'(x)|$ is not necessarily small.

PROBLEMS 3.1

1. For each of the following differential equations, consider a one-parameter family of solutions, determine a differential equation of first order $y' = \phi(x,y)$ that is

satisfied by this one-parameter family, and investigate if and where $(1,\phi(x,y))$ defines a field.

(a) $y'' + y = 0$ (b) $\frac{d}{dx}[(1 - x^2)y'] + 12y = 0$

(c) $\frac{d}{dx}\left(\frac{y'}{\sqrt{1 + y'^2}}\right) = 0$ (d) $x^2y'' + xy' + y = 0$

2. Show that the extremals of the following variational problems are embeddable in fields:

(a) $\int_0^{\pi/2} (y'^2(x) - y^2(x))\, dx \to \text{minimum}, \quad y(0) = 1, \quad y(\pi/2) = 0$

(b) $\int_0^1 (y'(x) - 1)^2(y'(x) + 1)^2\, dx \to \text{minimum}, \quad y(0) = 0, \quad y(1) = 1$

3. Find $\phi(x,y)$ for problem $2a$ and b.
4. Use the results of problem 3 to demonstrate that

$$\phi(x,Y(x,c)) = Y'(x,C(x,Y(x,c))) = Y'(x,c).$$

5. Show that $\Delta I \geq 0$, where

$$I[y] = \int_0^1 (1 + y'^2(x))\, dx,$$

$y(0) = 0,\ y(1) = 0,\ y_o(x) \equiv 0,\ \bar{y}(x) \equiv \sin n\pi x,\ n = 1, 2, 3, \ldots.$

*6. Consider the problem of minimal surfaces of revolution (Sec. 2.6), and assume that beginning point and endpoint both lie on either the ascending or the descending portion of the catenary $y = \alpha \cosh[(x - \beta)/\alpha]$. Show that such an extremal is embeddable in a field that may be obtained by translating the extremal in the x direction. Investigate how far to both sides such a field may be extended.
7. Consider the variational problem

$$I[y] = \int_0^1 \sqrt{1 + y'^2(x)}\, dx \to \text{minimum}, \quad y(0) = 0, \quad y(1) = 1.$$

Embed the extremal $y = x$, $x \in [0,1]$, in a field other than those that have been found in the text.

3.2 HILBERT'S INVARIANT INTEGRAL

We assume that the extremal $y = y_o(x)$ is embeddable in a field $\mathfrak{F}$; that is, a vector function $(1,\phi(x,y))$ is defined in $\mathfrak{A} \supset N_w^\delta(y_o)$, and $\phi(x,y)$ satisfies the requirements of Definitions 3.1.1 and 3.1.2.

Now, if $y = \bar{y}(x) \in N_w^\delta(y_o)$, then with every point $(x,\bar{y}(x))$ of $y = \bar{y}(x)$ there is associated the slope $y'(x) = \phi(x,\bar{y}(x))$ and the slope $\bar{y}'(x)$ of $y = \bar{y}(x)$. (See Fig. 3.3.) Therefore, the expression

$$h(x,\bar{y},\bar{y}') = f(x,\bar{y},\phi(x,\bar{y})) - \phi(x,\bar{y})f_{y'}(x,\bar{y},\phi(x,\bar{y})) + \bar{y}'f_{y'}(x,\bar{y},\phi(x,\bar{y})) \tag{3.2.1}$$

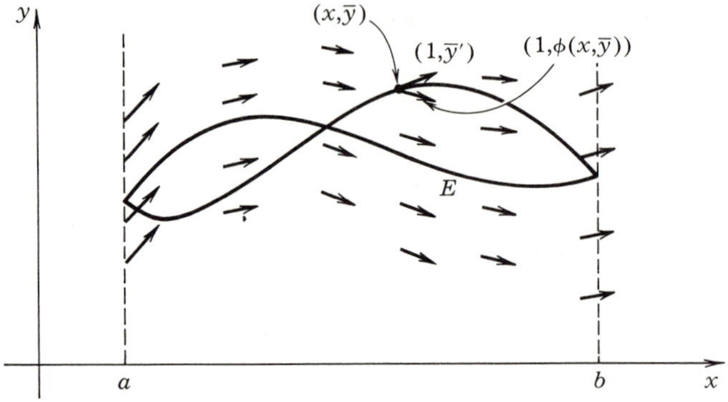

Figure 3.3

is defined for every point of the curve $y = \bar{y}(x) \in N_w^\delta(y_o)$. Since some curve $y = \bar{y}(x)$ can be put through any point (x,y) in $N_w^\delta(y_o)$, $h(x,\bar{y},\bar{y}')$ in (3.2.1) can be considered as defined for all points $(x,\bar{y})$ in a weak neighborhood of $y = y_o(x)$ and all slopes $\bar{y}'$.

In order to avoid confusion, we shall adopt the following convention on differentiation: $f_x, f_y, f_{y'}$ denote the partial derivatives of f with respect to the variables in the positions that are normally occupied by x, y, y' in $f(x,y,y')$. $(\partial f/\partial x) = f_x + f_{y'}\phi_x$, $(\partial f/\partial y) = f_y + f_{y'}\phi_y$ denote total derivatives with respect to x and y, respectively, if x and y are considered as independent variables and if $y' = \phi(x,y)$ is considered as a function of x and y. $(df/dx) = f_x + f_y y' + f_{y'}y''$ is used only if $y = y(x)$, $y' = y'(x)$ in $f(x,y,y')$.

These rules as stated for f apply just as well to the various partial derivatives of f, such as $f_y, f_{y'}, f_{yy'}, f_{y'y'}$.

The following theorem, due to *David Hilbert* (1862–1943), is the key to the transformation of the total variation:

Theorem 3.2 *If the extremal $y = y_o(x)$ is embeddable in a field $(1,\phi(x,y))$ that covers $N_w^\delta(y_o)$, then the integral*

$$U[\bar{y}] = \int_a^b h(x,\bar{y}(x),\bar{y}'(x))\,dx$$

$$= \int_a^b [f(x,\bar{y}(x),\phi(x,\bar{y}(x))) - \phi(x,\bar{y}(x))f_{y'}(x,\bar{y}(x),\phi(x,\bar{y}(x))) + \bar{y}'(x)f_{y'}(x,\bar{y}(x),\phi(x,\bar{y}(x)))]\,dx$$

is independent of $y = \bar{y}(x) \in C^1[a,b]$ and depends only on $\bar{y}(a)$ and $\bar{y}(b)$ so long as $\bar{y} \in N_w^\delta(y_o)$ and $f \in C^2(\mathcal{R})$, where $\mathcal{R}$ denotes the domain

$$\mathcal{R} = \{(x,y,y') \mid A < x < B, |y - y_o(x)| < \delta, -\infty < y' < \infty, [a,b] \subset (A,B)\}.$$

The integral $U[\bar{y}]$ is called *Hilbert's invariant integral.* The traditional notation $U[\bar{y}]$ stems from the German term *Unabhaengigkeitsintegral,* for "invariant integral."

Proof: In view of our assumptions on ϕ, $\bar{y}$, and f, the integral $U[\bar{y}]$ exists.

Let

$$P(x,y) = f(x,y,\phi(x,y)) - \phi(x,y)f_{y'}(x,y,\phi(x,y))$$
$$Q(x,y) = f_{y'}(x,y,\phi(x,y)).$$

Then, along any smooth curve $y = \bar{y}(x) \in N_w{}^\delta(y_o)$,

$$\int_a^b h(x,\bar{y}(x),\bar{y}'(x))\,dx = \int_a^b [P(x,\bar{y}(x)) + Q(x,\bar{y}(x))\bar{y}'(x)]\,dx$$
$$= \int_{(a,\bar{y}(a))}^{(b,\bar{y}(b))} (P(x,y)\,dx + Q(x,y)\,dy),$$

this integral being a line integral.

If we can show that

$$P_y(x,y) - Q_x(x,y) = 0†$$

for all (x,y) in $N_w{}^\delta(y_o)$, then the invariance of $U[\bar{y}]$ is established.

We have

$$P_y = f_y + \cancel{f_{y'}\phi_y} - \cancel{\phi_y f_{y'}} - \phi[f_{y'y} + f_{y'y'}\phi_y],$$
$$Q_x = f_{y'x} + f_{y'y'}\phi_x.$$

Hence,
$$P_y - Q_x = f_y - \phi[f_{y'y} + f_{y'y'}\phi_y] - f_{y'x} - f_{y'y'}\phi_x$$
$$= f_y - f_{y'x} - f_{y'y}\phi - f_{y'y'}[\phi\phi_y + \phi_x].$$

With every point (x,y) in $N_w{}^\delta(y_o)$ there is associated a slope $y' = \phi(x,y)$ so that $(x,y,\phi(x,y))$ is a lineal element of a uniquely determined extremal $y = y(x)$ which satisfies $y'(x) = \phi(x,y(x))$. Then

$$y''(x) = \phi_x(x,y(x)) + \phi_y(x,y(x))y'(x)$$
$$= \phi_x(x,y(x)) + \phi_y(x,y(x))\phi(x,y(x))$$

Since $y = y(x)$ is also a solution of the Euler-Lagrange equation, that is,

$$f_y(x,y(x),y'(x)) - f_{y'x}(x,y(x),y'(x)) - f_{y'y}(x,y(x),y'(x))y'(x)$$
$$- f_{y'y'}(x,y(x),y'(x))y''(x) = 0,$$

we have for every point (x,y) in $N_w{}^\delta(y_o)$ that

$$P_y(x,y) - Q_x(x,y) = 0.$$

† *Ibid.*, p. 440.

Hence there exists a function $W = W(x,y) \in C^1(N_w^\delta(y_o))$ such that

$$W_x(x,y) = P(x,y), \qquad W_y(x,y) = Q(x,y)$$

and consequently,

$$\int_a^b h(x,\bar{y}(x),\bar{y}'(x))\,dx = \int_{(a,\bar{y}(a))}^{(b,\bar{y}(b))} dW(x,\bar{y}(x)) = W(b,\bar{y}(b)) - W(a,\bar{y}(a)),$$

that is, $U[\bar{y}]$ is indeed independent of the path and depends only on the coordinates of the beginning point and the endpoint.

PROBLEMS 3.2

1. Given the variational problem

$$I[y] = \int_0^1 \sqrt{1 + y'^2(x)}\,dx \to \text{minimum}, \qquad y(0) = 0, \qquad y(1) = 1.$$

The extremal $y = x$ is embeddable in a field that is defined by $\mathfrak{F} = \{(1,1)\}$. (See Sec. 3.1.) Write down Hilbert's invariant integral for this case, and show directly that it is indeed independent of the choice of $y = \bar{y}(x)$ and depends only on $\bar{y}(0)$ and $\bar{y}(1)$.

2. Same as in problem 1 with a field defined by $(1,(y+1)/(x+1))$.
3. Consider the variational problem

$$\int_0^1 (1 + y'^2(x))\,dx \to \text{minimum}, \qquad y(0) = y(1) = 0.$$

Find the extremal, embed it in a field, write Hilbert's invariant integral, and prove directly that it is independent of the choice of $y = \bar{y}(x)$ so long as $\bar{y}(0) = \bar{y}(1) = 0$.

4. Same as in problem 3 for the variational problem

$$\int_{\pi/4}^{\pi/2} (y'^2(x) - y^2(x))\,dx \to \text{minimum}, \qquad y\left(\frac{\pi}{4}\right) = y\left(\frac{\pi}{2}\right) = 0.$$

5. Let $h(x,\bar{y},\bar{y}')$ be defined as in (3.2.1). Show that the Euler-Lagrange equation of

$$\int_a^b h(x,\bar{y}(x),\bar{y}'(x))\,dx \to \text{minimum}, \qquad \bar{y}(a) = y_a, \qquad \bar{y}(b) = y_b$$

is identically satisfied. Explain.

3.3 TRANSFORMATION OF THE TOTAL VARIATION

We saw in the preceding section that $U[\bar{y}]$ as defined in Theorem 3.2 is independent of $y = \bar{y}(x)$ so long as $y = \bar{y}(x)$ stays in $N_w^\delta(y_o)$, and depends only on the coordinates of the beginning point and the endpoint. In particular,

$$U[\bar{y}] = U[y_o] \qquad \text{if } \bar{y}(a) = y_o(a),\ \bar{y}(b) = y_o(b). \tag{3.3.1}$$

If we take $U[y]$ for $y = y_o(x)$, then $\bar{y}'(x) = y_o'(x)$, $\phi(x,y_o(x)) = y_o'(x)$, and we obtain

$$U[y_o] = \int_a^b [f(x,y_o(x),y_o'(x)) - y_o'(x)f_{y'}(x,y_o(x),y_o'(x)) + y_o'(x)f_{y'}(x,y_o(x),y_o'(x))]\,dx$$

$$= \int_a^b f(x,y_o(x),y_o'(x))\,dx$$

$$= I[y_o]. \tag{3.3.2}$$

In view of (3.3.1) and (3.3.2), we may transform the total variation ΔI as follows:

$$\Delta I = I[\bar{y}] - I[y_0] = I[\bar{y}] - U[y_o] = I[\bar{y}] - U[\bar{y}]$$

$$= \int_a^b [f(x,\bar{y},\bar{y}') - f(x,\bar{y},\phi(x,\bar{y})) + (\phi(x,\bar{y}) - \bar{y}')f_{y'}(x,\bar{y},\phi(x,\bar{y}))]\,dx, \tag{3.3.3}$$

where $\bar{y}$ stands for $\bar{y}(x)$ and $\bar{y}'$ stands for $\bar{y}'(x)$. Here it is assumed that $y = \bar{y}(x) \in N_w{}^\delta(y_o)$ and $\bar{y}(a) = y_o(a)$, $\bar{y}(b) = y_o(b)$.

If we introduce the *Weierstrass excess function*

$$\mathcal{E}(x,y,y',\bar{y}') = f(x,y,\bar{y}') - f(x,y,y') + (y' - \bar{y}')f_{y'}(x,y,y'), \tag{3.3.4}$$

we may write the total variation as follows:

$$\Delta I = \int_a^b \mathcal{E}(x,\bar{y}(x),\phi(x,\bar{y}(x)),\bar{y}'(x))\,dx. \tag{3.3.5}$$

(The peculiar-looking E which is customarily used to denote the excess function is supposed to resemble the E of Weierstrass' handwriting.)

We are now ready to state a sufficient condition for a strong relative minimum:

Theorem 3.3.1 *If the extremal* $y = y_o(x)$ *is embeddable in a field* $\mathfrak{F}$ *which covers* $N_w{}^\delta(y_o)$ *and if*

$$\mathcal{E}(x,\bar{y},\phi(x,\bar{y}),\bar{y}') \geq 0$$

for all $(x,\bar{y}) \in N_w{}^\delta(y_o)$ *and all* $-\infty < \bar{y}' < \infty$, *then* $y = y_o(x)$ *yields a strong relative minimum for* $I[y]$.

[For a strong relative maximum, the condition is $\mathcal{E}(x,\bar{y},\phi(x,\bar{y}),\bar{y}') \leq 0$.

Proof:

$$\Delta I = \int_a^b \mathcal{E}(x,\bar{y}(x),\phi(x,\bar{y}(x)),\bar{y}'(x))\,dx \geq 0 \qquad \text{for all } y = \bar{y}(x) \in N_w^\delta(y_o).$$

(The next section will be devoted to a detailed discussion of an example.)

The condition which we stated in Theorem 3.3.1 involves knowledge of the function ϕ. We shall see in Sec. 3.7 that it is possible to establish the embeddability of an extremal in a field without explicit knowledge of ϕ. Then the condition as stated in Theorem 3.3.1 is not very practical.

We note that if

$$\mathcal{E}(x,\bar{y},y',\bar{y}') \geq 0$$

for all $(x,\bar{y}) \in N_w^\delta(y_o)$ and all $-\infty < y' < \infty$, $-\infty < \bar{y}' < \infty$, then

$$\mathcal{E}(x,\bar{y},\phi(x,\bar{y}),\bar{y}') \geq 0$$

for all $(x,\bar{y}) \in N_w^\delta(y_o)$ and all $-\infty < \bar{y}' < \infty$. Hence, we may state Theorem 3.3.1 in the more practical but somewhat stronger form:

Theorem 3.3.2: The Weierstrass condition *If the extremal* $y = y_o(x)$, $x \in [a,b]$, *is embeddable in a field* $\mathfrak{F}$ *which covers* $N_w^\delta(y_o)$ *and if*

$$\mathcal{E}(x,\bar{y},y',\bar{y}') \geq 0 \qquad (\leq 0)$$

for all $(x,\bar{y}) \in N_w^\delta(y_o)$ *and all* $-\infty < y' < \infty$, $-\infty < \bar{y}' < \infty$, *then* $y = y_o(x)$ *yields a strong relative minimum (maximum) for* $I[y]$.

PROBLEMS 3.3

1. Find the excess function $\mathcal{E}(x,y,y',\bar{y}')$ for the following functionals:

   (a) $\int_a^b \sqrt{1 + y'^2(x)}\,dx$ (b) $\int_a^b (1 - y'^2(x))^2\,dx$ (c) $\int_a^b (y'^2(x) - y^2(x))\,dx$

2. Consider the variational problem in Prob. 3.1.2*b* and investigate whether or not the extremal yields a strong relative minimum.

3.4 AN EXAMPLE OF A STRONG MINIMUM

We consider the functional

$$I[y] = \int_0^b (y'^2(x) - y^2(x))\,dx, \qquad 0 < b < \pi, \tag{3.4.1}$$

for

$$\Sigma = \{y \mid y \in C^1[a,b],\, y(0) = 0,\, y(b) = 1\}.$$

From the Euler-Lagrange equation for (3.4.1),

$$y'' + y = 0,$$

we obtain the general solution

$$y = y(x,\alpha,\beta) \equiv \alpha \cos x + \beta \sin x. \tag{3.4.2}$$

In view of the boundary conditions,

$$\alpha = 0, \qquad \beta = \frac{1}{\sin b} \qquad (\sin b \neq 0 \text{ since } 0 < b < \pi).$$

Thus $$y = y_o(x) \equiv \frac{\sin x}{\sin b} \tag{3.4.3}$$

is an extremal of (3.4.1) in Σ.

In order to obtain a field into which $y = y_o(x)$ is embedded, we construct a one-parameter family of functions $y = Y(x,c)$ from $y = y(x,\alpha,\beta)$ that contains the extremal (3.4.3) as a member and covers a weak δ neighborhood of $y = y_o(x)$ simply; i.e., through every point of $N_w^{\delta}(y_o)$, there will be exactly one member of $y = Y(x,c)$.

Since $b < \pi$, there is a $\Delta > 0$ such that $\Delta < \pi - b$. The extremal E: $y = y_o(x)$ passes through the point $P_0(-\Delta, -(\sin \Delta)/(\sin b))$. (See Fig. 3.4.) We consider all members from (3.4.2) that pass through

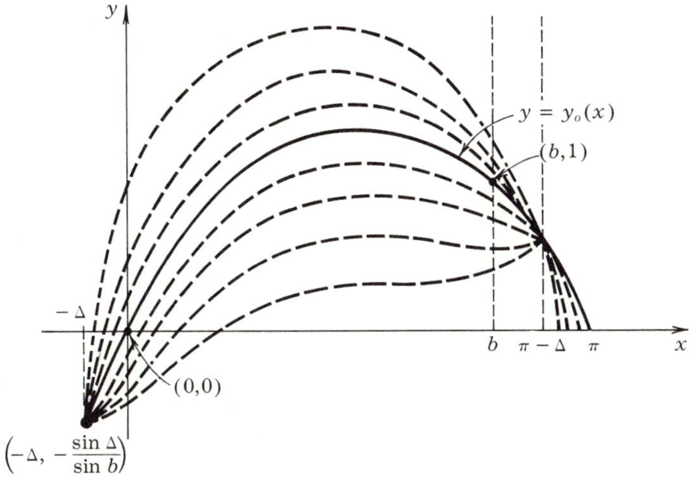

Figure 3.4

the point P_o:

$$y(-\Delta,\alpha,\beta) = \alpha \cos \Delta - \beta \sin \Delta = -\frac{\sin \Delta}{\sin b},$$

$$\alpha \cot \Delta - \beta = -\frac{1}{\sin b},$$

$$\alpha = \tan \Delta \left(\beta - \frac{1}{\sin b}\right).$$

Thus, if

$$Y(x,c) \equiv \tan \Delta \left(c - \frac{1}{\sin b}\right) \cos x + c \sin x, \tag{3.4.4}$$

then $y = Y(x,c)$ contains the extremal $y = y_o(x)$ for $c_o = 1/(\sin b)$. Further, $Y_c(x,c) \equiv \tan \Delta \cos x + \sin x$, and we see that

$$Y_c(x,c) = 0 \qquad \text{if and only if } \tan \Delta + \tan x = 0,$$

that is, $x = -\Delta + k\pi (k = 0, \pm 1, \pm 2, \ldots)$.

Hence, $Y_c(x,c)$ does not vanish in $[0,b]$, and it appears that (3.4.4) satisfies all the requirements of Lemma 3.1.

We obtain $\phi(x,y)$ by solving $y = Y(x,c)$ for c and substituting the result into $y' = Y'(x,c)$:

$$\phi(x,y) = \frac{y \sin b + \tan \Delta \cos x}{\tan \Delta \cos x + \sin x}\left(\frac{\cos x - \tan \Delta \sin x}{\sin b}\right) + \frac{\sin x}{\sin b} \tan \Delta.$$

(See Prob. 3.4.1.)

The next step in our investigation is a discussion of the sign of the excess function $\mathcal{E}$, as defined in (3.3.4). Since $f = y'^2 - y^2$, we have $f_{y'} = 2y'$, and hence

$$\begin{aligned} \mathcal{E}(x,\bar{y},y',\bar{y}') &= \bar{y}'^2 - \bar{y}^2 - y'^2 + \bar{y}^2 + (y' - \bar{y}')(2y') \\ &= (y' - \bar{y}')^2 \geq 0 \end{aligned}$$

for *all* $y',\bar{y}'$. Hence $y = y_o(x) \equiv (\sin x)/(\sin b)$ yields a strong relative minimum for (3.4.1). That it also yields an absolute minimum in $\Sigma = \{y \mid y \in C^1[0,b],\ y(0) = 0,\ y(b) = 1,\ 0 < b < \pi\}$ follows from the fact that the field is defined for all y and on the entire interval $[0,b]$ and that $\mathcal{E} \equiv 0$ in $[0,b]$ if and only if $\bar{y}'(x) = y_o'(x)$ for all $x \in [0,b]$.

Next, we consider the case $b = \pi$. Since $y(x,\alpha,\beta) = -y(x + \pi, \alpha, \beta)$ [see (3.4.2)], we see that it will be impossible in this case to join a given initial point $P_a(0,y_a)$ to just any given endpoint $P_b(\pi,y_b)$. The boundary conditions can then be satisfied only if $y_a = -y_b$, and in particular, $y_b = 0$

if $y_a = 0$. For reasons of simplicity, we discuss the case $y_a = y_b = 0$. Then we obtain from (3.4.2)

$$y = y_o(x) \equiv y(x,0,\beta) \equiv \beta \sin x,$$

a one-parameter family of extremals, all of which satisfy the given boundary conditions.

We obtain

$$I[y(x,0,\beta)] = \int_0^\pi (\beta^2 \cos^2 x - \beta^2 \sin^2 x)\, dx = \beta^2 \int_0^\pi \cos 2x\, dx = 0$$

for all β, that is, $I[y]$ has the same value, namely 0, for *all* extremals $y = y(x,0,\beta)$. (See also Prob. 3.4.3 and Sec. A3.13.)

That it does not suffice for a strong relative minimum (or maximum) to merely require $\mathcal{E} \geq 0$ (or ≤ 0) along the extremal and that the embedding condition is vital for this test will be obvious from the following discussion.

We consider the same variational problem as before, but we now drop the restriction $0 < b \leq \pi$. Only the case $\pi < b < 2\pi$ is really of significance.

Since $y(x,\alpha,\beta) \equiv \alpha \cos x + \beta \sin x$ vanishes every π units regardless of α,β, it is quite clear that a field cannot be constructed by means of Lemma 3.1. We shall see that in this case *no* extremal can possibly yield a minimum for $I[y]$.

Let beginning point and endpoint be given by $P_a(0,y_a)$, $P_b(b,y_b)$ and suppose the extremal y_o yields a minimum for $I[y]$. Since $\pi < b < 2\pi$, there exists a unique pair (α_o,β_o) such that

$$y = y_o(x) \equiv \alpha_o \cos x + \beta_o \sin x$$

satisfies the given boundary conditions regardless of what y_a and y_b are.

Let us now consider the curve

$$y = \bar{y}(x) = \begin{cases} \alpha_o \cos x + (\beta_o + \gamma) \sin x & \text{for } 0 \leq x \leq \pi \\ \alpha_o \cos x + \beta_o \sin x & \text{for } \pi \leq x \leq b \end{cases}$$

for $\gamma \neq 0$. $y = \bar{y}(x)$ is continuous at $x = \pi$ and hence continuous in $0 \leq x \leq b$, but $\bar{y}'$ has a discontinuity at $x = \pi$, namely,

$$\bar{y}'(\pi - 0) = -\beta_o - \gamma, \qquad \bar{y}'(\pi + 0) = -\beta_o.$$

Now, should $y = \bar{y}(x)$ yield a minimum for $I[y]$, then, by necessity, the Weierstrass-Erdmann corner conditions have to be satisfied at $x = \pi$ (see Theorem 2.9). Since

$$(f_{y'})_{\pi-0} = (2\bar{y}')_{\pi-0} = -2(\beta_o + \gamma), \qquad (f_{y'})_{\pi+0} = -2\beta_o,$$

we see that $(f_{y'})_{\pi-0} \neq (f_{y'})_{\pi+0}$, that is, not even the first of the two corner conditions is satisfied. Therefore, no matter what value $y = \bar{y}(x)$ yields for $I[y]$, it cannot be a relative minimum value.

We shall now show that $I[\bar{y}] = I[y_o]$: First we note that for any A,B and $y = A \cos x + B \sin x$, we have

$$\int_0^\pi [y'^2(x) - y^2(x)]\,dx = \int_0^\pi [(-A \sin x + B \cos x)^2 - (A \cos x + B \sin x)^2]\,dx$$

$$= \int_0^\pi [(B^2 - A^2) \cos 2x - 2AB \sin 2x]\,dx = 0.$$

Hence

$$I[\bar{y}] = \int_0^\pi [((\beta_o + \gamma) \cos x - \alpha_o \sin x)^2 - (\alpha_o \cos x + (\beta_o + \gamma) \sin x)^2]\,dx + \int_\pi^b (y_o'^2(x) - y_o^2(x))\,dx$$

$$= \int_\pi^b (y_o'^2(x) - y_o^2(x))\,dx$$

and

$$I[y_o] = \int_0^\pi [(\beta_o \cos x - \alpha_o \sin x)^2 - (\alpha_o \cos x + \beta_o \sin x)^2]\,dx + \int_\pi^b (y_o'^2(x) - y_o^2(x))\,dx$$

$$= \int_\pi^b (y_o'^2(x) - y_o^2(x))\,dx$$

and we indeed have

$$I[\bar{y}] = I[y_o].$$

We observe that we can choose for any $\delta > 0$ a γ so small that $\bar{y} \in N^{\delta/2}(y_o)$ and, by the same token, $y_o \in N^{\delta/2}(\bar{y})$. Since $\bar{y}$ does not yield a weak relative minimum, there is a $y = \eta(x) \in C_s^1[0,b]$, $\eta \in N^{\delta/2}(\bar{y})$, such that $I[\eta] < I[\bar{y}] = I[y_o]$. But, since we have $\eta \in N^\delta(y_o)$ and since $I[\eta] < I[y_o]$, we see that y_o does not yield a weak relative minimum for $I[y]$ relative to sectionally smooth functions.

Since $\eta \in N^\delta(y_o)$ and $I[\eta] < I[y_o]$, we can find by the construction that was indicated in the proof of Corollary 2 to Theorem 2.10 a function

$y = y(x) \in C^1[a,b]$, $y \in N^\delta(y_o)$ such that $I[y] < I[y_o]$, and we see that y_o does not yield a weak relative minimum for $I[y]$ relative to smooth functions either.

PROBLEMS 3.4

*1. Find $\phi(x,y)$ for the problem of this section with $0 < b < \pi$.

2. Show that

$$\int_0^b (y'^2(x) - y^2(x))\, dx \geq \frac{(y_a^2 + y_b^2)\cos b - 2y_a y_b}{\sin b}$$

for all $y \in C_s^1[0,b]$, $y(0) = y_a$, $y(b) = y_b$, $0 < b < \pi$.

3. Do the extremals $y = \beta \sin x$ yield a strong relative minimum for

$$\int_0^\pi (y'^2(x) - y^2(x))\, dx, \qquad y(0) = y(\pi) = 0?$$

3.5 FIELD CONSTRUCTION AND THE JACOBI EQUATION

In the example of the preceding section, we succeeded in constructing a field by considering a one-parameter family of extremals that emanated from one point (common center) located outside the field. Such fields are often referred to as *central fields*. We shall let our experience with the preceding example serve as a guideline in an attempt to formulate conditions under which a field may be constructed in this manner.

We assume our variational problem to be regular; that is,

$$f_{y'y'}(x,y,y') \neq 0 \qquad \text{for all } (x,y,y').$$

Let $\quad y = Y(x,c), \qquad c_1 - \varepsilon < c < c_1 + \varepsilon, \qquad a - \delta < x < b + \delta \quad (3.5.1)$

represent a one-parameter family of extremals, of which we assume that $y = y_o(x)$ is a member for $c = c_o$:

$$Y(x,c_o) = y_o(x), \qquad c_o \in (c_1 - \varepsilon, c_1 + \varepsilon).$$

We further assume that

$$Y(x,c),\ Y'(x,c),\ Y_c(x,c),\ Y_c'(x,c),\ Y''(x,c)$$

are continuous for all $x \in (a - \delta, b + \delta)$, $c \in (c_1 - \varepsilon, c_1 + \varepsilon)$.

We also assume that all the extremals in (3.5.1) pass through the same point $P_o(x_o,y_o)$ where $a - \delta < x_o < a$. Then

$$y_o = Y(x_o,c)$$

for all $c \in (c_1 - \varepsilon, c_1 + \varepsilon)$, and we have

$$Y_c(x_o,c) = 0.$$

It will be our task to find out if and when $Y_c(x,c)$ vanishes again, because we have seen that $Y_c \neq 0$ together with appropriate continuity assumptions will suffice to embed $y = y_0(x)$ in a field. (See Lemma 3.1.)

Since $y = Y(x,c)$ are solutions of the Euler-Lagrange equation for all $c \in (c_1 - \varepsilon, c_1 + \varepsilon)$, we have

$$f_y(x,Y(x,c),Y'(x,c)) - \frac{d}{dx} f_{y'}(x,Y(x,c),Y'(x,c)) = 0$$

for all $c \in (c_1 - \varepsilon, c_1 + \varepsilon)$. Hence

$$\frac{d}{dc}\left[f_y - \frac{d}{dx} f_{y'}\right]_{\substack{y=Y(x,c)\\ y'=Y'(x,c)}} = 0.$$

[The differentiation (d/dc) has to be carried out with respect to c wherever c occurs, namely, in $Y(x,c)$ and $Y'(x,c)$.] If we assume that the third partial derivatives of f with respect to all variables are continuous, then, in view of all our other assumptions, $Y_c''(x,c)$ is also continuous (see Probs. 3.5.2 and 3.5.3), and we may interchange the differentiation with respect to c and with respect to x:

$$\frac{d}{dc}\left[\frac{d}{dx} f_{y'}\right] = \frac{d}{dx}\left[\frac{d}{dc} f_{y'}\right].$$

We obtain

$$\frac{d}{dc} f_y - \frac{d}{dx}\left[\frac{d}{dc} f_{y'}\right] = 0.$$

Since

$$\frac{d}{dc} f_y(x,Y(x,c),Y'(x,c)) = f_{yy} Y_c + f_{yy'} Y_c'$$

and

$$\frac{d}{dc} f_{y'}(x,Y(x,c),Y'(x,c)) = f_{y'y} Y_c + f_{y'y'} Y_c',$$

we obtain

$$f_{yy} Y_c + f_{yy'} Y_c' - \frac{d}{dx}[f_{y'y} Y_c + f_{y'y'} Y_c'] = 0. \tag{3.5.2}$$

This is a second-order differential equation for Y_c. In order to simplify the notation, we denote

$$Y_c(x,c) \equiv \eta(x,c)$$

and, in particular,

$$Y_c(x,c_0) \equiv \eta(x).$$

We consider equation (3.5.2) for $c = c_o$. Then, $Y(x,c_o) = y_o(x)$ and $Y'(x,c_o) = y_o'(x)$.

Since

$$\frac{d}{dx}(f_{y'y}\eta) = \eta \frac{d}{dx}(f_{y'y}) + f_{y'y}\eta',$$

we obtain

$$\left[f_{yy} - \frac{d}{dx} f_{y'y}\right]\eta - \frac{d}{dx}[f_{y'y'}\eta'] = 0. \tag{3.5.3}$$

With

$$\alpha(x) = f_{y'y'}(x,y_o(x),y_o'(x))$$

$$\beta(x) = f_{yy}(x,y_o(x),y_o'(x)) - \frac{d}{dx} f_{y'y}(x,y_o(x),y_o'(x)),$$

we can write (3.5.3) for $c = c_o$ as follows:

$$\frac{d}{dx}[\alpha(x)\eta'] - \beta(x)\eta = 0. \tag{3.5.4}$$

Equation (3.5.3) is called the *Jacobi equation*.† (This equation will also play an important role in the theory of the second variation; see Sec. 7.3.)

In view of our assumptions on f, we see that $\alpha(x)$ and $\beta(x)$ are continuous and that, moreover, $\alpha(x)$ has a continuous derivative. The Jacobi equation is a self-adjoint differential equation of second order for η. We know that $\eta(x_o) = 0$. We want to find out if and where η vanishes again. This will then leave us with a certain interval where $\eta(x) \not\equiv 0$ and, in view of the continuity of $\eta(x,c)$, where also $\eta(x,c) \equiv Y_c(x,c) \not\equiv 0$ for $|c - c_o|$ sufficiently small, and we shall have made a major step toward the establishment of a sufficient condition for the existence of a field. The zeros of the solutions of the Jacobi equation, which appear to be the key to the whole problem, will be discussed in the next section.

PROBLEMS 3.5

1. Assume that $f \in C^3$, and show that the difference of two extremals for which $\max_{[a,b]} |y_1 - y_2| + \max_{[a,b]} |y_1' - y_2'| + \max_{[a,b]} |y_1'' - y_2''|$ is small satisfies the Jacobi equation, with coefficients evaluated for one of the extremals, except for terms that are small of higher than first order.

† Carl Gustav Jacob Jacobi, 1804–1851.

*2. Let $f_{y'y'} > 0$ (< 0). Assume that $f \in C^3$, and assume also that Y, Y', Y_c, Y'_c are continuous, where $y = Y(x,c)$ is a one-parameter family of solutions of the Euler-Lagrange equation. Show that Y''_c is also continuous.

*3. Show that under the conditions of problem 2,

$$\frac{d}{dc}\left[\frac{d}{dx}f_{y'}\right] = \frac{d}{dx}\left[\frac{d}{dc}f_{y'}\right].$$

4. Let $y = y(x,\alpha,\beta)$ denote a two-parameter family of solutions of the Euler-Lagrange equation such that $y = y(x,\alpha_o,\beta_o) \equiv y_o(x)$ represents the extremal of $I[y]$, $y(a) = y_a$, $y(b) = y_b$. Let

$$y = y(x,\alpha,\beta) \in C^1[[a,b] \times [\alpha_1,\alpha_2] \times [\beta_1,\beta_2]],$$

where $\alpha_o \in (\alpha_1,\alpha_2)$, $\beta_o \in (\beta_1,\beta_2)$, and let $[\partial(y,y')/\partial(\alpha,\beta)] \neq 0$ for all $(x,\alpha,\beta) \in [a,b] \times [\alpha_1,\alpha_2] \times [\beta_1,\beta_2]$. Prove that $\eta = (\partial y/\partial\alpha)_{\alpha_o,\beta_o}$, $\eta = (\partial y/\partial\beta)_{\alpha_o,\beta_o}$ are two linearly independent solutions of the Jacobi equation (3.5.3).

5. Utilize the result of problem 4 to find a solution $\eta = \eta(x)$ of the Jacobi equation that is associated with the extremal $y = \cosh x$ of the problem of minimal surfaces, $\int_0^b y(x)\sqrt{1 + y'^2(x)}\,dx \to$ minimum, such that $\eta(0) = 0$.

6. Show that $y = \cosh x$ yields a strong relative minimum for

$$\int_0^b y(x)\sqrt{1 + y'^2(x)}\,dx \to \text{minimum}, \qquad y(0) = 1, \qquad y(b) = \cosh b.$$

(See also problem 5.)

3.6 THE ZEROS OF THE SOLUTIONS OF THE JACOBI EQUATION—CONJUGATE POINTS

As we pointed out in the preceding section, the Jacobi equation

$$\frac{d}{dx}[\alpha(x)\eta'] - \beta(x)\eta = 0 \tag{3.6.1}$$

is a self-adjoint differential equation of second order.

We make the following assumptions: $y = y_o(x)$ will be a regular extremal on (A,B), where $[a,b] \subset (A,B)$ and $f \in C^3(\mathcal{R})$, where

$$\mathcal{R} = \{(x,y,y') \mid A < x < B, |y - y_o(x)| < \alpha_o, |y' - y'_o(x)| < \beta_o\}. \tag{3.6.2}$$

Then $\alpha(x) > 0$ (or <0) in (A,B), $\alpha,\beta \in C(A,B)$, and, moreover, $\alpha \in C^1(A,B)$.

It is possible to find out a great deal about the nature of the solutions of the Jacobi equation without ever having to solve it.

For the convenience of the reader, we list here two basic lemmas about the zeros of the solutions of the Jacobi equation. Our further investigation will be based on these lemmas:

Lemma 3.6.1 *If* $\eta = \eta(x)$ *is a solution of (3.6.1) and if for some* $x_o \in (A,B)$ *we have* $\eta(x_o) = \eta'(x_o) = 0$, *then* $\eta(x) \equiv 0$ *for all* $x \in (A,B)$; *that is, the zeros of a nontrivial solution of the Jacobi equation are simple.*†

Lemma 3.6.2 *If a solution* $\eta = \eta(x)$ *of the Jacobi equation has two consecutive zeros* x_1 *and* x_1^*, *then any other solution* $\eta = \bar{\eta}(x)$ *which is linearly independent of* $\eta = \eta(x)$ *has to vanish at a point* $\bar{x} \in (x_1,x_1^*)$; *that is, the zeros of the solutions of the Jacobi equation separate each other.*‡

Definition 3.6 *If* $x_1 < x_1^*$ *are two consecutive zeros of a solution of the Jacobi equation, then* x_1^* *is called* conjugate *to* x_1, *or the* conjugate point *to* x_1.

[Note that x_1^* is determined by x_1 independent of the particular solution $\eta = \eta(x)$ of the Jacobi equation for which $\eta(x_1) = 0$; any two nontrivial solutions of the Jacobi equation that vanish for the same x_1 differ by a multiplicative constant only. See Prob. 3.6.5.]

The preceding two lemmas will enable us to prove the following important theorem on conjugate points:

Theorem 3.6.1 *If* x_1^* *is conjugate to* x_1, *then* x_1^* *is a continuous, monotonically increasing function of* x_1.

Proof: Let η_1, η_2 denote two linearly independent solutions of the Jacobi equation. Then

$$\eta = \lambda_1\eta_1(x) + \lambda_2\eta_2(x)$$

represents the general solution of the Jacobi equation. In order to find a nontrivial solution that vanishes at $x = x_1$, we have to solve the equation

$$\lambda_1\eta_1(x_1) + \lambda_2\eta_2(x_1) = 0$$

for λ_1 and λ_2. Since $[\eta_1(x_1),\eta_2(x_1)] \neq [0,0]$ in view of the linear independence of these two solutions, we may assume without loss of generality that $\eta_1(x_1) \neq 0$ and solve for λ_1/λ_2:

$$\frac{\lambda_1}{\lambda_2} = -\frac{\eta_2(x_1)}{\eta_1(x_1)}.$$

Hence,

$$\lambda_1 = \lambda\eta_2(x_1), \qquad \lambda_2 = -\lambda\eta_1(x_1)$$

† H. Sagan, "Boundary and Eigenvalue Problems in Mathematical Physics," 3d printing, p. 154, John Wiley & Sons, Inc., New York, 1966.
‡ *Ibid.*, p. 155.

for some $\lambda \neq 0$, and

$$\eta = \lambda[\eta_2(x_1)\eta_1(x) - \eta_1(x_1)\eta_2(x)] = \lambda \begin{vmatrix} \eta_1(x) & \eta_2(x) \\ \eta_1(x_1) & \eta_2(x_1) \end{vmatrix}$$
$$= \lambda D(x,x_1)$$

represents all solutions of the Jacobi equation that vanish at x_1. Without loss of generality, we may assume $\lambda = 1$.

To find other zeros of $\eta = \eta(x)$—if they exist—we have to solve

$$D(x,x_1) = 0$$

for x. By hypothesis, there exists a conjugate point x_1^* to x_1. Hence, $D(x_1^*,x_1) = 0$. We shall now demonstrate that x_1^* is a continuous, monotonically increasing function of x_1.

(*a*) x_1^* is a continuous function of x_1: By hypothesis, we have $\eta(x_1^*) = D(x_1^*,x_1) = 0$, and by Lemma 3.6.1, we have $\eta'(x_1^*) = [\partial D\ (x,x_1)/\partial x]|_{x=x_1^*} \neq 0$ [otherwise, $\eta = \eta(x)$ would be the trivial solution]. Since $\eta \in C^1(A,B)$, we have that $D(x,x_1), \partial D\ (x,x_1)/\partial x \in C(A,B)^2$, and since $[\partial D\ (x,x_1)/\partial x]|_{x=x_1^*} \neq 0$, we may apply the theorem on implicit functions and solve $D(x,x_1) = 0$ for x in a neighborhood of x_1^*. We obtain $x_1^* = \varphi(x_1)$, where $\varphi(x)$ is continuous in a neighborhood of x_1. A repeated application of this argument will show that $\varphi(x)$ is continuous as long as x_1^* stays in (A,B).

(*b*) $\varphi(x)$ is monotonically increasing: We shall show that $x_1 < x_2$ implies $\varphi(x_1) < \varphi(x_2)$, that is, $x_1^* < x_2^*$. We first assume that, on the contrary, $x_2^* \leq x_1^*$. Then we have a solution $\eta = \bar{\eta}(x) = D(x,x_2)$, which is linearly independent of $\eta = \eta(x)$ (see Prob. 3.6.4), such that $\bar{\eta}(x_2) = \bar{\eta}(x_2^*) = 0$, where $x_1 < x_2 < x_2^* \leq x_1^*$. Hence, by Lemma 3.6.2, there has to exist an $\bar{x}$ such that $\eta(\bar{x}) = 0$, where $x_1 < x_2 < \bar{x} < x_2^* \leq x_1^*$, but this contradicts our hypothesis that x_1^* is the first zero of $\eta = \eta(x)$ that follows x_1.

We shall now utilize these findings to study conjugate points of the solutions of the Jacobi equation for a given extremal. We consider (3.6.1) for

$$\alpha(x) = f_{y'y'}(x,y_o(x),y_o'(x))$$

$$\beta(x) = f_{yy}(x,y_o(x),y_o'(x)) - \frac{d}{dx} f_{y'y}(x,y_o(x),y_o'(x)),$$

and we assume that all the conditions which we stated at the beginning of this section are satisfied.

Let $\eta = \eta(x)$ denote that nontrivial solution of (3.6.1) for which

$\eta(a) = 0$. Then there are two possibilities:

1. There is no conjugate point to a in $(a,b]$; that is, either $a^* > b$ or a^* does not exist.
2. The interval $(a,b]$ contains at least one conjugate point to a.

Only the first case is of interest. We have:

Theorem 3.6.2 *If $(a,b]$ does not contain a conjugate point to a, then there exists a $\Delta > 0$ such that $(a - \Delta, b]$ does not contain a conjugate point to $a - \Delta$ either.*

Proof (by contradiction): Suppose there is no such Δ. Then $\varphi(a - \Delta_n) \leq b$ for $\{\Delta_n\} \to 0$, and hence, by the continuity of φ, $\lim_{n\to\infty} \varphi(a - \Delta_n) = \varphi(a) \leq b$, which is contrary to our hypothesis that $a^* = \varphi(a) > b$.

As an illustration of this theorem, let us consider the following two examples:

A. $I[y] = \int_0^b (y'^2(x) - y^2(x))\, dx$, $y(0) = 0$, $y(b) = 1$, $0 < b < \pi$. Since $f_y = -2y$, $f_{y'} = 2y'$, $f_{yy'} = 0$, $f_{yy} = -2$, $f_{y'y'} = 2$, we obtain as the Jacobi equation of this problem

$$\eta'' + \eta = 0.$$

$\eta = \sin x$ is a nontrivial solution for which $\eta(0) = 0$. We see that $0^* = \pi$, and hence $(0,b]$ does not contain a conjugate point to 0. Neither does $(-\Delta,b]$ contain a conjugate point to $-\Delta$, provided that $0 < \Delta < \pi - b$. (See also Sec. 3.4.)

B. $I[y] = \int_0^1 \sqrt{1 + y'^2(x)}\, dx$, $y(0) = 0$, $y(1) = 1$. Since $f_y = 0$, $f_{yy'} = 0$, $f_{yy} = 0$, $f_{y'} = y'/\sqrt{1 + y'^2}$, $f_{y'y'} = 1/(1 + y'^2)^{3/2}$ and since $y = x$ is the extremal of this problem, the Jacobi equation reads

$$\eta'' = 0,$$

and $\eta = ax$ $(a \neq 0)$ is a nontrivial solution that vanishes at $x = 0$. Since $\eta(x) \equiv ax \neq 0$ for all $x > 0$, we see that a conjugate point does not exist. Nor, for that matter, does there exist a conjugate point to $-\Delta$ $(\Delta > 0)$, regardless of Δ.

PROBLEMS 3.6

1. Find the extremals of the following variational problems, solve the associated Jacobi equation, and investigate the possibility of conjugate points to $x = a$, where a is the lower integration limit.

   (*a*) $\displaystyle\int_0^2 (xy'(x) + y'^2(x))\, dx \to \text{minimum},\ y(0) = 1,\ y(2) = 0$

   (*b*) $\displaystyle\int_0^b (y'^2(x) + 2y(x)y'(x) - 16y^2(x))\, dx \to \text{minimum},\ y(0) = 0,\ y(b) = 0,$

   where $b > 0$

(c) $\int_1^2 y'(x)(1 + x^2y'(x))\,dx \to \text{minimum},\ y(1) = 3,\ y(2) = 5$

(d) $\int_0^{\pi/4} (4y^2(x) - y'^2(x) + 8y(x))\,dx \to \text{maximum},\ y(0) = -1,\ y\left(\frac{\pi}{4}\right) = 0$

(e) $\int_0^b y'^3(x)\,dx \to \text{minimum},\ y(0) = 0,\ y(b) = y_b,\ b > 0,\ y_b > 0$

2. Prove Lemmas 3.6.1 and 3.6.2.
3. Show that, if η_1, η_2 are two linearly independent solutions of the Jacobi equation, then $[\eta_1(x_1),\eta_2(x_1)] \neq [0,0]$.

*4. Show that the solutions η, $\bar{\eta}$ of the Jacobi equation in part b of the proof of Theorem 3.6.1 are linearly independent.

*5. Let $\eta = \eta_1(x)$, $\eta = \eta_2(x)$ represent two solutions of the Jacobi equation. Show: If $\eta_1(x_o) = \eta_2(x_o) = 0$, then $\eta_1(x) = C\eta_2(x)$ for some constant C.

3.7 CONJUGATE POINTS AND FIELD EXISTENCE

In view of Theorem 3.6.2 and some results which we obtained earlier in Theorem 2.5.1 and Lemma 3.1, we are now in a position to prove the following *embedding theorem*:

Theorem 3.7.1 *If $y = y_o(x) \in C^1(A,B)$ is a regular extremal on $(A,B) \supset [a,b]$, if $f \in C^3(\mathfrak{R})$, where $\mathfrak{R} = \{(x,y,y') \mid A < x < B, |y - y_o(x)| < \alpha_1, |y' - y_o'(x)| < \beta_1\}$, and if there is no conjugate point to a in $(a,b]$, then $y = y_o(x)$, $x \in [a,b]$, is embeddable in a field $\mathfrak{F}$.*

Proof: Let $p_o(x) = f_{y'}(x, y_o(x), y_o'(x))$. Then, by Theorem 2.5.1 there exists a domain $\mathfrak{D} = \{(x,y,p) \mid a_1 < x < b_1, |y - y_o(x)| < \delta, |p - p_o(x)| < \delta\}$, where $A < a_1 < a < b < b_1 < B$, $\delta > 0$, and where the Euler-Lagrange equation may be replaced by

$$\frac{dy}{dx} = \Phi(x,y,p)$$
$$\frac{dp}{dx} = \Psi(x,y,p) \tag{3.7.1}$$

such that $\Phi,\Psi \in C^1(\mathfrak{D})$.

Let $(x_o,\lambda,c) \in \mathfrak{D}$. Then (3.7.1) has a unique solution $y = y(x,\lambda,c)$, $p = p(x,\lambda,c)$ that satisfies the initial conditions

$$y(x_o,\lambda,c) = \lambda$$
$$p(x_o,\lambda,c) = c,$$

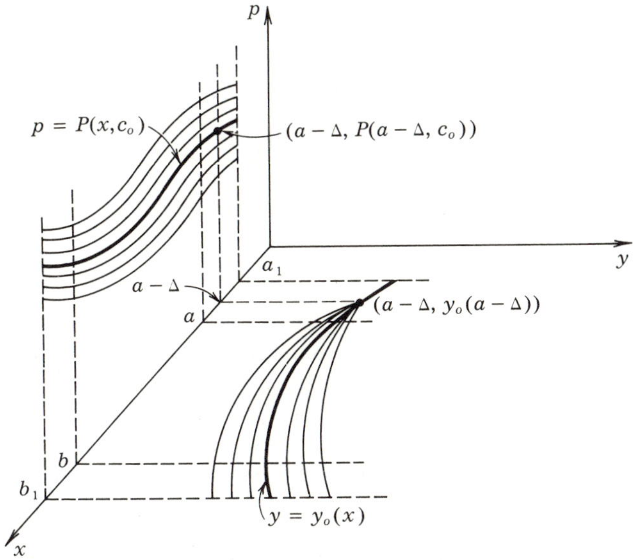

Figure 3.5

and $y', p', y_\lambda, y_c, p_\lambda, p_c$ are continuous so long as the solutions remain in $\mathfrak{D}$.†

Since $(a,b]$ does not contain a conjugate point to a, there exists a Δ, where $0 < \Delta < a - a_1$, such that $a - \Delta$ has no conjugate point in $(a,b]$. (See Theorem 3.6.2.)

We now determine those solutions of (3.7.1) for which

$$y(a - \Delta) = y_o(a - \Delta),$$
$$p(a - \Delta) = c$$

with $(a - \Delta, y_o(a - \Delta), c) \in \mathfrak{D}$, that is, $|c - p_o(a - \Delta)| < \delta$. We denote these solutions by

$$y = Y(x,c)$$
$$p = P(x,c).$$

We have thus obtained a one-parameter family of solutions of the Euler-Lagrange equation that pass through the point $(a - \Delta, y_o(a - \Delta))$ and contain—in view of the uniqueness theorem—the extremal $y = y_o(x)$ for $c_o = p_o(a - \Delta)$. (See also Fig. 3.5.)

† E. A. Coddington and N. Levinson, "Theory of Ordinary Differential Equations," p. 25, McGraw-Hill Book Company, New York, 1955.

Since

$$Y(a - \Delta, c) = y_o(a - \Delta)$$

for all c, we have

$$Y_c(x,c)|_{x=a-\Delta} = 0,$$

and, in particular,

$$Y_c(x,c_o)|_{x=a-\Delta} = 0.$$

Since $\eta = Y_c(x,c_o)$ is a solution of the Jacobi equation and since, by hypothesis, $(a - \Delta)^* > b$, we have

$$Y_c(x,c_o) \neq 0, \qquad \text{for all } a - \Delta < x \leq b \tag{3.7.2}$$

and, in particular, for all $x \in [a,b]$.

We shall now demonstrate that $y = Y(x,c)$ satisfies the hypotheses of Lemma 3.1 for $x \in (\alpha,\beta)$, where $[a,b] \subset (\alpha,\beta)$ and $c_1 < c < c_2$, where α,β,c_1,c_2 will be determined in the subsequent investigation.

Since $\Phi,\Psi \in C^1(\mathfrak{D})$, the solutions $y = Y(x,c)$, $p = P(x,c)$ are continuous functions of c so long as they stay in $\mathfrak{D}$, and, in particular,

$$|Y(x,c) - y_o(x)| + |P(x,c) - P(x,c_o)| \leq |c - c_o|\, e^{k(x-a+\Delta)},†$$

where $y_o(x) = Y(x,c_o)$. Let $0 < \Delta_1 < b_1 - b$, and let

$$|c - c_o| < \delta e^{-k(b_1-\Delta_1+\Delta-a)}.$$

Then $\quad |Y(x,c) - y_o(x)| < \delta \quad$ for all $x \in (a - \Delta, b_1 - \Delta_1)$,

and $\quad |P(x,c) - P(x,c_o)| < \delta \quad$ for all $x \in (a - \Delta, b_1 - \Delta_1)$.

Hence, if we take

$$c_1 = c_o - \delta e^{-k(b_1-\Delta_1-a+\Delta)}, \qquad c_2 = c_o + \delta e^{-k(b_1-\Delta_1-a+\Delta)},$$

then $\quad y = Y(x,c) \in \mathfrak{D}, \qquad p = P(x,c) \in \mathfrak{D}$

for all $a - \Delta < x < b_1 - \Delta_1$ and $c_1 < c < c_2$.

Let $\alpha_o = a - \Delta$, $\beta_o = b_1 - \Delta_1$, and $\mathcal{S} = \{(x,c) \mid \alpha_o < x < \beta_o, c_1 < c < c_2\}$. Then $Y,Y_c,P,P_c \in C(\mathcal{S})$,‡ and hence

$$Y_c' = \frac{\partial}{\partial c}[\Phi(x,Y(x,c),P(x,c))] = \Phi_y Y_c + \Phi_p P_c \in C(\mathcal{S})$$

$$Y'' = \frac{d}{dx}[\Phi(x,Y(x,c),P(x,c))] = \Phi_x + \Phi_y Y' + \Phi_p P'$$

$$= \Phi_x + \Phi_y \Phi + \Phi_p \Psi \in C(\mathcal{S}).$$

Thus $Y,Y'',Y_c,Y_c' \in C(\mathcal{S})$.

† *Ibid.*, p. 24.
‡ *Ibid.*

By (3.7.2), $Y_c(x,c_0) \neq 0$ for all $x \in [a,b]$, and hence, by Lemma 3.1, $y = y_0(x)$ is embeddable in a field $\mathfrak{F}$ on $[a,b]$.

In view of this theorem and the results which we obtained for examples A and B in Sec. 3.6, it follows that the extremals in these two examples are embeddable in fields. For the example in Sec. 3.4, we arrived at this result the hard way, namely, by actually constructing a field that was generated by a one-parameter family of extremals.

Theorem 3.7.1 enables us to state the sufficient condition which we formulated in Theorem 3.3.2 in a (in some respects) more practical form:

Theorem 3.7.2 *The regular extremal* $y = y_0(x) \in C^1(A,B)$, $[a,b] \subset (A,B)$ *yields a* strong *relative minimum (maximum) for* $I[y]$ *on* $[a,b]$ *if:*

1. *$(a,b]$ does not contain a conjugate point to* a.
2. $\mathcal{E}(x,\bar{y},y',\bar{y}') \geq 0 (\leq 0)$ *for all* $-\infty < y' < \infty$, $-\infty < \bar{y}' < \infty$ *and all* $(x,\bar{y}) \in N_w^{\delta}(y_0)$ *for some* $\delta > 0$.

[Hereby we assume that $f \in C^3(\mathcal{R})$, where $\mathcal{R}$ is a suitably chosen domain in (x,y,y') space.]

PROBLEMS 3.7

1. To which of the variational problems listed in Prob. 3.6.1*a* to *c* does Theorem 3.7.2 apply?
2. Consider the variational problem

$$\int_0^1 \sqrt{1 + y'^2(x)}\, dx \to \text{minimum}, \qquad y(0) = 0, \qquad y(1) = 1.$$

Follow the proof of Theorem 3.7.1 step by step and construct in this manner a central field into which $y = x$, $x \in [0,1]$, is embedded. If you choose $\Delta = 1$, then you will obtain the field that is defined by

$$\left(1, \frac{y+1}{x+1}\right).$$

3. Show that regular extremals of $I[y] = \int_a^b f(x,y'(x))\, dx \to$ minimum can always be embedded in a field, where it is assumed that $f \in C^3$.
4. Suppose that $f = f(y,y')$, $f_{y'y'} > 0$, and that $y = y_0(x) \equiv c$ is the extremal of a variational problem associated with this $f(y,y')$. Show that:
   (*a*) If $f_{yy} \geq 0$, then there are no conjugate points.
   (*b*) If $f_{yy} < 0$, then there are conjugate points $\pi\sqrt{-(f_{y'y'}/f_{yy})_{y=c,y'=0}}$ units apart.
5. The velocity field in a river with straight banks that are parallel to the x axis is given by

$$u = u(y)$$
$$v = 0.$$

A boat with the constant speed $c > |u(y)|$ in still water is to move from $A(a,y_a)$

to $B(b,y_a)$ in the shortest possible time. In general, the travel time is given by

$$t = \int_a^b \frac{\sqrt{c^2 + (c^2 - u^2)y'^2} - u}{c^2 - u^2}\,dx$$

(See also example D, Sec. 1.1, and note that x and y are interchanged.) Show that a straight line does provide the path of shortest travel time if:

(a) A, B lie in the line of (relative) maximum river velocity ($u'(y_a) = 0$, $u''(y_a) < 0$) and B is downstream from A, or A, B lie in the line of (relative) minimum river velocity ($u'(y_a) = 0$, $u''(y_a) > 0$) and B is upstream from A.

(b) A, B lie in the line of (relative) minimum river velocity and B is downstream from A, or A, B lie in the line of (relative) maximum river velocity and B is upstream from A, provided that A and B are sufficiently close together: $|b - a| < \pi(c + u(y_a))/\sqrt{cu''(y_a)}$.

3.8 A SUFFICIENT CONDITION FOR A WEAK MINIMUM

The excess function is defined by

$$\mathcal{E}(x,y,y',\bar{y}') = f(x,y,\bar{y}') - f(x,y,y') + (y' - \bar{y}')f_{y'}(x,y,y')$$

[see (3.3.4)].

By Taylor's formula,

$$f(x,y,\bar{y}') = f(x,y,y') + f_{y'}(x,y,y')(\bar{y}' - y') + f_{y'y'}(x,y,y' + \Theta(\bar{y}' - y'))\frac{(\bar{y}' - y')^2}{2},$$

where $0 \leq \Theta \leq 1$. Thus $\mathcal{E}$ may be represented as follows:

$$\mathcal{E}(x,y,y',\bar{y}') = f_{y'y'}(x,y,y' + \Theta(\bar{y}' - y'))\frac{(\bar{y}' - y')^2}{2}. \tag{3.8.1}$$

The following definition will simplify the terminology in the subsequent discussion:

Definition 3.8 *A lineal element* (x_0,y_0,y_0') *is called* strongly regular *if*

$$\mathcal{E}(x_0,y_0,y_0',\bar{y}') > 0 \qquad \text{for all } \bar{y}' \neq y_0'$$

and weakly regular *if*

$$f_{y'y'}(x_0,y_0,y_0') > 0.$$

Suppose now that the regular extremal $y = y_0(x) \in C^1[a,b]$ consists of weakly regular lineal elements only. Then

$$f_{y'y'}(x,y_0(x),y_0'(x)) > 0 \qquad \text{for all } x \in [a,b].$$

If $f \in C^2(\mathcal{R})$, where $\mathcal{R}$ denotes a simply connected domain that contains all lineal elements of $y = y_o(x)$, then for $y = y(x) \in C^1[a,b]$,

$$f_{y'y'}(x,y(x),y'(x)) > 0 \qquad \text{for all } x \in [a,b]$$

so long as $|y(x) - y_o(x)| < \delta_1$ and $|y'(x) - y_o'(x)| < \delta_2$ for some $\delta_1 > 0$, $\delta_2 > 0$. (See Prob. 3.8.3.)

Suppose now that $y = y_o(x)$ is embeddable in a field $\mathfrak{F}$. Then, with $|\Theta(x)| \leq 1$,

$$\begin{aligned} \mathcal{E}(x,\bar{y}(x),\phi(x,\bar{y}(x)),\bar{y}'(x)) &= f_{y'y'}(x,\bar{y}(x),\phi(x,\bar{y}(x)) \\ &\quad + \Theta(x)(\bar{y}'(x) - \phi(x,\bar{y}(x)))) \frac{(\bar{y}'(x) - \phi(x,\bar{y}(x)))^2}{2} \geq 0, \end{aligned}$$

provided that

$$|\bar{y}(x) - y_o(x)| < \delta_1$$

and $$|\phi(x,\bar{y}(x)) + \Theta(x)(\bar{y}'(x) - \phi(x,\bar{y}(x))) - y_o'(x)| < \delta_2.$$

We have

$$\begin{aligned} |\phi(x,\bar{y}(x)) + \Theta(x)(\bar{y}'(x) - \phi(x,\bar{y}(x))) - y_o'(x)| &\leq |\phi(x,\bar{y}(x)) - y_o'(x)| \\ &\quad + |\Theta(x)(\bar{y}'(x) - y_o'(x))| + |\Theta(x)(y_o'(x) - \phi(x,\bar{y}(x)))| \end{aligned}$$

[Note that $\bar{y}'(x)$ denotes the slope of $y = \bar{y}(x)$ while $\phi(x,\bar{y}(x))$ denotes the slope of the uniquely determined extremal through the point $(x,\bar{y}(x))$.]

Since

$$|y_o'(x) - \phi(x,\bar{y}(x))| = |\phi(x,y_o(x)) - \phi(x,\bar{y}(x))| < \frac{\delta_2}{3},$$

provided that $|y_o(x) - \bar{y}(x)| < \delta_3$ for some $\delta_3 > 0$, and if we also require that $|y_o'(x) - \bar{y}'(x)| < (\delta_2/3)$, then

$$|\phi(x,\bar{y}(x)) + \Theta(x)(\bar{y}'(x) - \phi(x,\bar{y}(x))) - y_o'(x)| < \delta_2,$$

and we obtain

$$\mathcal{E}(x,\bar{y}(x),\phi(x,\bar{y}(x)),\bar{y}'(x)) \geq 0 \qquad \text{for all } x \in [a,b]$$

and for all $y = \bar{y}(x) \in N^\delta(y_o)$, where $\delta = \min(\delta_1,\delta_2/3,\delta_3)$. Hence,

$$\Delta I = I[\bar{y}] - I[y_o] = \int_a^b \mathcal{E}(x,\bar{y}(x),\phi(x,\bar{y}(x)),\bar{y}'(x))\, dx \geq 0 \qquad (3.8.2)$$

for all $y = \bar{y}(x) \in N^\delta(y_o)$ for some $\delta > 0$.

We formulate this result as a theorem:

Theorem 3.8.1 *The regular extremal* $y = y_o(x) \in C^1[a,b]$ *yields a* weak relative minimum *for* $I[y]$, $f \in C^2(\mathcal{R})$, *if* $y = y_o(x)$ *is embeddable in a field and if* $y = y_o(x)$ *consists of* weakly regular lineal elements *only, that is,*

$$f_{y'y'}(x,y_o(x),y_o'(x)) > 0 \quad \text{for all } x \in [a,b].$$

(An analogous theorem for a weak relative maximum is obtained by reversing the inequality sign.)

With the new terminology introduced in Definition 3.8, we may now state Theorem 3.3.1 as follows:

Theorem 3.8.2 *If the extremal* $y = y_o(x)$ *is embeddable in a field and if all lineal elements of the* field *are* strongly regular, *then* $y = y_o(x)$ *yields a* strong relative minimum *for* $I[y]$. (See Probs. 3.8.1 and 3.8.4.)

The characteristic of f provides a simple geometric interpretation of weakly and strongly regular lineal elements. We suppose that $f_{y'y'}(x_o,y_o,y_o') > 0$ and consider the characteristic of f at (x_o,y_o):

$$\eta = f(x_o,y_o,\xi).$$

Then the equation of the tangent line T at $\xi = y_o'$ is given by

$$\eta_T - f(x_o,y_o,y_o') = f_{y'}(x_o,y_o,y_o')(\xi - y_o').$$

Hence, at $\xi = \bar{y}'$, we have

$$\begin{aligned} f(x_o,y_o,\bar{y}') - \eta_T(\bar{y}') &= f(x_o,y_o,\bar{y}') - f(x_o,y_o,y_o') + (y_o' - \bar{y}')f_{y'}(x_o,y_o,y_o') \\ &= \mathcal{E}(x_o,y_o,y_o',\bar{y}'). \end{aligned}$$

Thus (x_o,y_o,y_o') is a strongly regular lineal element if the graph of the characteristic $\eta = f(x_o,y_o,\xi)$ remains above the tangent line T for all $\xi = \bar{y}' \neq y_o'$. (See Fig. 3.6). If the graph of the characteristic dips below the tangent line for some $\xi = \bar{y}' \neq y_o'$, or touches it, then (x_o,y_o,y_o') is only weakly regular.

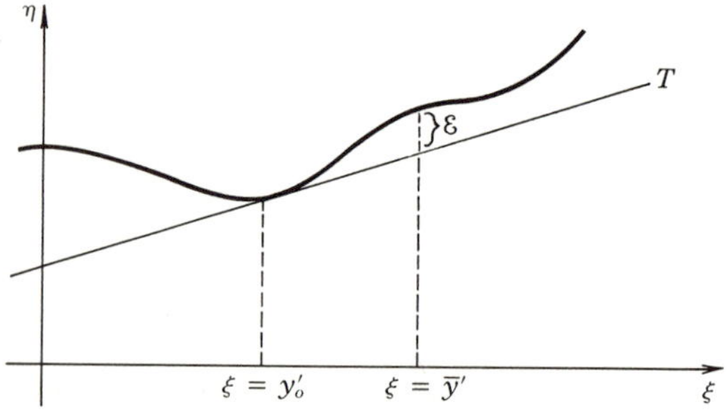

Figure 3.6

Since the tangent line to the characteristic at a point that corresponds to a strongly regular lineal element cannot ever touch the characteristic again, it follows that a corner cannot possibly occur at a point of an extremal that is associated with a strongly regular lineal element. (See also Sec. 2.9.)

PROBLEMS 3.8

*1. Prove Theorem 3.8.2.

2. Let (x_o,y_o,y'_o) represent a strongly regular lineal element of the extremal $y = y_o(x)$. Show that the extremal cannot possibly have a corner at the point (x_o,y_o).

*3. Prove: If $f \in C^2(\mathcal{R})$ and if $f_{y'y'}(x,y_o(x)y'_o(x)) > 0$ for all $x \in [a,b]$, where $y = y_o(x) \in C^1[a,b]$, then there exist a $\delta_1 > 0$ and a $\delta_2 > 0$ such that

$$f_{y'y'}(x,y(x),y'(x)) > 0 \qquad \text{for all } x \in [a,b],$$

provided that $|y(x) - y_o(x)| < \delta_1$, $|y'(x) - y'_o(x)| < \delta_2$ and $y \in C^1[a,b]$.

*4. Given the variational problem

$$I[y] = \int_0^1 y'^2(x)(1 + y'(x))^2\,dx \to \text{minimum}, \qquad y(0) = 0, \qquad y(1) = m.$$

Show:

(a) If $m \le -1$ or $m \ge 0$, then the extremal yields a strong relative minimum.

(b) If $-1 < m < -\frac{1}{2} - (\sqrt{3}/6)$ or if $-\frac{1}{2} + (\sqrt{3}/6) < m < 0$, then the extremal yields a weak relative minimum.

(c) If $-\frac{1}{2} - (\sqrt{3}/6) < m < -\frac{1}{2} + (\sqrt{3}/6)$, then the extremal yields a weak relative maximum.

3.9 A NECESSARY CONDITION FOR A STRONG RELATIVE MINIMUM

We consider again the simplest variational problem

$$I[y] = \int_a^b f(x,y(x),y'(x))\,dx \to \text{minimum},$$

where $\quad y(a) = y_a, \qquad y(b) = y_b$

are given.

We assume that $y = y_o(x) \in C^1[a,b]$ yields a *strong relative minimum* for $I[y]$ in $\Sigma_s = \{y \mid y \in C_s{}^1[a,b], y(a) = y_a, y(b) = y_b\}$, and we shall show that then, $\mathcal{E}(x,y_o(x),y'_o(x),\bar{y}') \ge 0$ for all $x \in [a,b]$ and all $-\infty < \bar{y}' < \infty$.

First we note that if $y = y_o(x)$ yields a strong relative minimum for $I[y]$ in Σ_s, then that portion of $y = y_o(x)$ that is defined in $[a,x_o]$, where x_o is an arbitrary point in $a < x \le b$, will yield a strong relative minimum for

$$I_1[y] = \int_a^{x_o} f(x,y(x),y'(x))\,dx$$

in $$\Sigma_s' = \{y \mid y \in C_s^1[a,x_o],\ y(a) = y_a,\ y(x_o) = y_o(x_o)\}.$$

[Note that the terminal value is the value that is assumed by $y = y_o(x)$ at $x = x_o$.]

Now, if this were not the case, then there would be, for any $\delta > 0$, a sectionally smooth function $y = y_1(x) \in C_s^1[a,x_o]$, $y_1(a) = y_a$, $y_1(x_o) = y_o(x_o)$, $y_1 \in N_w^\delta(y_o)$, such that

$$I_1[y_1] < I_1[y_o],$$

and hence we would have, for

$$y = \eta(x) \equiv \begin{Bmatrix} y_1(x) & \text{for } a \le x \le x_o \\ y_o(x) & \text{for } x_o \le x \le b \end{Bmatrix} \in N_w^\delta(y_o),$$

that $$I[\eta] = I_1[y_1] + \int_{x_o}^b f(x,y_o(x),y_o'(x))\,dx$$

$$< I_1[y_o] + \int_{x_o}^b f(x,y_o(x),y_o'(x))\,dx = I[y_o];$$

that is, $I[\eta] < I[y_o]$, and y_o does not yield a strong relative minimum for $I[y]$ in Σ_s.

We shall now demonstrate that if there exists an $x_o \in (a,b]$ and a specific value $\bar{y}'$ such that

$$\mathcal{E}(x_o,y_o(x_o),y_o'(x_o),\bar{y}') < 0, \tag{3.9.1}$$

then $y = y_o(x)$ cannot yield a strong relative minimum for $I_1[y]$ under the boundary conditions $y(a) = y_a$, $y(x_o) = y_o(x_o)$. Hence $y = y_o(x)$ cannot yield a strong relative minimum for $I[y]$ under the boundary conditions $y(a) = y_a$, $y(b) = y_b$ either. It will then follow that $\mathcal{E}(x,y_o(x),y_o'(x),\bar{y}') \ge 0$ is necessary if $y = y_o(x)$ is to yield a strong relative minimum for $I[y]$.

We assume without loss of generality (see Prob. 3.9.1) that

$$\bar{y}' < 0 \qquad \text{and} \qquad \bar{y}' < y_o'(x_o).$$

We then proceed to construct a function $y = \eta(x) \in C_s^1[a,x_o]$ as follows:

$$y = \eta_t(x) \equiv \begin{cases} y_o(x) + th(x) & \text{for } a \le x \le x_o + \Delta,\ t \ge 0 \\ (\bar{y}')(x - x_o) + y_o(x_o) & \text{for } x_o + \Delta < x \le x_o, \end{cases}$$

where we choose $\Delta < 0$ such that $\eta = \eta_t(x)$ is continuous and choose $h \in C^1[a,x_o]$, $h(x) > 0$ for $x > a$ and $h(a) = 0$. (See Fig. 3.7.)

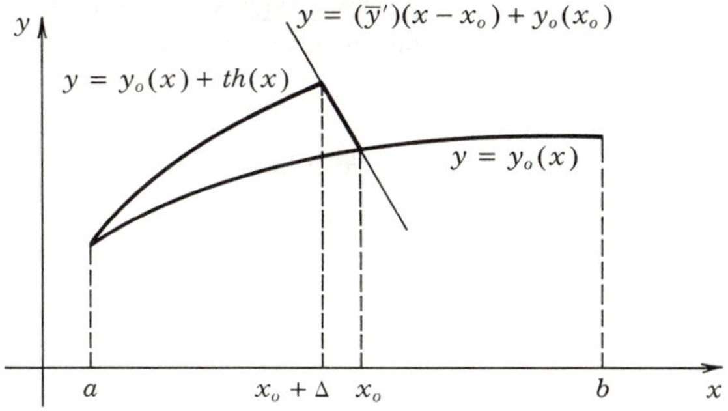

Figure 3.7

As in Sec. 2.8, we obtain

$$\Delta = \Delta(t) \qquad \text{and} \qquad \Delta'(0) = \frac{h(x_o)}{\bar{y}' - y_o'(x_o)}. \tag{3.9.2}$$

We shall now show that it is possible to choose t sufficiently small so that $\eta_t \in N_w{}^\delta(y_o)$ and $I_1[\eta_t] < I_1[y_o]$. With the understanding that in all the following integrals, y_o, y_o', h, h' stand for $y_o(x)$, $y_o'(x)$, $h(x)$, $h'(x)$, we have

$$I_1[\eta_t] - I_1[y_o] = \int_a^{x_o+\Delta} f(x, y_o + th, y_o' + th')\, dx$$

$$+ \int_{x_o+\Delta}^{x_o} f(x, (\bar{y}')(x - x_o) + y_o(x_o), \bar{y}')\, dx - \int_a^{x_o} f(x,y_o,y_o')\, dx. \tag{3.9.3}$$

As in Sec. 2.8, we see that the derivative of

$$\int_a^{x_o+\Delta} f(x, y_o + th, y_o' + th')\, dx$$

exists at $t = 0$ if we assume that $f \in C^1(\mathcal{R})$, where $\mathcal{R}$ is a domain that contains all lineal elements of $y = y_o(x)$, and is given by

$$\int_a^{x_o} [f_y(x,y_o y_o')h + f_{y'}(x,y_o,y_o')h']\, dx + f(x_o,y_o(x_o),y_o'(x_o))\Delta'(0).$$

Hence, by Theorem 1.5,

$$\int_a^{x_o+\Delta} f(x, y_o + th, y_o' + th')\, dx - \int_a^{x_o} f(x,y_o,y_o')\, dx$$

$$= t\int_a^{x_o} [f_y(x,y_o,y_o')h + f_{y'}(x,y_o,y_o')h']\, dx$$

$$+ tf(x_o,y_o(x_o), y_o'(x_o))\, \Delta'(0) + \varepsilon[th], \tag{3.9.4}$$

where $(\varepsilon[th]/t) \to 0$ as $t \to 0$.

We note that $\int_{x_o+\Delta}^{x_o} f(x,(\bar{y}')(x-x_o)+y_o(x_o),\bar{y}')\,dx$ in (3.9.3) is a function of t, and we convince ourselves easily that it is differentiable at $t=0$. We obtain

$$\frac{d}{dt}\int_{x_o+\Delta}^{x_o} f(x,\,(\bar{y}')(x-x_o)+y_o(x_o),\bar{y}')\,dx\Big|_{t=0} = -f(x_o,y_o(x_o),\bar{y}')\,\Delta'(0),$$

and hence, since $\int_{x_o+\Delta}^{x_o} f\,dx\big|_{t=0} = 0$,

$$\int_{x_o+\Delta}^{x_o} f(x,\,(\bar{y}')(x-x_o)+y_o(x_o),\bar{y}')\,dx = -tf(x_o,y_o(x_o),\bar{y}')\,\Delta'(0)+\beta(t), \tag{3.9.5}$$

where $[\beta(t)/t]\to 0$ as $t\to 0$. (See Definition 1.3.3.)

Now we obtain for (3.9.3), in view of (3.9.4) and (3.9.5),

$$I_1[\eta_t]-I_1[y_o] = t\int_a^{x_o}[(f_y)_o h+(f_{y'})_o h']\,dx + t[f(x_o,y_o(x_o),y_o'(x_o)) - f(x_o,y_o(x_o),\bar{y}')]\,\Delta'(0)+\varepsilon[th]+\beta(t),$$

where $\dfrac{\varepsilon[th]+\beta(t)}{t}\equiv\dfrac{\gamma(t)}{t}\to 0$ as $t\to 0$.

After integration by parts, as in Sec. 2.3, we obtain, because $y=y_o(x)$ is supposed to yield a strong relative minimum and is therefore, by necessity, a solution of the Euler-Lagrange equation, and in view of (3.9.2),

$$I_1[\eta_t]-I_1[y_o] = th(x_o)\left[f_{y'}(x,y_o(x_o),y_o'(x_o)) + (f(x_o,y_o(x_o),y_o'(x_o)) - f(x_o,y_o(x_o),\bar{y}'))\frac{1}{\bar{y}'-y_o'(x_o)}\right]+\gamma(t),$$

where $h(x_o)>0$ and $[\gamma(t)/t]\to 0$ as $t\to 0$.

By (3.3.4),

$$\mathcal{E}(x_o,y_o(x_o),y_o'(x_o),\bar{y}') = f(x_o,y_o(x_o),\bar{y}') - f(x_o,y_o(x_o),y_o'(x_o)) + (y_o'(x_o)-\bar{y}')f_{y'}(x_o,y_o(x_o),y_o'(x_o)).$$

Hence,

$$I_1[\eta_t]-I_1[y_o] = -\left[\frac{h(x_o)}{\bar{y}'-y_o'(x_o)}\,\mathcal{E}(x_o,y_o(x_o),y_o'(x_o),\bar{y}') - \frac{\gamma(t)}{t}\right]t.$$

By (3.9.1), $\mathcal{E}(x_o,y_o(x_o),y_o'(x_o),\bar{y}')<0$, and in view of our assumptions on $\bar{y}'$ and $y_o'(x_o)$, $\bar{y}'-y_o'(x_o)<0$. Also $h_o(x_o)>0$.

Since $[\gamma(t)/t]\to 0$ as $t\to 0$, we can choose a $t>0$ so small that

$$I_1[\eta_t]-I_1[y_o]<0,$$

and $\eta_t\in N_w{}^\delta(y_o)$.

Hence, if we denote

$$y = \eta(x) \equiv \begin{cases} \eta_t(x) & \text{for } a \leq x \leq x_o \\ y_o(x) & \text{for } x_o < x \leq b, \end{cases}$$

we have

$$I[\eta] < I[y_o],$$

whereby $\eta \in N_w^{\delta}(y_o)$, and we see that y_o does not yield a strong relative minimum in Σ_s.

By corollary 1 to Theorem 2.10, we can now find a function $y = y(x) \in \Sigma$ in *any* weak neighborhood of η and hence in any weak neighborhood of y_o such that $I[y] < I[y_o]$.

We formulate our result in the following:

Theorem 3.9.1 *If the extremal* $y = y_o(x) \in C^1[a,b]$ *yields a* strong relative minimum *for* $I[y]$ *in* $\Sigma_s = \{y \mid y \in C_s^1[a,b],\ y(a) = y_a,\ y(b) = y_b\}$*—or in* $\Sigma = \{y \mid y \in C^1[a,b], y(a) = y_a, y(b) = y_b\}$*, for that matter—whereby we assume that* $f \in C^1(\mathfrak{R})$*,* $\mathfrak{R}$ *denoting a domain that contains all lineal elements of* $y = y_o(x)$*, then it is necessary that*

$$\mathcal{E}(x,y_o(x),y_o'(x),\bar{y}') \geq 0 \tag{3.9.6}$$

for all $x \in [a,b]$ *and all* $-\infty < \bar{y}' < \infty$.

By the same token, every smooth portion of a broken extremal $y = y_o(x) \in C_s^1[a,b]$ has to satisfy (3.9.6) if $y = y_o(x)$ yields a strong relative minimum for $I[y]$ in Σ_s.

The derivation of this theorem was based on the possibility of constructing the function $y = \eta(x)$. This construction is only possible if $x_o > a$. However, it is easy to see that (3.9.6) has to hold at $x = a$ as well, since all the quantities involved are continuous.

In a similar manner and with the aid of the relationship between $\mathcal{E}$ and $f_{y'y'}$, as expressed in (3.8.1), one can obtain the so-called *Legendre*† *condition,* which we shall derive by an entirely different method in Sec. 7.2.

We express the Legendre condition in the following theorem:

Theorem 3.9.2 *For the extremal* $y = y_o(x) \in C^1[a,b]$ *to yield a weak relative minimum (maximum) for* $I[y]$ *in* $\Sigma_s = \{y \mid y \in C_s^1[a,b], y(a) = y_a, y(b) = y_b\}$*, where* $f \in C^2(\mathfrak{R})$*, it is necessary that*

$$f_{y'y'}(x,y_o(x),y_o'(x)) \geq 0 \qquad (\leq 0) \tag{3.9.7}$$

for all $x \in [a,b]$. (See Prob. 3.9.5.)

† Adrien Marie Legendre, 1752–1833.

PROBLEMS 3.9

*1. In the derivation of Theorem 3.9.1, we assumed that $\bar{y}' < 0$ and $\bar{y}' < y_o'(x_o)$. Construct suitable functions $y = \eta(x)$ for the other possibilities:
(*a*) $\bar{y}' < 0$ and $\bar{y}' > y_o'(x)$
(*b*) $\bar{y}' > 0$ and $\bar{y}' < y_o'(x_o)$
(*c*) $\bar{y}' > 0$ and $\bar{y}' > y_o'(x_o)$

2. Show that $\mathcal{E}(x,y_o(x),y_o'(x),\bar{y}') = 0$ for all $\bar{y}'$, for the extremal $y = y_o(x)$ of the variational problem

$$\int_0^1 y^2(x)(1 - y'(x))^2\,dx \to \text{minimum}, \qquad y(0) = 0, \qquad y(1) = 0.$$

3. Show that $y = x$ cannot yield a strong relative minimum for

$$\int_0^1 y'^3(x)\,dx \to \text{minimum}, \qquad y(0) = 0, \qquad y(1) = 1.$$

4. Given the variational problem

$$I[y] = \int_0^1 (y'^2(x) - 4y(x)y'^3(x) + 2xy'^4(x))\,dx \to \text{minimum}, \quad y(0) = y(1) = 0.$$

(*a*) Show that the extremal $y = y_o(x) \equiv 0$ is embeddable in a field.
(*b*) Show that $y = y_o(x) \equiv 0$ satisfies the Weierstrass necessary condition (3.9.6).
(*c*) Show that $y = y_o(x) \equiv 0$ satisfies the Legendre condition for a weak relative minimum, Theorem 3.8.1.
(*d*) Show that $y = y_o(x) \equiv 0$ does not yield a strong relative minimum.

*5. Prove: If the extremal $y = y_o(x) \in C^1[a,b]$ yields a weak relative minimum for $I[y]$, then it is necessary that

$$f_{y'y'}(x,y_o(x),y_o'(x)) \geq 0$$

for all $x \in [a,b]$.

6. Show that if $y = y_o(x)$ is an extremal of $I[y] = \int_a^b [\alpha(x,y) + \beta(x,y)y']\,dx \to$ minimum, $y(a) = y_a$, $y(b) = y_b$, then

$$\mathcal{E}(x,y_o(x),y_o'(x),\bar{y}') = 0$$

for all $x \in [a,b]$ and all $\bar{y}'$.

7. If $y = y_o(x) \in C^1[a,b]$ minimizes $I[y]$, then by (3.9.6),

$$\min_{(\bar{y}')} \mathcal{E}(x,y_o(x),y_o'(x),\bar{y}') = 0$$

for all $x \in [a,b]$. Then, by necessity, $(d\mathcal{E}/d\bar{y}')|_{\bar{y}'=y'_o} = 0$. Show that this is indeed the case.

8. Given the variational problem

$$I[y] = \int_0^1 y^2(x)(1 - y'^2(x))\,dx \to \text{minimum}, \qquad y(0) = 1, \qquad y(1) = 1.$$

Show that no extremal can possibly yield a relative minimum, either weak or strong.

9. Given the variational problem

$$I[y] = \int_0^1 (y'^2(x) - 4xy'^3(x) + 4x^2y'^4(x))\, dx \rightarrow \text{minimum},$$

$y(0) = 0$, $y(1) = 0$. Show that the extremal of the problem yields a strong relative minimum.

3.10 A SUFFICIENT CONDITION FOR THE PROBLEM IN n UNKNOWN FUNCTIONS

We consider the variational problem of Sec. 2.11:

$$I[\hat{y}] = \int_a^b f(x,\hat{y}(x),\hat{y}'(x))\, dx \rightarrow \text{minimum},$$

where we shall immediately consider the more general case in which beginning point and endpoint are free to move in the n-dimensional planes $x = a$ and $x = b$. [We remind the reader that $\hat{y} = (y_1, \ldots, y_n)$.]

We have seen in Sec. 2.11 that if $\hat{y} = \hat{y}_o(x) \in \mathfrak{C}^1[a,b]^n$ is the solution of this problem, then it has to satisfy the n Euler-Lagrange equations

$$f_{y_k} - \frac{d}{dx} f_{y'_k} = 0, \qquad k = 1, 2, \ldots, n$$

and the $2n$ natural boundary conditions

$$f_{y'_k}(a,\hat{y}_o(a),\hat{y}'_o(a)) = 0 \tag{3.10.1}$$
$$f_{y'_k}(b,\hat{y}_o(b),\hat{y}'_o(b)) = 0, \qquad k = 1, 2, \ldots, n.$$

If we wish to construct a field $(1,\hat{\phi}(x,\hat{y}))$ for this problem, we shall have to make sure that the solutions of $\hat{y}' = \hat{\phi}(x,\hat{y})$ satisfy not only the Euler-Lagrange equations but also the natural boundary conditions (3.10.1). These conditions alone will not suffice, however, when $n \geq 2$, to ensure the independence of the invariance integral as adapted to the problem that is now under investigation.

We give the following definition, reminding the reader of the differentiation convention which we adopted on page 138.

Definition 3.10.1 $\mathfrak{F} = \{(1,\hat{\phi}(x,\hat{y})) \mid (x,\hat{y}) \in \mathfrak{A}\}$ *defines a* Mayer field *in an open and simply connected point set* $\mathfrak{A}$ *in* $(x,y_1, \ldots, y_n)$ *space for the variational problem* $I[\hat{y}] \rightarrow$ *minimum, with beginning point and endpoint varying on the vertical planes* $x = a$ *and* $x = b$, *if*

$$\hat{\phi}, \hat{\phi}_x, \hat{\phi}_{y_k} \in C(\mathfrak{A})$$

and if:

1. *The solutions* $\hat{y} = \hat{Y}(x,\hat{c})$ *of* $\hat{y}' = \hat{\phi}(x,\hat{y})$ *are solutions of the Euler-Lagrange equations of* $I[\hat{y}]$.
2. $\left.\begin{aligned} f_{y'_k}(a,\hat{y}(a),\hat{\phi}(a,\hat{y}(a))) &= 0 \\ f_{y'_k}(b,\hat{y}(b),\hat{\phi}(b,\hat{y}(b))) &= 0 \end{aligned}\right\} k = 1, 2, \ldots, n.$
3. $(\partial/\partial y_i)\, f_{y'_k}(x,\hat{y},\hat{\phi}(x,\hat{y})) = (\partial/\partial y_k)\, f_{y'_i}(x,\hat{y},\hat{\phi}(x,\hat{y}))$ *for all* $i, k = 1, 2, \ldots, n$.

[Condition 3 is trivially satisfied for $n = 1$, namely, $(\partial/\partial y)\, f_{y'} = (\partial/\partial y)\, f_{y'}$. This is the reason why we have not encountered this condition before.]

The extremal $\hat{y} = \hat{y}_0(x)$, $x \in [a,b]$, *is embeddable in* $\mathfrak{F}$ *if* $(x,\hat{y}_0(x)) \in \mathfrak{A}$ *for all* $x \in [a,b]$ *and if* $\hat{y} = \hat{y}_0(x)$ *is a solution of* $\hat{y}' = \hat{\phi}(x,\hat{y})$.

Suppose we have such a Mayer field. Then we can prove:

Theorem 3.10.1 *If* $\hat{y} = \hat{\bar{y}}(x) \in C^1[a,b]$ *remains entirely in* $\mathfrak{A}$, *then*

$$U[\hat{\bar{y}}] = \int_a^b [f(x,\hat{\bar{y}},\hat{\phi}(x,\hat{\bar{y}})) + (\hat{\bar{y}}' - \hat{\phi}(x,\hat{\bar{y}}))f_{\hat{y}'}{}^T(x,\hat{\bar{y}},\hat{\phi}(x,\hat{\bar{y}}))]\, dx$$

is:

1. *Independent of the path* $\hat{y} = \hat{\bar{y}}(x)$.
2. *Independent of the terminal coordinates of* $\hat{y} = \hat{\bar{y}}(x)$ *at* $x = a$ *and* $x = b$.

In this formulation,

$$f_{\hat{y}'} = (f_{y'_1}, f_{y'_2}, \ldots, f_{y'_n}),$$

and hence

$$(\hat{\bar{y}}' - \hat{\phi})f_{\hat{y}'}{}^T = \sum_{i=1}^n (\bar{y}_i' - \phi_i)f_{y'_i}.$$

Proof: (*a*) First, we shall prove that $U[\hat{\bar{y}}]$ is independent of the path $\hat{y} = \hat{\bar{y}}(x)$, assuming that beginning point and endpoint remain fixed. We write $U[\hat{\bar{y}}]$ in greater detail by expressing all the dot products that are involved in terms of the vector components:

$$U[\hat{\bar{y}}] = \int_a^b ([f(x,\hat{\bar{y}},\hat{\phi}(x,\hat{\bar{y}})) - \sum_{i=1}^n \phi_i(x,\hat{\bar{y}})f_{y'_i}(x,\hat{\bar{y}},\hat{\phi}(x,\hat{\bar{y}}))]\, dx$$
$$+ \sum_{i=1}^n f_{y'_i}(x,\hat{\bar{y}},\hat{\phi}(x,\hat{\bar{y}}))\, d\bar{y}_i).$$

This line integral is independent of the path† if and only if

$$\frac{\partial(f - \hat{\phi} f_{\hat{y}'}{}^T)}{\partial y_k} = \frac{\partial f_{y'_k}}{\partial x}, \qquad k = 1, 2, 3, \ldots, n \tag{3.10.2}$$

$$\frac{\partial f_{y'_i}}{\partial y_k} = \frac{\partial f_{y'_k}}{\partial y_i}, \qquad i,k = 1, 2, 3, \ldots, n. \tag{3.10.3}$$

The latter condition is satisfied since $\hat{y}' = \hat{\phi}(x,\hat{y})$ defines a Mayer field.

It remains to check the condition (3.10.2):

$$\frac{\partial}{\partial x} f_{y'_k}(x,\hat{y},\hat{\phi}(x,\hat{y})) = f_{y'_k x} + \sum_{i=1}^{n} f_{y'_k y'_i} \phi_{ix}.$$

$$\frac{\partial}{\partial y_k}\Big[f(x,\hat{y},\hat{\phi}(x,\hat{y})) - \sum_{i=1}^{n} \phi_i(x,\hat{y}) f_{y'_i}(x,\hat{y},\hat{\phi}(x,\hat{y}))\Big] \tag{3.10.4}$$

$$= f_{y_k} + \sum_{i=1}^{n} f_{y'_i}\phi_{iy_k} - \sum_{i=1}^{n}\left[\phi_{iy_k} f_{y'_i} + \phi_i \frac{\partial f_{y'_i}}{\partial y_k}\right].$$

Since, by (3.10.3),

$$\frac{\partial}{\partial y_k} f_{y'_i} = \frac{\partial}{\partial y_i} f_{y'_k} = f_{y'_k y_i} + \sum_{j=1}^{n} f_{y'_k y'_j} \phi_{j y_i},$$

we have

$$\frac{\partial}{\partial y_k}[f - \hat{\phi} f_{\hat{y}'}{}^T] = f_{y_k} - \sum_{i=1}^{n} \phi_i \Big[f_{y'_k y_i} + \sum_{j=1}^{n} f_{y'_k y'_j} \phi_{j y_i}\Big]. \tag{3.10.5}$$

Equating (3.10.4) with (3.10.5),

$$f_{y_k} = f_{y'_k x} + \sum_{i=1}^{n} f_{y'_k y'_i}\phi_{ix} + \sum_{i=1}^{n} \phi_i f_{y'_k y_i} + \sum_{i=1}^{n}\sum_{j=1}^{n} \phi_i f_{y'_k y'_j} \phi_{j y_i}$$

$$= f_{y'_k x} + \sum_{i=1}^{n} f_{y'_k y'_i}\phi_{ix} + \sum_{i=1}^{n} \phi_i f_{y'_k y_i} + \sum_{j=1}^{n}\sum_{i=1}^{n} \phi_j f_{y'_k y'_i} \phi_{i y_j}$$

$$= f_{y'_k x} + \sum_{i=1}^{n} \phi_i f_{y'_k y_i} + \sum_{i=1}^{n} f_{y'_k y'_i}\Big(\phi_{ix} + \sum_{j=1}^{n} \phi_j \phi_{i y_j}\Big).$$

Since $\quad \hat{y}' = \hat{\phi}(x,\hat{y}),$

we have $\quad y''_i = \phi_{ix} + \sum_{j=1}^{n} \phi_{i y_j} y'_j = \phi_{ix} + \sum_{j=1}^{n} \phi_{i y_j} \phi_j.$

† T. M. Apostol, "Mathematical Analysis," pp. 292ff, Addison-Wesley Publishing Company, Inc., Reading, Mass., 1957.

Hence

$$f_{y_k} = f_{y'_k x} + \sum_{i=1}^{n} f_{y'_k y_i} y'_i + \sum_{i=1}^{n} f_{y'_k y'_i} y''_i = \frac{d}{dx} f_{y'_k},$$

which is true since the Euler-Lagrange equations are satisfied by all integral curves of $\hat{y}' = \hat{\phi}(x,\hat{y})$. Thus we have seen that $U[\hat{\bar{y}}]$ is indeed independent of the path.

(*b*) Now we have to show that $U[\hat{\bar{y}}]$ is also independent of the beginning and end ordinates of $\hat{y} = \hat{\bar{y}}(x)$.

In view of (3.10.2) and (3.10.3), there exists a function†

$$W = W(x,y_1, \ldots ,y_n)$$

such that

$$W_x = f - \hat{\phi} f_{\hat{y}'}{}^T \qquad \text{and} \qquad W_{\hat{y}} = f_{\hat{y}'},$$

and consequently,

$$U[\hat{\bar{y}}] = W(b,\hat{\bar{y}}(b)) - W(a,\hat{\bar{y}}(a)).$$

Since

$$W_{y_k}|_{x=a} = f_{y'_k}(a,\hat{y}(a),\hat{\phi}(a,\hat{y}(a))) = 0$$
$$W_{y_k}|_{x=b} = f_{y'_k}(b,\hat{y}(b),\hat{\phi}(b,\hat{y}(b))) = 0$$

because all solutions of $\hat{y}' = \hat{\phi}(x,\hat{y})$ satisfy the natural boundary conditions (3.10.1), it follows that $W(a,\hat{y}(a))$ and $W(b,\hat{y}(b))$ are independent of $\hat{y}$, and hence

$$W(a,\hat{y}_1(a)) = W(a,\hat{y}(a)), \qquad W(b,\hat{y}_1(b)) = W(b,\hat{y}(b)).$$

Hence, if $\hat{y} = \hat{y}(x)$ has the terminal ordinates $\hat{y}(a)$ and $\hat{y}(b)$ and another curve $\hat{y} = \hat{y}_1(x)$ has the terminal ordinates $\hat{y}_1(a)$ and $\hat{y}_1(b)$, then $U[\hat{y}] = U[\hat{y}_1]$.

Now that we have proved this theorem, we may transform the total variation as follows (see also Prob. 3.10.1):

$$I[\hat{\bar{y}}_1] - I[\hat{\bar{y}}_o] = I[\hat{\bar{y}}_1] - U[\hat{\bar{y}}_o] = I[\hat{\bar{y}}_1] - U[\hat{\bar{y}}_1]$$
$$= \int_a^b [f(x,\hat{\bar{y}},\hat{\bar{y}}') - f(x,\hat{\bar{y}},\hat{\phi}(x,\hat{\bar{y}}))$$
$$+ (\hat{\phi}(x,\hat{\bar{y}}) - \hat{\bar{y}}') f_{\hat{y}'}{}^T(x,\hat{\bar{y}},\hat{\phi}(x,\hat{\bar{y}}))]\, dx.$$

We can state:

Theorem 3. 10.2 *If the extremal* $\hat{y} = \hat{y}_o(x) \in C^1[a,b]^n$ *is embeddable in a*

† *Ibid.*, p. 296.

Mayer field and if

$$\mathcal{E}(x,\hat{\bar{y}},\hat{y}',\hat{\bar{y}}') = f(x,\hat{\bar{y}},\hat{\bar{y}}') - f(x,\hat{\bar{y}},\hat{y}') + (\hat{y}' - \hat{\bar{y}}')f_{\hat{y}'}{}^T(x,\hat{\bar{y}},\hat{y}') \geq 0 \quad (3.10.6)$$

for all $(x,\hat{\bar{y}}) \in N_w{}^\delta(\hat{y}_o)$ *and all possible* $-\infty < y_k' < \infty$, $-\infty < \bar{y}_k' < \infty$, $k = 1, 2, \ldots, n$, *then* $\hat{y} = \hat{y}_o(x)$ *yields a strong relative minimum for* $I[\hat{y}]$, *with beginning point and endpoint variable on* $x = a$ *and* $x = b$.

It is not at all clear that a field as defined in Definition 3.10.1 exists. Heuristically speaking, the natural boundary conditions at $x = a$ will essentially define the initial slopes of the field extremals, and these, in turn, will be uniquely determined by initial point and initial slope. That the natural boundary conditions at $x = b$ are also satisfied by these same field extremals will be merely accidental, if it happens at all. One speaks in this latter case of *consistent boundary conditions.* We shall formulate this idea now with greater precision:

Suppose that the extremal $\hat{y} = \hat{y}_o(x)$ is regular at the beginning point $x = a$. Then $\det|f_{y'_iy'_k}(x,\hat{y}_o(a),\hat{y}_o'(a))| > 0$ (or <0), and we can solve the natural boundary conditions at $x = a$,

$$f_{y'_k}(a,\hat{y}(a),\hat{y}'(a)) = 0,$$

for $\hat{y}'(a)$ in a neighborhood of $\hat{y}_o(a)$:

$$\hat{y}'(a) = \hat{Y}_a(a,\hat{y}(a)).$$

Similarly, if $\hat{y} = \hat{y}_o(x)$ is regular at the endpoint $x = b$, we can solve $f_{y'_k}(b,\hat{y}(b),\hat{y}'(b)) = 0$ in a neighborhood of $\hat{y}_o(b)$ for $\hat{y}'(b)$:

$$\hat{y}'(b) = \hat{Y}_b(b,\hat{y}(b)).$$

Now we can state that *the natural boundary conditions are consistent with* $I[\hat{y}]$ *if*

$$\hat{\phi}(a,\hat{y}) = \hat{Y}_a(a,\hat{y})$$

for all $\hat{y}$ *in a neighborhood of* $\hat{y}_o(a)$ *and*

$$\hat{\phi}(b,\hat{y}) = \hat{Y}_b(b,\hat{y})$$

for all $\hat{y}$ *in a neighborhood of* $\hat{y}_o(b)$, *where the solutions of* $\hat{y}' = \hat{\phi}(x,\hat{y})$ *are also solutions of the Euler-Lagrange equations of* $I[\hat{y}]$.

In order to obtain a field for the problem in n unknown functions with one endpoint fixed and the other varying on $x = a$ (or $x = b$) or for the problem with two fixed endpoints, one of the conditions 2 of Definition 3.10.1 may be omitted in the one case and both conditions may be omitted in the other case.

We leave it to the reader to convince himself that the transformation of the total variation by means of the invariance integral, when restricted to the problems mentioned above, can be carried out in such fields.

To facilitate later references, we shall formulate the definition of a Mayer field for the fixed-beginning-point, fixed-endpoint problem:

Definition 3.10.2 $\mathfrak{F} = \{(1,\hat{\phi}(x,\hat{y})) \mid (x,\hat{y}) \in \mathfrak{a}\}$ *defines a* Mayer field *in a simply connected domain* $\mathfrak{a}$ *in x,y space for the variational problem* $I[\hat{y}] \to$ *minimum, with fixed beginning point and fixed endpoint, if*

$$\hat{\phi},\hat{\phi}_x,\hat{\phi}_{y_k} \in C(\mathfrak{a}^n)$$

and if:

1. *The solutions* $\hat{y} = \hat{Y}(x,\hat{c})$ *of* $\hat{y}' = \hat{\phi}(x,\hat{y})$ *are also solutions of the Euler-Lagrange equations of* $I[\hat{y}]$.
2. $(\partial/\partial y_i)\ f_{y'_k}(x,\hat{y},\hat{\phi}(x,\hat{y})) = (\partial/\partial y_k)\ f_{y'_i}(x,\hat{y},\hat{\phi}(x,\hat{y}))$ *for all* i, $k = 1, 2, \ldots, n$.

As in the case of the simplest problem in one unknown function, we can now obtain a sufficient condition for a weak relative minimum by exploring the relationship between the excess function and the $f_{y'_i y'_k}(i, k = 1, 2, \ldots, n)$. We shall postpone this problem, however, for the time being and take it up in Chap. 7, where we shall have the appropriate terminology at our disposal. (See Prob. 7.9.8.)

PROBLEMS 3.10

*1. Show that $U[\hat{y}_o] = I[\hat{y}_o]$, where $U[\hat{y}]$ is defined in Theorem 3.10.1.
2. Impose suitable conditions so that $f_{y'_k}(x,\hat{y},\hat{y}') = 0$ yields $\hat{y}' = \hat{\phi}(x,\hat{y})$, and $(1,\hat{\phi}(x,\hat{y}))$ defines a field in the sense of Definition 3.10.1.
3. Suppose that $f = f(x,\hat{y}')$ and that $\det|f_{y'_k y'_i}| \neq 0$. Show that under suitable conditions imposed on f, the natural boundary conditions associated with this variational problem are consistent.
4. Verify the result of problem 3 for $f = xy_1' + x^2y_1'y_2' + x^3y_2'^2$, $x > 0$.

BRIEF SUMMARY

Given the variational problem

$$I[y] = \int_a^b f(x,y(x),y'(x))\,dx \to \text{minimum}, \qquad f \in C^2,$$

with fixed endpoints or variable endpoints. The extremal $y = y_o(x) \in C^1[a,b]$ is embeddable in a field that is defined by $(1,\phi(x,y))$ if the solutions of $y' = \phi(x,y)$, $\phi,\phi_x,\phi_y \in C$, are solutions of the Euler-Lagrange equation and if $y = y_o(x)$ is one of them (Sec. 3.1).

For $y = y_o(x) \in C^1[a,b]$ to yield a *strong* relative minimum for $I[y]$, it is sufficient that $y = y_o(x)$ be embeddable in a field and that

$$\mathcal{E}(x,y,y',\bar{y}') = f(x,y,\bar{y}') - f(x,y,y') + (y' - \bar{y}')f_{y'}(x,y,y') \geq 0$$

for all $(x,y) \in N_w^\delta(y_o)$ and all $y',\bar{y}'$ (Secs. 3.3 and 3.10.)

For $y = y_o(x) \in C^1[a,b]$ to be embeddable in a field, provided $f \in C^3$, it suffices that any solution of the Jacobi equation

$$\frac{d}{dx}(f_{y'y'}(x,y_o(x),y_o'(x))\eta') - \left(f_{yy}(x,y_o(x),y_o'(x)) - \frac{d}{dx}f_{y'y}(x,y_o(x),y_o'(x))\right)\eta = 0$$

that vanishes at $x = a$ does not vanish again in $(a,b]$; that is, there is no conjugate point to a in $(a,b]$ (Sec. 3.7).

$y = y_o(x) \in C^1[a,b]$ yields a *weak* relative minimum for $I[y]$ if it is embeddable in a field and if $f_{y'y'}(x,y_o(x),y_o'(x)) > 0$ for all $x \in [a,b]$ (Sec. 3.8).

For $y = y_o(x) \in C^1[a,b]$ to yield a *strong* relative minimum for $I[y]$, it is necessary that

$$\mathcal{E}(x,y_o(x),y_o'(x),\bar{y}') \geq 0$$

for all $x \in [a,b]$ and all $\bar{y}'$ (Sec. 3.9); to yield a *weak* relative minimum for $I[y]$, it is necessary that

$$f_{y'y'}(x,y_o(x),y_o'(x)) \geq 0$$

for all $x \in [a,b]$ (Sec. 3.9).

The concepts of field and excess function have obvious and straightforward generalizations for the problem in n unknown functions (Sec. 3.10).

APPENDIX

A3.11 SUFFICIENT CONDITIONS FOR THE VARIABLE-ENDPOINT PROBLEM

In Sec. 2.8, we derived necessary conditions for the minimizing function of the variational problem

$$I[y] = \int_a^b f(x,y(x),y'(x))\,dx \rightarrow \text{minimum},$$

where beginning point and endpoint were free to move on given curves $y = \psi(x)$ and $y = \varphi(x)$, $\psi,\varphi \in C^1(-\infty,\infty)$.

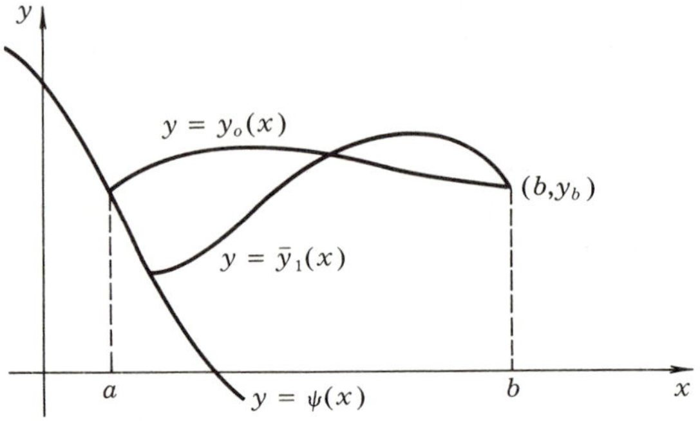

Figure 3.8

At that time, we studied the problem with a variable endpoint in great detail and then found that an analogous result has to hold if the beginning point is variable. To provide variety, we shall now study the case of a variable beginning point and a fixed endpoint and worry later about the case where beginning point and endpoint are both variable.

To summarize the content of Sec. 2.8: We found that the minimizing function $y = y_o(x)$ has to be an extremal (i.e., satisfy the Euler-Lagrange equation) and has to intersect the beginning curve $y = \psi(x)$ [terminal curve $y = \varphi(x)$] in such a manner that the transversality condition (2.8.5)

$$f_{y'}(a,y(a),y'(a))(\psi'(a) - y'(a)) + f(a,y(a),y'(a)) = 0$$

is satisfied. [For the endpoint, b is to be substituted for a and $\varphi'(b)$ for $\psi'(a)$.]

In order to arrive at a sufficient condition for a strong minimum in terms of the excess function, we need again a transformation of the total variation by means of Hilbert's invariant integral, which in turn presupposes the existence of a field and the embeddability of the extremal in a field. Since we now have to compare the value of the functional for $y = y_o(x)$ with its value for some other function $y = \bar{y}_1(x)$ that does not necessarily emanate from the same beginning point, our results of Sec. 3.3 are not immediately applicable. If $U[y]$ denotes the invariant integral, then we would like to argue again that $U[\bar{y}_1] = U[y_o]$, but this time, $y = y_o(x)$ and $y = \bar{y}_1(x)$ have different initial values. (See Fig. 3.8.)

We shall show that the argument still applies provided that a field is properly constructed. Then the invariant integral will have the same value, no matter which beginning point on $y = \psi(x)$ is chosen.

In Sec. 3.2, we showed that

$$P_y(x,\bar{y}) = Q_x(x,\bar{y})$$

if

$$P(x,\bar{y}) = f(x,\bar{y},\phi(x,\bar{y})) - \phi(x,\bar{y})f_{y'}(x,\bar{y},\phi(x,\bar{y})),$$

$$Q(x,\bar{y}) = f_{y'}(x,\bar{y},\phi(x,\bar{y})),$$

where $(1,\phi(x,y))$ defines a field $\mathfrak{F}$ in a weak neighborhood of $y = y_o(x)$. If P and Q are chosen in this manner, then there exists a function $W(x,y)$ such that

$$W_x(x,y) = P(x,y), \qquad W_y(x,y) = Q(x,y)\dagger \tag{A3.11.1}$$

and we have

$$U[\bar{y}] = \int_a^b [P(x,\bar{y}) + Q(x,\bar{y})\bar{y}']\,dx = W(b,\bar{y}(b)) - W(a,\bar{y}(a)), \tag{A3.11.2}$$

where $y = \bar{y}(x)$ emanates from $(a,\bar{y}(a))$ and terminates at $(b,\bar{y}(b))$.

Now, if we could replace $W(a,\bar{y}(a))$ by $W(a - \Delta, \psi(a - \Delta))$ for any sufficiently small $|\Delta|$ in the above expression without changing the value, we would be all set. But what does this mean? Apparently, we can make such a replacement if $W(x,y)$ remains constant along the beginning curve $y = \psi(x)$:

$$W(x,\psi(x)) = \text{constant},$$

which is true if and only if $[dW\,(x,\psi(x))]/dx = 0$.

Since

$$\frac{dW\,(x,\psi(x))}{dx} = W_x(x,\psi(x)) + W_y(x,\psi(x))\psi'(x),$$

or in view of (A3.11.1),

$$\frac{dW\,(x,\psi(x))}{dx} = P(x,\psi(x)) + Q(x,\psi(x))\psi'(x),$$

we obtain in view of the definition of P and Q,

$$\frac{d}{dx}W(x,\psi(x)) = f_{y'}(x,\psi(x),\phi(x,\psi(x)))(\psi'(x) - \phi(x,\psi(x))) + f(x,\psi(x),\phi(x,\psi(x))) = 0.$$

This is the transversality condition for each point on the curve $y = \psi(x)$. So, if the integral curves of the field $(1,\phi(x,y))$ penetrate the initial curve *transversally*, i.e., satisfy the transversality condition along $y = \psi(x)$, then the integral $U[y]$ will indeed be independent of the coordinates of the beginning point so long as the beginning point remains on $y = \psi(x)$. If we are able to construct such a field, then everything will turn out all right.

† A. E. Taylor, "Advanced Calculus," p. 440, Ginn and Company, Boston, 1955.

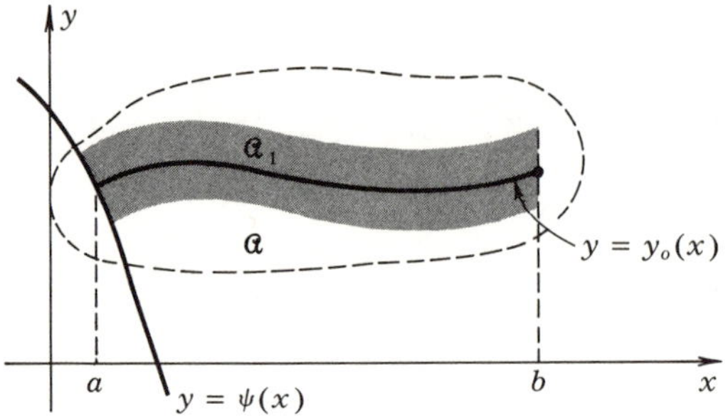

Figure 3.9

We shall concern ourselves with the field construction later. At the moment, we are more interested in the transformation of the total variation, which will be based on the following definition of a field:

Definition A3.11.1 $(1,\phi(x,y))$ *defines a* transversal field $\mathfrak{F}_T$ *for* $I[y]$ *and the initial curve* $y = \psi(x)$ *in a simply connected domain* $\mathcal{A}$ *of the* x,y *plane if* $\phi,\phi_x,\phi_y \in C(\mathcal{A})$ *and if the solutions* $y = Y(x,c)$ *of* $y' = \phi(x,y)$:

1. *Are solutions of the Euler-Lagrange equation of* $I[y]$, *and*
2. *Satisfy the transversality condition on that portion of* $y = \psi(x)$ *that lies in* $\mathcal{A}$.

Definition A3.11.2 *The extremal* $y = y_o(x)$ *which satisfies the transversality condition at* $x = a$, $y = \psi(a)$ *with respect to* $y = \psi(x)$ *is embeddable in a transversal field* $\mathfrak{F}_T$ *if:*

1. $\mathfrak{F}_T$ *is defined in a simply connected domain* $\mathcal{A}$ *that properly contains a weak neighborhood of* $y = y_o(x)$.
2. $y = y_o(x)$ *is a solution of* $y' = \phi(x,y)$.

Now, suppose we do have a transversal field $\mathfrak{F}_T$, and suppose that y_o is embeddable in $\mathfrak{F}_T$. Since $\mathcal{A}$ is supposed to properly contain a weak neighborhood of $y = y_o(x)$, there exist a $\Delta_1 > 0$ and a $\Delta_2 > 0$ such that $(x,\psi(x))$ for $a - \Delta_1 < x < a + \Delta_2$ lies entirely in $\mathcal{A}$.

Since a transversal field is also a field, by Definition 3.1.1, and since $y = y_o(x)$ is embeddable in $\mathfrak{F}_T$, we have, by (3.3.1), that

$$U[\bar{y}] = U[y_o] \qquad \text{if } \bar{y}(a) = y_o(a), \bar{y}(b) = y_o(b). \qquad \text{(A3.11.3)}$$

By (A3.11.2),

$$U[\bar{y}] = W(b,\bar{y}(b)) - W(a,\bar{y}(a)).$$

$\mathfrak{F}_T$ is a transversal field with respect to $y = \psi(x)$, that is,

$$f_{y'}(x,\psi(x),\phi(x,\psi(x)))(\psi'(x) - \phi(x,\psi(x))) + f(x,\psi(x),\phi(x,\psi(x))) = 0$$

for all $a - \Delta_1 < x < a + \Delta_2$. Hence, if $\bar{a} \in (a - \Delta_1, a + \Delta_2)$, we have

$$W(a,\bar{y}(a)) = W(a,\psi(a)) = W(\bar{a},\psi(\bar{a})),$$

and therefore, if $y = \bar{y}_1(x)$ is any curve that emanates from $y = \psi(x)$ at $x = \bar{a}$ and terminates at $x = b$, $y = y_b$, all the time remaining in $\mathfrak{A}$, we have

$$U[\bar{y}] = W(b,\bar{y}(b)) - W(a,\bar{y}(a)) = W(b,y_b) - W(\bar{a},\psi(\bar{a})) = U[\bar{y}_1].$$

Hence, by (A3.11.2),

$$U[\bar{y}_1] = U[\bar{y}] = U[y_o],$$

and we can again write [see (3.3.4)]

$$I[\bar{y}_1] - I[y_o] = I[\bar{y}_1] - U[y_o] = I[\bar{y}_1] - U[\bar{y}_1]$$
$$= \int_a^b \mathcal{E}(x,\bar{y}_1(x),\phi(x,\bar{y}_1(x)),\bar{y}_1'(x))\,dx.$$

If $\mathfrak{A}_1$ is such a subset of $\mathfrak{A}$, as indicated by the shaded region in Fig. 3.9, then we can say that $y = y_o(x)$ yields a strong minimum for the variable-beginning-point, fixed-endpoint problem, as formulated in the beginning of this section, if

$$\mathcal{E}(x,\bar{y},y',\bar{y}') \geq 0$$

for all $(x,\bar{y}) \in \mathfrak{A}_1$ and all $-\infty < y' < \infty$, $-\infty < \bar{y}' < \infty$.

Let us discuss a simple example:

$$I[y] = \int_a^3 \sqrt{1 + y'^2(x)}\,dx, \qquad \psi(x) \equiv x^2, \qquad y(3) = 0$$

(the shortest distance from the point (3,0) to the parabola $y = x^2$; see Fig. 3.10).

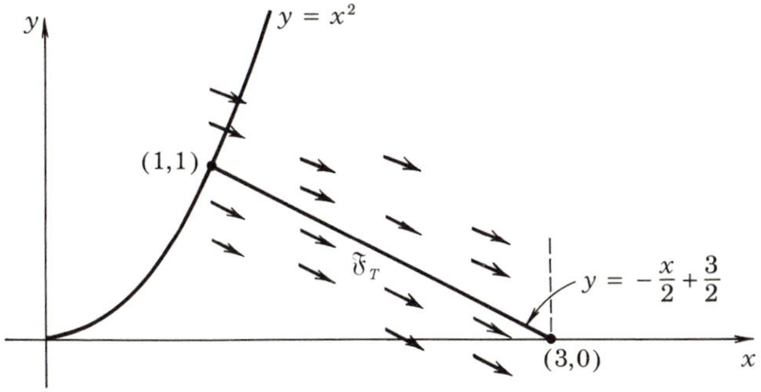

Figure 3.10

The Euler-Lagrange equation of this problem has the general solution

$$y = Ax + B.$$

Since this is a Fermat-type problem—see Sec. 2.8—transversality means orthogonality, and we have at the point (a,a^2) where the extremal penetrates the parabola $y = x^2$, which has slope $2a$ at that point, the condition

$$y'(a) = -\frac{1}{2a}.$$

We obtain then $A = -(1/2a)$, $B = (3/2a)$, and $a = 1$. Thus

$$y = -\frac{x}{2} + \frac{3}{2} \tag{A3.11.4}$$

appears to be the solution to our problem. That it is indeed the solution will follow from the fact that the function in (A3.11.4) is embeddable in a transversal field and that the excess function is positive, as we now show:

The transversal field will be generated by all straight lines that are orthogonal to $y = x^2$, $x > 0$:

$$y = Y(x,c) \equiv -\frac{x}{2c} + \frac{1}{2} + c^2,$$

where

$$y_0(x) \equiv Y(x,1).$$

Finally,

$$\mathcal{E}(x,\bar{y},y',\bar{y}') = \sqrt{1 + \bar{y}'^2} - \frac{1 + y'\bar{y}'}{\sqrt{1 + y'^2}} \geq 0 \qquad \text{for all } y', \bar{y}'.$$

(See Prob. A3.11.1.)

Let us now turn our attention to the case where the beginning point varies on a curve $y = \psi(x)$ and the endpoint varies on a curve $y = \varphi(x)$. It seems reasonable to expect that any two solutions of the Euler-Lagrange equation which satisfy the transversality conditions at $y = \psi(x)$ as well as at $y = \varphi(x)$ would yield the same value for the functional $I[y]$. That this is indeed the case under certain circumstances will be shown now.

Let $y = Y(x,c)$ represent a one-parameter family of solutions of the Euler-Lagrange equation that defines a transversal field

$$(1,Y'(x,C(x,y)))$$

with respect to $y = \psi(x)$ and $y = \varphi(x)$; that is, every solution $y = Y(x,c)$ satisfies the transversality condition on $y = \varphi(x)$ as well as on $y = \psi(x)$. Let $y = y_1(x) \equiv Y(x,c_1)$, where $y_1(a_1) = \psi(a_1)$, $y_1(b_1) = \varphi(b_1)$, and let $y = y_2(x) \equiv Y(x,c_2)$, where $y_2(a_2) = \psi(a_2)$ and $y_2(b_2) = \varphi(b_2)$. (See Fig. 3.11.)

Since

$$W(x,\psi(x)) = \text{constant}, \qquad W(x,\varphi(x)) = \text{constant},$$

Figure 3.11

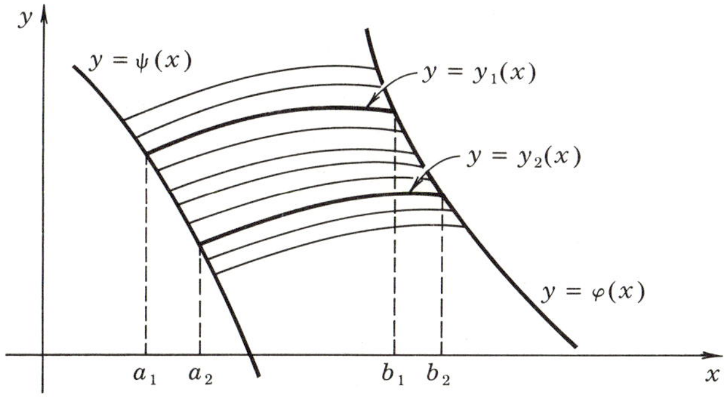

we have

$$W(b_1,\varphi(b_1)) = W(b_2,\varphi(b_2))$$
$$W(a_1,\psi(a_1)) = W(a_2,\psi(a_2))$$

and hence,

$$\begin{aligned} I[y_1] = U[y_1] &= W(b_1,\varphi(b_1)) - W(a_1,\psi(a_1)) \\ &= W(b_2,\varphi(b_2)) - W(a_2,\psi(a_2)) = U[y_2] = I[y_2] \end{aligned}$$

Therefore we have: If the one-parameter family of extremals $y = Y(x,c)$, $c_1 < c < c_2$, generates a field, in the sense of Lemma 3.1, and if this field is transversally cut by the two curves $y = \psi(x)$ and $y = \varphi(x)$, then

$$I[Y] = \int_{a(c)}^{b(c)} f(x,Y(x,c),Y'(x,c))\,dx$$

is independent of c. Here, $a(c)$ and $b(c)$ are the solutions of $Y(x,c) = \psi(x)$ and $Y(x,c) = \varphi(x)$, respectively.

The following example and, even more convincingly, the investigations of the next section will show that, in general, it is unreasonable to expect a field that is transversal to one given curve, say $y = \psi(x)$, to be also transversal to the other given curve $y = \varphi(x)$.

Let

$$I[y] = \int_a^b \sqrt{1 + y'^2(x)}\,dx, \qquad \psi(x) \equiv x, \qquad \varphi(x) \equiv \sqrt{x - 1}.$$

The solutions of the Euler-Lagrange equation are

$$y = Ax + B,$$

and the only field that is transversal to $y = \psi(x) \equiv x$ is given by

$$(1,\phi(x,y)) = (1,-1).$$

Among the integral curves of this field there is *only one*, namely,

$$y = -x + \tfrac{7}{4},$$

that is also transversal to the terminal curve $y = \sqrt{x-1}$. (It terminates at $x = \frac{5}{4}$, $y = \frac{1}{2}$.)

PROBLEMS A3.11

*1. Show that

$$\sqrt{1+\bar{y}'^2} - \frac{1+y'\bar{y}'}{\sqrt{1+y'^2}} \geq 0$$

for all $y',\bar{y}'$.

2. Verify that $I[Y(x,c)] = \int_{a(c)}^{b(c)} f(x,Y(x,c),Y'(x,c))\,dx$ is independent of c, where the family $y = Y(x,c)$ is defined by a field and is transversally cut by $y = \psi(x)$, $y = \varphi(x)$, for the example

$$I[y] = \int_a^b \sqrt{1+y'^2(x)}\,dx \to \text{minimum},$$

$y(a) = a$, $y(b) = b - 1$.

A3.12 EXISTENCE OF A TRANSVERSAL FIELD

In this section, we shall sketch an argument that will lead to the construction of a transversal field with respect to a (beginning) curve $y = \psi(x)$.

We assume that $y = y_o(x) \in C^1[a,b]$ is the solution of the problem

$$I[y] = \int_a^b f(x,y(x),y'(x))\,dx \to \text{minimum},$$

where the beginning point is allowed to vary on $y = \psi(x)$ while the end-point is considered as fixed. As the above notation suggests, the penetration of $y = \psi(x)$ by $y = y_o(x)$ takes place at $x = a$.

By necessity, $y = y_o(x)$ satisfies the transversality condition at $x = a$:

$$f_{y'}(a,y_o(a),y_o'(a))(\psi'(a) - y_o'(a)) + f(a,y_o(a),y_o'(a)) = 0. \quad \text{(A3.12.1)}$$

This condition determines the initial slope $y_o'(a)$ in terms of the slope of $y = \psi(x)$ at $x = a$.

From here on, x will play a dual role, that of the independent variable in the representation of $y = y_o(x)$ and that of the independent variable in the representation of $y = \psi(x)$. To avoid unnecessary confusion, we shall consider the initial curve from now on as represented in parameter form in terms of a parameter c:

$$x = c$$
$$y = \psi(c).$$

In order to construct a transversal field, we need the initial slopes of extremals to both sides of $y = y_o(x)$ which satisfy the transversality condition. For that purpose, we shall have to solve (A3.12.1) for y' in terms of ψ'.

We observe that if $f_{y'y'}(a,y_o(a),y_o'(a)) \neq 0$, if $y_o'(a) \neq \psi'(a)$, and finally, if $f \in C^2$ in some suitably chosen domain, then

$$f_{y'}(c,\psi,y')(\psi' - y') + f(c,\psi,y') \in C^1,$$

$$f_{y'}(a,\psi(a),y'(a))(\psi'(a) - y'(a)) + f(a,\psi(a),y'(a)) = 0, \quad y'(a) = y_o'(a),$$

$$\frac{\partial}{\partial y'}[f_{y'}(\psi' - y') + f]_{c=a} = f_{y'y'}(a,y_o(a),y_o'(a))(\psi'(a) - y_o'(a)) \neq 0.$$

Hence, by the theorem on implicit functions, we can solve

$$f_{y'}(c,\psi,y')(\psi' - y') + f(c,\psi,y') = 0 \tag{A3.12.2}$$

in a neighborhood of the beginning point of $y = y_o(x)$ uniquely for y',

$$y' = F(c,\psi,\psi'),$$

such that (A3.12.2) is identically satisfied and $F \in C^1$.

Since $\psi = \psi(c)$, we obtain y' as a function of c:

$$y'(c) = F(c,\psi(c),\psi'(c)) \equiv G(c)$$

for all c in some neighborhood of $c = a$. Thus it would appear that the construction of a transversal field amounts to a solution of the one-parameter family of initial-value problems

$$f_y - \frac{d}{dx} f_{y'} = 0$$

$$y(c) = \psi(c)$$
$$y'(c) = G(c)$$

for all c in some neighborhood of $c = a$.

We have seen in Sec. 2.5 that under suitable conditions on f, the Euler-Lagrange equation is equivalent to the following system of first-order equations:

$$\begin{aligned} y' &= \Phi(x,y,p) \\ p' &= \Psi(x,y,p), \end{aligned} \tag{A3.12.3}$$

$\Phi,\Psi \in C^1$, where $p = f_{y'}(x,y,y')$.

The initial conditions will be transformed into

$$y(c) = \psi(c)$$
$$p(c) = f_{y'}(c,\psi(c),G(c)).$$

These initial conditions will, by the standard theorems on the existence and uniqueness of the solutions of systems of differential equations† and in view of $\Phi,\Psi \in C^1$, determine a unique solution for each c in some neighborhood of $c = a$:

$$y = Y(x,c)$$
$$p = P(x,c),$$

where $Y(x,a) = y_o(x)$.

In order to show that $y = Y(x,c)$ generates a field—at least locally—we have to demonstrate that $Y_c(x,c) \neq 0$ in some neighborhood of $c = a$, $x = a$. (See also Sec. 3.1.)

The system of differential equations (A3.12.3) together with the initial conditions is equivalent to the system of integral equations‡

$$Y(x,c) = \psi(c) + \int_c^x \Phi(t,Y(t,c),P(t,c))\,dt$$

$$P(x,c) = f_{y'}(c,\psi(c),G(c)) + \int_c^x \Psi(t,Y(t,c),P(t,c))\,dt.$$

We obtain from the first equation,

$$Y_c(x,c) = \psi'(c) + \int_c^x [\Phi_y Y_c + \Phi_p P_c]\,dt$$
$$- \Phi(c,Y(c,c),P(c,c)).$$

Hence, $Y_c(x,c)\Big|_{\substack{x=a\\c=a}} = \psi'(a) - \Phi(a,Y(a,a),P(a,a)).$

By (A3.12.3), $\Phi(a,Y(a,a),P(a,a)) = y_o'(a)$, and we have

$$Y_c(x,c)\Big|_{\substack{x=a\\c=a}} = \psi'(a) - y_o'(a) \neq 0.$$

We now see that $y = Y(x,c)$ can be solved uniquely for c in some neighborhood of $x = a$, $c = a$. Thus the existence of a transversal field is guaranteed—at least locally, that is, in a neighborhood of the beginning point of the extremal $y = y_o(x)$.

Some of the pitfalls that are encountered in the construction of transversal fields will be discussed in the next section.

† G. Birkhoff and G. C. Rota, "Ordinary Differential Equations," pp. 99ff, Ginn and Company, Boston, 1962.
‡ *Ibid.*, p. 111.

PROBLEMS A3.12

1. Given the variable-beginning-point problem

$$\int_a^{\pi/2} (y'^2(x) - y^2(x))\,dx \to \text{minimum},$$

$\psi(x) \equiv x$, $y(\pi/2) = 0$. Construct a transversal field and investigate how far it can be extended. (What happens outside $-1 < x < 1$ on the beginning curve?) Show that $y = 0$ yields a strong relative minimum.
2. Explore the possibility of extending the local transversal field that was constructed in this section up to the endpoint $x = b$.
3. Formulate the appropriate conditions on f, $y = y_o(x)$, $\psi(c)$, under which the implicit-function theorem may be applied to the solution of (A3.12.2) for y'. Describe all the domains involved.

A3.13 FOCAL POINTS IN TRANSVERSAL FIELDS

We shall now discuss some of the phenomena that might occur when extending a transversal field, the existence of which near the beginning point was guaranteed essentially by $f_{y'y'} > 0$ and $\psi'(a) \neq y_o'(a)$, as we saw in the preceding section.

We shall assume for the subsequent discussion that

$$f_{y'y'} > 0 \text{ in } \mathfrak{R} \qquad \text{and} \qquad f \in C^3(\mathfrak{R}),$$

where $\mathfrak{R}$ is a domain of the (x,y,y') space that consists of a simply connected pointset $\mathfrak{A}$ in the x,y plane and all $y' \in (-\infty,\infty)$. We shall also assume that $y = y_o(x) \in C^2[a,b]$ is a member of a family of extremals

$$y = Y(x,c)$$

for $c = c_o$, where $Y \in C^2((A,B) \times (c_o - \Delta, c_o + \Delta))$, $A < a < b < B$.

Then, when $\eta = \eta(x,c)$ is considered a function of c,

$$\eta = \eta(x,c) \equiv Y_c(x,c) \in C^1(c_o - \Delta, c_o + \Delta),$$

and when it is considered a function of x,

$$\eta = \eta(x,c) \equiv Y_c(x,c) \in C^2(A,B).$$

(See Prob. 3.5.2.) Hereby, $\eta = \eta(x,c)$ is that solution of the Jacobi equation that corresponds to the extremal $y = Y(x,c)$. In particular,

$$\eta = \eta(x,c_o) \equiv \eta(x)$$

is the solution of the Jacobi equation that corresponds to the extremal $y = y_o(x)$, which is to be investigated.

Suppose now that $\eta(b) = \eta(b,c_o) = 0$. This could be caused by all extremals of $y = Y(x,c)$ passing through the same endpoint (b,y_b), because then $Y(b,c) = y_b$ for all c, and hence $Y_c(b,c) = \eta(b,c) = 0$. We shall investigate this case first. We expect that all extremals from $y = Y(x,c)$ will yield the same value for $I[y]$, and this is indeed the case. To show this, we have to find first the x coordinate of the point at which the extremals $y = Y(x,c)$ emanate from the beginning curve $y = \psi(x) \in C^1(-\infty,\infty)$. At that point, we must have

$$Y(x,c) = \psi(x).$$

We shall have to make the restricting assumption that

$$Y'(a,c_o) \neq \psi'(a), \tag{A3.13.1}$$

that is, that the extremal $y = y_o(x)$ does not leave the initial curve tangentially. Then we can solve

$$Y(x,c) - \psi(x) = 0$$

for x in a δ_1 neighborhood of $c = c_o$ and obtain

$$x = X_1(c) \in C^1(N^{\delta_1}(c_o)).$$

Hence $y = Y(x,c)$ will leave the initial curve $y = \psi(x)$ at

$$\left.\begin{aligned} x &= X_1(c) \\ y &= Y(X_1(c),c) = Y_1(c) \end{aligned}\right\} c \in N^{\delta_1}(c_o). \tag{A3.13.2}$$

Thus we obtain for $I[y]$ taken along one of the extremals $y = Y(x,c)$ between the initial curve $y = \psi(x)$ and the endpoint (b,y_b),

$$I[Y(x,c)] = I(c) = \int_{X_1(c)}^{b} f(x,Y(x,c),Y'(x,c))\,dx,$$

which is a function of c, and consequently,

$$I'(c) = \int_{X_1(c)}^{b} (f_y Y_c + f_{y'} Y_c')\,dx - f(X_1(c),Y_1(c),Y'(X_1(c),c))X_1'(c).$$

We obtain $X_1'(c)$ by implicit differentiation from

$$Y(X_1(c),c) = \psi(X_1(c)).$$

We have

$$Y'(X_1(c),c)X_1'(c) + Y_c(X_1(c),c) = \psi'(X_1(c))X_1'(c),$$

and hence

$$X_1'(c) = \frac{Y_c(X_1(c),c)}{\psi'(X_1(c)) - Y'(X_1(c),c)}. \tag{A3.13.3}$$

If we also apply integration by parts, as in Sec. 2.4, to the integral in the expression for $I'(c)$—note that this is permissible because we have $f \in C^3$ and $Y(x,c) \in C^2$—we obtain, in view of the fact that $y = Y(x,c)$

is a solution of the Euler-Lagrange equation and because of $Y_c(b,c) = 0$,

$$I'(c) = -\left(f_{y'} + f\frac{1}{\psi' - Y'}\right) Y_c \Big|_{\substack{x=X_1(c)\\ y=Y_1(c)}} = 0$$

because $y = Y(x,c)$, by hypothesis, satisfies the transversality condition (2.8.5) on $y = \psi(x)$.

So we see that the value of $I(c)$ is independent of c, that is, it does not depend on which curve from the family $y = Y(x,c)$ we select, as long as $y = Y(x,c)$ is transversal to $y = \psi(x)$ but not tangential and all curves $y = Y(x,c)$ terminate at the same point (b,y_b). This is also true, as the reader can easily convince himself, if all curves from the family $y = Y(x,c)$ emanate from the same point (a,y_a) and terminate at the same point (b,y_b). (See Prob. A3.13.1 and the case $a = 0$, $b = \pi$ in Sec. 3.4.)

Next, we shall consider a more interesting case, where $\eta(b,c_o) = 0$ but where the extremals $y = Y(x,c)$ do not pass through one point at $x = b$.

Since the zeros of the solutions of the Jacobi equation are simple (see Lemma 3.6.1), we have $\eta'(b,c_o) \neq 0$, and we may solve

$$\eta(x,c) = 0$$

in a δ_2 neighborhood of c_o for x:

$$x = X_2(c) \in C^1(N^{\delta_2}(c_o)).$$

The corresponding y value on $y = Y(x,c)$ we obtain from

$$y = Y(X_2(c),c) = Y_2(c).$$

Hence,

$$\left.\begin{aligned} x &= X_2(c) \\ y &= Y_2(c) \end{aligned}\right\} \quad c \in N^{\delta_2}(c_o) \tag{A3.13.4}$$

is the locus of all points for which $\eta(x,c) = Y_c(x,c) = 0$. We call these points *focal points* to $y = \psi(x)$ on $y = Y(x,c)$.

In order to find the slope of the locus of focal points, we need

$$Y_2'(c) = Y'(X_2(c),c)X_2'(c) + Y_c(X_2(c),c) = Y'(X_2(c),c)X_2'(c).$$

[Note that $Y_c(X_2(c),c) = 0$ by the definition of $X_2(c)$.] If $X_2'(c) \equiv 0$, then $Y_2'(c) \equiv 0$ also, and we have $X_2(c) =$ constant, $Y_2(c) =$ constant; that is, all extremals pass through the same point (b,y_b), and we have the case which we discussed at the beginning of this section. If $X_2'(c_o) = 0$, then $Y_2'(c_o) = 0$ and the locus of focal points has a singular point at (b,y_b).

We shall only discuss the case where $X_2'(c_o) \neq 0$, which is true if and only if $\eta_c(X_2(c_o),c_o) \neq 0$. (See Prob. A3.13.5.) Then $X_2'(c) \neq 0$ for $c \in N^\delta(c_o)$ for some $\delta > 0$, which we choose to be less than δ_1 and less than δ_2.

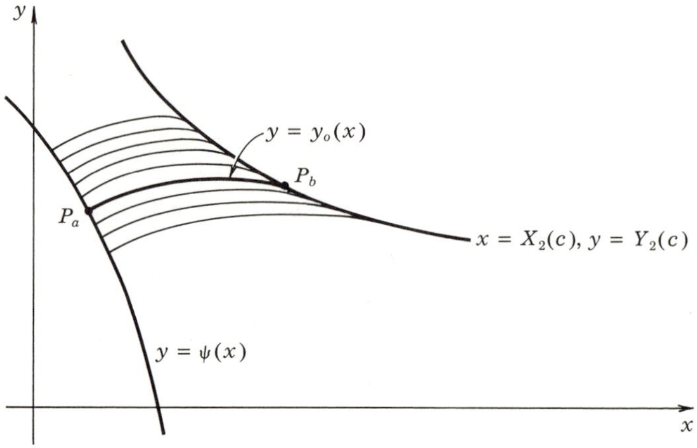

Figure 3.12

We obtain for the slope of the locus of focal points

$$\frac{dy}{dx} = \frac{Y_2'(c)}{X_2'(c)} = Y'(X_2(c),c), \qquad c \in N^\delta(c_o),$$

and we see that *the locus of focal points to* $y = \psi(x)$ *on* $y = Y(x,c)$ *is an envelope of* $y = Y(x,c)$. (See Fig. 3.12.) This is also true if all extremals $y = Y(x,c)$ emanate from the same point (a,y_a). Then $Y_c(a,c) = 0$ and the locus of focal points becomes the locus of conjugate points to (a,y_a) on $y = Y(x,c)$, and we have that the locus of conjugate points to (a,y_a) on $y = Y(x,c)$ is an envelope of $y = Y(x,c)$. (See Prob. A3.13.2.) Conversely, if $y = Y(x,c)$ has an envelope, then $Y_c(x,c) = 0$ for all (x,y) on the envelope. (See Prob. A3.13.8.)

The value of $I[y]$ between the beginning curve $y = \psi(x)$ and the locus of focal points is a function of c:

$$I(c) = \int_{X_1(c)}^{X_2(c)} f(x,Y(x,c),Y'(x,c))\,dx.$$

If we take this integral for two different curves $y = Y(x,c_1)$ and $y = Y(x,c_2)$ from $y = Y(x,c)$, then we shall see that the difference of the values of the integral is equal to the integral taken for the envelope (locus of focal points) between the endpoints of $y = Y(x,c_1)$ and $y = Y(x,c_2)$ on the envelope.

In order to find $I(c_2) - I(c_1)$, we first take $I'(c)$ and then obtain the desired result from

$$I(c_2) - I(c_1) = \int_{c_1}^{c_2} I'(c)\,dc. \qquad \text{(A3.13.5)}$$

We have

$$I'(c) = \int_{X_1(c)}^{X_2(c)} (f_y Y_c + f_{y'} Y'_c)\, dx + f\left(X_2(c), Y_2(c), \frac{Y'_2(c)}{X'_2(c)}\right) X'_2(c)$$
$$- f(X_1(c), Y_1(c), Y'(X_1(c),c)) X'_1(c).$$

After having carried out Lagrange's integration by parts, we see that the remaining integral vanishes because $y = Y(x,c)$ satisfies the Euler-Lagrange equation. One integrated term together with the last term in the above expression vanishes as before on account of the transversality condition (2.8.5) that has to be satisfied on $y = \psi(x)$; the other integrated term vanishes because of $Y_c(X_2(c),c) = 0$, and all that remains is

$$I'(c) = f\left(X_2(c), Y_2(c), \frac{Y'_2(c)}{X'_2(c)}\right) X'_2(c).$$

Hence we have, from (A3.13.5),

$$I(c_2) - I(c_1) = \int_{c_1}^{c_2} f\left(X_2(c), Y_2(c), \frac{Y'_2(c)}{X'_2(c)}\right) X'_2(c)\, dc$$
$$= \int_{X_2(c_1)}^{X_2(c_2)} f\left(X_2, Y_2, \frac{dY_2}{dX_2}\right) dX_2.$$

Therefore, we can state with reference to Fig. 3.13 that

$$I_{P_2P'_2} - I_{P_1P'_1} = I_{P'_1P'_2}, \qquad \text{(A3.13.6)}$$

where $I_{P_iP'_i} = \int_{X_1(c_i)}^{X_2(c_i)} f(x, Y(x,c_i), Y'(x,c_i))\, dx, \qquad i = 1, 2,$

and $I_{P'_1P'_2} = \int_{X_2(c_1)}^{X_2(c_2)} f\left(X_2, Y_2, \frac{dY_2}{dX_2}\right) dX_2,$

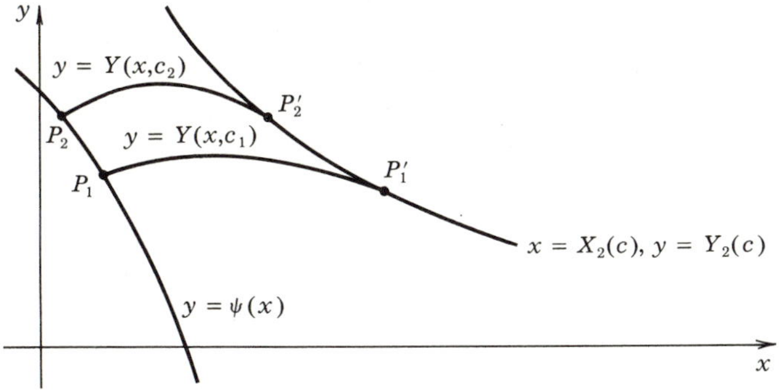

Figure 3.13

whereby the latter integral is to be taken along the envelope $x = X_2(c)$, $y = Y_2(c)$. Equation (A3.13.6) is the content of *Kneser's* so-called *envelope theorem.*

If we apply the formula (A3.13.6) to $y = y_o(x) \equiv Y(x,c_o)$ and $y = Y(x,c_2)$, then we obtain with the notation

$$\bar{y}(x) = \begin{cases} Y(x,c_2),\ X_1(c_2) \leq x \leq X_2(c_2) \\ Y_o(x),\ X_2(c_2) \leq x \leq X_2(c_o), \end{cases}$$

where $y = Y_o(x)$ is the explicit representation of the envelope (A3.13.4) (see Prob. A3.13.3),

$$\int_{X_1(c_2)}^{b} f(x,\bar{y}(x),\bar{y}'(x))\,dx = \int_a^b f(x,y_o(x),y_o'(x))\,dx, \qquad \text{(A3.13.7)}$$

that is, the curve $y = \bar{y}(x)$, which consists partly of an extremal and partly of a portion of the envelope, yields the same value for $I[y]$ between $y = \psi(x)$ and (b,y_b) as does $y = y_o(x)$.

We observe that we can choose c_2 in $y = Y(x,c_2)$ sufficiently close to c_o so that $\bar{y} \in N^{\delta/2}(y_o)$. We also observe that $\bar{y}$ cannot possibly yield a minimum because it is not an extremal. (It is pieced together from a portion of an extremal and a portion of the envelope. The envelope cannot be an extremal because if it were, we would have a contradiction to Theorem 2.5.2, according to which the extremals of a regular variational problem are uniquely determined by point and slope.) Hence there exists a function $y = \eta_o(x) \in C_s^1[a,b]$, $\eta_o \in N^{\delta/2}(\bar{y})$ such that $I[\eta_o] < I[\bar{y}] = I[y_o]$. Since $\eta_o \in N^{\delta}(y_o)$ and $I[\eta_o] < I[y_o]$, we can find, by a process that was explained in the proof of Corollary 2 to Theorem 2.10, a smooth function $y = y(x) \in C^1[a,b]$, $y \in N^{\delta}(y_o)$ such that $I[y] < I[y_o]$, and we see that $y = y_o(x)$ most certainly ceases to yield a weak relative minimum if it is continued up to the focal point.

This is also true if all extremals $y = Y(x,c)$ emanate from the same point (a,y_a). So we can say that if $X_2'(c_o) \neq 0$, *the extremal* $y = y_o(x)$ *ceases to yield a weak relative minimum if it is continued up to the focal point or the conjugate point,* depending on whether we have a variable-beginning-point problem or a fixed-beginning-point problem. (See Prob. A3.13.9 and Sec. 7.3.)

As an example, let us discuss the problem of finding the shortest distance from a point (b,y_b), the coordinates of which are to be specified later, to the parabola $y = x^2$:

$$I[y] = \int_a^b \sqrt{1 + y'^2}\,dx \to \text{minimum}$$

$$y(a) = a^2, \qquad y(b) = y_b.$$

This is a Fermat-type problem, and hence transversality means orthogonality (see Sec. 2.8). The extremals are straight lines, and

$$y = Y(x,c) \equiv -\frac{x}{2c} + \frac{1}{2} + c^2$$

represents a one-parameter family of extremals that are orthogonal (transversal) to $y = x^2$. (These lines intersect $y = x^2$ at $x = c$, $y = c^2$, where the parabola has the slope $2c$.)

We obtain the envelope (locus of focal points) (A3.13.4) by solving

$$Y_c(x,c) \equiv \frac{x}{2c^2} + 2c = 0$$

for x in terms of c and substituting the result into $y = Y(x,c)$:

$$x = X_2(c) \equiv -4c^3$$
$$y = Y_2(c) \equiv 3c^2 + \tfrac{1}{2}.$$

We note that $x = 0$, $y = \frac{1}{2}$ (for $c = 0$) is a singular point of the envelope because there, $X_2'(c) \equiv -12c^2 = 0$, $Y_2'(c) \equiv 6c = 0$. Elimination of c leads to the following explicit representation of the envelope:

$$y = \tfrac{3}{4}(2x)^{2/3} + \tfrac{1}{2},$$

which we recognize now as *Neil's parabola.* (See Fig. 3.14.)

Let us consider the endpoint

$$(b,y_b) = (\tfrac{1}{2},\tfrac{5}{4}),$$

which lies on the envelope. There are, of course, two extremals through this point, one that leads to that portion of the parabola that lies in the right half-plane and one that touches the envelope at this point and leads to that branch of the parabola that lies in the left half-plane. (The extensions of these straight lines in the opposite direction also intersect the parabola but not transversally, i.e., at right angles, and can be ignored.)

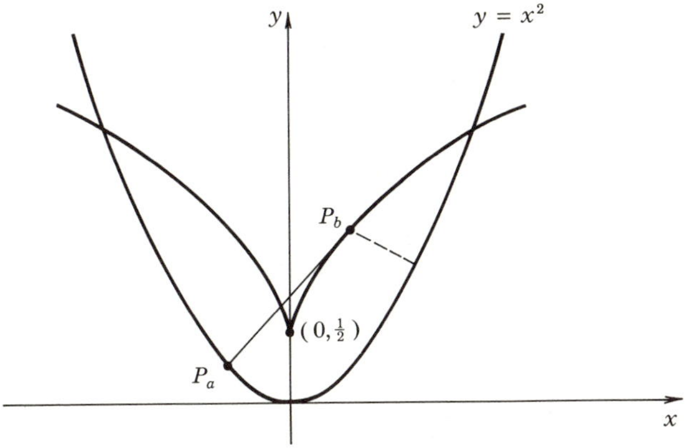

Figure 3.14

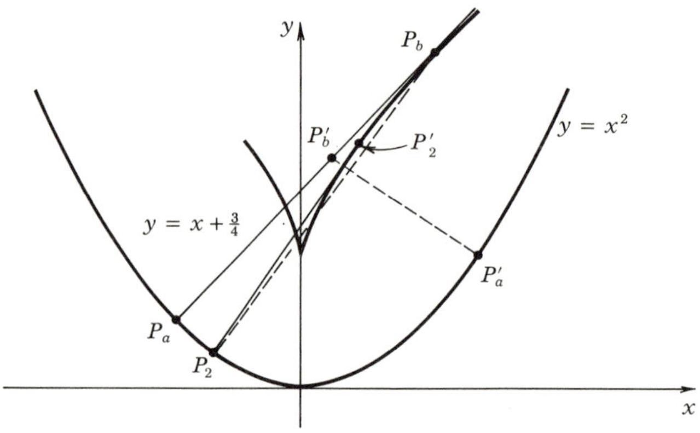

Figure 3.15

We shall consider the extremal from P_a to P_b that is indicated in Fig. 3.14 by a solid line. We obtain its equation by finding that value of c for which $-(1/2c)$ is the slope of the envelope at $P_b(\frac{1}{2},\frac{5}{4})$. Since

$$y'(x) = (2x)^{-(1/3)}$$

is the slope of the envelope $(x \neq 0)$, we obtain for its slope at $x = \frac{1}{2}$,

$$y'(\tfrac{1}{2}) = 1,$$

and hence $c = -\frac{1}{2}$. The corresponding extremal $y = x + \frac{3}{4}$ will intersect the parabola $y = x^2$ at $x = -\frac{1}{2}$, $y = \frac{1}{4}$, and we obtain for $I[y]$ taken along this extremal

$$I[y] = \int_{-1/2}^{1/2} \sqrt{1 + y'^2}\, dx = \int_{-1/2}^{1/2} \sqrt{2}\, dx = \sqrt{2}.$$

From (A3.13.7) and with reference to Fig. 3.15, we have also

$$I_{P_2P'_2} + I_{P'_2P_b} = \sqrt{2},$$

where the latter integral is to be taken along the envelope and where P_2' may be chosen as close to P_b as we please. But clearly,

$$I_{P_2P'_2} + I_{P'_2P_b} > d(P_2,P_b),$$

where $d(P_2,P_b)$ denotes the shortest distance from P_2 to P_b.

So we see that the extremal between $(-\frac{1}{2},\frac{1}{4})$ and $(\frac{1}{2},\frac{5}{4})$ cannot yield a weak *relative minimum*. (See also Prob. A3.13.6.) As a matter of fact, there is no weak minimum in a strong neighborhood of $\overline{P_aP_b}$, as one can easily convince oneself.

Between P_a and any point P_b' on $y = x + \frac{3}{4}$ to the left of P_b but in the right half-plane, the line segment $\overline{P_aP_b'}$ does yield a *strong relative minimum* (embeddable in a transversal field, excess function nonnegative) but, of course, *not* an *absolute minimum* because the absolute minimum is rendered by the line segment $\overline{P_a'P_b'}$ (see Fig. 3.15).

Obviously, for any point in the right half-plane and inside the parabola, the absolute minimum is rendered by that line from $y = Y(x,c)$ that leads to the portion of the parabola lying in the right half-plane, and for any point in the left half-plane by the one that leads to the portion of the parabola in the left half-plane. If $P_b(0,y_b)$ and if $y_b > \frac{1}{2}$, then there are two extremals, one going to the left and one going to the right, that both yield the same absolute minimum; and if $0 < y_b \leq \frac{1}{2}$, then the y axis between 0 and y_b yields the absolute minimum. (See also Prob. A3.13.7.)

PROBLEMS A3.13

*1. Show that $I[Y(x,c)]$, taken between a and b, is independent of c if all the extremals $y = Y(x,c)$ pass through the same beginning point (a,y_a) and the same endpoint (b,y_b).

*2. Make suitable assumptions and show that the locus of conjugate points to (a,y_a) is an envelope to $y = Y(x,c)$.

*3. State suitable conditions under which the envelope (A3.13.4) can be represented in the form $y = Y(x)$.

4. Show that (A3.13.6) is also valid if all extremals $y = Y(x,c)$ emanate from the same point (a,y_a).

*5. Show that $X_2'(c_o) \neq 0$ if and only if $\eta_c(X_2(c_o),c_o) \neq 0$.

*6. Consider the example of the shortest distance from the point $P_b(\frac{1}{2},\frac{5}{4})$ to the parabola $y = x^2$, discussed at the end of this section. Take the point P_2' with the x coordinate $\frac{1}{2}(1 - \varepsilon)$ on the envelope and find the intersection point P_2 of the extremal that touches the envelope at P_2' with the parabola $y = x^2$. Find $d(P_2,P_b)$, and show that $d(P_2,P_b) < \sqrt{2}$. (See also Fig. 3.15.)

*7. Show that the shortest distance from $(0,y_b)$, $y_b > \frac{1}{2}$, to the parabola $y = x^2$ is rendered by two extremals from the family

$$y = -\frac{x}{2c} + \frac{1}{2} + c^2,$$

one intersecting the parabola in the right half-plane and the other symmetric to the first one with respect to the y axis. If $0 < y_b \leq \frac{1}{2}$, then the minimum is rendered by the y axis itself between 0 and y_b.

*8. Suppose that the one-parameter family of curves $y = Y(x,c)$ has an envelope; i.e., that there is a curve $x = X_o(c)$, $y = Y_o(c)$ such that $Y(X_o(c),c) = Y_o(c)$, $Y'(X_o(c),c) = Y_o'(c)/X_o'(c)$. Impose suitable conditions and show that $[\partial Y(x,c)/\partial c]|_{x=X_o(c)} = 0$.

*9. Assume that $y_o(x) \equiv Y(x,c_o)$ and $\eta_c(x,c_o) \neq 0$. Show that the extremal $y = y_o(x)$ ceases to yield a weak relative minimum between the beginning point and the point that is conjugate to the beginning point.

10. Does the theorem in problem 9 apply to the variational problem with $f = y'^2 - y^2$, $(a,y_a) = (0,0)$?

11. Find the envelope of the one-parameter family of curves

$$y = cx + \frac{1 + c^2}{4} x^2.$$

12. Show that the extremal of

$$\int_0^1 \sqrt{y(x) + 1}\sqrt{1 + y'^2(x)}\, dx \to \text{minimum}$$

$y(0) = 0$, $y(1) = -\frac{3}{4}$, does not yield a weak relative minimum. (See also problem 11.)

A3.14 FIELD, INVARIANT INTEGRAL, AND EXCESS FUNCTION FOR THE PROBLEM IN TWO INDEPENDENT VARIABLES

We shall briefly sketch a method by which a sufficient condition for a strong relative minimum can be obtained for the problem in two independent variables formulated in Sec. A2.17.

The problem is to find a function $u = u(x,y) \in C^2[\mathcal{a}]$ such that

$$\iint_{\mathcal{a}} f(x,y,u,u_x,u_y)\, dx\, dy \to \text{minimum}$$

under the boundary condition

$$u(x(t),y(t)) = z(t), \qquad t \in [a,b],$$

whereby $x = x(t), y = y(t)$ is the parameter representation of the boundary curve J of $\mathcal{a}$. We retain all the assumptions on f, $\mathcal{a}$, J, z that we imposed on the problem in Sec. A2.17 for the purpose of deriving the Euler-Lagrange equation, with the exception that we require now that f and its partial derivatives of second order be continuous for *all* u_x, u_y.

We saw in earlier sections of this chapter that the embeddability of an extremal in a field is a prerequisite for the application of the Weierstrass test. The availability of universal existence and uniqueness theorems for the solutions of systems of ordinary first-order differential equations was essential to any attempts at field construction or any proof of field existence. Such universal theorems are not available in the theory of partial differential equations, and hence certain difficulties arise at the very outset of any discussion of sufficient conditions for a strong relative minimum of double integrals (and multiple integrals in general).

A number of theories have been proposed and developed to various extents to overcome these difficulties. We shall outline here the basic idea of a field theory that was proposed by H. Weyl.†

For any two functions

$$P = P(x,y,u), Q = Q(x,y,u) \in C^1(N^\delta(u_0))$$

for some $\delta > 0$, where $u = u_0(x,y)$ is the extremal surface that is to be investigated, the integral

$$U[u] = \iint_{\mathcal{a}} \left[\frac{\partial P(x,y,u(x,y))}{\partial x} + \frac{\partial Q(x,y,u(x,y))}{\partial y}\right] dx\, dy \qquad \text{(A3.14.1)}$$

depends only on the values $u(x,y)$ on the boundary J of $\mathcal{a}$ and not on the

† H. Weyl, Geodesic Fields in the Calculus of Variations for Multiple Integrals, *Ann. Math.*, ser. 2, vol. 36, pp. 607–629, 1935.

values $u(x,y)$ in the interior of $\mathfrak{A}$ because, by Green's theorem,

$$\iint\limits_{\mathfrak{A}} \left[\frac{\partial P}{\partial x} + \frac{\partial Q}{\partial y}\right] dx\,dy = \int_J P\,dy - Q\,dx$$

$$= \int_a^b \left[P(x(t),y(t),u(x(t),y(t)))\,\frac{dy}{dt} - Q(x(t),y(t),u(x(t),y(t)))\,\frac{dx}{dt}\right] dt.$$

We again call $U[u]$ the *invariant integral* because it is independent of the choice of $u = u(x,y)$ and depends only on the boundary values of $u = u(x,y)$.

Weyl's theory is based on the use of an invariant integral of the type (A3.14.1). [C. Carathéodory, in his field theory,† proposed the use of

$$\iint\limits_{\mathfrak{A}} \left[\frac{\partial P}{\partial x}\frac{\partial Q}{\partial y} - \frac{\partial P}{\partial y}\frac{\partial Q}{\partial x}\right] dx\,dy$$

as an invariant integral. (See also Prob. A3.14.1.)]

We now have to relate the invariant integral (A3.14.1) by proper choice of $P(x,y,u)$ and $Q(x,y,u)$ to our variational problem. For this purpose, we define a field by the following system of first-order partial differential equations:

$$\begin{aligned} u_x &= \phi(x,y,u) \\ u_y &= \psi(x,y,u) \end{aligned} \qquad \text{(A3.14.2)}$$

[which plays the role of $y' = \phi(x,y)$ for the problem in one independent variable]. We choose the functions $P(x,y,u)$, $Q(x,y,u)$, $\phi(x,y,u)$, and $\psi(x,y,u)$ in such a manner that the extremal surface $u = u_o(x,y)$ is embedded in the field defined by $(1,\phi,\psi)$, that is,

$$\begin{aligned} u_{ox}(x,y) &= \phi(x,y,u_o(x,y)) \\ u_{oy}(x,y) &= \psi(x,y,u_o(x,y)) \end{aligned} \qquad \text{(A3.14.3)}$$

and $$f(x,y,u,\phi,\psi) = \left[\frac{\partial P(x,y,u(x,y))}{\partial x} + \frac{\partial Q(x,y,u(x,y))}{\partial y}\right]_{u_x=\phi,\ u_y=\psi}$$

$$\begin{aligned} f_{u_x}(x,y,u,\phi,\psi)\Big|_{\substack{\phi=\phi(x,y,u)\\ \psi=\psi(x,y,u)}} &= P_u(x,y,u) \\ f_{u_y}(x,y,u,\phi,\psi)\Big|_{\substack{\phi=\phi(x,y,u)\\ \psi=\psi(x,y,u)}} &= Q_u(x,y,u). \end{aligned} \qquad \text{(A3.14.4)}$$

[Note that $(\partial P/\partial x) + (\partial Q/\partial y) = P_x + Q_y + P_u u_x + Q_u u_y$.]

† C. Carathéodory, Ueber die Variationsrechnung bei mehrfachen Integralen, *Acta Szeged, Sect. Math.*, vol. 4, p. 103, 1929.

Then we can rewrite (A3.14.1) for a function $u = \bar{u}(x,y)$ as follows:

$$U[\bar{u}] = \iint_{\mathcal{A}} [f(x,y,\bar{u},\phi(x,y,\bar{u}),\psi(x,y,\bar{u})) + (\bar{u}_x - \phi(x,y,\bar{u}))f_{u_x}(x,y,\bar{u},\phi(x,y,\bar{u}),\psi(x,y,\bar{u})) + (\bar{u}_y - \psi(x,y,\bar{u}))f_{u_y}(x,y,\bar{u},\phi(x,y,\bar{u}),\psi(x,y,\bar{u}))]\,dx\,dy. \quad \text{(A3.14.5)}$$

We see that if $u = u_o(x,y)$ is embedded in the field $(1,\phi,\psi)$ so that (A3.14.3) holds, then

$$U[u_o] = I[u_o].$$

Hence, we can transform the total variation as follows:

$$I[\bar{u}] - I[u_o] = I[\bar{u}] - U[u_o] = I[\bar{u}] - U[\bar{u}]$$
$$= \iint_{\mathcal{A}} [f(x,y,\bar{u},\bar{u}_x,\bar{u}_y) - f(x,y,\bar{u},\phi,\psi) + (\phi - \bar{u}_x)f_{u_x}(x,y,\bar{u},\phi,\psi) + (\psi - \bar{u}_y)f_{u_y}(x,y,\bar{u},\phi,\psi)]\,dx\,dy,$$

and we see that again

$$\mathcal{E}(x,y,\bar{u},u_x,u_y,\bar{u}_x,\bar{u}_y) = f(x,y,\bar{u},\bar{u}_x,\bar{u}_y) - f(x,y,\bar{u},u_x,u_y) + (u_x - \bar{u}_x)f_{u_x}(x,y,\bar{u},u_x,u_y) + (u_y - \bar{u}_y)f_{u_y}(x,y,\bar{u},u_x,u_y) \geq 0,$$

for all $(x,y,\bar{u}) \in N_w^{\delta}(u_o)$ and for *all* u_x, u_y, $\bar{u}_x$, $\bar{u}_y$, is, in conjunction with the embeddability into a field, sufficient for $u = u_o(x,y)$ to yield a strong relative minimum for $I[u]$.

The reader has, no doubt, noticed that at no point in our definition of a field (A3.14.2) and in the embeddability conditions (A3.14.3) and (A3.14.4) have we required that the solutions of (A3.14.2) be extremal surfaces of the variational problem, i.e., solutions of the Euler-Lagrange equation. This seems to deviate sharply from our approach to the problem in one independent variable, where we required that the solutions of $y' = \phi(x,y)$ also be solutions of the Euler-Lagrange equation. The fact is that had we taken the same approach to the problem in one independent variable as we have taken here, we would have been led automatically to the fact that the Euler-Lagrange equation is satisfied by the solutions of $y' = \phi(x,y)$. (See Prob. A3.14.2.) We can show here only that any surface $u = u(x,y) \in C^2[\mathcal{A}]$ which is *embedded* in a field by (A3.14.3), so that all the conditions (A3.14.4) are satisfied, also satisfies the Euler-Lagrange equation.

This may be seen as follows: First of all, we note that since $U[u]$ depends only on the boundary values of $u = u(x,y)$ on J, every surface $u = u(x,y) \in C^2[\mathcal{A}]$ which satisfies the boundary conditions $u(x(t),y(t)) =$

$z(t)$ is an extremal of the variational problem

$$U[u] \to \text{minimum}, \qquad u(x(t),y(t)) = z(t),$$

that is, satisfies the Euler-Lagrange equation associated with the functional $U[u]$. If we denote the integrand of $U[u]$ in (A3.14.5) by

$$\begin{aligned} F(x,y,u,u_x,u_y) = {} & f(x,y,u,\phi(x,y,u),\psi(x,y,u)) \\ & + (u_x - \phi(x,y,u))f_{u_x}(x,y,u,\phi(x,y,u),\psi(x,y,u)) \\ & + (u_y - \psi(x,y,u))f_{u_y}(x,y,u,\phi(x,y,u),\psi(x,y,u)), \end{aligned}$$

then

$$F_u - \frac{\partial}{\partial x} F_{u_x} - \frac{\partial}{\partial y} F_{u_y} = 0 \qquad \text{(A3.14.6)}$$

is satisfied by all $u = u(x,y) \in C^2[\mathfrak{a}]$. Hence this equation is, in particular, satisfied by that function $u = u(x,y)$ that is embedded in the field defined by (A3.14.2).

Secondly, we note that if $u = u(x,y)$ is embedded in the field defined by (A3.14.2), then, because of (A3.14.3) and (A3.14.4), we have

$$F(x,y,u,u_x,u_y) = f(x,y,u,u_x,u_y),$$

and we obtain from (A3.14.6) that

$$f_u - \frac{\partial}{\partial x} f_{u_x} - \frac{\partial}{\partial y} f_{u_y} = 0,$$

which is the Euler-Lagrange equation of the original variational problem $I[u] \to \text{minimum}$.

That (A3.14.6) is identically satisfied for all $u = u(x,y) \in C^2[\mathfrak{a}]$ may also be seen directly. In view of (A3.14.4), we can rewrite the integrand of (A3.14.5) as

$$\begin{aligned} F(x,y,u,u_x,u_y) = {} & f(x,y,u,\phi(x,y,u),\psi(x,y,u)) \\ & + P_u(x,y,u)(u_x - \phi(x,y,u)) + Q_u(x,y,u)(u_y - \psi(x,y,u)). \end{aligned}$$

Then,

$$\begin{aligned} F_u &= f_u + \cancel{f_{u_x}\phi_u} + \cancel{f_{u_y}\psi_u} + P_{uu}(u_x - \phi) \\ &\qquad + Q_{uu}(u_y - \psi) - \cancel{P_x\phi_u} - \cancel{Q_u\psi_u} \\ &= f_u + P_{uu}(u_x - \phi) + Q_{uu}(u_y - \psi), \end{aligned}$$

and $\qquad F_{u_x} = P_u, \qquad F_{u_y} = Q_u.$

Hence, $\quad \dfrac{\partial}{\partial x} F_{u_x} = P_{ux} + P_{uu}u_x, \qquad \dfrac{\partial}{\partial y} F_{u_y} = Q_{uy} + Q_{uu}u_y.$

Therefore,

$$F_u - \frac{\partial}{\partial x} F_{u_x} - \frac{\partial}{\partial y} F_{u_y} = f_u - P_{uu}\phi - Q_{uu}\psi - P_{ux} - Q_{uy}.$$

But, by the first equation of (A3.14.4),

$$f_u = P_{xu} + Q_{yu} + P_{uu}\phi + Q_{uu}\psi,$$

and hence

$$F_u - \frac{\partial}{\partial x} F_{u_x} - \frac{\partial}{\partial y} F_{u_y} = 0.$$

We have said nothing about the existence of a field, as postulated by (A3.14.2) to (A3.14.4)—or about the construction of such a field, for that matter. In regard to these questions, we refer the reader to the literature.†

PROBLEMS A3.14

1. Impose suitable conditions and show that

$$\iint_{\mathfrak{a}} \left(\frac{\partial P}{\partial x} \frac{\partial Q}{\partial y} - \frac{\partial P}{\partial y} \frac{\partial Q}{\partial x} \right) dx\, dy,$$

where $P = P(x,y,u(x,y))$, $Q = Q(x,y,u(x,y))$ is an invariant integral, i.e., depends on the boundary values of $u(x,y)$ only.

*2. Postulate the existence of a field and the embeddability of an extremal for the problem in one independent variable by simulating the approach which we used in this section for the problem in two independent variables. Show that the solutions of the field-defining equation are also solutions of the Euler-Lagrange equation.

† Weyl, *loc. cit.*

CHAPTER 4

THE HOMOGENEOUS PROBLEM†

4.1 PARAMETER INVARIANCE OF INTEGRAL

The preceding three chapters were devoted to a discussion of the problem of minimizing the functional

$$I[y] = \int_a^b f(x,y(x),y'(x))\,dx,$$

where beginning point and endpoint were fixed or variable, and to obvious generalizations of this problem to more unknown functions and to more independent variables.

The function $y = y(x)$ to be found was assumed to be at least sectionally smooth and, *eo ipso*, continuous. This assumption ruled out multiple-valued functions and functions that possess vertical tangent lines. This imposes, of course, some unnatural restrictions on the problem. If we

† This chapter may be omitted entirely. We shall not make use of the material discussed here in later chapters except in Secs. A7.10 and A7.11, which in turn may be omitted.

think for a moment of the isoperimetric problem (problem C, Sec. 1.1) in its more general formulation, where among all closed curves of a given length the one is sought that will encompass the largest possible region, then the solution to this problem, the circle, can obviously not be represented by a continuous function with an everywhere finite derivative: $y = \pm\sqrt{R^2 - x^2}$ is double-valued, and the derivative fails to exist (in the proper sense) at the points $x = \pm R$. In the problem of navigation (problem D, Sec. 1.1), to mention another case, it is entirely feasible that for certain velocity fields, the course of shortest crossing time will include points where the vessel moves momentarily parallel to the banks, i.e., has a vertical tangent line.

Similar difficulties arise in dealing with endpoints that vary on given curves. There, we had to assume that the tangent line to these curves is nowhere vertical. At times we may be very hard put to comply with such restrictions.

In order to circumvent such difficulties, it is often of advantage to formulate the problem in parameter form. Instead of seeking a function $y = y(x) \in C_s^1[a,b]$ that minimizes the functional $I[y]$, it is frequently better to look for an arc γ, given in parameter form $x = x(t)$, $y = y(t)$ of a later-to-be-decided-upon quality, that minimizes

$$I[\gamma] = \int_{P_a}^{P_b} f\left(x(t), y(t), \frac{\dot{y}(t)}{\dot{x}(t)}\right)\dot{x}(t)\, dt,$$

where the dot denotes derivatives with respect to t.

In general, we can formulate a fixed-endpoint problem as follows: To find an arc

$$\gamma\begin{cases} x = x(t) \\ y = y(t) \end{cases}$$

such that

$$I[\gamma] = \int_{P_a}^{P_b} F(x(t),y(t),\dot{x}(t),\dot{y}(t))\, dt \tag{4.1.1}$$

becomes a minimum, where P_a and P_b have the given coordinates (a,y_a) and (b,y_b), which are to be assumed by $x(t)$, $y(t)$ for some parameter values t_a and t_b.

Several points require clarification before we can proceed with an analysis of this problem. First, we have to impose some reasonable restrictions on γ. For the time being, we shall assume that γ is a *simple smooth arc*, and later, when the need arises, we shall relent somewhat.

Definition 4.1.1 *γ is a* simple smooth arc *if it is possible to find a parameter representation of γ: $x = x(t)$, $y = y(t)$, $t \in [t_a,t_b]$, such that $x,y \in C^1[t_a,t_b]$,*

$(x(t_1),y(t_1)) \neq (x(t_2),y(t_2))$ *whenever* $t_1 \neq t_2$ *for any* $t_1,t_2 \in [t_a,t_b]$ *and* $\dot{x}^2(t) + \dot{y}^2(t) > 0$, *that is,* $\dot{x}(t)$ *and* $\dot{y}(t)$ *will not vanish simultaneously.*

γ is a simple closed smooth arc if $(x(t_a),y(t_a)) = (x(t_b),y(t_b))$, but all the other conditions stated above remain valid except the second one which has the weaker form $t_1,t_2 \in [t_a,t_b)$.

We can see quite easily that if γ is a simple smooth arc, then not only is there one parameter representation of γ that satisfies the conditions of Definition 4.1.1 but there are infinitely many:

Lemma 4.1 *Let* $\{x = x(t), y = y(t)\}$, $t \in [t_a,t_b]$, *represent the simple smooth arc* γ. *If a new parameter* τ *is introduced by virtue of* $t = t(\tau)$, $t(\tau_a) = t_a$, $t(\tau_b) = t_b$, $t = t(\tau) \in C^1[\tau_a,\tau_b]$, $(dt/d\tau) > 0$ *for all* $\tau \in [\tau_a,\tau_b]$, *then*

$$\left.\begin{aligned} x &= x(t(\tau)) = \bar{x}(\tau) \\ y &= y(t(\tau)) = \bar{y}(\tau) \end{aligned}\right\} \tau \in [\tau_a,\tau_b]$$

is also a parameter representation of γ *that satisfies the conditions of Definition 4.1.1.*

Proof: Since $t \in C^1[\tau_a,\tau_b]$ and since $x \in C^1[t_a,t_b]$, $t(\tau_a) = t_a$, $t(\tau_b) = t_b$, we have $x(t(\tau)) = \bar{x}(\tau) \in C^1[\tau_a,\tau_b]$. An analogous argument holds for $y = \bar{y}(\tau)$.

With $t(\tau_1) = t_1$, $t(\tau_2) = t_2$,

$$(\bar{x}(\tau_1),\bar{y}(\tau_1)) = (x(t(\tau_1)),y(t(\tau_1))) = (x(t_1),y(t_1))$$
$$(\bar{x}(\tau_2),\bar{y}(\tau_2)) = (x(t(\tau_2)),y(t(\tau_2))) = (x(t_2),y(t_2)).$$

Since $(dt/d\tau) > 0$, it follows from $\tau_1 \neq \tau_2$ that $t_1 \neq t_2$, and hence

$$(\bar{x}(\tau_1),\bar{y}(\tau_1)) \neq (\bar{x}(\tau_2),\bar{y}(\tau_2)).$$

Finally,

$$\left(\frac{d\bar{x}}{d\tau}\right)^2 + \left(\frac{d\bar{y}}{d\tau}\right)^2 = \dot{x}^2\left(\frac{dt}{d\tau}\right)^2 + \dot{y}^2\left(\frac{dt}{d\tau}\right)^2 = \left(\frac{dt}{d\tau}\right)^2(\dot{x}^2 + \dot{y}^2) > 0.$$

As an example, let us consider

$$\gamma\left\{\begin{aligned} x &= t^2 \\ y &= t \end{aligned}\right\} t \in [-\tfrac{1}{2},\tfrac{1}{2}].$$

We introduce a new parameter by

$$t = -\cos\tau, \qquad \tau \in \left[\frac{\pi}{3}, \frac{2\pi}{3}\right]$$

$$\frac{dt}{d\tau} = \sin\tau > 0,$$

and we have another representation of γ:

$$\left.\begin{aligned} x &= \cos^2 \tau \\ y &= -\cos \tau \end{aligned}\right\} \tau \in \left[\frac{\pi}{3}, \frac{2\pi}{3}\right].$$

If we would permit parameter transformations with $(dt/d\tau) < 0$, then the orientation of the arc would be reversed. We make the following *loose* agreement about the orientation of γ: For increasing parameter t, the point $(x(t),y(t))$ on γ will move from the beginning point of γ to the endpoint of γ. The beginning point will always be the point on the left, or if the one point is vertically above the other, then the beginning point will be the one on the bottom. In the rare case where γ is a closed curve, $(x(t),y(t))$ will move along γ in the positive (counterclockwise) direction. Occasionally, we shall deviate from this agreement.

The fact that the same arc may be represented in such a variety of different— but for all practical purposes, equivalent— ways raises a serious question: What if the integral $I[\gamma]$ changes its value as the parameter representation of γ is changed? Then the arc γ may yield the minimum of $I[\gamma]$ for one parameter representation of γ but perhaps not for another one. This would hardly be in the spirit of our problem. Therefore, we have to require, in the following precise sense, that the value of $I[\gamma]$ not depend on the choice of parameter representation.

Definition 4.1.2 *$I[\gamma]$ is parameter invariant in $\mathcal{A}$, where $\mathcal{A}$ represents a simply connected domain in the x,y plane, if for all simple smooth arcs γ that lie entirely in $\mathcal{A}$,*

$$\int_P^Q F(x(t),y(t),\dot{x}(t),\dot{y}(t))\,dt = \int_P^Q F\left(\bar{x}(\tau),\bar{y}(\tau),\frac{d\bar{x}(\tau)}{d\tau},\frac{d\bar{y}(\tau)}{d\tau}\right)d\tau$$

for any parameter transformation $t = t(\tau)$ that satisfies the requirements of Lemma 4.1 *and for any two points P, Q on γ.*

Hereby it is assumed that $F \in C(\mathcal{R})$, where $\mathcal{R}$ consists of all $(x,y,\dot{x},\dot{y})$ for which $(x,y) \in \mathcal{A}$ and $(\dot{x},\dot{y}) \neq (0,0)$.

The following theorem, which is due to Weierstrass, settles the question of parameter invariance once and for all:

Theorem 4.1 *$I[\gamma]$ is parameter invariant in $\mathcal{A}$ in the sense of Definition* 4.1.2 *if and only if for any real $\lambda > 0$ and for all $x,y \in \mathcal{A}$,*

$$F(x,y,\lambda\dot{x},\lambda\dot{y}) = \lambda F(x,y,\dot{x},\dot{y}),$$

that is, F is positive homogeneous *of the first degree in $\dot{x}$, $\dot{y}$.*

Proof: (*a*) The condition is sufficient. We consider the parameter transformation $t = t(\tau)$ that will satisfy the conditions of Lemma 4.1. Then, if we denote $x(t(\tau)) = \bar{x}(\tau)$, $y(t(\tau)) = \bar{y}(\tau)$, and since $(d\tau/dt) > 0$ because of $(dt/d\tau) > 0$,

$$\int_P^Q F(x(t),y(t),\dot{x}(t),\dot{y}(t))\,dt = \int_{t_a}^{t_b} F(x(t),y(t),\dot{x}(t),\dot{y}(t))\,dt$$

$$= \int_{\tau_a}^{\tau_b} F\left(\bar{x}(\tau),\bar{y}(\tau),\frac{d\bar{x}(\tau)}{d\tau}\frac{d\tau}{dt},\frac{d\bar{y}(\tau)}{d\tau}\frac{d\tau}{dt}\right)\frac{dt}{d\tau}\,d\tau$$

$$= \int_{\tau_a}^{\tau_b} F\left(\bar{x}(\tau),\bar{y}(\tau),\frac{d\bar{x}(\tau)}{d\tau},\frac{d\bar{y}(\tau)}{d\tau}\right)d\tau$$

$$= \int_P^Q F\left(\bar{x}(\tau),\bar{y}(\tau),\frac{d\bar{x}(\tau)}{d\tau},\frac{d\bar{y}(\tau)}{d\tau}\right)d\tau.$$

(*b*) The condition is necessary. We have

$$\int_P^Q F(x(t),y(t),\dot{x}(t),\dot{y}(t))\,dt = \int_P^Q F\left(\bar{x}(\tau),\bar{y}(\tau),\frac{d\bar{x}(\tau)}{d\tau},\frac{d\bar{y}(\tau)}{d\tau}\right)d\tau$$

for all possible transformations $t = t(\tau)$ that satisfy the conditions of Lemma 4.1, all possible simple smooth arcs γ in $\mathfrak{a}$, and all possible pairs of points P, Q on γ.

Hence,

$$\int_{\tau_a}^{\tau_b} F\left(\bar{x}(\tau),\bar{y}(\tau),\frac{d\bar{x}(\tau)}{d\tau}\frac{d\tau}{dt},\frac{d\bar{y}(\tau)}{d\tau}\frac{d\tau}{dt}\right)\frac{dt}{d\tau}\,d\tau$$

$$= \int_{\tau_a}^{\tau_b} F\left(\bar{x}(\tau),\bar{y}(\tau),\frac{d\bar{x}(\tau)}{d\tau},\frac{d\bar{y}(\tau)}{d\tau}\right)d\tau.$$

We keep P fixed and vary Q on γ; that is, we keep τ_a fixed and let τ_b vary and differentiate with respect to the upper limit τ_b:

$$F\left(\bar{x}(\tau_b),\bar{y}(\tau_b),\frac{d\bar{x}(\tau_b)}{d\tau}\left(\frac{d\tau}{dt}\right),\frac{d\bar{y}(\tau_b)}{d\tau}\left(\frac{d\tau}{dt}\right)\right)\left(\frac{dt}{d\tau}\right)$$

$$= F\left(\bar{x}(\tau_b),\bar{y}(\tau_b),\frac{d\bar{x}(\tau_b)}{d\tau},\frac{d\bar{y}(\tau_b)}{d\tau}\right),$$

where $(d\tau/dt)$ and $(dt/d\tau)$ are to be taken at $\tau = \tau_b$.

By Definition 4.1.2, this has to hold for any point Q in $\mathfrak{a}$, for all $(\dot{x},\dot{y}) \neq (0,0)$, and for any transformation $t = t(\tau)$ that satisfies the requirements of Lemma 4.1. Hence it has to hold, in particular, for the transformation $t = (1/\lambda)\tau$, where $\lambda > 0$ is a constant, that is,

$$F(x,y,\lambda\dot{x},\lambda\dot{y}) = \lambda F(x,y,\dot{x},\dot{y})$$

for all $\lambda > 0$, all $(x,y) \in \mathcal{A}$, and all $(\dot{x},\dot{y}) \neq (0,0)$.

Because of this homogeneity property of the integrand, the variational problem in parameter form is referred to as the *homogeneous problem.*

It is now easy to see that the problem which we dealt with in the preceding two chapters, and to which we shall henceforth refer as the *x problem,* satisfies the requirements of Theorem 4.1 when a parameter representation is introduced:

$$F(x,y,\dot{x},\dot{y}) = f\left(x,\, y,\, \frac{\dot{y}}{\dot{x}}\right)\dot{x}.$$

Hence

$$F(x,y,\lambda\dot{x},\lambda\dot{y}) = f\left(x,\, y,\, \frac{\lambda\dot{y}}{\lambda\dot{x}}\right)\lambda\dot{x} = f\left(x,\, y,\, \frac{\dot{y}}{\dot{x}}\right)\lambda\dot{x} = \lambda F(x,y,\dot{x},\dot{y}).$$

On the other hand, if we have a homogeneous problem to begin with, then we may change it to an x problem, provided that $\dot{x} \neq 0$ and that $y = y(t^{-1}(x)) \equiv y(x)$ is single-valued and satisfies the customary differentiability conditions:

$$\int_{t_a}^{t_b} F(x,y,\dot{x},\dot{y})\,dt = \int_{t_a}^{t_b} \dot{x}F\left(x,\, y,\, 1,\, \frac{\dot{y}}{\dot{x}}\right)dt = \int_a^b F(x,y,1,y')\,dx.$$

This clearly demonstrates that even though we have to impose the restriction of positive homogeneity of the first order on the integrand, we have not lost anything with the new formulation of our problems. On the contrary, we have gained greater generality inasmuch as we can now admit vertical tangents and also curves that cross vertical lines more than once.

PROBLEMS 4.1

1. Formulate the following variational problems as homogeneous problems:
   (*a*) The Brachistochrone problem (problem A, Sec. 1.1)
   (*b*) The problem of minimal surfaces of revolution (problem B, Sec. 1.1)
   (*c*) The problem of navigation with fixed endpoints (problem D, Sec. 1.1)
   (*d*) $\int_a^b y'^2(x)\,dx \to \text{minimum}, \qquad y(a) = y_a, \qquad y(b) = y_b$
   (*e*) $\int_a^b n(x,y(x))\sqrt{1 + y'^2(x)}\,dx \to \text{minimum}, \qquad y(a) = y_a, \qquad y(b) = y_b$

2. Find equivalent representations of the following arcs by introducing the arc length as parameter:
   (a) $x = t$, $y = t$, $t \in [0,1]$
   (b) $x = 2\cos t$, $y = 2\sin t$, $t \in [0,\pi]$
   (c) $x = x(t)$, $y = y(t)$, $t \in [t_a,t_b]$, $x,y \in C^1[t_a,t_b]$, $\dot{x}^2 + \dot{y}^2 > 0$, $(x(t_1),y(t_1)) \neq (x(t_2),y(t_2))$ for $t_1 \neq t_2$
3. Evaluate

$$\int_0^{2\pi} [\sqrt{\dot{x}^2 + \dot{y}^2} + \tfrac{1}{2}(x\dot{y} - y\dot{x})]\, dt$$

   for $x = a + \cos t$, $y = b + \sin t$, $t \in [0,2\pi]$.
4. Consider the variational problem $I[y] = \int_a^b f(y(x),y'(x))\, dx \to \text{minimum}$, $y(a) = y_a$, $y(b) = y_b$. Suppose that $f(y,\lambda y') = \lambda f(y,y')$ for all $\lambda > 0$, for all y and all $y' \neq 0$.
   (a) Show that $f_{y'y'} = 0$.
   (b) Show that $I[y]$ is invariant under a transformation $x = x(\xi)$, $x'(\xi) > 0$, $x' \in C$.

4.2 PROPERTIES OF HOMOGENEOUS FUNCTIONS

For the following, we shall assume that $F \in C^2(\mathcal{R})$, where

$$\mathcal{R} = \{(x,y,\dot{x},\dot{y}) \mid (x,y) \in \mathcal{A},\ (\dot{x},\dot{y}) \neq (0,0)\},$$

whereby $\mathcal{A}$ is a simply connected domain in the x,y plane. We shall also assume that $F = F(x,y,\dot{x},\dot{y})$ is positive homogeneous of the first degree in $\dot{x}$, $\dot{y}$ in $\mathcal{R}$.

Then F satisfies *Euler's identity*

$$\dot{x}F_{\dot{x}}(x,y,\dot{x},\dot{y}) + \dot{y}F_{\dot{y}}(x,y,\dot{x},\dot{y}) = F(x,y,\dot{x},\dot{y}). \tag{4.2.1}$$

The proof is quite simple: We differentiate

$$F(x,y,\lambda\dot{x},\lambda\dot{y}) = \lambda F(x,y,\dot{x},\dot{y}) \tag{4.2.2}$$

with respect to λ:

$$F_{\dot{x}}(x,y,\lambda\dot{x},\lambda\dot{y})\dot{x} + F_{\dot{y}}(x,y,\lambda\dot{x},\lambda\dot{y})\dot{y} = F(x,y,\dot{x},\dot{y}).$$

We let $\lambda = 1$, and (4.2.1) follows.

$F_{\dot{x}}$, $F_{\dot{y}}$ are positive homogeneous of the zeroth degree:

$$\begin{aligned} F_{\dot{x}}(x,y,\lambda\dot{x},\lambda\dot{y}) &= F_{\dot{x}}(x,y,\dot{x},\dot{y}) \\ F_{\dot{y}}(x,y,\lambda\dot{x},\lambda\dot{y}) &= F_{\dot{y}}(x,y,\dot{x},\dot{y}). \end{aligned} \tag{4.2.3}$$

We see this by differentiating (4.2.2) with respect to $\dot{x}$,

$$F_{\dot{x}}(x,y,\lambda\dot{x},\lambda\dot{y})\lambda = \lambda F_{\dot{x}}(x,y,\dot{x},\dot{y}),$$

and cancelling λ. (Analogously for $F_{\dot{y}}$.)

Finally, $F_{\dot{x}\dot{x}}$, $F_{\dot{x}\dot{y}}$, $F_{\dot{y}\dot{y}}$ are positive homogeneous of the (-1)st degree:

$$
\begin{aligned}
F_{\dot{x}\dot{x}}(x,y,\lambda\dot{x},\lambda\dot{y}) &= \frac{1}{\lambda}F_{\dot{x}\dot{x}}(x,y,\dot{x},\dot{y}) \\
F_{\dot{x}\dot{y}}(x,y,\lambda\dot{x},\lambda\dot{y}) &= \frac{1}{\lambda}F_{\dot{x}\dot{y}}(x,y,\dot{x},\dot{y}) \\
F_{\dot{y}\dot{y}}(x,y,\lambda\dot{x},\lambda\dot{y}) &= \frac{1}{\lambda}F_{\dot{y}\dot{y}}(x,y,\dot{x},\dot{y}).
\end{aligned}
\tag{4.2.4}
$$

We can see this by differentiating (4.2.3) with respect to $\dot{x}$ and $\dot{y}$, respectively.

If we differentiate Euler's identity (4.2.1) with respect to $\dot{x}$ and then with respect to $\dot{y}$, we obtain

$$
\begin{aligned}
\dot{x}F_{\dot{x}\dot{x}}(x,y,\dot{x},\dot{y}) + \dot{y}F_{\dot{y}\dot{x}}(x,y,\dot{x},\dot{y}) &= 0 \\
\dot{x}F_{\dot{x}\dot{y}}(x,y,\dot{x},\dot{y}) + \dot{y}F_{\dot{y}\dot{y}}(x,y,\dot{x},\dot{y}) &= 0.
\end{aligned}
\tag{4.2.5}
$$

These relations indicate that the second partial derivatives of F with respect to $\dot{x},\dot{y}$ are not entirely independent of each other. We can show that all three may be expressed in terms of the function

$$F_1 = \frac{F_{\dot{x}\dot{x}} + F_{\dot{y}\dot{y}}}{\dot{x}^2 + \dot{y}^2}. \tag{4.2.6}$$

(Remember that $\dot{x}^2 + \dot{y}^2 > 0$!) We obtain from (4.2.5),

$$F_{\dot{x}\dot{x}} + F_{\dot{y}\dot{y}} = -F_{\dot{x}\dot{y}}\left(\frac{\dot{y}}{\dot{x}} + \frac{\dot{x}}{\dot{y}}\right),$$

and hence

$$F_{\dot{x}\dot{y}} = -\dot{x}\dot{y}F_1. \tag{4.2.7a}$$

Substitution of (4.2.7*a*) into (4.2.5) yields

$$F_{\dot{x}\dot{x}} = \dot{y}^2F_1 \tag{4.2.7b}$$

$$F_{\dot{y}\dot{y}} = \dot{x}^2F_1. \tag{4.2.7c}$$

Since $F_{\dot{x}\dot{x}}$, $F_{\dot{y}\dot{y}}$ are positive homogeneous of the (-1)st degree and since $\dot{x}^2 + \dot{y}^2$ is homogeneous of the second degree, it follows that F_1 is homogeneous of the (-3)rd degree:

$$F_1(x,y,\lambda\dot{x},\lambda\dot{y}) = \frac{1}{\lambda^3}F_1(x,y,\dot{x},\dot{y}). \tag{4.2.8}$$

PROBLEMS 4.2

1. Show that $F_{\dot y}(x,y,\lambda\dot x,\lambda\dot y) = F_{\dot y}(x,y,\dot x,\dot y)$ if F is positive homogeneous of the first degree in $\dot x$, $\dot y$.
2. Show that $F_{\dot x\dot x}$, $F_{\dot x\dot y}$, $F_{\dot y\dot y}$ are positive homogeneous of the (-1)st degree in $\dot x$, $\dot y$ if F is positive homogeneous of the first degree in $\dot x$, $\dot y$.
3. Prove: If $F(x,y,\dot x,\dot y) \in C^1(\mathcal{R})$, where $\mathcal{R} = \{(x,y,\dot x,\dot y) \mid (x,y) \in \mathcal{A},\ (\dot x,\dot y) \neq (0,0)\}$ and if $\dot xF_{\dot x}(x,y,\dot x,\dot y) + \dot yF_{\dot y}(x,y,\dot x,\dot y) = F(x,y,\dot x,\dot y)$, then F is positive homogeneous of the first degree in $\dot x$, $\dot y$. (This is the converse of Euler's identity.)
4. Find F_1 as defined in (4.2.6) for the following functions F:
   (a) $F = n(x,y)\sqrt{\dot x^2 + \dot y^2}$
   (b) $F = \sqrt{\dot x^2 + \dot y^2} + \frac{1}{2}(x\dot y - y\dot x)$
   (c) $F = \dfrac{\dot y^2}{\dot x}$
   (d) $F = \dfrac{\dot x^2 + \dot y^2}{\sqrt{2(\dot x^2 + \dot y^2)} + \dot x}$
5. Find the most general form of a function $F(x,y,\dot x,\dot y)$ which is positive homogeneous of the first degree in $\dot x$, $\dot y$ and for which $F_1 = 0$ for all $(x,y,\dot x,\dot y)$.

4.3 WEAK AND STRONG RELATIVE EXTREMA

From now on, we shall admit simple, sectionally smooth arcs for consideration, i.e., arcs for which there is a parameter representation $x = x(t)$, $y = y(t)$, $t \in [t_a,t_b]$ such that $x,y \in C_s^1[t_a,t_b]$ while all the other conditions of Definition 4.1.1 are retained with the understanding that $\dot x^2(t) + \dot y^2(t) > 0$ will hold wherever $\dot x$, $\dot y$ exist.

Before deriving a necessary condition for a simple, sectionally smooth arc to yield a relative minimum for

$$I[\gamma] = \int_{P_a}^{P_b} F(x,y,\dot x,\dot y)\,dt,$$

where $P_a(a,y_a)$, $P_b(b,y_b)$ are given, we have to define what we mean by a relative minimum of $I[\gamma]$. Again we shall distinguish between strong and weak relative minima, but we shall see that in this respect there is no strict analogy between the x problem and the homogeneous problem.

Let us first define what we mean when we say that an arc lies in a weak neighborhood of another arc:

Definition 4.3.1 *The simple, sectionally smooth arc γ lies in a weak δ neighborhood of the simple, sectionally smooth arc γ_o: $\gamma \in N_w^\delta(\gamma_o)$ if there exists a parameter representation such that*

$$\left.\begin{aligned} x &= x(t) \\ y &= y(t) \end{aligned}\right\} t \in [t_a,t_b]$$

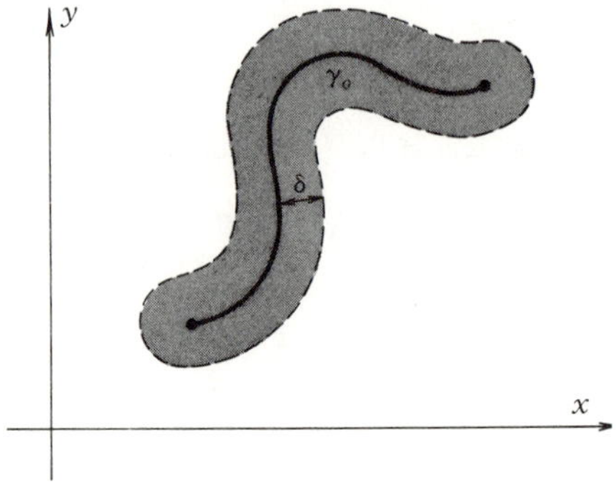

Figure 4.1

represents γ, *and*

$$\left.\begin{aligned} x &= x_o(t) \\ y &= y_o(t) \end{aligned}\right\} t \in [t_a,t_b]$$

represents γ_o, *such that*

$$[x(t) - x_o(t)]^2 + [y(t) - y_o(t)]^2 < \delta^2$$

for all $t \in [t_a,t_b]$. (See Fig. 4.1.)

Note that it is very important to include the condition, "there exists a parameter representation such that..." in this definition. Without this, we would run into paradoxical situations, such as that arcs are not even in a weak neighborhood of themselves.

Take, for example,

$$\left.\begin{aligned} x &= t \\ y &= t \end{aligned}\right\} t \in [0,1],$$

which may also be represented by

$$\left.\begin{aligned} x &= t^2 \\ y &= t^2 \end{aligned}\right\} t \in [0,1].$$

Then $$(t - t^2)^2 + (t - t^2)^2 = 2t^2(1 - t)^2 = \tfrac{1}{8} \qquad \text{for } t = \tfrac{1}{2}$$

and *cannot* be made less than δ for all $t \in [0,1]$ and any $\delta > 0$.

It is not difficult to see that if an arc γ lies in a weak neighborhood of an arc γ_o in the sense of Definition 4.3.1, then $(x,y) = (x(t),y(t))$ will lie in a weak neighborhood of $(x,y) = (x_o(t),y_o(t))$ as defined in Sec. 2.11. (See Prob. 4.3.1.)

We can now define a strong minimum as follows:

Definition 4.3.2 *The simple, sectionally smooth arc γ_o yields a* strong relative minimum *for* $I[\gamma] = \int_{P_a}^{P_b} F(x,y,\dot{x},\dot{y})\,dt$ *if there is a* $\delta > 0$ *such that*

$$I[\gamma] - I[\gamma_o] \geq 0$$

for all sectionally smooth arcs $\gamma \in N_w^\delta(\gamma_o)$ *which join the same beginning point* P_a *to the same endpoint* P_b.

We observe that a strong minimum when the problem is formulated as an x problem is not necessarily equivalent to a strong minimum when the same problem is formulated as a homogeneous problem. The following example will show this quite clearly. For

$$I[y] = \int_0^1 y'^2(x)\,dx \to \text{minimum}, \qquad y(0) = 0, \qquad y(1) = 1,$$

$y = x$ yields a strong relative minimum. (See Prob. 4.3.2.)

Now we shall formulate this problem as a homogeneous problem. We have

$$I[\gamma] = \int_{(0,0)}^{(1,1)} \frac{\dot{y}^2}{\dot{x}}\,dt \to \text{minimum}.$$

We see that the parameter representation

$$\left.\begin{aligned} x &= t \\ y &= t \end{aligned}\right\} t \in [0,1]$$

of $y = x$, $x \in [0,1]$, yields the value 1 for this functional.

However, we shall see now that one can find in any weak δ neighborhood of $y = x$, a simple, sectionally smooth arc γ such that $I[\gamma] = -M$, where $M > 1$ may be made as large as we please.

Given any $\delta > 0$, we can choose n sufficiently large so that the zigzag line in Fig. 4.2 lies in a weak δ neighborhood of $y = x$. (See Prob. 4.3.3.) We assume the slopes of the nonhorizontal portions to be $-M$, where $M > 1$, but otherwise quite arbitrary. Since the horizontal portions will not contribute to the integral on account

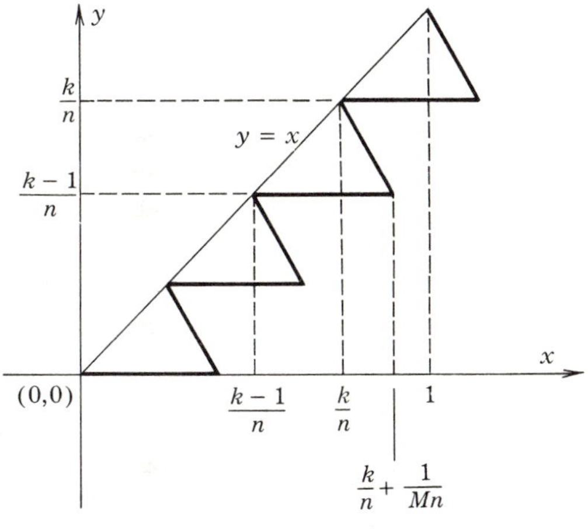

Figure 4.2

of $\dot{y} = 0$, $\dot{x} \neq 0$, we need only establish a parameter representation of the nonhorizontal portions. It is easily verified (see Prob. 4.3.4) that

$$\left.\begin{aligned} x &= -\frac{1}{M}t + \frac{k(M+1)}{nM} \\ y &= t \end{aligned}\right\} t \in \left[\frac{k-1}{n}, \frac{k}{n}\right], \qquad k = 1, 2, \ldots, n,$$

is such a parameter representation. Then

$$\int_0^1 \frac{\dot{y}^2}{\dot{x}}\,dt = -\sum_{k=1}^{n} \int_{(k-1)/n}^{k/n} M\,dt = -\sum_{k=1}^{n} M\frac{1}{n} = -M,$$

and we see that $x = t$, $y = t$ cannot yield a strong relative minimum. This should not surprise us too much, considering that we admit in the homogeneous formulation of a variational problem a much wider class of competing functions. (See also Prob. 4.3.5.)

When trying to define a strong neighborhood of an arc, one has to take into account that $(\dot{x},\dot{y})$ and $(\lambda\dot{x},\lambda\dot{y})$, $\lambda > 0$, represent the same direction of the tangent line to the arc that is given by $x = x(t)$, $y = y(t)$. Also, when considering $x = x_o(t) + h(t)$, $y = y_o(t) + k(t)$ for the purpose of deriving the Euler-Lagrange equations, as in Sec. 2.11, one wishes to prevent $(\dot{x}_o(t_o) + \dot{h}(t_o), \dot{y}_o(t_o) + \dot{k}(t_o)) = (0,0)$ from happening for some $t = t_o$, since F is, in general, not defined for $(\dot{x},\dot{y}) = (0,0)$.

We propose here the following definition:

Definition 4.3.3 *The simple, sectionally smooth arc γ lies in a* strong δ neighborhood *of the simple, sectionally smooth arc γ_o: $\gamma \in N^\delta(\gamma_o)$ if there exists a parameter representation such that*

$$\left.\begin{aligned} x &= x(t) \\ y &= y(t) \end{aligned}\right\} t \in [t_a,t_b]$$

represents γ, and

$$\left.\begin{aligned} x &= x_o(t) \\ y &= y_o(t) \end{aligned}\right\} t \in [t_a,t_b]$$

represents γ_o, such that

$$[x(t) - x_o(t)]^2 + [y(t) - y_o(t)]^2 < \delta^2$$

for all $t \in [t_a,t_b]$, and

$$[\dot{x}(t) - \dot{x}_o(t)]^2 + [\dot{y}(t) - \dot{y}_o(t)]^2 < \delta^2[\dot{x}_o^2(t) + \dot{y}_o^2(t)]$$

for all $t \in [t_a,t_b]$, where $\dot{x}$, $\dot{x}_o$, $\dot{y}$, $\dot{y}_o$ exist.

Again, it is not difficult to relate this concept of a strong neighborhood to the one that is based on (2.11.3).

We note that this definition is invariant under parameter transformations such as the ones in Lemma 4.1.

If we introduce the arc length s on γ_o as a new parameter by means of

$$s = \int_{t_a}^{t} \sqrt{\dot{x}_o^2(u) + \dot{y}_o^2(u)}\, du,$$

we obtain with

$$x(t(s)) = \bar{x}(s), \quad y(t(s)) = \bar{y}(s), \quad x_o(t(s)) = \bar{x}_o(s), \quad y_o(t(s)) = \bar{y}_o(s),$$

and because of $s(t_a) = 0$, $s(t_b) = l$, where l is the length of γ_o,

$$[\bar{x}(s) - \bar{x}_o(s)]^2 + [\bar{y}(s) - \bar{y}_o(s)]^2 < \delta^2$$

and

$$\left(\frac{d\bar{x}}{ds} - \frac{d\bar{x}_o}{ds}\right)^2 + \left(\frac{d\bar{y}}{ds} - \frac{d\bar{y}_o}{ds}\right)^2 < \delta^2$$

for all $s \in [0,l]$ because $(d\bar{x}_o/ds)^2 + (d\bar{y}_o/ds)^2 = 1$ and because $(ds/dt)^2 > 0$ appears as a common factor on both sides and can be cancelled. The latter condition means geometrically that $|\theta| < \sin^{-1}\delta$, whereby θ denotes the angle between the two tangent vectors to γ and γ_o at corresponding points. (See Prob. 4.3.7.)

In terms of this concept of a strong neighborhood, we give the following definition of a weak relative minimum:

Definition 4.3.4 *The simple, sectionally smooth arc γ_o yields a* weak relative minimum *for*

$$I[\gamma] = \int_{P_a}^{P_b} F(x,y,\dot{x},\dot{y})\, dt$$

if there exists a $\delta > 0$ such that

$$I[\gamma] - I[\gamma_o] \geq 0$$

for all simple, sectionally smooth arcs $\gamma \in N^\delta(\gamma_o)$ which join the same beginning point P_a to the same endpoint P_b.

We note that if $\gamma \in N^\delta(\gamma_o)$, then $\gamma \in N_w^\delta(\gamma_o)$. Hence, if γ_o yields a strong relative minimum for $I[\gamma]$, then it yields necessarily a weak relative minimum for $I[\gamma]$.

We can now show that if $\delta > 0$ is chosen sufficiently small and if γ_h: $x = x_o(t) + h(t)$, $y = y_o(t)$, $t \in [t_a,t_b]$, $h \in C_s^1[t_a,t_b]$, $h(t_a) = h(t_b) = 0$, lies in a strong δ neighborhood of γ_o: $x = x_o(t)$, $y = y_o(t)$, $t \in [t_a,t_b]$, then

$$(\dot{x}(t),\dot{y}(t)) = (\dot{x}_o(t) + \dot{h}(t), \dot{y}_o(t)) \neq (0,0)$$

for all $t \in [t_a,t_b]$. [An analogous statement holds for γ_k: $x = x_o(t)$, $y = y_o(t) + k(t)$.]

Let $m = \inf_{[t_a,t_b]} (\dot{x}_o^2(t) + \dot{y}_o^2(t)) > 0$, and choose h so that

$$||h|| = \max_{[t_a,t_b]} |h(t)| + \sup_{[t_a,t_b]} |\dot{h}(t)| < \delta^* = \min\left(\delta,\delta\sqrt{m},\frac{\sqrt{m}}{2}\right),$$

which implies that $|h(t)| < \delta^*$, $|\dot{h}(t)| < \delta^*$. Then

$$[x(t) - x_o(t)]^2 + [y(t) - y_o(t)]^2 = h^2(t) < \delta^2$$

and
$$[\dot{x}(t) - \dot{x}_o(t)]^2 + [\dot{y}(t) - \dot{y}_o(t)]^2 = \dot{h}^2(t) < \delta^2 m$$
$$< \delta^2[\dot{x}_o^2(t) + \dot{y}_o^2(t)],$$

that is, $\gamma_h \in N^\delta(\gamma_o)$. Also, $(\dot{x}_o(t) + \dot{h}(t), \dot{y}_o(t)) \neq (0,0)$ because, whenever $\dot{y}_o(t) = 0$, then $\dot{x}_o(t) \geq \sqrt{m}$, or $\dot{x}_o(t) \leq -\sqrt{m}$, and hence $\dot{x}_o(t) + \dot{h}(t) \geq \sqrt{m} - (\sqrt{m}/2)$, or $\dot{x}_o(t) + \dot{h}(t) \leq -\sqrt{m} + (\sqrt{m}/2)$.

With this, we have laid the groundwork for the application of the method that was developed in Sec. 2.11 for the purpose of deriving the Euler-Lagrange equations, which we shall do in the next section.

PROBLEMS 4.3

*1. Show that if an arc γ lies in a weak neighborhood of an arc γ_o, in the sense of Definition 4.3.1, then $(x(t),y(t))$ lies in a weak neighborhood of $(x_o(t),y_o(t))$, as defined in Sec. 2.11.

*2. Show that $y = x$ yields a strong relative minimum for

$$\int_0^1 y'^2(x)\,dx \to \text{minimum}, \qquad y(0) = 0, \qquad y(1) = 1.$$

*3. Show that the zigzag line in Fig. 4.2 lies in a weak δ neighborhood of $x = t$, $y = t$, $t \in [0,1]$, if n is chosen properly.

*4. Verify that

$$\left.\begin{aligned} x &= -\frac{1}{M}t + \frac{k(M+1)}{nM} \\ y &= t \end{aligned}\right\} \quad t \in \left[\frac{k-1}{n}, \frac{k}{n}\right]$$

is a parameter representation of the nonhorizontal portions of the zigzag line in Fig. 4.2.

*5. Consider the variational problem

$$\int_0^1 \frac{y'^2(x)}{1 + y'^2(x)}\,dx \to \text{minimum}, \qquad y(0) = y(1) = 0.$$

(a) Show that $y = 0$ yields a strong relative minimum.

(b) Formulate the problem as a homogeneous problem and show that $x = t$, $y = 0$, $t \in [0,1]$, does *not* yield a strong relative minimum. [*Hint*: Consider a path Γ that consists of the line segment that joins $(0,0)$ to $(1 + \varepsilon, \varepsilon)$ and the line segment that joins $(1 + \varepsilon, \varepsilon)$ to $(1,0)$.]

6. Consider the following definition of a strong neighborhood: There shall exist a parameter representation of γ and γ_o such that $x = x_o(t)$, $y = y_o(t)$, $t \in [t_a,t_b]$ represents γ_o and $x = x(t)$, $y = y(t)$, $t \in [t_a,t_b]$, represents γ, so that

$$[x(t) - x_o(t)]^2 + [y(t) - y_o(t)]^2 < \delta^2$$

and $[\dot{x}(t) - \dot{x}_o(t)]^2 + [\dot{y}(t) - \dot{y}_o(t)]^2 < \delta^2\sqrt{\dot{x}_o^2(t) + \dot{y}_o^2(t)}\sqrt{\dot{x}^2(t) + \dot{y}^2(t)}$

for all $t \in [t_a,t_b]$. Show that this definition is invariant under a parameter transformation $t = t(\tau)$, $(dt/d\tau) > 0$ (if carried out simultaneously for γ and γ_o) and is symmetric in γ and γ_o.

7. Show that if

$$\left(\frac{dx}{ds} - \frac{dx_o}{ds}\right)^2 + \left(\frac{dy}{ds} - \frac{dy_o}{ds}\right)^2 < \delta^2,$$

where

$$\left(\frac{dx_o}{ds}\right)^2 + \left(\frac{dy_o}{ds}\right)^2 = 1,$$

then the following inequality holds for the angle θ between the vectors $(dx/ds, dy/ds)$ and $(dx_o/ds, dy_o/ds)$:

$$|\theta| < \sin^{-1}\delta.$$

Is the converse also true?

4.4 THE EULER-LAGRANGE EQUATIONS FOR THE HOMOGENEOUS PROBLEM

We consider the homogeneous variational problem

$$I[\gamma] = \int_{P_a}^{P_b} F(x,y,\dot{x},\dot{y})\, dt \to \text{minimum},$$

where $P_a(a,y_a)$ and $P_b(b,y_b)$ are given. We assume that $F \in C^1(\mathcal{R})$, where $\mathcal{R} = \{(x,y,\dot{x},\dot{y}) \mid (x,y) \in \mathcal{A},\ (\dot{x},\dot{y}) \neq (0,0)\}$, whereby $\mathcal{A}$ denotes a simply connected domain in the x,y plane.

Suppose that the simple, sectionally smooth arc γ_o: $x = x_o(t)$, $y = y_o(t)$, $t \in [t_a,t_b]$, yields a weak relative minimum for $I[\gamma]$. Then

$$I[\gamma_h] \geq I[\gamma_o],$$

where γ_h: $x = x_o(t) + h(t)$, $y = y_o(t)$, $h \in C_s^1[t_a,t_b]$, $h(t_a) = h(t_b) = 0$, for all $\gamma_h \in N^\delta(\gamma_o)$ for sufficiently small $\delta > 0$. Similarly,

$$I[\gamma_k] \geq I[\gamma_o],$$

where γ_k: $x = x_o(t)$, $y = y_o(t) + k(t)$, $k \in C_s^1[t_a,t_b]$, $k(t_a) = k(t_b) = 0$, $\gamma_k \in N^\delta(\gamma_o)$.

With this, our problem is reduced to the one in Sec. 2.11 with $n = 2$, and we obtain as a necessary condition the Euler-Lagrange equations in integrated form:

$$F_{\dot{x}}(x_o(t),y_o(t),\dot{x}_o(t),\dot{y}_o(t)) = \int_{t_a}^{t} F_x(x_o(u),y_o(u),\dot{x}_o(u),\dot{y}_o(u))\,du + C_1 \tag{4.4.1}$$

$$F_{\dot{y}}(x_o(t),y_o(t),\dot{x}_o(t),\dot{y}_o(t)) = \int_{t_a}^{t} F_y(x_o(u),y_o(u),\dot{x}_o(u),\dot{y}_o(u))\,du + C_2,$$

which have to be satisfied for all $t \in [t_a,t_b]$ except at those points where $\dot{x} = \dot{x}_o(t)$ and/or $\dot{y} = \dot{y}_o(t)$ have a jump discontinuity.

The *Weierstrass-Erdmann corner conditions* follow readily: By the same argument as in Sec. 2.9, we conclude that for any parameter value $t_c \in (t_a,t_b)$ for which $\dot{x} = \dot{x}_o(t)$ and/or $\dot{y} = \dot{y}_o(t)$ have a jump discontinuity, we have to have

$$\begin{aligned}(F_{\dot{x}})_{t_c-0} &= (F_{\dot{x}})_{t_c+0}\\ (F_{\dot{y}})_{t_c-0} &= (F_{\dot{y}})_{t_c+0}.\end{aligned} \tag{4.4.2}$$

(See also Prob. 4.4.1.) For any smooth portion of the arc γ_o, we again obtain by differentiation of (4.4.1), the Euler-Lagrange equations

$$\begin{aligned}F_x(x,y,\dot{x},\dot{y}) - \frac{d}{dt}F_{\dot{x}}(x,y,\dot{x},\dot{y}) &= 0\\ F_y(x,y,\dot{x},\dot{y}) - \frac{d}{dt}F_{\dot{y}}(x,y,\dot{x},\dot{y}) &= 0\end{aligned} \tag{4.4.3}$$

γ_o: $x = x_o(t)$, $y = y_o(t)$, $t \in [t_a,t_b]$, is called an *extremal arc* if $x = x_o(t)$, $y = y_o(t)$ are smooth solutions of the Euler-Lagrange equations (4.4.3).

In view of the results of Sec. 4.2 concerning positive homogeneous functions, the two equations (4.4.3) depend on each other to a much greater extent than the equations (2.11.5). We shall see that both equations may actually be replaced by one equation.

To demonstrate this, we assume that $F \in C^2(\mathcal{R})$ and carry out the differentiations with respect to t. We obtain, provided that $\ddot{x}$, $\ddot{y}$ exist,

$$\begin{aligned}F_x - F_{\dot{x}x}\dot{x} - F_{\dot{x}y}\dot{y} - F_{\dot{x}\dot{x}}\ddot{x} - F_{\dot{x}\dot{y}}\ddot{y} &= 0\\ F_y - F_{\dot{y}x}\dot{x} - F_{\dot{y}y}\dot{y} - F_{\dot{y}\dot{x}}\ddot{x} - F_{\dot{y}\dot{y}}\ddot{y} &= 0.\end{aligned}$$

From Euler's identity (4.2.1),

$$\begin{aligned}F_x &= \dot{x}F_{x\dot{x}} + \dot{y}F_{x\dot{y}}\\ F_y &= \dot{x}F_{y\dot{x}} + \dot{y}F_{y\dot{y}}.\end{aligned}$$

If we replace F_x, F_y by these expressions and replace $F_{\dot{x}\dot{x}}$, $F_{\dot{x}\dot{y}}$, $F_{\dot{y}\dot{y}}$ by the appropriate expressions in terms of F_1 [see (4.2.6)], we obtain

$$\dot{x}F_{x\dot{x}} + \dot{y}F_{x\dot{y}} - F_{x\dot{x}}\dot{x} - F_{\dot{x}y}\dot{y} - F_1\dot{y}^2\ddot{x} + F_1\dot{x}\dot{y}\ddot{y} = 0$$
$$\dot{x}F_{\dot{x}y} + \dot{y}F_{y\dot{y}} - F_{x\dot{y}}\dot{x} - F_{\dot{y}y}\dot{y} + F_1\dot{x}\dot{y}\ddot{x} - F_1\dot{x}^2\ddot{y} = 0,$$

or

$$\dot{y}[(F_{x\dot{y}} - F_{y\dot{x}}) + F_1(\dot{x}\ddot{y} - \dot{y}\ddot{x})] = 0$$
$$-\dot{x}[(F_{x\dot{y}} - F_{y\dot{x}}) + F_1(\dot{x}\ddot{y} - \dot{y}\ddot{x})] = 0. \tag{4.4.4}$$

Since $\dot{x}^2 + \dot{y}^2 > 0$, it follows that

$$F_{x\dot{y}} - F_{y\dot{x}} + F_1(\dot{x}\ddot{y} - \dot{y}\ddot{x}) = 0. \tag{4.4.5}$$

We see that if the Euler-Lagrange equations (4.4.3) are satisfied, then either (4.4.5) is satisfied or $\dot{x} = \dot{y} = 0$. Since $\dot{x}^2 + \dot{y}^2 > 0$ by hypothesis, it follows that (4.4.5) has to be satisfied. On the other hand, suppose that (4.4.5) is satisfied. Then, by (4.4.4) the Euler-Lagrange equations have to be satisfied also.

Equation (4.4.5) is called the *Weierstrass equation* of the homogeneous problem.

We state our results as a theorem:

Theorem 4.4 *If the simple, sectionally smooth arc γ_o: $x = x_o(t)$, $y = y_o(t)$, $t \in [t_a,t_b]$, yields a (weak or strong) relative minimum for the functional*

$$I[\gamma] = \int_{P_a}^{P_b} F(x,y,\dot{x},\dot{y})\, dt,$$

where $F \in C^1(\mathfrak{R})$, then $x = x_o(t)$, $y = y_o(t)$ have to satisfy the Euler-Lagrange equations in integrated form,

$$F_{\dot{x}}(x_o(t),y_o(t),\dot{x}_o(t),\dot{y}_o(t)) = \int_{t_a}^{t} F_x(x_o(u),y_o(u),\dot{x}_o(u),\dot{y}_o(u))\, du + C_1$$

$$F_{\dot{y}}(x_o(t),y_o(t),\dot{x}_o(t),\dot{y}_o(t)) = \int_{t_a}^{t} F_y(x_o(u),y_o(u),\dot{x}_o(u),\dot{y}_o(u))\, du + C_2$$

for all $t = [t_a,t_b]$ except where $\dot{x}_o$, $\dot{y}_o$ have jump discontinuities.

Every smooth portion of γ_o has to satisfy the Euler-Lagrange equations in differentiated form,

$$F_x - \frac{d}{dt}F_{\dot{x}} = 0, \qquad F_y - \frac{d}{dt}F_{\dot{y}} = 0.$$

If $F \in C^2(\mathfrak{R})$ and if $\ddot{x}_o(t)$, $\ddot{y}_o(t)$ exist for all $t \in [t_a,t_b]$, then the latter *two equations may be replaced by the Weierstrass equation*

$$F_{x\dot{y}} - F_{y\dot{x}} + F_1(\dot{x}\ddot{y} - \dot{y}\ddot{x}) = 0$$

where

$$F_1 = \frac{F_{\dot{x}\dot{x}} + F_{\dot{y}\dot{y}}}{\dot{x}^2 + \dot{y}^2}.$$

We see quite clearly from (4.4.4) that if $F_x - (d/dt)F_{\dot{x}} = 0$ is satisfied, then $F_y - (d/dt)F_{\dot{y}} = 0$ is *not necessarily* satisfied, because from the first equation it follows only that either $\dot{y} = 0$ or $F_{x\dot{y}} - F_{y\dot{x}} + F_1(\dot{x}\ddot{y} - \dot{y}\ddot{x}) = 0$ (or both).

This situation is best explained by discussing a simple example: Let

$$F = \sqrt{\dot{x}^2 + \dot{y}^2} + \tfrac{1}{2}(x\dot{y} - y\dot{x})$$

(which is obviously positive homogeneous of the first degree). Then

$$F_x = \tfrac{1}{2}\dot{y}, \qquad F_{\dot{x}} = \frac{\dot{x}}{\sqrt{\dot{x}^2 + \dot{y}^2}} - \frac{y}{2},$$

and we have

$$F_x - \frac{d}{dt}F_{\dot{x}} = \tfrac{1}{2}\dot{y} - \frac{d}{dt}\left(\frac{\dot{x}}{\sqrt{\dot{x}^2 + \dot{y}^2}} - \frac{y}{2}\right) = -\frac{d}{dt}\left(\frac{\dot{x}}{\sqrt{\dot{x}^2 + \dot{y}^2}} - y\right) = 0,$$

or

$$\frac{\dot{x}}{\sqrt{\dot{x}^2 + \dot{y}^2}} - y = A.$$

This equation is clearly satisfied by $y = 1 - A$, $\dot{y} = 0$. Now, let us look at the second equation:

$$F_y = -\tfrac{1}{2}\dot{x}, \qquad F_{\dot{y}} = \frac{\dot{y}}{\sqrt{\dot{x}^2 + \dot{y}^2}} + \tfrac{1}{2}x,$$

$$F_y - \frac{d}{dt}F_{\dot{y}} = \frac{d}{dt}\left(\frac{\dot{y}}{\sqrt{\dot{x}^2 + \dot{y}^2}} + x\right) = 0$$

or

$$\frac{\dot{y}}{\sqrt{\dot{x}^2 + \dot{y}^2}} + x = B.$$

This equation is *not* satisfied by $\dot{y} = 0$, but it is satisfied by $\dot{x} = 0$ $(x = B - 1)$, which, in turn, does not satisfy the first equation.

Thus the problem has neither vertical nor horizontal extremals. In order to find the extremals, it is best to introduce the arc length s as parameter. Then

$$\left(\frac{dx}{ds}\right)^2 + \left(\frac{dy}{ds}\right)^2 = 1,$$

and the two Euler-Lagrange equations assume the simple form

$$\frac{dx}{ds} = y + A$$

$$\frac{dy}{ds} = -x + B,$$

and we obtain

$$(x - B)^2 + (y + A)^2 = 1,$$

that is, all circles with radius 1.

The same result is obtained, of course, if we use (4.4.5) instead of the Euler-Lagrange equations. We have

$$F_{x\dot{y}} = \tfrac{1}{2}, \qquad F_{y\dot{x}} = -\tfrac{1}{2}, \qquad F_1 = \frac{1}{(\dot{x}^2 + \dot{y}^2)^{3/2}}.$$

Hence, (4.4.5) reads

$$1 + \frac{\dot{x}\ddot{y} - \dot{y}\ddot{x}}{(\dot{x}^2 + \dot{y}^2)^{3/2}} = 0.$$

$$\left| \frac{\dot{x}\ddot{y} - \dot{y}\ddot{x}}{(\dot{x}^2 + \dot{y}^2)^{3/2}} \right| = \kappa$$

represents the curvature of $x = x(t)$, $y = y(t)$,† equation (4.4.5) as applied to this case states that the extremals have constant curvature 1, that is, are circles of radius 1.

PROBLEMS 4.4

*1. Derive the Weierstrass-Erdmann corner conditions for the x problem (2.9.1) and (2.9.4) from (4.4.2).

2. Find the extremal arc of the Brachistochrone problem in homogeneous formulation. (See also Prob. 4.1.1*a*.)

3. Same as in problem 2 for the problem of minimal surfaces of revolution. (See also Prob. 4.1.1*b*.)

4.5 DISCUSSION OF THE EULER-LAGRANGE EQUATIONS

In Sec. 2.11, we introduced the concept of a regular lineal element for the x problem in more than one unknown function. (See Theorem 2.12.) According to the definition that was given there, and in the notation of the homogeneous problem, a lineal element $(t,x(t),y(t),\dot{x}(t),\dot{y}(t))$ would be regular if

$$\begin{vmatrix} F_{\dot{x}\dot{x}} & F_{\dot{x}\dot{y}} \\ F_{\dot{y}\dot{x}} & F_{\dot{y}\dot{y}} \end{vmatrix} \neq 0.$$

However, this definition does not make sense in the context of the homogeneous problem.

First, due to the interdependence of the second partial derivatives of F with respect to $\dot{x},\dot{y}$, we have, in view of (4.2.7*a*, *b*, *c*), that

$$\begin{vmatrix} F_{\dot{x}\dot{x}} & F_{\dot{x}\dot{y}} \\ F_{\dot{y}\dot{x}} & F_{\dot{y}\dot{y}} \end{vmatrix} = F_{\dot{x}\dot{x}}F_{\dot{y}\dot{y}} - F_{\dot{x}\dot{y}}^2 = \dot{y}^2F_1\dot{x}^2F_1 - \dot{x}^2\dot{y}^2F_1^2 = 0,$$

that is, all lineal elements would appear to be singular.

† R. Courant and F. John, "Introduction to Calculus and Analysis," vol. 1, p. 355, Interscience Publishers, Inc., New York, 1965.

Second, the value of the parameter t in $(t,x(t),y(t),\dot{x}(t),\dot{y}(t))$ cannot possibly be of any consequence since it can be replaced by any other value by introduction of a new parameter representation that changes neither the coordinates nor the direction of the lineal element.

Since the essence of the concept of a lineal element are the coordinates of a point and the direction that is associated with this point, we shall henceforth refer to $(x,y,\dot{x},\dot{y})$ as a lineal element, with the understanding that $(x,y,\lambda\dot{x},\lambda\dot{y})$ for any $\lambda > 0$ is an equivalent representation of the same lineal element in view of the fact that $(\lambda\dot{x},\lambda\dot{y})$, $\lambda > 0$, represents the same direction as $(\dot{x},\dot{y})$.

To obtain a lead in our search for a meaningful concept of a regular lineal element, we recall our motive for introducing that concept: to make it possible to replace the Euler-Lagrange equations by a system of first-order differential equations so that the general existence and uniqueness theorems become applicable.

The following heuristic argument will put us on the right track: Let us suppose that we have chosen s, the arc length, as the parameter. Then

$$\left(\frac{dx}{ds}\right)^2 + \left(\frac{dy}{ds}\right)^2 = 1,$$

and we can write

$$\frac{dx}{ds} = \cos\alpha, \qquad \frac{dy}{ds} = \sin\alpha,$$

where α is the angle between the tangent to the curve and the positive x axis. Hence

$$\frac{d\alpha}{ds} = \frac{\dot{x}\ddot{y} - \dot{y}\ddot{x}}{(\dot{x}^2 + \dot{y}^2)^{3/2}} \dagger$$

Since $\dot{x}^2 + \dot{y}^2 = 1$, we have from (4.4.5) that

$$\frac{d\alpha}{ds} = \frac{F_{y\dot{x}} - F_{x\dot{y}}}{F_1},$$

and we see that if $F_1 \neq 0$, it appears to be possible to replace the Euler-Lagrange equations by three first-order equations in normal form, namely,

$$\frac{dx}{ds} = \cos\alpha, \qquad \frac{dy}{ds} = \sin\alpha, \qquad \frac{d\alpha}{ds} = \frac{F_{y\dot{x}} - F_{x\dot{y}}}{F_1}.$$

Thus it appears that the role of $f_{y'y'}$, or $\det|f_{y'_iy'_k}|$, is now assumed by the function F_1.

† *Ibid.*

We see that if $F_1(x,y,\dot{x},\dot{y}) \neq 0$ for a given lineal element $(x,y,\dot{x},\dot{y})$, then $F_1(x,y,\lambda\dot{x},\lambda\dot{y}) \neq 0$, $\lambda > 0$, for all equivalent representations $(x,y,\lambda\dot{x},\lambda\dot{y})$ of the same lineal element because F_1 is positive homogeneous of the (-3)rd degree in $(\dot{x},\dot{y})$.

Definition 4.5 *A lineal element* $(x,y,\dot{x},\dot{y})$ *is called* regular *if* $F_1(x,y,\dot{x},\dot{y}) \neq 0$. *An extremal arc is called* regular *if it consists of regular lineal elements only.*

We assume, for what is to follow, that γ_o: $x = x_o(t)$, $y = y_o(t)$, $t \in [t_a,t_b]$, is a regular extremal arc. We also assume that a parameter has been chosen such that

$$G(x(t),y(t),\dot{x}(t),\dot{y}(t)) = 1,$$

where G is a given function that is positive homogeneous of the first degree in $(\dot{x},\dot{y})$, $G \in C^2(\mathcal{R})$, and $G(x,y,\dot{x},\dot{y}) > 0$ in $\mathcal{R}$. [For example, $\sqrt{\dot{x}^2(t) + \dot{y}^2(t)} = 1$, arc length.] This is always possible. No matter what the original parameter representation in terms of a parameter τ, we can always introduce a new parameter t by means of $t = t(\tau) \in C^1[\tau_a,\tau_b]$, $t(\tau_a) = t_a$, $t(\tau_b) = t_b$, $dt(\tau)/d\tau > 0$, so that

$$G(x(t),y(t),\dot{x}(t),\dot{y}(t)) = G\left(x(\tau(t)),\, y(\tau(t)),\, \frac{dx(\tau(t))}{d\tau}\frac{d\tau}{dt},\, \frac{dy(\tau(t))}{d\tau}\frac{d\tau}{dt}\right)$$

$$= G\left(x(\tau(t)),\, y(\tau(t)),\, \frac{dx(\tau(t))}{d\tau},\, \frac{dy(\tau(t))}{d\tau}\right)\frac{d\tau}{dt} = 1,$$

if we choose only a parameter transformation for which

$$\frac{dt}{d\tau} = G\left(\bar{x}(\tau),\, \bar{y}(\tau),\, \frac{d\bar{x}(\tau)}{d\tau},\, \frac{d\bar{y}(\tau)}{d\tau}\right).$$

[For example, $(ds/d\tau) = \sqrt{\dot{x}^2(\tau) + \dot{y}^2(\tau)}$.]

Since F, G are positive homogeneous of the first degree in $\dot{x}$, $\dot{y}$,

$$\Phi(x,y,\dot{x},\dot{y},\mu) = F(x,y,\dot{x},\dot{y}) + \mu G(x,y,\dot{x},\dot{y})$$

is also positive homogeneous of the first degree in $(\dot{x},\dot{y})$, and all the formulas of Sec. 4.2 that were derived for F apply to Φ as well. In particular, with

$$\Phi_1 = \frac{\Phi_{\dot{x}\dot{x}} + \Phi_{\dot{y}\dot{y}}}{\dot{x}^2 + \dot{y}^2},$$

we have from (4.2.7*a*, *b*, *c*) that

$$\Phi_{\dot{x}\dot{x}} = \dot{y}^2\Phi_1, \qquad \Phi_{\dot{x}\dot{y}} = -\dot{x}\dot{y}\Phi_1, \qquad \Phi_{\dot{y}\dot{y}} = \dot{x}^2\Phi_1.$$

As in Sec. 2.12, we introduce new variables p, q by means of

$$p = \Phi_{\dot{x}}(x,y,\dot{x},\dot{y},\mu), \qquad q = \Phi_{\dot{y}}(x,y,\dot{x},\dot{y},\mu)$$

and try to solve

$$\begin{aligned} \Phi_{\dot{x}}(x,y,\dot{x},\dot{y},\mu) &= p \\ \Phi_{\dot{y}}(x,y,\dot{x},\dot{y},\mu) &= q \\ G(x,y,\dot{x},\dot{y}) &= 1 \end{aligned} \tag{4.5.1}$$

for $\dot{x}$, $\dot{y}$, μ.

Since G is positive homogeneous of the first degree in $(\dot{x},\dot{y})$ and, as such, satisfies Euler's identity (4.2.1), we have

$$\dot{x}G_{\dot{x}} + \dot{y}G_{\dot{y}} = G,$$

and hence

$$\frac{\partial(\Phi_{\dot{x}},\Phi_{\dot{y}},G)}{\partial(\dot{x},\dot{y},\mu)} = \begin{vmatrix} \Phi_{\dot{x}\dot{x}} & \Phi_{\dot{x}\dot{y}} & G_{\dot{x}} \\ \Phi_{\dot{x}\dot{y}} & \Phi_{\dot{y}\dot{y}} & G_{\dot{y}} \\ G_{\dot{x}} & G_{\dot{y}} & 0 \end{vmatrix} = -\Phi_1 G^2.$$

Since $\Phi(x,y,\dot{x},\dot{y},0) = F(x,y,\dot{x},\dot{y})$, we have $\Phi_1(x,y,\dot{x},\dot{y},0) = F_1(x,y,\dot{x},\dot{y})$, and consequently,

$$\left.\frac{\partial(\Phi_{\dot{x}},\Phi_{\dot{y}},G)}{\partial(\dot{x},\dot{y},\mu)}\right|_{\mu=0} = -F_1(x,y,\dot{x},\dot{y}) \neq 0$$

for any regular lineal element $(x,y,\dot{x},\dot{y})$—and hence for all lineal elements in a neighborhood of $(x,y,\dot{x},\dot{y})$ and all the equivalent lineal elements.

Since γ_o: $x = x_o(t)$, $y = y_o(t)$, $t \in [t_a,t_b]$, is assumed to be a regular extremal arc, we can solve (4.5.1) in a strong neighborhood of γ_o (for all lineal elements of arcs in a strong neighborhood of γ_o) and of $\mu = 0$ for $\dot{x}$, $\dot{y}$, μ and obtain

$$\begin{aligned} \dot{x} &= \varphi(x,y,p,q) \\ \dot{y} &= \psi(x,y,p,q) \\ \mu &= H(x,y,p,q) \end{aligned} \tag{4.5.2}$$

where $\varphi,\psi,H \in C^1$ in some domain of the (x,y,p,q) space that contains $x = x_o(t)$, $y = y_o(t)$, $p = \Phi_{\dot{x}}(x_o(t),y_o(t),\dot{x}_o(t),\dot{y}_o(t))$, $q = \Phi_{\dot{y}}(x_o(t),y_o(t),\dot{x}_o(t),\dot{y}_o(t))$ for all $t \in [t_a,t_b]$.

Since $\Phi(x,y,\dot{x},\dot{y},0) = F(x,y,\dot{x},\dot{y})$, we can write the Euler-Lagrange equations (4.4.3) as

$$\Phi_x(x,y,\dot{x},\dot{y},\mu) - \frac{d}{dt}\Phi_{\dot{x}}(x,y,\dot{x},\dot{y},\mu) = 0$$

$$\Phi_y(x,y,\dot{x},\dot{y},\mu) - \frac{d}{dt}\Phi_y(x,y,\dot{x},\dot{y},\mu) = 0$$

with the understanding that $\mu = 0$ and $G(x,y,\dot{x},\dot{y}) = 1$.

These four equations are, in view of (4.5.1) and (4.5.2), equivalent to

$$\begin{aligned}
\dot{x} &= \varphi(x,y,p,q) \\
\dot{y} &= \psi(x,y,p,q) \\
\dot{p} &= \Phi_x(x,y,\varphi(x,y,p,q),\psi(x,y,p,q),H(x,y,p,q)) \\
\dot{q} &= \Phi_y(x,y,\varphi(x,y,p,q),\psi(x,y,p,q),H(x,y,p,q)) \\
H(x,y,p,q) &= 0 \\
G(x,y,\varphi(x,y,p,q),\psi(x,y,p,q)) &= 1.
\end{aligned} \tag{4.5.3}$$

We have

$$d\Phi = \Phi_x\,dx + \Phi_y\,dy + \Phi_{\dot{x}}\,d\dot{x} + \Phi_{\dot{y}}\,d\dot{y} + \Phi_\mu\,d\mu$$

and since $\Phi_\mu = G = 1$ and $d\mu = dH$,

$$d\Phi = \Phi_x\,dx + \Phi_y\,dy + \Phi_{\dot{x}}\,d\dot{x} + \Phi_{\dot{y}}\,d\dot{y} + dH. \tag{4.5.4}$$

On the other hand, Φ, being positive homogeneous of the first degree in $(\dot{x},\dot{y})$, satisfies Euler's identity (4.2.1):

$$\Phi = \Phi_{\dot{x}}\dot{x} + \Phi_{\dot{y}}\dot{y}.$$

Hence
$$d\Phi = \Phi_{\dot{x}}\,d\dot{x} + \Phi_{\dot{y}}\,d\dot{y} + \dot{x}\,d\Phi_{\dot{x}} + \dot{y}\,d\Phi_{\dot{y}},$$

and since $d\Phi_{\dot{x}} = dp$, $d\Phi_{\dot{y}} = dq$,

$$d\Phi = p\,d\dot{x} + q\,d\dot{y} + \dot{x}\,dp + \dot{y}\,dq. \tag{4.5.5}$$

Equating the right sides of (4.5.4) and (4.5.5) yields, along the extremal arc,

$$dH = \dot{x}\,dp + \dot{y}\,dq - \Phi_x\,dx - \Phi_y\,dy.$$

Hence $\quad H_p = \dot{x}, \qquad H_q = \dot{y}, \qquad H_x = -\Phi_x, \qquad H_y = -\Phi_y.$

Since we have already utilized $G = 1$, we can now write (4.5.3) in the *canonical form*:

$$\begin{aligned}
\dot{x} &= H_p(x,y,p,q) \\
\dot{y} &= H_q(x,y,p,q) \\
\dot{p} &= -H_x(x,y,p,q) \\
\dot{q} &= -H_y(x,y,p,q),
\end{aligned} \tag{4.5.6}$$

to which we have to add

$$H(x,y,p,q) = 0. \tag{4.5.7}$$

We observe that the standard existence and uniqueness theorems can now be applied to (4.5.6).

Since, for any solution $(x,y,p,q) = (x(t),y(t),p(t),q(t))$ of (4.5.6),

$$\frac{dH}{dt} = H_x\dot{x} + H_y\dot{y} + H_q\dot{q} + H_p\dot{p} = H_xH_p + H_yH_q - H_qH_y - H_pH_x = 0,$$

we see that $H =$ constant is a first integral of (4.5.6). Only $H = 0$ can be utilized in view of (4.5.7), and this means that only three integration constants will be available. This is, in general, quite sufficient to solve the initial-value problem of finding a solution of the Euler-Lagrange equations that passes through a given point (x_o,y_o) with the given direction $(\dot{x},\dot{y})$. First, we determine $\lambda > 0$ so that $G(x_o,y_o,\dot{x}_o,\dot{y}_o) = 1$, where $\dot{x}_o = \lambda\dot{x}$, $\dot{y}_o = \lambda\dot{y}$. Then we obtain the initial values of p and q by

$$p_o = \Phi_{\dot{x}}(x_o,y_o,\dot{x}_o,\dot{y}_o,0), \qquad q_o = \Phi_{\dot{y}}(x_o,y_o,\dot{x}_o,\dot{y}_o,0)$$

and determine t, a, b, c in the solutions $x = x(t,a,b,c)$, $y = y(t,a,b,c)$, $p = p(t,a,b,c)$, $q = q(t,a,b,c)$ of [(4.5.6), (4.5.7)] so that the initial conditions $x(t_o,a_o,b_o,c_o) = x_o$, $y(t_o,a_o,b_o,c_o) = y_o$, $p(t_o,a_o,b_o,c_o) = p_o$, $q(t_o,a_o,b_o,c_o) = q_o$ are satisfied.

In order to understand the transformation of the Euler-Lagrange equations better, we consider the following example:

$$F(x,y,\dot{x},\dot{y}) = n(x,y)\sqrt{\dot{x}^2 + \dot{y}^2}.$$

Before we carry out the transformation of this section and discuss the resulting system of equations (4.5.6), let us first look at the Euler-Lagrange equations, or equivalently, at equation (4.4.5):

$$F_{x\dot{y}} - F_{y\dot{x}} + F_1(\dot{x}\ddot{y} - \dot{y}\ddot{x}) = 0.$$

We have

$$F_x = n_x\sqrt{\dot{x}^2 + \dot{y}^2}, \qquad F_{x\dot{y}} = \frac{n_x\dot{y}}{\sqrt{\dot{x}^2 + \dot{y}^2}}, \qquad F_{\dot{x}} = \frac{n\dot{x}}{\sqrt{\dot{x}^2 + \dot{y}^2}}$$

$$F_y = n_y\sqrt{\dot{x}^2 + \dot{y}^2}, \qquad F_{y\dot{x}} = \frac{n_y\dot{x}}{\sqrt{\dot{x}^2 + \dot{y}^2}}, \qquad F_{\dot{y}} = \frac{n\dot{y}}{\sqrt{\dot{x}^2 + \dot{y}^2}}$$

$$F_1 = \frac{n}{(\dot{x}^2 + \dot{y}^2)^{3/2}}.$$

Hence,
$$\frac{n_x\dot{y} - n_y\dot{x}}{\sqrt{\dot{x}^2 + \dot{y}^2}} + \frac{n(\dot{x}\ddot{y} - \dot{y}\ddot{x})}{(\dot{x}^2 + \dot{y}^2)^{3/2}} = 0.$$

We choose the arc length s as parameter: $\sqrt{\dot{x}^2 + \dot{y}^2} = 1$. Then

$$\frac{\dot{x}}{\sqrt{\dot{x}^2 + \dot{y}^2}} = \cos\alpha, \qquad \frac{\dot{y}}{\sqrt{\dot{x}^2 + \dot{y}^2}} = \sin\alpha,$$

and since

$$\frac{d\alpha}{ds} = \frac{\dot{x}\ddot{y} - \dot{y}\ddot{x}}{(\dot{x}^2 + \dot{y}^2)^{3/2}},$$

we obtain for (4.4.5),

$$n_y \cos \alpha - n_x \sin \alpha = n \frac{d\alpha}{ds}. \tag{4.5.8}$$

Let us now go through the procedure that was developed in this section: We choose the arc length as parameter, that is,

$$G(x,y,\dot{x},\dot{y}) = \sqrt{\dot{x}^2 + \dot{y}^2} = 1,$$

and we obtain for Φ:

$$\Phi(x,y,\dot{x},\dot{y},\mu) = (n(x,y) + \mu)\sqrt{\dot{x}^2 + \dot{y}^2}.$$

Then

$$p = \Phi_{\dot{x}} = (n + \mu)\frac{\dot{x}}{\sqrt{\dot{x}^2 + \dot{y}^2}},$$

$$q = \Phi_{\dot{y}} = (n + \mu)\frac{\dot{y}}{\sqrt{\dot{x}^2 + \dot{y}^2}}.$$

We observe that

$$\Phi_1 = \frac{n + \mu}{(\dot{x}^2 + \dot{y}^2)^{3/2}} \quad \text{and hence} \quad \Phi_1 \big|_{\mu=0} \neq 0.$$

Thus we can solve

$$\Phi_{\dot{x}} = p$$
$$\Phi_{\dot{y}} = q$$
$$\sqrt{\dot{x}^2 + \dot{y}^2} = 1$$

for $\dot{x}$, $\dot{y}$, μ and obtain

$$\dot{x} = \frac{p}{\sqrt{p^2 + q^2}}$$

$$\dot{y} = \frac{q}{\sqrt{p^2 + q^2}}$$

$$\mu = \sqrt{p^2 + q^2} - n(x,y) \equiv H(x,y,p,q),$$

and we have for (4.5.6),

$$\frac{dx}{ds} = \frac{p}{\sqrt{p^2 + q^2}}$$

$$\frac{dy}{ds} = \frac{q}{\sqrt{p^2 + q^2}}$$

$$\frac{dp}{ds} = n_x$$

$$\frac{dq}{ds} = n_y.$$

We want only those solutions for which

$$H = \sqrt{p^2 + q^2} - n(x,y) = 0,$$

that is, $p^2 + q^2 = n^2$. Since s is the parameter ($G = 1$), we have

$$\frac{dx}{ds} = \cos \alpha, \qquad \frac{dy}{ds} = \sin \alpha$$

and hence

$$p = n \cos \alpha, \qquad q = n \sin \alpha,$$

and we obtain for $(dp/ds) = n_x$, $(dq/ds) = n_y$,

$$\frac{dp}{ds} = (n_x \cos \alpha + n_y \sin \alpha) \cos \alpha - n \sin \alpha \frac{d\alpha}{ds} = n_x$$

$$\frac{dq}{ds} = (n_x \cos \alpha + n_y \sin \alpha) \sin \alpha + n \cos \alpha \frac{d\alpha}{ds} = n_y.$$

Since $(\cos \alpha, \sin \alpha) \neq (0,0)$, both equations imply

$$n_y \cos \alpha - n_x \sin \alpha = n \frac{d\alpha}{ds},$$

which is again (4.5.8).

After division by n, this equation may also be written as

$$\frac{\partial \log n}{\partial y} \cos \alpha - \frac{\partial \log n}{\partial x} \sin \alpha = \frac{d\alpha}{ds}.$$

This equation relates the curvature $(d\alpha/ds)$ of the extremal to every point (x,y) and every angle α of the extremal.

PROBLEMS 4.5

1. Impose suitable conditions and show:
   (a) A regular extremal arc is uniquely determined by an initial point and an initial direction.
   (b) If γ_o is a regular extremal arc, then it can be represented by $x = x_o(t)$, $y = y_o(t)$, $t \in [t_a,t_b]$, so that $x_o,y_o \in C^2[t_a,t_b]$.
2. Write the canonical system (4.5.6) for the following homogeneous variational problems, using the arc length as parameter:

   (a) $$\int_{P_a}^{P_b} \sqrt{\dot{x}^2 + \dot{y}^2}\, dt \to \text{minimum}$$

   (b) $$\int_{P_a}^{P_b} y\sqrt{\dot{x}^2 + \dot{y}^2}\, dt \to \text{minimum}$$

   (c) $$\int_{P_a}^{P_b} \frac{\sqrt{\dot{x}^2 + \dot{y}^2}}{\sqrt{y}}\, dt \to \text{minimum}$$

4.6 TRANSVERSALITY CONDITION

We now consider the problem of minimizing the functional

$$I[\gamma] = \int_{P_a}^{P_\epsilon} F(x,y,\dot{x},\dot{y})\, dt,$$

where P_a is fixed with the coordinates (a,y_a) and where P_ϵ lies somewhere on the simple smooth curve

$$C\begin{cases} x = \varphi(\varepsilon) \\ y = \psi(\varepsilon). \end{cases}$$

It is *not* possible to freely vary one coordinate at a time, because, unless very complicated restrictions are placed on the variation, the varied curve will in general *not* terminate on the curve C—it is most probable that it will miss it entirely!

In order to solve our problem with as little effort as possible, we shall now depart somewhat from our pattern that was based on the concept of the Gâteaux variation of a functional.

We assume that the simple, sectionally smooth arc γ_o: $x = x_o(t)$, $y = y_o(t)$ yields the minimum for $I[\gamma]$ and that γ_o terminates on C at the point $P_{\epsilon_o}(\varphi(\varepsilon_o),\psi(\varepsilon_o))$. Further, t_a corresponds to P_a, and t_{ϵ_o} corresponds to P_{ϵ_o}.

We note that γ_o has to yield a minimum also relative to all other simple, smooth arcs that join the same two points P_a and P_{ϵ_o}, and hence $x = x_o(t)$, $y = y_o(t)$ have to satisfy the Euler-Lagrange equations in integrated form (4.4.1).

Now let us consider the following family of nonlinear variations γ_ϵ of γ_o (see Fig. 4.3):

$$\left.\begin{aligned} x = x(t,\varepsilon) &\equiv x_o(t) + \frac{t - t_a}{t_{\epsilon_o} - t_a}\,[\varphi(\varepsilon) - x_o(t_{\epsilon_o})] \\ y = y(t,\varepsilon) &\equiv y_o(t) + \frac{t - t_a}{t_{\epsilon_o} - t_a}\,[\psi(\varepsilon) - y_o(t_{\epsilon_o})] \end{aligned}\right\}\gamma_\epsilon \qquad (4.6.1)$$

By hypothesis, $x_o(t_{\epsilon_o}) = \varphi(\varepsilon_o)$, $y_o(t_{\epsilon_o}) = \psi(\varepsilon_o)$. Therefore, $\gamma_{\epsilon_o} = \gamma_o$. Since $x(t_a,\varepsilon) = x_o(t_a) = a$, $y(t_a,\varepsilon) = y_o(t_a) = y_a$ and $x(t_{\epsilon_o},\varepsilon) = \varphi(\varepsilon)$,

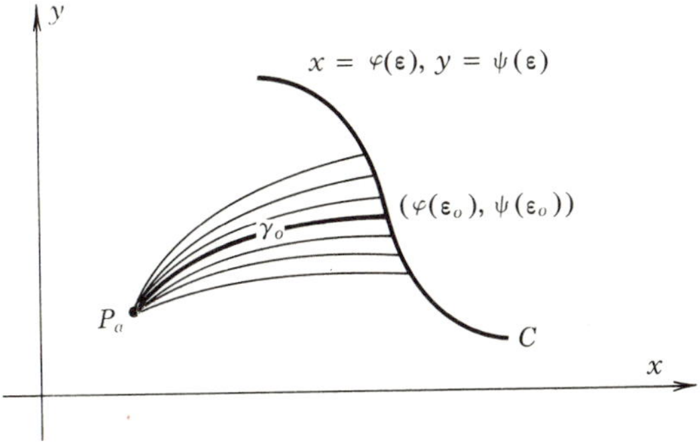

Figure 4.3

$y(t_{\varepsilon_o},\varepsilon) = \psi(\varepsilon)$, all arcs γ_ε emanate from P_a for $t = t_a$ and terminate on C for $t = t_{\varepsilon_o}$.

Let us now consider

$$I[\gamma_\varepsilon] = \int_{P_a}^{P_\varepsilon} F(x(t,\varepsilon),y(t,\varepsilon),\dot{x}(t,\varepsilon),\dot{y}(t,\varepsilon))\, dt.$$

Since $x = x(t,\varepsilon)$, $y = y(t,\varepsilon)$ are known functions of t and ε, the functional $I[\gamma_\varepsilon]$ thus appears as a *function* of ε: $I[\gamma_\varepsilon] = I(\varepsilon)$. By hypothesis, $I(\varepsilon)$ assumes a minimum for $\varepsilon = \varepsilon_o$. If we assume that $I'(\varepsilon)$ exists at $\varepsilon = \varepsilon_o$—which is ensured by our standing assumption that $F \in C^2(\mathfrak{R})$ in a suitably chosen domain $\mathfrak{R}$—then we have as necessary condition for a minimum that

$$I'(\varepsilon_o) = 0. \tag{4.6.2}$$

We obtain

$$I'(\varepsilon) = \int_{t_a}^{t_{\varepsilon_o}} \left(F_x \frac{\partial x}{\partial \varepsilon} + F_y \frac{\partial y}{\partial \varepsilon} + F_{\dot{x}} \frac{\partial \dot{x}}{\partial \varepsilon} + F_{\dot{y}} \frac{\partial \dot{y}}{\partial \varepsilon}\right) dt, \tag{4.6.3}$$

where the partial derivatives of F in this integrand all have the argument $(x(t,\varepsilon),y(t,\varepsilon),\dot{x}(t,\varepsilon),\dot{y}(t,\varepsilon))$. (Note that the upper integration limit is fixed.)

We consider (4.6.3) at $\varepsilon = \varepsilon_o$ and apply integration by parts to the first two terms under the integral. With

$$\phi(x_o(t),y_o(t),\dot{x}_o(t),\dot{y}_o(t)) = \int_{t_a}^{t} F_x(x_o(u),y_o(u),\dot{x}_o(u),\dot{y}_o(u))\, du,$$

$$\psi(x_o(t),y_o(t),\dot{x}_o(t),\dot{y}_o(t)) = \int_{t_a}^{t} F_y(x_o(u),y_o(u),\dot{x}_o(u),\dot{y}_o(u))\, du,$$

we obtain for $I'(\varepsilon_o) = 0$:

$$\left[\phi \frac{\partial x}{\partial \varepsilon} + \psi \frac{\partial y}{\partial \varepsilon}\right]_{t_a}^{t_{\varepsilon_o}} + \int_{t_a}^{t_{\varepsilon_o}} \left[\frac{\partial \dot{x}}{\partial \varepsilon}(F_{\dot{x}} - \phi) + \frac{\partial \dot{y}}{\partial \varepsilon}(F_{\dot{y}} - \psi)\right] dt = 0. \tag{4.6.4}$$

Since $x_o(t)$, $y_o(t)$ have to satisfy the Euler-Lagrange equations in integrated form, we have from (4.4.1),

$$F_{\dot{x}} - \phi = C_1, \qquad F_{\dot{y}} - \psi = C_2$$

and also

$$\phi\big|_{t_{\varepsilon_o}} = F_{\dot{x}}\big|_{t_{\varepsilon_o}} - C_1, \qquad \psi\big|_{t_{\varepsilon_o}} = F_{\dot{y}}\big|_{t_{\varepsilon_o}} - C_2,$$

while

$$\left.\frac{\partial x}{\partial \varepsilon}\right|_{t_a} = \left.\frac{\partial y}{\partial \varepsilon}\right|_{t_a} = 0$$

because $x(t_a,\varepsilon) = a$, $y(t_a,\varepsilon) = y_a$ for all ε.

Hence we obtain from (4.6.4),

$$\left[F_{\dot{x}}\frac{\partial x}{\partial \varepsilon}+F_{\dot{y}}\frac{\partial y}{\partial \varepsilon}\right]_{t=t_{\epsilon_o},\epsilon=\epsilon_o}=0.$$

From (4.6.1),

$$\frac{\partial x(t,\varepsilon)}{\partial \varepsilon}=\frac{t-t_a}{t_{\epsilon_o}-t_a}\varphi'(\varepsilon),\qquad \frac{\partial y(t,\varepsilon)}{\partial \varepsilon}=\frac{t-t_a}{t_{\epsilon_o}-t_a}\psi'(\varepsilon),$$

and hence

$$\left.\frac{\partial x(t,\varepsilon)}{\partial \varepsilon}\right|_{t_{\epsilon_o},\epsilon_o}=\varphi'(\varepsilon_o),\qquad \left.\frac{\partial y(t,\varepsilon)}{\partial \varepsilon}\right|_{t_{\epsilon_o},\epsilon_o}=\psi'(\varepsilon_o),$$

and we obtain the *transversality condition* at the endpoint that corresponds to the parameter value t_{ϵ_o}:

$$F_{\dot{x}}(x(t_{\epsilon_o}),y(t_{\epsilon_o}),\dot{x}(t_{\epsilon_o}),\dot{y}(t_{\epsilon_o}))\varphi'(\varepsilon_o)$$
$$+F_{\dot{y}}(x(t_{\epsilon_o}),y(t_{\epsilon_o}),\dot{x}(t_{\epsilon_o}),\dot{y}(t_{\epsilon_o}))\psi'(\varepsilon_o)=0. \quad (4.6.5)$$

(An analogous condition is obtained for a variable beginning point.)

It may be of interest to note that the transversality condition (2.8.5) for the x problem can easily be deduced from (4.6.5): With $\dot{x}\neq 0$,

$$F(x,y,\dot{x},\dot{y})=\dot{x}F\left(x,y,1,\frac{\dot{y}}{\dot{x}}\right)=\dot{x}f(x,y,y').$$

Then $$F_{\dot{x}}=F\left(x,y,1,\frac{\dot{y}}{\dot{x}}\right)-\frac{\dot{x}\dot{y}}{\dot{x}^2}F_{\dot{y}}\left(x,y,1,\frac{\dot{y}}{\dot{x}}\right)=f(x,y,y')-y'f_{y'}(x,y,y'),$$

$$F_{\dot{y}}=\frac{\dot{x}}{\dot{x}}F_{\dot{y}}\left(x,y,1,\frac{\dot{y}}{\dot{x}}\right)=f_{y'}(x,y,y').$$

Since, for $\varphi'(\varepsilon_o)\neq 0$,

$$\frac{\psi'(\varepsilon_o)}{\varphi'(\varepsilon_o)}=\phi'(b),$$

where $y=\phi(x)$ represents the terminal curve, we have

$$F_{\dot{x}}\varphi'(\varepsilon_o)+F_{\dot{y}}\psi'(\varepsilon_o)=\varphi'(\varepsilon_o)[f-y'f_{y'}+\phi'f_{y'}]_b,$$

and (2.8.5) follows readily.

Since $\varphi'(\varepsilon_o)$ may now vanish, as long as $\psi'(\varepsilon_o)$ does not vanish at the same time, the *natural boundary condition* for the problem where the endpoint may vary along the vertical line $\varphi(\varepsilon)=b$ appears now as a special case of (4.6.5), namely,

$$F_{\dot y}\big|_{x=b} = 0. \tag{4.6.6}$$

If the extremal is to terminate on a horizontal line $\psi(\varepsilon) = c$, then we obtain from (4.6.5)

$$F_{\dot x}\big|_{y=c} = 0. \tag{4.6.7}$$

As an illustration of condition (4.6.6), we shall now discuss the *navigation problem* which we discussed earlier, in Sec. 1.1, problem D.

Let us assume that we have a body of water, the surface velocity of which is given by

$$\begin{cases} u = u(x,y) \\ v = v(x,y). \end{cases}$$

A vessel with constant speed c in still water is moving across this body of water. We assume that the surface velocity of the water is retained to a sufficient depth so that we can treat this problem as a two-dimensional problem and that $u^2 + v^2 < c^2$.

We obtain for the equations of motion of the vessel,

$$\left.\begin{aligned} \frac{dx}{dt} &= u(x,y) + c\cos\alpha \\ \frac{dy}{dt} &= v(x,y) + c\sin\alpha \end{aligned}\right\},$$

where t denotes the time and where the angle α depends on the course of the vessel. Elimination of α yields

$$\left(\frac{dx}{dt} - u\right)^2 + \left(\frac{dy}{dt} - v\right)^2 = c^2.$$

We introduce a new parameter τ by $t = t(\tau)$, $(dt/d\tau) > 0$, $t(0) = 0$, and denote the derivatives with respect to τ by a dot. Then

$$\left(\dot x\frac{d\tau}{dt} - u\right)^2 + \left(\dot y\frac{d\tau}{dt} - v\right)^2 = c^2.$$

Multiplication by $\dot t^2$ yields

$$(\dot x - u\dot t)^2 + (\dot y - v\dot t)^2 = c^2\dot t^2,$$

and we obtain

$$\dot t = \frac{-(\dot x u + \dot y v) + \sqrt{(\dot x u + \dot y v)^2 + (c^2 - u^2 - v^2)(\dot x^2 + \dot y^2)}}{c^2 - u^2 - v^2}.$$

After some simplifications, we obtain for the traveling time,

$$t = \int_{\tau_a}^{\tau_b} \frac{\sqrt{c^2(\dot x^2 + \dot y^2) - (\dot x v - \dot y u)^2} - (\dot x u + \dot y v)}{c^2 - u^2 - v^2}\, d\tau.$$

We note that the integrand is positive homogeneous of the first degree in $(\dot x,\dot y)$. Hence the problem is a legitimate homogeneous problem.

Figure 4.4

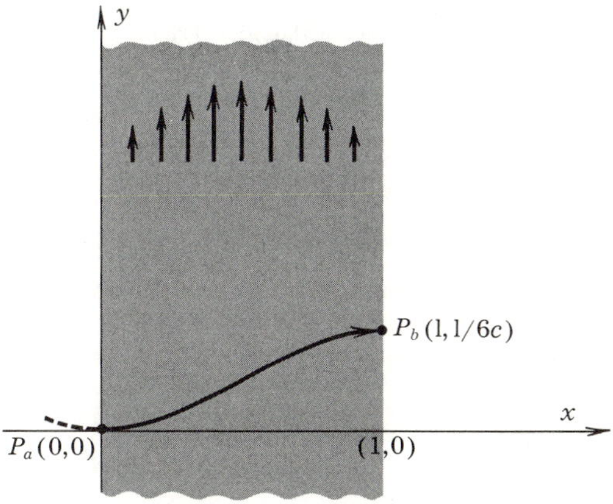

Let us now specialize and consider the concrete problem of finding the course of shortest crossing time of a river of width 1 with banks parallel to the y axis and a parabolic velocity profile

$$\begin{cases} u(x,y) = 0 \\ v(x,y) = x - x^2. \end{cases}$$

As a point of departure, we may choose without loss of generality the point $P_a(0,0)$. (See Fig. 4.4.) In this special case, we have

$$t = \int_0^{\tau_b} \frac{\sqrt{c^2(\dot{x}^2 + \dot{y}^2) - \dot{x}^2v^2} - \dot{y}v}{c^2 - v^2}\, d\tau.$$

We have $F_1 = 1/[c^2(\dot{x}^2 + \dot{y}^2) - \dot{x}^2v^2]^{3/2} > 0$ for all $(x,y,\dot{x},\dot{y})$, and hence we may immediately consider the Euler-Lagrange equations in differentiated form.

Since F is independent of y, $F_y = 0$ and the second of the Euler-Lagrange equations (4.4.3) simplifies to

$$F_{\dot{y}} = \text{constant}.$$

At the bank opposite P_a, the natural boundary condition (4.6.6)

$$F_{\dot{y}}\,|_{x=1} = 0$$

has to hold. Hence

$$F_{\dot{y}} = 0.$$

Now,
$$F_{\dot{y}} = \frac{1}{c^2 - v^2}\left[\frac{c^2\dot{y}}{\sqrt{c^2(\dot{x}^2 + \dot{y}^2) - \dot{x}^2v^2}} - v\right].$$

Hence $F_{\dot{y}} = 0$ yields

$$c^2\dot{y} - v\sqrt{c^2(\dot{x}^2 + \dot{y}^2) - \dot{x}^2v^2} = 0,$$

or after simplifications,

$$c\dot{y} = \pm v\dot{x}.$$

We eliminate τ and obtain

$$y = \pm\frac{1}{c}\int v\, dx = \pm\frac{1}{c}\int (x - x^2)\, dx = \pm\frac{1}{c}\left(\frac{x^2}{2} - \frac{x^3}{3}\right) + A.$$

Since x = constant is *not* a solution of this equation, the solution which we just obtained is also a solution of the other Euler-Lagrange equation [see (4.4.4)].

$P_a = (0,0)$, hence $A = 0$, and we have as the course of shortest crossing time,

$$y = \pm\frac{1}{c}\left(\frac{x^2}{2} - \frac{x^3}{3}\right)$$

(see Fig. 4.4). For the shortest crossing time itself, we have, with

$$\left.\begin{aligned} x &= \tau \\ y &= \frac{1}{c}\left(\frac{\tau^2}{2} - \frac{\tau^3}{3}\right) \end{aligned}\right\} 0 \le \tau \le 1$$

and $v = \tau - \tau^2$,

$$t = \int_0^1 \frac{\sqrt{c^2(1 + (1/c^2)(\tau - \tau^2)^2) - (\tau - \tau^2)^2} \pm (1/c)(\tau - \tau^2)^2}{c^2 - (\tau - \tau^2)^2}\, d\tau = \frac{1}{c}.$$

(The crossing time varies inversely with the speed of the vessel. See Prob. 4.5.5.)

PROBLEMS 4.6

1. Formulate and solve the navigation problem for a river of width 1, the banks of which are parallel to the x axis. Assume again a parabolic velocity profile of the river.
2. In the navigation problem of this section, assume that $u = 0$ and $v(0) = v(1)$. Show that the angle of departure is equal to the angle of arrival.
3. Consider the navigation problem of crossing a river, the banks of which are parallel to the y axis, with variable endpoint and variable beginning point. Assume that $v = v(x)$, $u = u(x)$. Show that if the boat were steered tangential to its path along a finite portion of the crossing path, then the resulting crossing would not occur in minimum time, unless $u = v = 0$.

*4. Deduce the Weierstrass-Erdmann corner conditions from (4.6.5).

*5. Show that $1/c$ is the value of the integral which represents the shortest crossing time and which is stated at the end of this section.

4.7 CARATHÉODORY'S INDICATRIX

The indicatrix is to the homogeneous problem what the characteristic is to the x problem. (See Secs. 2.8, 2.9, and 3.8.)

We shall assume that

$$F(x,y,\dot{x},\dot{y}) > 0$$

in some neighborhood N of (x_o,y_o) and for *all* $\dot{x}^2 + \dot{y}^2 > 0$. A homogeneous problem with integrand which satisfies this condition is called *positive definite* in a neighborhood of x_o, y_o.

The curve

$$F(x_o,y_o,\xi,\eta) = 1 \tag{4.7.1}$$

in the ξ,η plane is called *Carathéodory's indicatrix*† *of F at* (x_o,y_o).

† Constantin Carathéodory, 1873–1950.

In order to find out some of the basic properties of the indicatrix, it is of advantage to introduce polar coordinates

$$\xi = r\cos\theta$$
$$\eta = r\sin\theta.$$

Then, in view of the positive homogeneity of F of the first degree in ξ, η, we obtain, instead of (4.7.1),

$$r = \frac{1}{F(x_o, y_o, \cos\theta, \sin\theta)}, \qquad 0 \le \theta \le 2\pi, \tag{4.7.2}$$

as the equation of the indicatrix in polar coordinates. Since $r(\theta) = r(\theta + 2\pi)$, we see that the indicatrix is a *closed curve.*

Since $F(x_o, y_o, \cos\theta, \sin\theta) > 0$ for all $0 \le \theta \le 2\pi$ and since F is a continuous function of θ in a closed interval, it follows that F assumes on $[0,2\pi]$ a positive minimum m and a positive maximum M. Hence, $0 < (1/M) \le r \le (1/m)$. It also follows that the indicatrix contains the origin of the ξ,η coordinate system in its interior.

From (4.7.1),

$$\operatorname{grad} F(x_o,y_o,\xi,\eta) = (F_{\dot{x}},F_{\dot{y}}),$$

that is, $(F_{\dot{x}},F_{\dot{y}})$ represents the direction orthogonal to the indicatrix at the point ξ, η. (See Fig. 4.5.)

Suppose now that the extremal γ_o terminates on the curve

$$C\begin{cases} x = \varphi(\varepsilon) \\ y = \psi(\varepsilon) \end{cases}$$

at the point $x_o = \varphi(\varepsilon_o)$, $y_o = \psi(\varepsilon_o)$. Then the transversality condition (4.6.5)

$$F_{\dot{x}}(x_o,y_o,\xi,\eta)\varphi'(\varepsilon_o) + F_{\dot{y}}(x_o,y_o,\xi,\eta)\psi'(\varepsilon_o) = 0$$

has to hold. This means that

$$(F_{\dot{x}},F_{\dot{y}}) \perp (\varphi'(\varepsilon_o),\psi'(\varepsilon_o)).$$

Since $(F_{\dot{x}},F_{\dot{y}})$ is orthogonal to the tangent line T to the indicatrix at the point (ξ,η), this means, in turn, that $(\varphi'(\varepsilon_o),\psi'(\varepsilon_o))$ has to be parallel to T.

So we see that the indicatrix yields, for every terminal direction $\dot{x} = \xi$, $\dot{y} = \eta$ of the extremal (which is represented by the polar angle θ) the corresponding transversal direction as the direction of the tangent line at the point (ξ,η).

Conversely, if the direction of the terminal curve C at the point of impact is given, then the terminal direction of the extremal is given by the coordinates ξ, η of that point on the indicatrix where a line with the direction numbers $(\varphi'(\varepsilon_o),\psi'(\varepsilon_o))$ is tangent to the indicatrix.

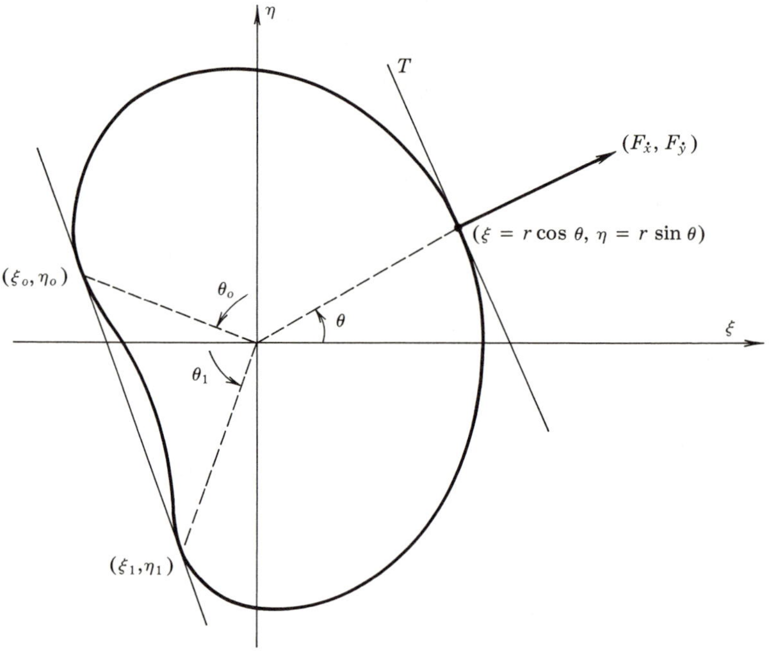

Figure 4.5

For a Fermat-type problem with

$$F = n(x,y)\sqrt{\dot{x}^2 + \dot{y}^2}, \qquad n(x,y) > 0,$$

the indicatrix is a circle,

$$\xi^2 + \eta^2 = \frac{1}{n^2(x,y)}$$

for every point (x,y). Hence we see again that transversality for such problems means orthogonality.

Let us now turn to a discussion of the corner conditions (4.4.2). In particular, we shall investigate the possibility of a corner at the point (x_o,y_o).

The tangent line to the indicatrix (4.7.1) at the point (ξ_o,η_o) is given by

$$(\xi - \xi_o)F_{\dot{x}}(x_o,y_o,\xi_o,\eta_o) + (\eta - \eta_o)F_{\dot{y}}(x_o,y_o,\xi_o,\eta_o) = 0.$$

By Euler's identity (4.2.1) and by (4.7.1),

$$\xi_o F_{\dot{x}}(x_o,y_o,\xi_o,\eta_o) + \eta_o F_{\dot{y}}(x_o,y_o,\xi_o,\eta_o) = 1,$$

and the equation of the tangent line simplifies to

$$\xi F_{\dot{x}}(x_o,y_o,\xi_o,\eta_o) + \eta F_{\dot{y}}(x_o,y_o,\xi_o,\eta_o) = 1,$$

or, since $F_{\dot{x}}$, $F_{\dot{y}}$ are positive homogeneous of the zeroth degree in ξ, η,

$$\xi F_{\dot{x}}(x_o, y_o, \cos\theta_o, \sin\theta_o) + \eta F_{\dot{y}}(x_o, y_o, \cos\theta_o, \sin\theta_o) = 1. \quad (4.7.3)$$

Suppose now that there is a corner at (x_o,y_o) and that the incoming direction is given by θ_o and the outgoing direction by θ_1. Then, by the corner conditions (4.4.2),

$$F_{\dot{x}}(x_o, y_o, \cos\theta_o, \sin\theta_o) = F_{\dot{x}}(x_o, y_o, \cos\theta_1, \sin\theta_1)$$

$$F_{\dot{y}}(x_o, y_o, \cos\theta_o, \sin\theta_o) = F_{\dot{y}}(x_o, y_o, \cos\theta_1, \sin\theta_1),$$

and it follows that the tangent lines to the indicatrix at the point (ξ_o,η_o) corresponding to the direction θ_o and at (ξ_1,η_1) corresponding to the direction θ_1 have to have the same equation, i.e., have to be the same line. So we have to have a double tangent that touches the indicatrix at the point (ξ_o,η_o) as well as at the point (ξ_1,η_1). (See Fig. 4.5.)

It follows readily that if the indicatrix for a given point (x_o,y_o) is convex, then a double tangent is not possible and hence a corner cannot occur. (Since the indicatrix for a Fermat-type problem is a circle for all points of the x,y plane, it follows that a Fermat-type problem cannot have corner solutions.)

We shall see in a moment that the convexity of the indicatrix is intimately related to the sign of the function F_1 in a manner similar to that with which the convexity of the characteristic is related to the sign of $f_{y'y'}$. (See also the end of Sec. 2.9.) In order to study this relationship, we shall discuss the change of the angle of the tangent line to the indicatrix as we move along the indicatrix in the positive direction (θ increases from 0 to 2π), or equivalently, the change of the angle of the normal to the indicatrix.

The normal direction to the indicatrix is given by $(F_{\dot{x}},F_{\dot{y}})$. Hence

$$\tan\nu = \frac{F_{\dot{y}}}{F_{\dot{x}}},$$

or

$$\nu = \arctan\frac{F_{\dot{y}}(x_o, y_o, \cos\theta, \sin\theta)}{F_{\dot{x}}(x_o, y_o, \cos\theta, \sin\theta)},$$

if ν is the angle of the normal with the positive ξ axis. Then

$$\frac{d\nu}{d\theta} = \frac{(-F_{\dot{y}\dot{x}}\sin\theta + F_{\dot{y}\dot{y}}\cos\theta)F_{\dot{x}} - (-F_{\dot{x}\dot{x}}\sin\theta + F_{\dot{x}\dot{y}}\cos\theta)F_{\dot{y}}}{F_{\dot{x}}^2 + F_{\dot{y}}^2}.$$

By (4.2.7*a*, *b*, *c*),

$$F_{\dot{x}\dot{x}}(x_o, y_o, \cos\theta, \sin\theta) = \sin^2\theta F_1(x_o, y_o, \cos\theta, \sin\theta)$$

$$F_{\dot{x}\dot{y}}(x_o, y_o, \cos\theta, \sin\theta) = -\sin\theta\cos\theta F_1(x_o, y_o, \cos\theta, \sin\theta)$$

$$F_{\dot{y}\dot{y}}(x_o, y_o, \cos\theta, \sin\theta) = \cos^2\theta F_1(x_o, y_o, \cos\theta, \sin\theta).$$

Hence
$$\frac{d\nu}{d\theta} = \frac{F_1(F_{\dot{x}}\cos\theta + F_{\dot{y}}\sin\theta)}{F_{\dot{x}}^2 + F_{\dot{y}}^2}.$$

Finally, from Euler's identity (4.2.1),

$$F_{\dot{x}}\cos\theta + F_{\dot{y}}\sin\theta = F(x, y, \cos\theta, \sin\theta) = \frac{1}{r},$$

and we have

$$\frac{d\nu}{d\theta} = \frac{F_1}{r(F_{\dot{x}}^2 + F_{\dot{y}}^2)}. \tag{4.7.4}$$

So we see that:

If $F_1 > 0$, then $(d\nu/d\theta) > 0$ and the indicatrix is *convex*.
If $F_1 < 0$, then $(d\nu/d\theta) < 0$ and the indicatrix is *concave*.
If $F_1 = 0$ and $F_1 > 0$ on the one side and $F_1 < 0$ on the other side, then $(d\nu/d\theta)$ changes sign and the indicatrix has a point of inflection.

[Observe that $(d\nu/d\theta)\cdot(d\theta/ds)$, where s represents the arc length, is the curvature of the indicatrix.]

We can see from this discussion that a corner solution is possible only if F_1 changes sign somewhere.

The following example—which is quite remarkable in many respects—will demonstrate this quite clearly. Let

$$F(x,y,\dot{x},\dot{y}) = \frac{\dot{x}^2 + \dot{y}^2}{\sqrt{2(\dot{x}^2 + \dot{y}^2)} + \dot{x}} \tag{4.7.5}$$

We see that

$$F > 0 \qquad \text{for } all\ (x,y) \text{ and for all } \dot{x}^2 + \dot{y}^2 > 0.$$

The indicatrix, which is, for this case, independent of (x,y), is given by

$$\xi^2 + \eta^2 = \sqrt{2(\xi^2 + \eta^2)} + \xi,$$

or, in polar coordinates $\xi = r\cos\theta$, $\eta = r\sin\theta$,

$$r = \sqrt{2} + \cos\theta.$$

This represents a special case of a *Pascal snail* and is depicted in Fig. 4.6.

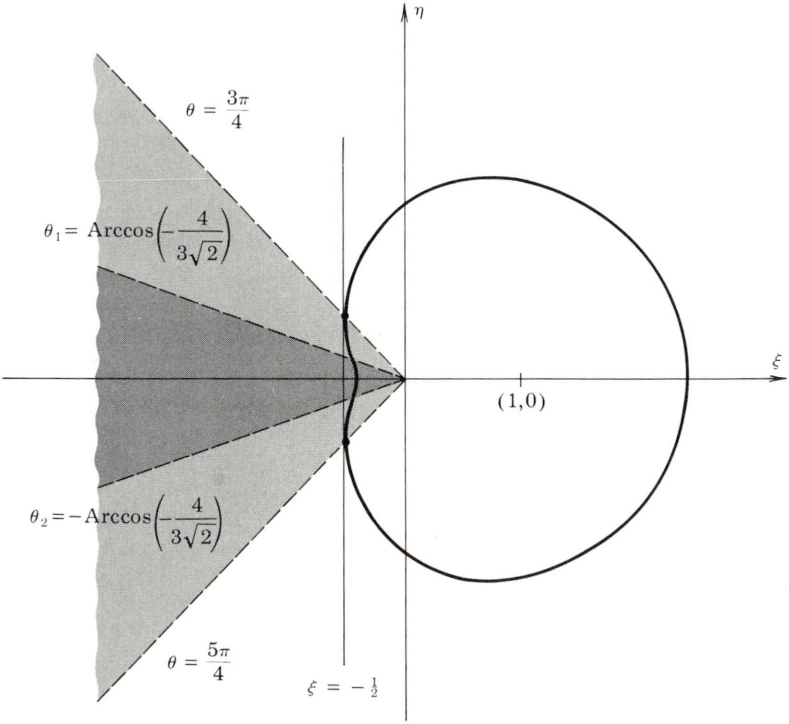

Figure 4.6

Apparently, there is a vertical double tangent

$$r = \frac{b}{\cos \theta}$$

for $b < 0$. In order to find b and the angle θ for which the line touches the indicatrix, we intersect the line with the Pascal snail:

$$\frac{b}{\cos \theta} = \sqrt{2} + \cos \theta,$$

or

$$\cos^2 \theta + \sqrt{2} \cos \theta - b = 0$$

$$(\cos \theta)_{1,2} = \frac{-\sqrt{2} \pm \sqrt{2 + 4b}}{2}.$$

Hence $b = -\frac{1}{2}$ and $\cos \theta = (\sqrt{2}/2)$, that is, $\theta_1 = 3\pi/4$, $\theta_2 = 5\pi/4$. (Note that another, simple vertical tangent is obtained for $b = 1 + \sqrt{2}$, $\theta = 0$.)

Thus we see that corners are possible at every point of the x,y plane, provided that the extremal goes into the corner at an angle $3\pi/4$ with the positive x axis and leaves the corner at an angle of $5\pi/4$, or vice versa.

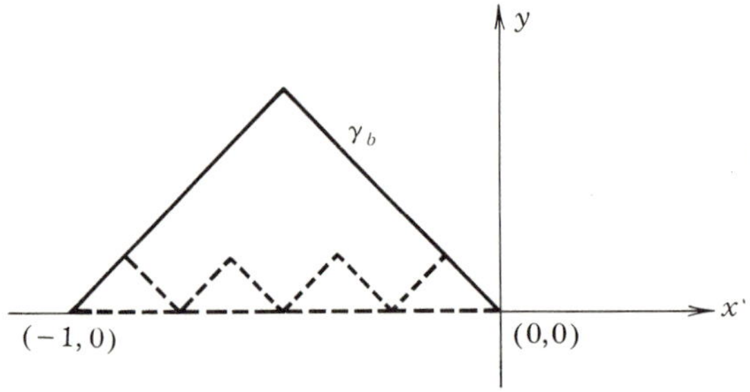

Figure 4.7

Before solving the Euler-Lagrange equations or the equivalent, equation (4.4.5), let us investigate F_1. From the shape of the indicatrix, it is obvious that F_1 will change sign twice. In order to compute F_1, we shall use one of the formulas (4.2.7*a*,*b*,*c*). Since F contains fewer $\dot{y}$'s than $\dot{x}$'s, it seems most convenient to use formula (4.2.7*c*):

$$F_1 = \frac{F_{\dot{y}\dot{y}}}{\dot{x}^2}.$$

We obtain, after lengthy and cumbersome manipulations and after introducing polar coordinates,

$$F_1 = \frac{4 + 3\sqrt{2}\cos\theta}{(2 + \cos\theta)^3},$$

and we see that

$$F_1 = 0 \text{ for } \cos\theta = -\frac{4}{3\sqrt{2}},$$

or

$$\theta = \begin{cases} \operatorname{Arccos}\left(-\frac{4}{3\sqrt{2}}\right) = \theta_1 \\ -\operatorname{Arccos}\left(-\frac{4}{3\sqrt{2}}\right) = \theta_2. \end{cases}$$

We see also that

$$F_1 > 0 \qquad \text{for } \theta_2 < \theta < \theta_1$$
$$F_1 < 0 \qquad \text{for } \theta_1 < \theta < 2\pi + \theta_2.$$

Since $F_{x\dot{y}} = F_{y\dot{x}} = 0$, Weierstrass' equation (4.4.5) simplifies to

$$F_1(\dot{x}\ddot{y} - \dot{y}\ddot{x}) = 0.$$

For $F_1 \neq 0$, we have therefore $\dot{x}\ddot{y} - \dot{y}\ddot{x} = 0$; that is, the extremals have curvature zero (are straight lines).

Let us now consider the variational problem that is associated with the integrand given in (4.7.5), and let us first assume that the beginning point P_a is given by (0,0) and that the endpoint P_b is located so that it can be reached by a line the slope of which lies in the "forbidden zone" (darkly shaded area in Fig. 4.6). For reasons of simplicity, we take P_b to be the point $(-1,0)$. Then, if we take the extremal

$$\gamma_f \begin{cases} x = -t \\ y = 0 \end{cases} 0 \leq t \leq 1,$$

we obtain

$$I[\gamma_f] = \int_0^1 \frac{1}{\sqrt{2} - 1}\, dt = \sqrt{2} + 1 = 2.4142\cdots.$$

On the other hand, if we take the broken extremal

$$\gamma_b \begin{cases} x = -t \\ y = t \end{cases} 0 \le t \le \tfrac{1}{2}, \qquad \begin{cases} x = -t \\ y = -t + 1 \end{cases} \tfrac{1}{2} \le t \le 1$$

instead—it satisfies the corner condition at $(-\frac{1}{2},\frac{1}{2})$—we obtain

$$I[\gamma_b] = \int_0^1 \frac{2}{2 - 1}\, dt = 2 < I[\gamma_f].$$

As a matter of fact, one can see quite easily that any zigzag extremal, as indicated in Fig. 4.7 by a dotted line, will yield the same value 2 for the functional.

In Sec. 4.10, we shall see that γ_f cannot possibly yield a minimum.

Next, we take an endpoint that can be reached by a line the slope of which lies in the "restricted zone" (lightly shaded area in Fig. 4.6). To be specific, we choose the point $(-1,\frac{1}{2})$. The extremal

$$\gamma_r \begin{cases} x = -t \\ y = t/2 \end{cases} 0 \le t \le 1$$

yields

$$I[\gamma_r] = \int_0^1 \frac{1 + \frac{1}{4}}{\sqrt{\frac{5}{2}} - 1}\, dt = \tfrac{5}{6}\left(\sqrt{\tfrac{5}{2}} + 1\right) = 2.1509\cdots.$$

On the other hand, the broken extremal

$$\gamma_b \begin{cases} x = -t \\ y = t \end{cases} 0 \le t \le \tfrac{3}{4}, \qquad \begin{cases} x = -t \\ y = -t + \frac{3}{2} \end{cases} \tfrac{3}{4} \le t \le 1,$$

which satisfies the corner condition at $(-\frac{3}{4},\frac{3}{4})$ (see Fig. 4.8), yields

$$I[\gamma_b] = \int_0^1 \frac{2}{2 - 1}\, dt = 2.$$

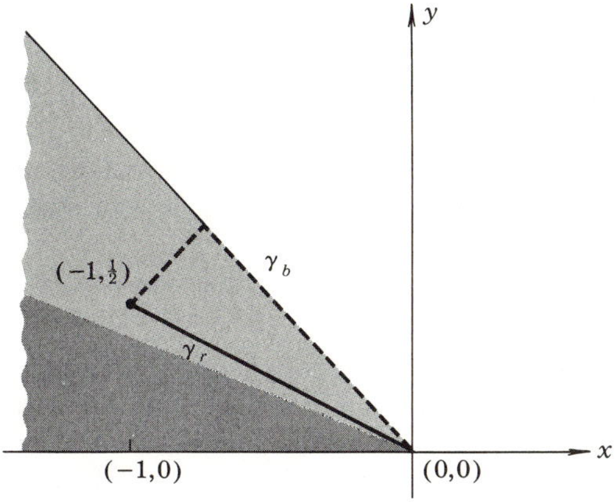

Figure 4.8

We shall see in Sec. 4.10 that γ_r yields only a weak minimum. [It turns out that the reason for this is that the tangent line to the indicatrix at the point that corresponds to the direction of γ_r intersects the indicatrix again—and this is, of course, true for all points in the restricted zone that are joined to (0,0) by a straight line.]

PROBLEMS 4.7

1. Show that the homogeneous variational problem that is associated with $F = \sqrt{\dot{x}^2 + \dot{y}^2} - \frac{1}{2}(x\dot{y} - y\dot{x})$ is positive definite for all $x^2 + y^2 < 4$. Show that the indicatrix for any point (x_o,y_o) for which $x_o^2 + y_o^2 < 4$ is an ellipse.
2. Show that the zigzag line with slopes alternating between 1 and -1 of Fig. 4.7 yields the value 2 for the integral

$$I[\gamma] = \int_{(0,0)}^{(-1,0)} \frac{\dot{x}^2 + \dot{y}^2}{\sqrt{2(\dot{x}^2 + \dot{y}^2)} + \dot{x}}\, dt.$$

4.8 INTEGRALS OF THE EULER-LAGRANGE EQUATIONS

We saw in Sec. 2.6 [see (2.6.1) and Prob. 2.6.3] that when $f = f(y,y')$, then

$$f - y'f_{y'} = \text{constant}$$

is a first integral of the Euler-Lagrange equation (2.3.8) for the x problem, and when $f = f(x,y')$, then

$$f_{y'} = \text{constant}$$

is a first integral of the Euler-Lagrange equation.

Similarly, in case of the homogeneous problem, when $F = F(y,\dot{x},\dot{y})$, then

$$F_{\dot{x}} = \text{constant}$$

is a first integral of the Euler-Lagrange equations (4.4.3), and when $F = F(x,\dot{x},\dot{y})$, then

$$F_{\dot{y}} = \text{constant}$$

is such a first integral.

In the one case, F is invariant under all translations in the x direction, and in the other case, F is invariant under all translations in the y direction. Consequently, the functional that is associated with F, namely,

$$I[\gamma]_{1,2} = \int_{t_1}^{t_2} F\, dt,$$

will be invariant under such transformations, regardless of t_1, t_2.

As we have shown in Sec. A2.16 for the x problem, first integrals of the Euler-Lagrange equation can always be obtained whenever the integral is invariant under a certain family of transformations. A similar result can be obtained for the homogeneous problem.

Toward this end, we shall assume that $I[\gamma]$ is invariant under the family of transformations

$$T_\epsilon \begin{cases} \bar{x} = \bar{x}(x,y,\varepsilon) \\ \bar{y} = \bar{y}(x,y,\varepsilon), \end{cases} \tag{4.8.1}$$

where we assume that

$$\bar{x}(x,y,0) = x$$
$$\bar{y}(x,y,0) = y,$$

that is, T_ϵ contains the identity transformation as a member for $\varepsilon = 0$. We shall further assume that $[\partial(\bar{x},\bar{y})]/[\partial(x,y)] \neq 0$ and that $\bar{x},\bar{y} \in C^1$ in x, y, and ε.

Let

$$\gamma_o \begin{cases} x = x(t) \\ y = y(t) \end{cases}$$

represent an extremal of $I[\gamma]$. Then, if we subject γ_o to the transformations (4.8.1), we obtain a one-parameter family of curves

$$\gamma_\epsilon \begin{cases} \bar{x} = \bar{x}(x(t),y(t),\varepsilon) = \bar{x}(t,\varepsilon) \\ \bar{y} = \bar{y}(x(t),y(t),\varepsilon) = \bar{y}(t,\varepsilon), \end{cases}$$

and we note that γ_o is contained in this family for $\varepsilon = 0$.

By hypothesis,

$$I[\gamma_\epsilon]_{1,2} = I[\gamma_o]_{1,2}$$

for any pair (t_1,t_2), that is, $I[\gamma_\epsilon]_{1,2} = I(\varepsilon)_{1,2}$ is independent of ε:

$$\frac{dI(\varepsilon)_{1,2}}{d\varepsilon} = 0.$$

We find $[dI(\varepsilon)]/d\varepsilon$, as in Sec. 4.6, when we observe that in the light of our assumptions on T_ϵ and under the additional assumption that $F \in C^2$, all the operations that lead to (4.6.3) and (4.6.4) are permissible. We obtain

$$\frac{dI(\varepsilon)_{1,2}}{d\varepsilon} = \int_{t_1}^{t_2} \left[F_x \frac{\partial \bar{x}}{\partial \varepsilon} + F_y \frac{\partial \bar{y}}{\partial \varepsilon} + F_{\dot{x}} \frac{\partial \dot{\bar{x}}}{\partial \varepsilon} + F_{\dot{y}} \frac{\partial \dot{\bar{y}}}{\partial \varepsilon} \right] dt$$

$$= \left[\phi \frac{\partial \bar{x}}{\partial \varepsilon} + \psi \frac{\partial \bar{y}}{\partial \varepsilon} \right]_{t_1}^{t_2} + \int_{t_1}^{t_2} \left[\frac{\partial \bar{x}}{\partial \varepsilon} (F_{\dot{x}} - \phi) + \frac{\partial \bar{y}}{\partial \varepsilon} (F_{\dot{y}} - \psi) \right] dt.$$

Since $I'(\varepsilon)_{1,2} = 0$ for *all* ε, we have, in particular, that

$$\left. \frac{dI(\varepsilon)_{1,2}}{d\varepsilon} \right|_{\epsilon=0} = 0.$$

Since γ_o is, by hypothesis, an extremal, the Euler-Lagrange equations in integrated form (4.4.1) are satisfied, and we obtain the condition

$$\left.\frac{dI(\varepsilon)_{1,2}}{d\varepsilon}\right|_{\epsilon=0} = \left.\left[F_{\dot{x}}\frac{\partial\bar{x}}{\partial\varepsilon} + F_{\dot{y}}\frac{\partial\bar{y}}{\partial\varepsilon}\right]_{t_1}^{t_2}\right|_{\epsilon=0} = 0. \tag{4.8.2}$$

We denote the "initial velocities" of the transformations T_ϵ by

$$\left(\frac{\partial\bar{x}}{\partial\varepsilon}\right)_{\epsilon=0} = \xi(x,y)$$

$$\left(\frac{\partial\bar{y}}{\partial\varepsilon}\right)_{\epsilon=0} = \eta(x,y).$$

Then we can write (4.8.2) as

$$(F_{\dot{x}}\xi + F_{\dot{y}}\eta)_{t_1} = (F_{\dot{x}}\xi + F_{\dot{y}}\eta)_{t_2}$$

for all pairs t_1, t_2. This means that $F_{\dot{x}}\xi + F_{\dot{y}}\eta$ is independent of t along an extremal:

$$F_{\dot{x}}\xi + F_{\dot{y}}\eta = \text{constant}. \tag{4.8.3}$$

This is a *first integral* of the Euler-Lagrange equation.

The two cases which we mentioned initially now appear as special cases of (4.8.3), namely, if $F = F(y,\dot{x},\dot{y})$, then the integral is invariant under

$$T_\epsilon\begin{cases}\bar{x} = x + \varepsilon\\ \bar{y} = y,\end{cases}$$

and we have $\xi = 1, \eta = 0$. Hence, $F_{\dot{x}} = $ constant. If $F = F(x,\dot{x},\dot{y})$, then the integral is invariant under

$$T_\epsilon\begin{cases}\bar{x} = x\\ \bar{y} = y + \varepsilon,\end{cases}$$

and we have $\xi = 0$, $\eta = 1$; hence $F_{\dot{y}} = $ constant.

From (4.8.3), we can easily obtain the first integral of the x problem in one unknown function, namely,

$$(f - y'f_{y'})\xi + f_{y'}\eta = \text{constant}, \tag{4.8.4}$$

which is formula (A2.16.8) for $n = 1$. (See Prob. 4.8.1.)

PROBLEMS 4.8

*1. Derive the first integral (4.8.4) of the Euler-Lagrange equation for the x problem from (4.8.3).

2. Find first integrals of the Euler-Lagrange equations of the variational problem

$$\int_{P_a}^{P_b} \frac{\sqrt{\dot{x}^2 + \dot{y}^2}}{y}\, dt \to \text{minimum.}$$

3. Find the extremals of the variational problem in problem 2, utilizing the result of problem 2.

*4. Supply all the details in the derivation of condition (4.8.2).

4.9 FIELD AND EXCESS FUNCTION

We shall now adapt the ideas of Chap. 3 to fit the homogeneous problem. Again, the concept of a field will loom in the foreground. To provide some variety, we shall now take a shortcut and define a field immediately in terms of a one-parameter family of extremals that covers a certain portion of the plain simply:

Definition 4.9 *The one-parameter family $x = X(t,c)$, $y = Y(t,c)$ of simple, smooth arcs that are solutions of the Euler-Lagrange equations (4.4.3) defines a field $\mathfrak{F}$ in the domain $t_1 < t < t_2$, $c_1 < c < c_2$ if*

1. $X_t, Y_t, X_c, Y_c, X_{tc}, Y_{tc} \in C([t_1,t_2] \times [c_1,c_2])$
2. $[\partial(X,Y)]/\partial(t,c) \neq 0$ *for* $(t,c) \in ([t_1,t_2] \times [c_1,c_2])$

The extremal

$$\gamma_o \begin{cases} x = x_o(t) \\ y = y_o(t) \end{cases} t \in [t_a,t_b]$$

is embeddable in a field $\mathfrak{F}$ if such a field exists with $t_1 < t_a < t_b < t_2$ and if there exists a $c_o \in (c_1,c_2)$ such that

$$x_o(t) = X(t,c_o)$$
$$y_o(t) = Y(t,c_o)$$

for all $t \in [t_a,t_b]$.

It follows from the theorem on implicit functions that one can solve $x = X(t,c)$, $y = Y(t,c)$ uniquely for t, c:

$$t = T(x,y), \qquad c = C(x,y)$$

in $\mathfrak{A}$, where $\mathfrak{A}$ is a domain that is encompassed by the curves

$$\left.\begin{matrix} x = X(t,c_1) \\ y = Y(t,c_1) \end{matrix}\right\}, \quad \left.\begin{matrix} x = X(t,c_2) \\ y = Y(t,c_2) \end{matrix}\right\}, \quad \left.\begin{matrix} x = X(t_1,c) \\ y = Y(t_1,c) \end{matrix}\right\}, \quad \left.\begin{matrix} x = X(t_2,c) \\ y = Y(t_2,c) \end{matrix}\right\}.$$

(See Fig. 4.9.)

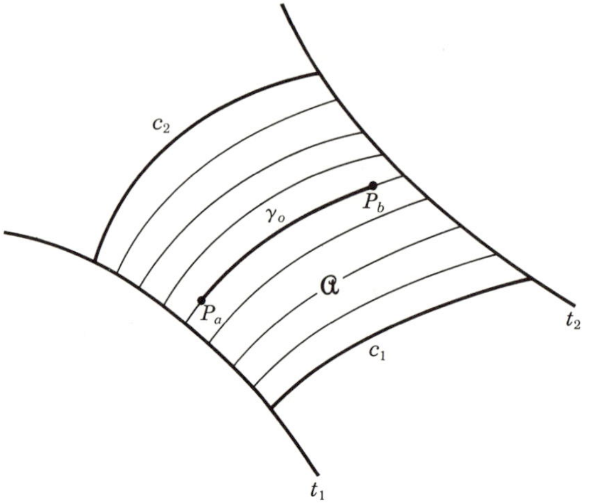

Figure 4.9

Lemma 4.9 *Under the conditions formulated above and if* $F = F(x,y,\dot{x},\dot{y}) \in C^1(\mathfrak{a})$ *and for all* $(\dot{x},\dot{y}) \neq (0,0)$, *there exists a function* $W = W(x,y) \in C^1(\bar{\mathfrak{a}})$ *such that*

$$W_x = F_{\dot{x}}(x,y,X_t(T(x,y),C(x,y)),Y_t(T(x,y),C(x,y))),$$
$$W_y = F_{\dot{y}}(x,y,X_t(T(x,y),C(x,y)),Y_t(T(x,y),C(x,y))).$$

Proof: First, we shall demonstrate that there is a function $W = W(t,c)$ such that

$$\begin{aligned} W_t &= F(X(t,c),Y(t,c),X_t(t,c),Y_t(t,c)), \\ W_c &= X_cF_{\dot{x}}(X(t,c),Y(t,c),X_t(t,c),Y_t(t,c)) \\ &\quad + Y_cF_{\dot{y}}(X(t,c),Y(t,c),X_t(t,c),Y_t(t,c)). \end{aligned} \tag{4.9.1}$$

We have

$$W_{tc} - W_{ct} = \frac{\partial}{\partial c} F - \frac{\partial}{\partial t} (X_cF_{\dot{x}} + Y_cF_{\dot{y}})$$

$$= X_c\left(F_x - \frac{d}{dt} F_{\dot{x}}\right) + Y_c\left(F_y - \frac{d}{dt} F_{\dot{y}}\right) = 0$$

because $x = X(t,c)$, $y = Y(t,c)$ are, by hypothesis, solutions of the Euler-Lagrange equations (4.4.3). Hence a function $W = W(t,c)$ for which (4.9.1) holds exists.†

We now consider W as function of x and y by letting $t = T(x,y)$, $c = C(x,y)$. We note that $X(T(x,y),C(x,y)) = x$ and $Y(T(x,y),C(x,y)) = y$

† A. E. Taylor, "Advanced Calculus," p. 440, Ginn and Company, Boston, 1955.

hold identically in x and y, and we have

$$X_tT_x + X_cC_x = 1, \qquad Y_tT_y + Y_cC_y = 1,$$
$$X_tT_y + X_cC_y = 0, \qquad Y_tT_x + Y_cC_x = 0. \tag{4.9.2}$$

Then we have, by (4.9.1), because of Euler's identity $F = \dot{x}F_{\dot{x}} + \dot{y}F_{\dot{y}}$ and in view of (4.9.2), for $W(T(x,y),C(x,y)) = W(x,y)$:

$$\begin{aligned} W_x = W_tT_x + W_cC_x &= FT_x + (X_cF_{\dot{x}} + Y_cF_{\dot{y}})C_x \\ &= (\dot{x}F_{\dot{x}} + \dot{y}F_{\dot{y}})T_x + (X_cF_{\dot{x}} + Y_cF_{\dot{y}})C_x \\ &= F_{\dot{x}}(X_tT_x + X_cC_x) + F_{\dot{y}}(Y_tT_x + Y_cC_x) \\ &= F_{\dot{x}}(x,y,X_t(T(x,y),C(x,y)),Y_t(T(x,y),C(x,y))) \end{aligned}$$

and, similarly,

$$W_y = F_{\dot{y}}(x,y,X_t(T(x,y),C(x,y)),Y_t(T(x,y),C(x,y))).$$

On the basis of this lemma, it is very easy to prove:

Theorem 4.9 *The integral* (invariant integral)

$$\begin{aligned} U[\Gamma] = \int_{P_a}^{P_b} [&\dot{\bar{x}}F_{\dot{x}}(\bar{x},\bar{y},X_t(T(\bar{x},\bar{y}),C(\bar{x},\bar{y})),Y_t(Y(\bar{x},\bar{y}),C(\bar{x},\bar{y}))) \\ &+ \dot{\bar{y}}F_{\dot{y}}(\bar{x},\bar{y},X_t(T(\bar{x},\bar{y}),C(\bar{x},\bar{y})),Y_t(T(\bar{x},\bar{y}),C(\bar{x},\bar{y})))]\,dt \end{aligned}$$

is independent of the path Γ: $x = \bar{x}(t)$, $y = \bar{y}(t)$ *so long as* Γ *joins the same endpoints* $(\bar{x}_o(t_a),\ \bar{y}_o(t_a)) = (a,y_a)$, $(\bar{x}_o(t_b),\bar{y}_o(t_b)) = (b,y_b)$, *is a simple, smooth arc, and remains entirely in* $\mathfrak{A}$.

Proof: By Lemma 4.9, there exists in $\mathfrak{A}$ a function $W = W(x,y)$ such that

$$W_x = F_{\dot{x}}, \qquad W_y = F_{\dot{y}},$$

where $F_{\dot{x}}$, $F_{\dot{y}}$ are to be considered as functions of $\bar{x}$ and $\bar{y}$. Hence,

$$U[\Gamma] = \int_{P_a}^{P_b} [\dot{\bar{x}}W_x + \dot{\bar{y}}W_y]\,dt = \int_{P_a}^{P_b} \frac{dW}{dt}\,dt = W(b,y_b) - W(a,y_a).$$

We can now transform the total variation

$$\Delta I = I[\Gamma] - I[\gamma_o],$$

where Γ is any simple, smooth arc that joins the same endpoints as γ_o and lies in a strong neighborhood of γ_o. (Obviously, $\mathfrak{A}$ contains such a strong neighborhood of γ_o.) We observe that

$$U[\gamma_o] = \int_{P_a}^{P_b} [\dot{x}_o F_{\dot{x}}(x_o,y_o,\dot{x}_o,\dot{y}_o) + \dot{y}_o F_{\dot{y}}(x_o,y_o,\dot{x}_o,\dot{y}_o)]\,dt$$

$$= \int_{P_a}^{P_b} F(x_o,y_o,\dot{x}_o,\dot{y}_o)\,dt = I[\gamma_o].$$

Hence,

$$\Delta I = I[\Gamma] - I[\gamma_o] = I[\Gamma] - U[\gamma_o] = I[\Gamma] - U[\Gamma]$$

$$= \int_{P_a}^{P_b} [F(\bar{x},\bar{y},\dot{\bar{x}},\dot{\bar{y}}) - \dot{\bar{x}}F_{\dot{x}}(\bar{x},\bar{y},\dot{x},\dot{y}) - \dot{\bar{y}}F_{\dot{y}}(\bar{x},\bar{y},\dot{x},\dot{y})]\,dt,$$

where $\dot{x}$ and $\dot{y}$ have the argument $(T(\bar{x},\bar{y}),C(\bar{x},\bar{y}))$. The integrand

$$\mathcal{E}(\bar{x},\bar{y},\dot{x},\dot{y},\dot{\bar{x}},\dot{\bar{y}}) = F(\bar{x},\bar{y},\dot{\bar{x}},\dot{\bar{y}}) - \dot{\bar{x}}F_{\dot{x}}(\bar{x},\bar{y},\dot{x},\dot{y}) - \dot{\bar{y}}F_{\dot{y}}(\bar{x},\bar{y},\dot{x},\dot{y}) \qquad (4.9.3)$$

is again called the *Weierstrass excess function.*

With this notation, we can write the total variation as

$$\Delta I = \int_{P_a}^{P_b} \mathcal{E}(\bar{x},\bar{y},\dot{x},\dot{y},\dot{\bar{x}},\dot{\bar{y}})\,dt. \qquad (4.9.4)$$

We note that $\mathcal{E}$ is positive homogeneous of the first degree in $\dot{\bar{x}}$, $\dot{\bar{y}}$ and positive homogeneous of the zeroth degree in $\dot{x}$, $\dot{y}$. Thus, if we substitute

$$\dot{x} = r\cos\theta, \qquad \dot{\bar{x}} = \rho\cos\bar{\theta},$$
$$\dot{y} = r\sin\theta, \qquad \dot{\bar{y}} = \rho\sin\bar{\theta},$$

then

$$\mathcal{E}(\bar{x},\bar{y},\dot{x},\dot{y},\dot{\bar{x}},\dot{\bar{y}}) = \rho\mathcal{E}(\bar{x},\bar{y},\cos\theta,\sin\theta,\cos\bar{\theta},\sin\bar{\theta})$$
$$= \rho\bar{\mathcal{E}}(\bar{x},\bar{y},\theta,\bar{\theta}), \qquad \rho > 0, \qquad (4.9.5)$$

where

$$\bar{\mathcal{E}}(x,y,\theta,\bar{\theta}) = F(x,y,\cos\bar{\theta},\sin\bar{\theta}) - \cos\bar{\theta}F_{\dot{x}}(x,y,\cos\theta,\sin\theta)$$
$$- \sin\bar{\theta}F_{\dot{y}}(x,y,\cos\theta,\sin\theta). \qquad (4.9.6)$$

PROBLEMS 4.9

1. Given the variable-beginning-point, fixed-endpoint problem

$$\int_{P(x,y)}^{P_b(b,y_b)} F(x,y,\dot{x},\dot{y})\,dt \to \text{minimum},$$

where the beginning point lies on a simple smooth curve C. Extend Definition 4.9 of a field to cover this contingency and show that a field can be constructed, at least locally, if $F_1 \neq 0$ at the beginning point, provided that the initial direction of the extremal arc is different from the direction of the beginning curve C at the point of departure.

*2. Show that for $x = x_o(t)$, $y = y_o(t)$ to yield a strong relative minimum for $I[\gamma]$, it is necessary that

$$\bar{\mathcal{E}}\,(x_o(t),y_o(t),\theta_o(t),\bar{\theta}) \geq 0$$

for all $0 \leq \bar{\theta} \leq 2\pi$ and all $t \in [t_a,t_b]$, where $\dot{x}_o = r\cos\theta_o$, $\dot{y}_o = r\sin\theta_o$.

*3. Suppose that $x = X(t,c)$, $y = Y(t,c)$ is a one-parameter family of solutions of the Euler-Lagrange equations, where X, Y, $\dot{X}$, $\dot{Y}$, X_c, $\dot{Y}_c$ are continuous for $t_1 \leq t \leq t_2$ $(t_1 < t_a < t_b < t_2)$ and $c_1 \leq c \leq c_2$. Let the extremal arc be represented by $x = X(t,c_o)$, $y = Y(t,c_o)$, $t \in [t_a,t_b]$, $c_o \in (c_1,c_2)$, and assume that the extremal arc is simple and smooth. Assume further that $[\partial(X,Y)]/[\partial(t,c)] \neq 0$ for $c = c_o$ and all $t \in [t_a,t_b]$. Prove: $x = X(t,c)$, $y = Y(t,c)$ maps the region $t_a - \delta \leq t \leq t_b + \delta$, $c_o - \delta \leq c \leq c_o + \delta$, for some $\delta > 0$, biuniquely into a region of the x,y plane whereby interior is mapped into interior and boundary is mapped into boundary.

4. Show that

$$\mathcal{E}\,(x,y,\dot{x},\dot{y},\dot{\bar{x}},\dot{\bar{y}}) = F(x,y,\dot{\bar{x}},\dot{\bar{y}}) - F(x,y,\dot{x},\dot{y}) + (\dot{x} - \dot{\bar{x}})F_{\dot{x}}(x,y,\dot{x},\dot{y}) + (\dot{y} - \dot{\bar{y}})F_{\dot{y}}(x,y,\dot{x},\dot{y}),$$

and note the similarity to the representation of the excess function for the x problem in two unknown functions.

*5. Given $x = c(t - \sin t) + (c_o - c)\pi$, $y = c(1 - \cos t)$, $t \in [0,2\pi]$, $c > 0$. Show that $[\partial(x,y)]/[\partial(t,c)] \neq 0$ for all $t \in (0,2\pi)$, $c > 0$.

4.10 STRONG AND WEAK EXTREMA

If the extremal arc γ_o is embeddable in a field $\mathfrak{F}$, then there exists a weak neighborhood of γ_o in $\mathfrak{A}$ such that the transformation of the total variation (4.9.4) applies to all simple, smooth curves Γ that lie in that weak neighborhood of γ_o.

In view of (4.9.4) and (4.9.5), we can state:

Theorem 4.10.1 *If the extremal arc γ_o is embeddable in a field $\mathfrak{F}$ and if*

$$\bar{\mathcal{E}}\,(\bar{x},\bar{y},\theta,\bar{\theta}) \geq 0 \qquad (\leq 0)$$

for all $(\bar{x},\bar{y})$ in a weak neighborhood of γ_o and all $0 \leq \theta \leq 2\pi$, $0 \leq \bar{\theta} \leq 2\pi$, then γ_o yields a strong relative minimum (maximum) for $I[\gamma]$.

The theory as developed here and in the preceding section does not immediately apply to the Brachistochrone problem

$$\int_{P_0(0,0)}^{P_b(b,y_b)} \frac{\sqrt{\dot{x}^2 + \dot{y}^2}}{\sqrt{y}}\,dt \to \text{minimum},$$

because $F(x,y,\dot{x},\dot{y}) = \sqrt{\dot{x}^2 + \dot{y}^2}/\sqrt{y}$ is not continuous for $y = 0$ and hence not continuous at the beginning point $P_o(0,0)$. (This difficulty cannot be resolved simply by a translation of the beginning point in the y direction, say by a units, because then the denominator of the integrand would have to be changed to $\sqrt{y - a}$ and we would be back where we were.) The following argument, however, will reveal that the theory can be extended to apply to this exceptional case.

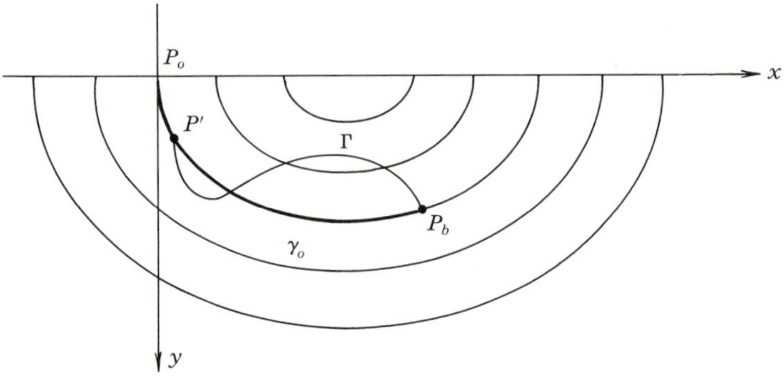

Figure 4.10

We obtain as an extremal arc a uniquely determined cycloid (see Prob. 2.6.1), provided that $b \neq 0$ and $y_b > 0$, namely,

$$x = c_o(t - \sin t)$$
$$y = c_o(1 - \cos t).$$

This can be embedded into the one-parameter family of extremals

$$x = X(t,c) \equiv c(t - \sin t) + (c_o - c)\pi$$
$$y = Y(t,c) \equiv c(1 - \cos t)$$

(see Fig. 4.10), which defines a field in the entire half-plane $y > 0$ for $0 < t < 2\pi$ and all $c > 0$. (See Prob. 4.9.5.)

We obtain, after some cumbersome manipulations,

$$F(X(t,c),Y(t,c),\dot{X}(t,c),\dot{Y}(t,c)) = \sqrt{2c},$$

$$\frac{\partial X}{\partial c} F_{\dot{x}} + \frac{\partial Y}{\partial c} F_{\dot{y}} = (t - \pi)\frac{1}{\sqrt{2c}}.$$

Hence the function

$$W(t,c) = (t - \pi)\sqrt{2c}$$

has the property that

$$\frac{\partial W}{\partial t} = \sqrt{2c} = F(X(t,c),Y(t,c),\dot{X}(t,c),\dot{Y}(t,c)),$$

$$\frac{\partial W}{\partial c} = \frac{t - \pi}{\sqrt{2c}} = \left(\frac{\partial X}{\partial c} F_{\dot{x}} + \frac{\partial Y}{\partial c} F_{\dot{y}}\right)_{x=X(t,c),y=Y(t,c)}.$$

Since $[\partial(X,Y)]/[\partial(t,c)] \neq 0$ for $0 < t < 2\pi$ and $c > 0$ (see Prob. 4.9.5), we can express t,c as unique functions of x,y, and we see that the invariant integral $U[\Gamma]$ is defined in the half-plane $y > 0$.

If we now consider a point P' on the extremal arc (see Fig. 4.10) that is close to the beginning point P_o, we then obtain between P' and P_b, in consideration of

$$\bar{\mathcal{E}}(x,y,\theta,\bar{\theta}) = (1/\sqrt{y})(1 - \cos(\bar{\theta} - \theta)) \geq 0,$$

$$I[\Gamma] - I[\gamma_o] = I[\Gamma] - U[\Gamma] = \int_{P'}^{P_b} \mathcal{E}\, dt \geq 0.$$

If we let $P' \to P_o$ along the extremal arc, we see that this integral exists as an improper integral, and we have between P_o and P_b

$$I[\Gamma] - I[\gamma_o] = \int_{P_0}^{P_b} \mathcal{E}\, dt \geq 0,$$

that is, the cycloid yields a strong relative minimum.

In order to rule out curves Γ that touch the x axis between $x = 0$ and $x = b$, we shall show that such curves can always be replaced by other curves Γ' that do *not* touch the x axis between $x = 0$ and $x = b$ and that yield a smaller value for $I[\gamma]$ than does Γ. In Fig. 4.11, let the solid curve represent Γ. Then Γ', the dotted curve, is obtained from Γ by moving the portion of Γ between P' and Q' to the left until it assumes a position between P_o and Q and moving that part of Γ that lies between P_o and P' down and to the right until it comes to rest between Q and Q'.

We observe that

$$I[\Gamma] = I[\Gamma_1] + I[\Gamma_2] + I[\Gamma_3]$$

and that

$$I[\Gamma_2] = I[\Gamma_2']$$

since the integrand is independent of x. Since Γ_1' is obtained from Γ_1 by increasing y, we have

$$I[\Gamma_1'] < I[\Gamma_1]$$

and hence

$$I[\Gamma'] < I[\Gamma].$$

Now, one can smooth out the corners of Γ' by a suitable adaptation of the fairing theorem and obtain a smooth arc that yields a smaller value for the integral than does Γ.

In Sec. 3.8, we established a relationship between the excess function of the x problem and $f_{y'y'}$. In a similar manner, we shall now establish a

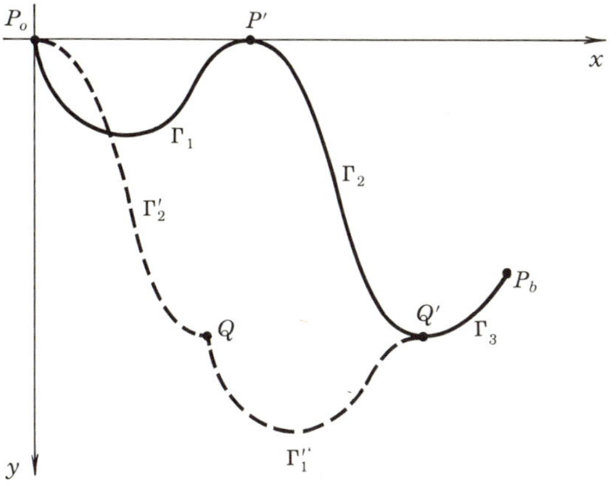

Figure 4.11

relationship between the excess function of the homogeneous problem and the function F_1.

From (4.9.6),

$$\frac{\partial \bar{\mathcal{E}}}{\partial \theta} = -\cos\bar{\theta}(-F_{\dot{x}\dot{x}}\sin\theta + F_{\dot{x}\dot{y}}\cos\theta) - \sin\bar{\theta}(-F_{\dot{x}\dot{y}}\sin\theta + F_{\dot{y}\dot{y}}\cos\theta).$$

By (4.2.7*a*, *b*, *c*), we may introduce F_1, thus eliminating the second partials of F, and we obtain

$$\begin{aligned}\frac{\partial \bar{\mathcal{E}}}{\partial \theta} &= -\cos\bar{\theta}(-\sin^3\theta - \sin\theta\cos^2\theta)F_1 - \sin\bar{\theta}(\sin^2\theta\cos\theta + \cos^3\theta)F_1\\ &= (\sin\theta\cos\bar{\theta} - \cos\theta\sin\bar{\theta})F_1(\bar{x}, \bar{y}, \cos\theta, \sin\theta)\\ &= \sin(\theta - \bar{\theta})F_1(\bar{x}, \bar{y}, \cos\theta, \sin\theta).\end{aligned}$$

We can now recover $\bar{\mathcal{E}}$ by integrating again with respect to θ and noting that $\bar{\mathcal{E}}(\bar{x},\bar{y},\bar{\theta},\bar{\theta}) = 0$. Then we obtain

$$\bar{\mathcal{E}}(\bar{x},\bar{y},\theta,\bar{\theta}) = \int_{\bar{\theta}}^{\theta} F_1(\bar{x}, \bar{y}, \cos\sigma, \sin\sigma)\sin(\sigma - \bar{\theta})\,d\sigma. \qquad (4.10.1)$$

(This representation is due to *H. A. Schwarz.*)

From this representation we obtain:

Corollary to Theorem 4.10.1 *If γ_o is embeddable in a field $\mathfrak{F}$ and if*

$$F_1(\bar{x}, \bar{y}, \cos\theta, \sin\theta) \geq 0 \qquad (\leq 0) \qquad (4.10.2)$$

for all $(\bar{x},\bar{y})$ in a weak neighborhood of γ_o and all $0 \leq \theta \leq 2\pi$, then γ_o yields a strong relative minimum (maximum) for $I[\gamma]$.

Proof: Since θ, $\bar{\theta}$ are determined but for additive multiples of 2π, we may assume without loss of generality that

$$|\theta - \bar{\theta}| \leq \pi.$$

We have to distinguish now between $\theta \geq \bar{\theta}$ and $\theta < \bar{\theta}$:

(*a*) $\theta \geq \bar{\theta}$. Then we have, in view of $|\theta - \bar{\theta}| \leq \pi$, that $0 \leq \theta - \bar{\theta} \leq \pi$, and since $\bar{\theta} \leq \sigma \leq \theta$, we also have $0 \leq \sigma - \bar{\theta} \leq \theta - \bar{\theta} \leq \pi$ and hence $\sin(\sigma - \bar{\theta}) \geq 0$. By (4.10.1) and (4.10.2), we have then $\bar{\mathcal{E}}(\bar{x},\bar{y},\theta,\bar{\theta}) \geq 0$.

(*b*) $\theta < \bar{\theta}$. Then $0 < \bar{\theta} - \theta \leq \pi$, and with $\theta \leq \sigma \leq \bar{\theta}$, we have $-\pi \leq \sigma - \bar{\theta} < 0$ and hence $\sin(\sigma - \bar{\theta}) \leq 0$. We interchange the integration limits in (4.10.1), thus compensating for the negative sign of $\sin(\sigma - \bar{\theta})$, and we obtain again $\bar{\mathcal{E}}(\bar{x},\bar{y},\theta,\bar{\theta}) \geq 0$.

(For an alternate way of proving this corollary, see Prob. 4.10.5.)

As in Sec. 3.8, we now introduce the concepts of strongly and weakly regular lineal elements:

Definition 4.10 *The lineal element* $(x,y,\dot{x},\dot{y})$ *is called* weakly regular *if* $F_1(x,y,\dot{x},\dot{y}) > 0$ *and* strongly regular *if* $F_1(x,y,\dot{x},\dot{y}) > 0$ and $\mathcal{E}(x,y,\dot{x},\dot{y},\dot{\bar{x}},\dot{\bar{y}}) > 0$ *for all* $(\dot{\bar{x}},\dot{\bar{y}}) \neq (\lambda\dot{x},\lambda\dot{y})$ *for all* $\lambda > 0$.

With $\dot{x} = r\cos\theta$, $\dot{y} = r\sin\theta$, $\dot{\bar{x}} = \rho\cos\bar{\theta}$, $\dot{\bar{y}} = \rho\sin\bar{\theta}$, we may put this, in view of the homogeneity of all the functions that are involved, as follows: The lineal element (x,y,θ) is *weakly regular* if $F_1(x, y, \cos\theta, \sin\theta) > 0$ and *strongly regular* if $F_1(x, y, \cos\theta, \sin\theta) > 0$ *and* $\bar{\mathcal{E}}(x,y,\theta,\bar{\theta}) > 0$ for all $\bar{\theta} \neq \theta + 2n\pi$.

Clearly, if the extremal arc is embeddable in a field and if all lineal elements of the field, including the ones of the extremal arc, are strongly regular, then, by the transformation of the total variation (4.9.4), we can say that the extremal arc yields a strong relative minimum. This is quite analogous to Theorem 3.8.2, which pertains to the x problem.

However, because of the particular structure of the homogeneous problem as induced by the positive homogeneity of the integrand, we can go further than that and guarantee a strong relative minimum under somewhat weaker conditions.

We obtain from (4.10.1) by application of the first mean value theorem (see Prob. 4.10.5) the following representation of the excess function:

$$\bar{\mathcal{E}}(x,y,\theta,\bar{\theta}) = F_1(x, y, \cos\bar{\sigma}, \sin\bar{\sigma})(1 - \cos(\bar{\theta} - \theta)).$$

We define

$$\mathcal{E}_1(x,y,\theta,\bar{\theta}) = \begin{cases} \dfrac{\bar{\mathcal{E}}(x,y,\theta,\bar{\theta})}{1 - \cos(\bar{\theta} - \theta)} = F_1(x, y, \cos\bar{\sigma}, \sin\bar{\sigma}), & \bar{\theta} \neq \theta + 2n\pi \\ F_1(x, y, \cos\theta, \sin\theta), & \bar{\theta} = \theta + 2n\pi. \end{cases} \tag{4.10.3}$$

Clearly, the function $\mathcal{E}_1$ is continuous since

$$\lim_{\bar{\theta}\to\theta+2n\pi} \mathcal{E}_1(x,y,\theta,\bar{\theta}) = F_1(x, y, \cos\theta, \sin\theta).$$

(See Prob. 4.10.1.)

We see now that our definition of a strongly regular lineal element $(x,y,\dot{x},\dot{y})$ [or (x,y,θ)] is equivalent to stating that $\mathcal{E}_1(x,y,\theta,\bar{\theta}) > 0$ for *all* $\bar{\theta} \in [0,2\pi]$.

Suppose that the extremal arc γ_o consists of strongly regular lineal elements only and is embeddable in a field that is defined by $x = X(t,c)$, $y = Y(t,c)$. Then $\mathcal{E}_1(X(t,c),Y(t,c),\theta(t,c),\bar{\theta})$ is a continuous function of

t, c, θ for $t_a \le t \le t_b$, $c_1 \le c \le c_2$, $0 \le \bar{\theta} \le 2\pi$ and hence is uniformly continuous so long as (x,y) remains in the field:

$$\mathcal{E}_1(X(t,c),Y(t,c),\theta(t,c),\bar{\theta}) = \phi(t,c,\bar{\theta}).$$

γ_o is given by $x = X(t,c_o)$, $y = Y(t,c_o)$, where $c_o \in (c_1,c_2)$, and we have by hypothesis

$$\phi(t,c_o,\bar{\theta}) > 0$$

for all $t_a \le t \le t_b$, $0 \le \bar{\theta} \le 2\pi$. Hence ϕ will assume on the extremal arc a positive minimum, and we have

$$\phi(t,c_o,\bar{\theta}) \ge m > 0$$

for all $t \in [t_a,t_b]$ and all $\bar{\theta} \in [0,2\pi]$. Since ϕ is uniformly continuous in this rectangle, we have that

$$\phi(t,c,\bar{\theta}) > 0$$

so long as $|c - c_o| < \delta$ for some $\delta > 0$. Hence the lineal elements of the field-defining extremals remain strongly regular in some weak neighborhood of γ_o, and we have:

Theorem 4.10.2 *If γ_o is embeddable in a field and if all lineal elements of γ_o are strongly regular, then γ_o yields a strong relative minimum for $I[\gamma]$.*

If all lineal elements of γ_o are weakly regular, i.e., if $F_1(x_o(t),y_o(t),\dot{x}_o(t),\dot{y}_o(t)) > 0$ for all $t \in [t_a,t_b]$, then, by continuity, F_1 will remain positive for all (x,y) in a weak neighborhood of γ_o and for all $\dot{x}(t)$, $\dot{y}(t)$ that are sufficiently close to $\dot{x}_o(t)$, $\dot{y}_o(t)$. Hence we have from (4.10.1) that $\mathcal{E}(x,y,\theta,\bar{\theta}) > 0$ for all field extremals that are sufficiently close to γ_o and so long as $|\theta - \bar{\theta}|$ remains sufficiently small. Hence:

Theorem 4.10.3 *If γ_o is embeddable in a field and if all lineal elements of γ_o are weakly regular, then γ_o yields a weak relative minimum for $I[\gamma]$.*

Remark: Theorems 4.10.2 and 4.10.3 apply equally well to relative maxima if the inequality signs in the definition of weak and strongly regular lineal elements are reversed.

Let us now assume that the variational problem under consideration is positive definite, and let us consider the indicatrix $F(x,y,\xi,\eta) = 1$ at a point (x_o,y_o) on the extremal arc γ_o. The equation of the tangent line T to the indicatrix at $\dot{x}_o = \xi_o$, $\dot{y}_o = \eta_o$ is given by

$$\xi F_{\dot{x}}(x_o,y_o,\xi_o,\eta_o) + \eta F_{\dot{y}}(x_o,y_o,\xi_o,\eta_o) = 1. \qquad (4.10.4)$$

[See (4.7.3).]

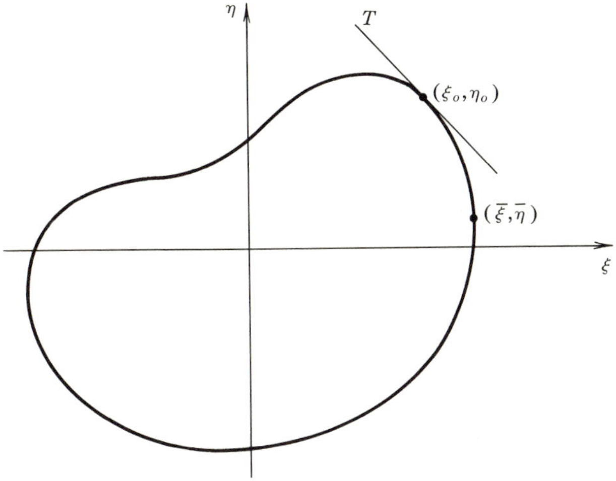

Figure 4.12

Now, if (ξ_o,η_o) and $(\bar{\xi},\bar{\eta})$ are two points on the indicatrix (see Fig. 4.12), we have

$$\begin{aligned}\mathcal{E}(x_o,y_o,\xi_o,\eta_o,\bar{\xi},\bar{\eta}) &= F(x_o,y_o,\bar{\xi},\bar{\eta}) - \bar{\xi}F_{\dot{x}}(x_o,y_o,\xi_o,\eta_o) - \bar{\eta}F_{\dot{y}}(x_o,y_o,\xi_o,\eta_o)\\ &= 1 - \bar{\xi}F_{\dot{x}}(x_o,y_o,\xi_o,\eta_o) - \bar{\eta}F_{\dot{y}}(x_o,y_o,\xi_o,\eta_o)\end{aligned}$$

(since $F = 1$ for all points $(\bar{\xi},\bar{\eta})$ on the indicatrix).

We see from (4.10.4) that

$$\mathcal{E}(x_o,y_o,\xi_o,\eta_o,\bar{\xi},\bar{\eta}) = 0$$

if and only if $(\bar{\xi},\bar{\eta})$ lies also on the tangent line T. Hence $\mathcal{E} > 0$ for all $(\bar{\xi},\bar{\eta})$ on one side of the tangent line T and $\mathcal{E} < 0$ for all $(\bar{\xi},\bar{\eta})$ on the other side of the tangent line T. Since $\mathcal{E}(x_o,y_o,\xi_o,\eta_o,0,0) = 1$, it follows that $\mathcal{E} > 0$ if $(\bar{\xi},\bar{\eta})$ lies on the same side of T as the origin (0,0). Thus, if all points of the indicatrix lie on the same side of the tangent as the origin, i.e., if the indicatrix lies on the same side of T as the origin, then $\mathcal{E}(x_o,y_o,\xi_o,\eta_o,\bar{\xi},\bar{\eta}) \geq 0$ for all $(\bar{\xi},\bar{\eta})$ on the indicatrix and, since $\mathcal{E}$ is positive homogeneous of the first degree in $\bar{\xi}$, $\bar{\eta}$, *for all possible* $(\bar{\xi},\bar{\eta})$. [For every pair $(\bar{\xi},\bar{\eta})$, there is a $\lambda > 0$ such that $(\lambda\bar{\xi},\lambda\bar{\eta})$ lies on the indicatrix.]

We can now say that $(x_o,y_o,\dot{x}_o,\dot{y}_o)$ is *strongly regular* (with $\mathcal{E} > 0$) if the indicatrix for the point (x_o,y_o) lies entirely on that side of the tangent line to the indicatrix at $(\dot{x}_o.\dot{y}_o)$ on which the origin is situated and if it does not touch the tangent line anywhere except at the point of tangency. On the other hand, $(x_o,y_o,\dot{x}_o,\dot{y}_o)$ is *weakly regular* (with $F_1 > 0$) if the indicatrix for the point (x_o,y_o) lies on that side of the tangent line to the indicatrix at

$(\dot{x}_o,\dot{y}_o)$ on which the origin is situated through some distance to the left and to the right of the point of tangency. (See Prob. 4.10.4.)

To illustrate this situation, let us return to the problem of Sec. 4.7, where the indicatrix turned out to be a *Pascal snail.* In that example, the indicatrix is fortunately the same for all points (x,y) since the integrand F does not depend on x and y. We see immediately from Fig. 4.6 that all points on the indicatrix that correspond to the directions $-3\pi/4 < \theta < 3\pi/4$ are strongly regular, all points that correspond to the directions $-\text{Arccos}\,[-(4/3\sqrt{2})] < \theta \leq -3\pi/4$ and $3\pi/4 < \theta \leq \text{Arccos}\,[-(4/3\sqrt{2})]$ are weakly regular with $F_1 > 0$, and all points that correspond to the directions $\text{Arccos}\,[-(4/3\sqrt{2})] < \theta < 2\pi - \text{Arccos}\,[-(4/3\sqrt{2})]$ are weakly regular with $F_1 < 0$. The points that correspond to the directions $\pm\text{Arccos}\,[-(4/3\sqrt{2})]$ itself are singular.

Hence, with reference to Fig. 4.8 any point in the white zone may be joined to the origin by a straight line which will yield a strong minimum. The straight lines that join the origin to points in the lightly shaded zone yield only weak relative minima, and the straight lines that join the origin to points in the darkly shaded zone yield weak relative maxima.

In order to investigate the nature of the extrema that are obtained from broken extremals, it is necessary to extend the field concept so that it also embraces the case of fields of broken extremals. For such a study, we refer the reader to the literature.†

PROBLEMS 4.10

*1. Show that the function $\mathcal{E}_1(x,y,\theta,\bar{\theta})$ as defined in (4.10.3) is continuous in all variables under the customary continuity assumptions on F.

2. Does the solution obtained for the navigation problem in Sec. 4.6 yield a weak or a strong relative minimum?

3. Show that $F_1(x_o(t),y_o(t),\dot{x}_o(t),\dot{y}_o(t)) \geq 0$ for all $t \in [t_a,t_b]$ is necessary for $x = x_o(t)$, $y = y_o(t)$ to yield a strong relative minimum for $I[\gamma]$.

*4. Show: If the indicatrix of the point (x_o,y_o) lies on the same side of the tangent at $(\dot{x}_0,\dot{y}_0)$ as the origin, for some distance to the left and to the right of the point of tangency, then

$$F_1(x_o,y_o,\dot{x}_o,\dot{y}_o) > 0.$$

*5. Prove the corollary to Theorem 4.10.1 by applying the first mean value theorem to the integral in (4.10.1).

6. Show that a statement analogous to Theorem 4.10.2 cannot possibly hold for the x problem. (*Hint*: consider $\int_0^1 (y'^2(x) - 4y(x)y'^3(x) + 2xy'^4(x))\,dx \to \text{minimum}$, $y(0) = y(1) = 0$.)

† O. Bolza, "Variationsrechnung," 2d ed., pp. 381ff, Chelsea Publishing Company, New York, 1962.

BRIEF SUMMARY

If the variational problem

$$I[\gamma] = \int_{P_a}^{P_b} F(x,y,\dot{x},\dot{y})\, dt \rightarrow \text{minimum}$$

is to make sense regardless of the choice of the parameter (within reason), then $F(x,y,\dot{x},\dot{y})$ has to be positive homogeneous of the first degree in $\dot{x}$, $\dot{y}$: $F(x,y,\lambda\dot{x},\lambda\dot{y}) = \lambda F(x,y,\dot{x},\dot{y})$, $\lambda > 0$ (Sec. 4.1).

For the simple, sectionally smooth arc γ_o: $x = x_o(t)$, $y = y_o(t)$, $t \in [t_a,t_b]$, to yield a weak relative minimum (and *eo ipso* a strong relative minimum), it is necessary that every smooth portion satisfy the Euler-Lagrange equations

$$F_x - \frac{d}{dt} F_{\dot{x}} = 0, \qquad F_y - \frac{d}{dt} F_{\dot{y}} = 0,$$

which are, due to the positive homogeneity of first degree of F in $\dot{x}$, $\dot{y}$, equivalent to the Weierstrass equation

$$F_{x\dot{y}} - F_{y\dot{x}} + F_1(\dot{x}\ddot{y} - \dot{y}\ddot{x}) = 0,$$

where $F_1 = (F_{\dot{x}\dot{x}} + F_{\dot{y}\dot{y}})/(\dot{x}^2 + \dot{y}^2)$ (Sec. 4.4).

At every corner, the Weierstrass-Erdmann corner conditions

$$(F_{\dot{x}})_{t_c-0} = (F_{\dot{x}})_{t_c+0}$$
$$(F_{\dot{y}})_{t_c-0} = (F_{\dot{y}})_{t_c+0}$$

have to be satisfied (Sec. 4.4).

If beginning point and/or endpoint are free to vary on a simple, smooth curve $x = \varphi(\varepsilon)$, $y = \psi(\varepsilon)$, then at the point of departure P (or impact, or both), the transversality condition

$$[F_{\dot{x}}\varphi' + F_{\dot{y}}\psi']_P = 0$$

has to be satisfied (Sec. 4.6).

For the simple, smooth extremal arc γ_o to yield a strong relative minimum, it suffices that γ_o be embeddable in a field and that

$$\mathcal{E}(x,y,\theta,\bar{\theta}) = F(x, y, \cos\bar{\theta}, \sin\bar{\theta}) - \cos\bar{\theta} F_{\dot{x}}(x, y, \cos\theta, \sin\theta) - \sin\bar{\theta} F_{\dot{y}}(x, y, \cos\theta, \sin\theta) \geq 0$$

for all (x,y) in a weak neighborhood of γ_o and all $0 \leq \bar{\theta} < 2\pi$ (Sec. 4.10).

For γ_o to yield a weak relative minimum, it suffices that γ_o be embeddable in a field and that

$$F_1(x_o(t),y_o(t),\dot{x}_o(t),\dot{y}_o(t)) > 0$$

for all $t \in [t_a,t_b]$ (Sec. 4.10).

If $I[\gamma]$ is invariant under the transformations

$$\bar{x} = \bar{x}(x,y,\varepsilon)$$
$$\bar{y} = \bar{y}(x,y,\varepsilon),$$

which contain the identity transformation for $\varepsilon = 0$ and where

$$\frac{\partial(\bar{x},\bar{y})}{\partial(x,y)} \neq 0,$$

then

$$F_{\dot{x}}\xi + F_{\dot{y}}\eta = \text{constant}$$

is a first integral of the Euler-Lagrange equations, where $\xi = (\partial\bar{x}/\partial\varepsilon)_{\epsilon=0}$, $\eta = (\partial\bar{y}/\partial\varepsilon)_{\epsilon=0}$ (Sec. 4.8).

CHAPTER 5

THE HAMILTON–JACOBI THEORY AND THE MINIMUM PRINCIPLE OF PONTRYAGIN

APPENDIX

5.1 A FUNDAMENTAL LEMMA OF CARATHÉODORY

We consider the simplest variational problem that is associated with the functional

$$I[y] = \int_a^b f(x,y(x),y'(x))\,dx.$$

In the preceding chapters, we have been primarily interested in finding the function, or functions, $y = y(x)$ for which this functional assumes a relative minimum value, given certain boundary conditions. We now shift the emphasis to the minimum value of the functional itself, and we shall investigate it as a function of the coordinates of one of the two endpoints $P_a(a,y_a)$ or $P_b(b,y_b)$.

In all our previous investigations, no restrictions were ever placed on the values of y' other than those imposed by the domain within which the integrand was defined and satisfied suitable differentiability conditions.

Frequently, the formulation of a problem calls for certain restrictions on the range of values of y', restrictions that are dictated by practical considerations. This is fairly plausible when one realizes that y' may stand for velocity, rate of fuel consumption, direction of thrust, or what have you. Only velocities, rates of fuel consumption, and so on, within a certain range are technologically feasible, and a solution for which y' attains values outside this range is, for all practical purposes, useless. For this reason, we shall change the formulation of our problem in order to take such restrictions on y' into account.

Specifically, we shall now study the problem of finding the *minimum value* of the functional

$$I[y] = \int_{P_a(a,y_a)}^{P(x_o,y_o)} f(x,y(x),y'(x))\, dx$$

as a function of the endpoint coordinates (x_o,y_o) whereby the functions $y = y(x)$ that are admitted are assumed to be sectionally smooth and are such that $y'(x) \in U$, wherever y' exists, where U is some given subset of the set of all reals, e.g., an open interval, a closed interval, a union of intervals, or some discrete pointset.

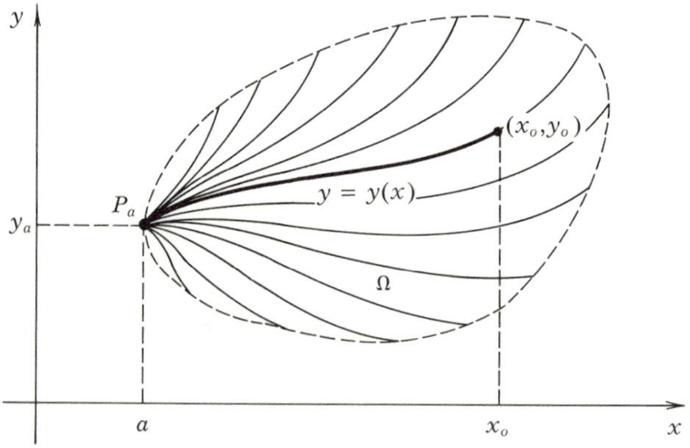

Figure 5.1

Clearly, if U stands for all the reals, then this is a reformulation of the original problem of the preceding chapters. This problem, where y' is *not* restricted, may now be viewed as a special case of the new, more general problem which we just formulated.

The symbols x, y will appear in two different roles in the sequel. While we shall continue to denote the independent variable by x and the dependent variable and the function by y, we shall also use (x,y) to denote the variable coordinates of a point $P(x,y)$ in the x,y plane. Whenever x, y appear in the same formula in the two different roles, we shall make a distinction by appropriate use of subscripts. When x, y appear in a formula in one role only, it will be clear from the context how they are to be interpreted.

In order to analyze this problem, we shall have to make a number of assumptions:

First, we assume that there exists a point set Ω which consists of a simply connected domain in the x,y plane *and* the point $P_a(a,y_a)$ which shall be a boundary point of that domain such that $x > a$ for all $(x,y) \in \Omega$, $y \neq y_a$, and such that $P_a(a,y_a)$ may be joined to every point $P(x_o,y_o) \in \Omega$ by *exactly one* sectionally smooth function $y = y(x)$, $(x,y(x)) \in \Omega$ for all $x \in [a, x_o]$, $y'(x) \in U$, which yields a minimum for $I[y]$ between $P_a(a,y_a)$ and $P(x_o,y_o)$, as compared with the values that are rendered for $I[y]$ by all other sectionally smooth functions with $y'(x) \in U$ that join P_a to P and remain entirely in Ω. We shall call a function $y = y(x)$ which renders $I[y]$ a minimum an *optimal trajectory*, and we shall call a pointset Ω with the stated properties an *admissible set of termination*. Note that all points in Ω other than $P_a(a,y_a)$ are interior points. (See Fig. 5.1.)

Similarly, we can define an *admissible set of inception* for the problem where the endpoint is considered as fixed and where the minimum of the functional $I[y]$ is considered as a function of the coordinates (x,y) of the beginning point.

Second, we assume that $f(x,y,y')$ is continuous for all $(x,y) \in \Omega$ and all $y' \in U$. Under these conditions, the following function, which we call *Hamilton's characteristic function*† (and which Hamilton called *optical distance*), is defined in Ω as a single-valued function of (x_o,y_o):

$$S(x_o,y_o) = \text{minimum value of} \int_{P_a(a,y_a)}^{P(x_o,y_o)} f(x,y(x),y'(x))\,dx \text{ as rendered by}$$

a uniquely defined optimal trajectory from P_a to P.

For example, for the variational problem $\int_{(o,o)}^{(x_o,y_o)} \sqrt{1 + y'^2(x)}\,dx \to \text{minimum}$, $U = (-\infty, \infty)$, the entire right half-plane including (0,0) is an admissible set of termination and $S(x_o,y_o) = \sqrt{x_o^2 + y_o^2}$.

† Not to be confused with the *hamiltonian*, which we introduced in Sec. 2.12.

Finally, we shall assume that $S \in C^1(\Omega)$.

Whether or not Hamilton's characteristic function S satisfies this condition in some Ω depends largely on the integrand $f(x,y,y')$, on the boundary conditions, and on other constraints that are imposed on the problem. We have already encountered a problem where S does not even exist (Prob. 1.7.2).

For instance, in the example above, we have

$$S_x = \frac{x}{\sqrt{x^2+y^2}}, \qquad S_y = \frac{y}{\sqrt{x^2+y^2}},$$

and we see that $S \notin C^1(\Omega)$ if Ω represents the right half-plane including $(0,0)$.

Let $y = \bar{y}(x)$, $\bar{y}'(x) \in U$, denote any sectionally smooth curve that joins $P_a(a, y_a)$ to $P(x_o, y_o) \in \Omega$ and remains entirely in Ω. If we note that Ω is, by hypothesis, simply connected and that $S(a, y_a) = 0$, then we have

$$S(x_o,y_o) = \int_{(a,y_a)}^{(x_o,y_o)} dS\,(x, \bar{y}(x))$$

$$= \int_{(a,y_a)}^{(x_o,y_o)} [S_x(x,\bar{y}(x)) + S_y(x,\bar{y}(x))\bar{y}'(x)]\,dx. \tag{5.1.1}$$

In view of the definition of $S(x_o, y_o)$ as the minimum of $I[y]$ between $P_a(a,y_a)$ and $P(x_o,y_o)$, we have

$$I[\bar{y}] - S(x_o,y_o)$$

$$= \int_{(a,y_a)}^{(x_o,y_o)} [f(x,\bar{y}(x),\bar{y}'(x)) - S_x(x,\bar{y}(x)) - S_y(x,\bar{y}(x))\bar{y}'(x)]\,dx \geq 0 \tag{5.1.2}$$

for all such curves $y = \bar{y}(x)$. More precisely, if $y = y(x)$ is the optimal trajectory of $I[y]$ between $P_a(a, y_a)$ and $P(x_o,y_o)$ which is, by hypothesis, uniquely determined, then

$$\int_{(a,y_a)}^{(x_o,y_o)} [f(x,y(x),y'(x)) - S_x(x,y(x)) - S_y(x,y(x))y'(x)]\,dx = 0, \tag{5.1.3}$$

while we have, for all other sectionally smooth functions $\bar{y} \neq y$, $\bar{y}'(x) \in U$, that also join $P_a(a,y_a)$ to $P(x_o,y_o)$ and remain entirely in Ω,

$$\int_{(a,y_a)}^{(x_o,y_o)} [f(x,\bar{y}(x),\bar{y}'(x)) - S_x(x,\bar{y}(x)) - S_y(x,\bar{y}(x))\bar{y}'(x)]\,dx > 0. \tag{5.1.4}$$

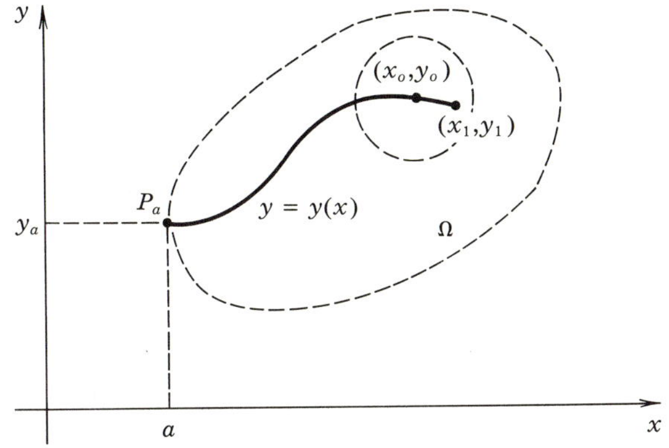

Figure 5.2

Lemma 5.1.1 If $f \in C$ *for all* $(x,y) \in \Omega$, *and all* $y' \in U$ *and if* $S \in C^1(\Omega)$, *then*

$$f(x,y,y') - S_x(x,y) - S_y(x,y)y' \geq 0$$

for all lineal elements (x,y,y'), *where* $(x,y) \in \Omega$ and $y' \in U$.

Proof (by contradiction): Suppose there exists a lineal element (x_o,y_o,y'_o), $(x_o,y_o) \in \Omega$, $x_o > a$, $y'_o \in U$, such that

$$f(x_o,y_o,y'_o) - S_x(x_o,y_o) - S_y(x_o,y_o)y'_o < 0.$$

Then there exists an $N^\delta(x_o,y_o) \subset \Omega$ such that

$$f(x,y,y'_o) - S_x(x,y) - S_y(x,y)y'_o < 0$$

for all $(x,y) \in N^\delta(x_o,y_o)$.

Hence we can find a point $Q(x_1,y_1) \in \Omega$ as close to $P(x_o,y_o)$ as we please on the line

$$y = (y'_o)(x - x_o) + y_o$$

that emanates from (x_o,y_o) with slope y'_o (see Fig. 5.2) such that

$$\int_{(x_o,y_o)}^{(x_1,y_1)} [f(x, (y'_o)(x - x_o) + y_o,y'_o) - S_x(x, (y'_o)(x - x_o) + y_o) \\ - S_y(x, (y'_o)(x - x_o) + y_o)y'_o]\, dx < 0.$$

Hence, if $y = y(x)$ is the optimal trajectory from $P_a(a,y_a)$ to $P(x_o,y_o)$, we obtain for

$$y = Y(x) = \begin{cases} y(x) & \text{for } a \leq x \leq x_o \\ (y'_o)(x - x_o) + y_o & \text{for } x > x_o \end{cases}$$

that

$$\int_{(a,y_a)}^{(x_1,y_1)} [f(x,Y(x),Y'(x)) - S_x(x,Y(x)) - S_y(x,Y(x))Y'(x)]\,dx < 0.$$

This is a contradiction of (5.1.2) if we also observe that $y = Y(x)$ is a sectionally smooth curve which joins $P(a,y_a)$ to $Q(x_1,y_1)$, that it remains entirely in Ω, and that $Y'(x) \in U$. By continuity, the lemma is also true for $(x,y) = (a,y_a)$.

Lemma 5.1.2 *If $f \in C$ for all $(x,y) \in \Omega$ and all $y' \in U$ and if $S \in C^1(\Omega)$, then for all lineal elements (x,y,y') of an optimal trajectory $y = y(x)$,*

$$f(x,y,y') - S_x(x,y) - S_y(x,y)y' = 0$$

wherever y' exists.

Proof: By Lemma 5.1.1,

$$f(x,y,y') - S_x(x,y) - S_y(x,y)y' \geq 0.$$

Suppose that for some lineal element $(x_o,y(x_o),y'(x_o))$ of the optimal trajectory $y = y(x)$,

$$f(x_o,y(x_o),y'(x_o)) - S_x(x_o,y(x_o)) - S_y(x_o,y(x_o))y'(x_o) > 0.$$

Then this is also true in a neighborhood of x_o and we obtain in view of Lemma 5.1.1 a contradiction to (5.1.3).

Combining Lemmas 5.1.1 and 5.1.2 yields the following equivalent statement:

Theorem 5.1 Carathéodory's lemma *If Hamilton's characteristic function $S = S(x, y)$ is defined on an admissible set of termination Ω and if $S \in C^1(\Omega)$,* then every lineal element (x_o,y_o,y'_o) *of an optimal trajectory that lies entirely in Ω is characterized by*

$$f(x_o,y_o,y'_o) - S_x(x_o,y_o) - S_y(x_o,y_o)y'_o = \min_{y' \in U} [f(x_o,y_o,y') - S_x(x_o,y_o) - S_y(x_o,y_o)y'] = 0.$$

R. E. Bellman, who calls this relation the *fundamental partial differential equation*, derived it by a different method.†

In order to use this lemma for the determination of an optimal trajectory, it appears that a knowledge of Hamilton's characteristic function

† R. E. Bellman and S. E. Dreyfus, "Applied Dynamic Programming," p. 191, Princeton University Press, Princeton, N.J., 1962; and S. E. Dreyfus, "Dynamic Programming and the Calculus of Variations," p. 78, Academic Press Inc., New York, 1965.

or, at the very least, a knowledge of S_x and S_y along the optimal trajectory is required.

In Sec. 5.2 we shall discuss a method for finding approximate values of S at discrete points and, at the same time, a polygonal approximation to the optimal trajectory. In Secs. 5.3 and 5.4 we shall demonstrate when and how Hamilton's characteristic function S may be found in a domain of the x,y plane for the case where $U = (-\infty,\infty)$. In Prob. 5.1.2 the reader is required to eliminate S_x, S_y from Carathéodory's lemma under the assumptions that $S \in C^2(\Omega)$ and $U = (-\infty,\infty)$. This elimination will lead to the Euler-Lagrange equation. In Sec. 5.7, we shall show in a much more general setting how S_x and S_y may be found as functions of the independent variable along the optimal trajectory. How this can be accomplished for the case at hand is indicated in Probs. 5.1.3 and 5.1.4.

PROBLEMS 5.1

1. State and prove Theorem 5.1 for a variational problem in n unknown functions that is associated with the functional

$$I[\hat{y}] = \int_{P_a}^{P_b} f(x,y_1(x), \ldots ,y_n(x),y_1'(x), \ldots ,y_n'(x))\, dx.$$

*2. Assume that $U = (-\infty,\infty)$ and that $f \in C^2$ for all $(x,y) \in \Omega$, $-\infty < y' < \infty$, $S \in C^2(\Omega)$, and derive the Euler-Lagrange equation from Theorem 5.1.

*3. Let $Q(x,y,y') = f(x,y,y') - S_x(x,y) - S_y(x,y)y'$. Then, by Carathéodory's lemma,

$$Q(x_o,y_o,y_o') = \min_{y'\epsilon U} Q(x_o,y_o,y') = 0$$

for every lineal element (x_o,y_o,y_o') of an optimal trajectory. If $\Delta x,\Delta y$ are sufficiently small, so that $(x_o + \Delta x,\ y_o + \Delta y) \in \Omega$, then

$$Q(x_o + \Delta x,\ y_o + \Delta y,\ y_o') \geq Q(x_o,y_o,y_o') = 0$$

and hence, by necessity,

$$Q_x(x_o,y_o,y_o') = 0, \qquad Q_y(x_o,y_o,y_o') = 0,$$

provided that Q_x, Q_y exist. Assume that $S \in C^2(\Omega)$ and that $y = y_o(x)$ is the optimal trajectory. Show that $\varphi(x) = S_x(x,y_o(x))$, $\psi(x) = S_y(x,y_o(x))$, that is, S_x, S_y, as functions of x along the optimal trajectory, satisfy the system of differential equations

$$\varphi'(x) = f_x(x,y_o(x),y_o'(x)), \qquad \psi'(x) = f_y(x,y_o(x),y_o'(x)).$$

*4. Show: If $y = y_o(x)$ is the optimal trajectory of the problem $I[y] = \int_a^b f(x,y(x),y'(x))\, dx \to$ minimum, $y(a) = y_a$, $y(b) = y_b$, $y'(x) \in U$, $x \in [a,b]$, and if the hypotheses of problem 3 are met, then it is necessary that

for every $x \in [a,b]$,

$$f(x,y_o(x),y_o'(x)) - \varphi(x) - \psi(x)y_o'(x) = \min_{y'\epsilon U}[f(x,y_o(x),y') - \varphi(x) - \psi(x)y'] = 0,$$

whereby $(\varphi,\psi) = (\varphi(x),\psi(x))$ are solutions of

$$\varphi'(x) = f_x(x,y_o(x),y_o'(x))$$
$$\psi'(x) = f_y(x,y_o(x),y_o'(x)).$$

5.2 DYNAMIC PROGRAMMING†

We consider again the problem of the preceding section, namely, the problem of minimizing the integral

$$I[y] = \int_{P_a(a,y_a)}^{P_b(b,y_b)} f(x,y(x),y'(x))\,dx \tag{5.2.1}$$

by means of a sectionally smooth function $y = y(x)$, where $y'(x) \in U$ for all x, which joins P_a to P_b. U is a given subset of the set of all reals.

R. E. Bellman arrived at Theorem 5.1 (Carathéodory's lemma) by means of a heuristic principle which he calls the *principle of optimality*‡ and by a method which is vaguely and generally referred to as *dynamic programming*.

Bellman's method for the derivation of Carathéodory's lemma will be discussed in some detail and in a more general setting in Sec. 5.7. Here, we shall demonstrate how Bellman utilizes an intermediate result in his derivation (which we shall recover from Carathéodory's lemma by a simple step backward) to solve, by a numerical procedure, variational problems such as those we have stated.

First, to achieve agreement with Bellman's formulas, we consider the beginning point in (5.2.1) as variable and the endpoint as fixed. Consequently, we shall work with an *admissible set of inception* rather than an admissible set of termination. All the conditions imposed in the preceding section are still assumed to hold.

If $S = S(x_o,y_o)$ now represents Hamilton's characteristic function for the problem with a fixed endpoint $P_b(b,y_b)$ and a variable beginning point $P(x_o,y_o)$, then instead of (5.1.1) we obtain

$$S(x_o,y_o) = -\int_{P(x_o,y_o)}^{P_b(b,y_b)} [S_x(x,\bar{y}(x)) + S_y(x,\bar{y}(x))\bar{y}'(x)]\,dx.$$

Consequently, the relation in Theorem 5.1 will assume the form

$$f(x_o,y_o,y_o') + S_x(x_o,y_o) + S_y(x_o,y_o)y_o'$$
$$= \min_{y'\in U} [f(x_o,y_o,y') + S_x(x_o,y_o) + S_y(x_o,y_o)y'] = 0 \tag{5.2.2}$$

for every lineal element (x_o,y_o,y_o') of an optimal trajectory $y = y(x)$,

† This section may be omitted without jeopardizing the continuity of the material.
‡ Bellman and Dreyfus, *op. cit.*, pp. 15 and 180ff; and Dreyfus, *op. cit.*, pp. 69ff.

$y'(x) \in U$, that joins P to P_b and remains entirely in Ω, the admissible set of inception.

If we assume that $\Delta x > 0$ is sufficiently small so that $(x_o + \Delta x, y_o + y' \Delta x) \in \Omega$ for all $y' \in U$, U bounded, then we have

$$S(x_o + \Delta x, y_o + y' \Delta x) - S(x_o,y_o)$$
$$= S_x(x_o,y_o) \Delta x + S_y(x_o,y_o)y' \Delta x + o(\Delta x),$$

where $\lim_{\Delta x \to 0} [o(\Delta x)/\Delta x] = 0$.

Accordingly, we replace $S_x(x_o,y_o) + S_y(x_o,y_o)y'$ in (5.2.2) by

$$\frac{S(x_o + \Delta x, y_o + y' \Delta x) - S(x_o,y_o)}{\Delta x},$$

neglect $[o(\Delta x)]/\Delta x$, multiply the resulting relation by $\Delta x > 0$, shift $S(x_o,y_o)$ to the left, and obtain the *functional equation*

$$S(x_o,y_o) = \min_{y' \epsilon U} [f(x_o,y_o,y') \Delta x + S(x_o + \Delta x, y_o + y' \Delta x)]. \quad (5.2.3)$$

This functional equation has been used by R. E. Bellman as the basis for a numerical approximation procedure for the solution of variational problems.†

Suppose we want to solve the fixed-beginning-point, variable-endpoint problem

$$I[y] = \int_a^b f(x,y(x),y'(x)) \, dx \rightarrow \text{minimum},$$

where the endpoint is allowed to slide freely on the vertical line $x = b$ and where the beginning point has the coordinates (a,y_a).

In order to utilize the functional equation (5.2.3) to obtain an approximate solution of this problem, we embed the problem in a family of problems with fixed endpoints $P(b,y)$ and variable beginning points. Hereby it is assumed that for all y in some neighborhood of the yet unknown y coordinate of the termination point, $P(b,y)$ admits an admissible set of inception.

We introduce grid-coordinates (i,k) by subdividing the interval $[a,b]$ into N equal subintervals of length Δ and by subdividing an interval of the line $x = b$ containing the anticipated endpoint into n equal subintervals of length δ. (This grid may have to be expanded when it turns out during the subsequently described process of computations that it does not cover

† R. E. Bellman, "Adaptive Control Processes," pp. 85ff, Princeton University Press, Princeton, N.J., 1961. See also Prob. 5.7.5.

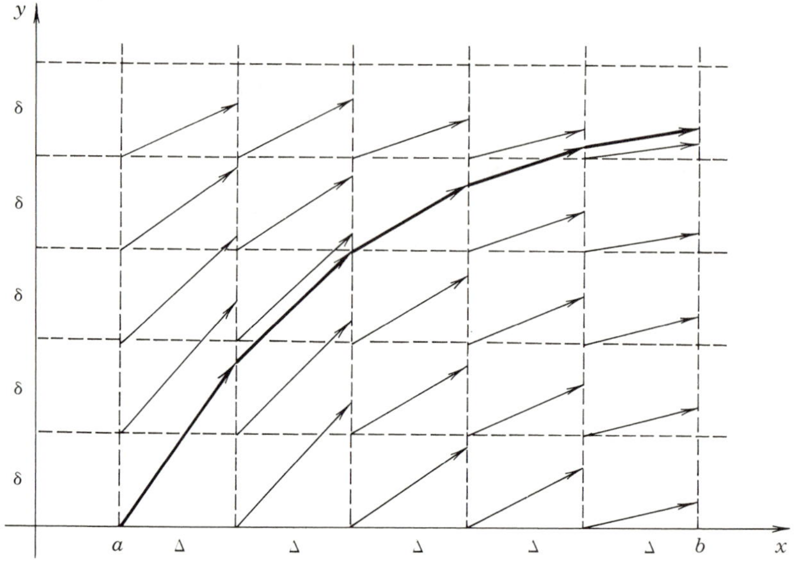

Figure 5.3

a sufficiently large range of y values.) For the purpose of this demonstration and for reasons of simplicity, we assume that the beginning point has the coordinates $(a,0)$. We start the latter subdivision at $y = 0$ and proceed in the positive direction. (See Fig. 5.3.)

The grid-point with the coordinates

$$x = a + (N - i)\Delta, \qquad i = 0, 1, \ldots, N$$

$$y = k\delta, \qquad k = 0, 1, \ldots, n$$

is characterized by the grid-coordinates (i,k). In particular, we have for $x = a$ the grid-abscissa $i = N$ and for $x = b$ the grid-abscissa $i = 0$.

Now we apply the functional equation (5.2.3) to each grid-point, whereby we adopt the notation

$$f(a + (N - i)\Delta, k\delta, y') = f_{i,k}(y')$$

$$S(a + (N - i)\Delta, y) = S_i(y).$$

Then (5.2.3) appears as

$$S_i(k\delta) = \min_{y' \epsilon U} [f_{i,k}(y')\Delta + S_{i-1}(k\delta + y'\Delta)]. \tag{5.2.4}$$

Since $S(b,y) = \int_b^b f(x,y(x),y'(x))\,dx = 0$, we have the "initial condition" $S(b,y) = 0$ for all y, or as applied to our grid,

$$S_o(k\delta + y'\Delta) = 0 \qquad \text{for all } k \text{ and all } y' \in U.$$

Hence,

$$S_1(k\delta) = \min_{y' \in U} [f_{1,k}(y')\Delta + S_o(k\delta + y'\Delta)]$$

$$= \min_{y' \in U} f(a + (N-1)\Delta, k\delta, y')\Delta.$$

This is, for every fixed k, a minimum-value problem for a function of the one variable y', and it may be solved numerically by a search ranging over discrete values of $y' \in U$. [Note that if U is the entire real axis, then the y' for which the minimum is achieved is the one for which $f_{y'}(a + (N-1)\Delta, k\delta, y') = 0$ and, as $\Delta \to 0$, $f_{y'}(b,k\delta,y') = 0$, that is, the natural boundary condition—see (2.7.4)—is satisfied at the endpoint.]

After we have solved this minimum problem for every $k = 0, 1, \ldots, n$, we know the values of $S_1(k\delta)$ for all k and we also know at each grid-point on $x = a + (N-1)\Delta$ the minimizing slope y' (see arrows in Fig. 5.3). Note that it is possible that several values of y' may yield the minimum at each step. (See Prob. 5.2.1.)

To prepare for the next step, we interpolate $S_1(k\delta)$ between grid-points on $x = a + (N-1)\Delta$ and assume for what is to follow that $S_1(y)$ is known, that is, $S(a + (N-1)\Delta, y)$ is known.

Now we apply formula (5.2.4) to $S_2(k\delta)$:

$$S_2(k\delta) = \min_{y' \in U} [f_{2,k}(y')\Delta + S_1(k\delta + y'\Delta)].$$

Since $S_1(k\delta + y'\Delta)$ is known for all k,y', we again have a minimum problem for a function of one variable y', which we may proceed to solve as before for all $k = 0, 1, \ldots, n$. We then obtain $S_2(k\delta)$ at all grid-points on the line $x = a + (N-2)\Delta$ and then, by interpolation, on the whole segment of the line between 0 and $n\delta$. We also obtain the minimizing slopes y' at all grid-points on this line. We continue to repeat this process, finding one S_i after the other in terms of its predecessor S_{i-1}, until we finally arrive at $S_N(k\delta) = S(a,k\delta)$.

We are now ready to reap the harvest from our labors. By construction, an arrow will emanate from the beginning point $(a,0)$. (See Fig. 5.3.) This arrow may or may not terminate at a grid-point on the line $x = a + \Delta$. If it does not, we have to find by interpolation an arrow on $x = a + \Delta$ that emanates from the termination point of the preceding arrow. We repeat this process until we arrive at $x = b$. (See bold polygon in Fig. 5.3.)

The polygon that has been put together from all these arrows is (hopefully) an approximation to the optimal trajectory of our problem, and the value $S_N(0)$ thus obtained is (hopefully) an approximate value of $S(a,0)$,† the minimum value of $I[y]$ between $(a,0)$ and $x = b$. (See also Prob. 5.2.2.)

PROBLEMS 5.2

1. Utilize formula (5.2.4) to solve the problem

$$\int_{P_a}^{P_b} (1 - y'^2(x))^2\,dx \to \text{minimum}, \qquad |y'| \leq \tfrac{1}{2}$$

   (*a*) For the case $P_a(0,y)$, $P_b(1,0)$
   (*b*) For the case $P_a(0,0)$, $P_b(1,0)$
   In both cases, use $N = 2$, $\Delta = \frac{1}{2}$, $\delta = \frac{1}{4}$.
2. Consider the set of all approximating polygonal solutions $y = P_N(x)$ of the fixed-beginning-point, variable-endpoint problem that were obtained from repeated application of the functional equation (5.2.4) for increasing N (decreasing Δ), as outlined in the preceding section. Assume that U is bounded above and below and show that $\{P_N(x)\}$ contains a uniformly convergent subsequence.

5.3 THE HAMILTON-JACOBI EQUATION

We shall now return to the problem of Sec. 5.1 with the beginning point fixed and the endpoint variable in an admissible set of termination Ω. Moreover, *we shall now free ourselves from the restriction $y' \in U$ and allow y' to assume any value between $-\infty$ and ∞.* Then, Carathéodory's lemma assumes the form

$$\begin{aligned} f(x_o,y_o,y_o') &- S_x(x_o,y_o) - S_y(x_o,y_o)y_o' \\ &= \min_{-\infty<y'<\infty} [f(x_o,y_o,y') - S_x(x_o,y_o) - S_y(x_o,y_o)y'] = 0, \end{aligned} \tag{5.3.1}$$

where we shall have to assume that $f(x,y,y')$ is defined and continuous for all $x,y \in \Omega$ and all $-\infty < y' < \infty$.

On this basis, it is now very easy to establish a fundamental relationship between Hamilton's characteristic function $S = S(x,y)$ and the integrand $f(x,y,y')$ of $I[y]$:

Theorem 5.3 *If $f \in C^1$ for all $(x,y) \in \Omega$ and all $-\infty < y' < \infty$ and if $S \in C^1(\Omega)$, then for every lineal element (x,y,y') of an optimal trajectory $y = y(x)$, except where $y' = y'(x)$ has a jump discontinuity, the following*

† See R. E. Bellman, Functional Equations in the Theory of Dynamic Programming—VI, A Direct Convergence Proof, *Ann. Math.*, vol. 65, no. 2, pp. 215–223, March, 1957.

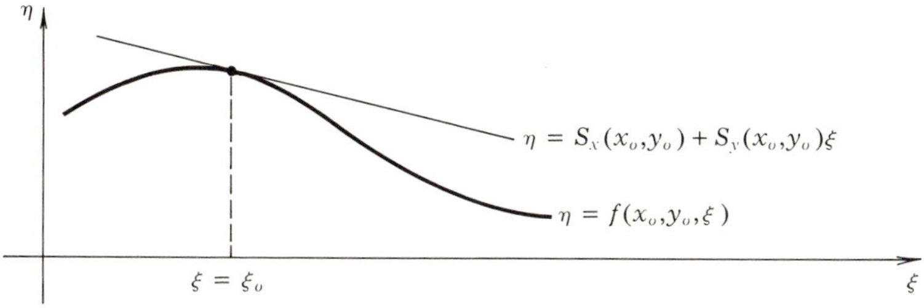

Figure 5.4

two relations have to hold:

$$S_x(x,y) = f(x,y,y') - y'f_{y'}(x,y,y'),$$
$$S_y(x,y) = f_{y'}(x,y,y'). \tag{5.3.2}$$

Here, y' is the uniquely determined slope of the optimal trajectory at the point (x,y) and is to be viewed as a function of x and y.

Proof: Let $P_o(x_o,y_o)$ represent any point on the optimal trajectory $y = y(x)$, where y' exists. We consider the characteristic of f at the point (x_o,y_o) (see Sec. 2.8)

$$\eta = f(x_o,y_o,\xi) \tag{5.3.3}$$

and the line

$$\eta = S_x(x_o,y_o) + S_y(x_o,y_o)\xi. \tag{5.3.4}$$

We see from Lemma 5.1.2 that the line (5.3.4) has to intersect the characteristic curve (5.3.3) at the point $\xi = \xi_o \equiv y'(x_o)$, which is the slope of the optimal trajectory at $x = x_o$. (See Fig. 5.4.)

From Lemma 5.1.1, we have that the line (5.3.4) has to remain on one side of the curve (5.3.3) for all $\xi \neq \xi_o$. Hence (5.3.4) has to be tangent line to (5.3.3) at $\xi = \xi_o$. But this tangent line is also given by

$$\eta = f_{y'}(x_o,y_o,\xi_o)(\xi - \xi_o) + f(x_o,y_o,\xi_o). \tag{5.3.5}$$

Comparison of (5.3.5) with (5.3.4) yields

$$S_x(x_o,y_o) = f(x_o,y_o,\xi_o) - \xi_o f_{y'}(x_o,y_o,\xi_o),$$
$$S_y(x_o,y_o) = f_{y'}(x_o,y_o,\xi_o).$$

Since (x_o,y_o) is any point on the optimal trajectory and since ξ_o represents the slope of the optimal trajectory at that point, formulas (5.3.2) follow immediately.

Suppose now that $f \in C^2$ for all $(x, y) \in \Omega$ and all y' and that our variational problem is regular in Ω and for all $y' \in (-\infty, \infty)$. Then we may replace y' and f by p and H by means of the Legendre transformation (2.12.9)

$$p = f_{y'}(x,y,y')$$
$$H(x,y,p) = y'f_{y'}(x,y,y') - f(x,y,y'),$$

and we can write equations (5.3.2) in the form

$$\begin{aligned} S_x(x,y) &= -H(x,y,p) \\ S_y(x,y) &= p. \end{aligned} \tag{5.3.6}$$

Elimination of p yields the *Hamilton-Jacobi equation*

$$S_x + H(x,y,S_y) = 0. \tag{5.3.7}$$

Note that, had we considered the endpoint as fixed and the beginning point as variable—as we did in the preceding section—then, for S as a function of the beginning point, defined on an *admissible set of inception*, we would have

$$S_x - H(x,y,-S_y) = 0.$$

As an illustration, we consider

$$I[y] = \int_{(0,0)}^{(x,y)} \sqrt{1 + y'^2(x)}\, dx \to \text{minimum},$$

where the endpoint is considered variable.

We have

$$f_{y'} = \frac{y'}{\sqrt{1 + y'^2}} = p$$

and hence $$y' = \Phi(x,y,p) = \frac{p}{\sqrt{1 - p^2}}.$$

Thus $$H(x,y,p) = [y'p - f(x,y,y')]_{y'=\Phi(x,y,p)} = -\sqrt{1 - p^2},$$

and we obtain as the Hamilton-Jacobi equation

$$S_x - \sqrt{1 - S_y^2} = 0,$$

or $$(S_x)^2 + (S_y)^2 = 1.$$

This is a (nonlinear) partial differential equation of the first order of the type

$$F(x,S_x) + G(y,S_y) = 0, \tag{5.3.8}$$

which can be solved by quadratures in the following manner:† For an arbitrary pa-

† I. N. Sneddon, "Elements of Partial Differential Equations," p. 72, McGraw-Hill Book Company, New York, 1957.

rameter α, solve the two ordinary differential equations

$$F(x,S_x) = \alpha, \qquad G(y,S_y) = -\alpha$$

by expressing S_x, S_y in terms of (x,α) and (y,α), respectively, and integrating the result, noting that the integration "constants" will depend on y and x, respectively. Then, match up the integration "constants" and obtain, after adding an arbitrary constant β, a two-parameter family of functions

$$S = S(x,y,\alpha) + \beta,$$

a so-called *complete integral*† of (5.3.8).

In our particular case, we have

$$F(x,S_x) = (S_x)^2, \qquad G(y,S_y) = (S_y)^2 - 1.$$

We proceed as outlined above:

$$(S_x)^2 = \alpha,$$

or

$$S_x = \sqrt{\alpha}.$$

Then

$$S = \sqrt{\alpha}x + A(y).$$

From

$$(S_y)^2 - 1 = -\alpha,$$

or

$$S_y = \sqrt{1-\alpha},$$

$$S = \sqrt{1-\alpha}y + B(x).$$

Matching up the constants: $A(y) = \sqrt{1-\alpha}y$, $B(x) = \sqrt{\alpha}x$. Hence

$$S = \sqrt{\alpha}x + \sqrt{1-\alpha}y + \beta. \tag{5.3.9}$$

For every β, (5.3.9) represents a one-parameter family of planes in (x,y,S) space that all pass through the point $(0,0,\beta)$.

In our problem $S(x,y)$ represents the minimum distance from (0,0) to the point (x,y). Therefore, if the endpoint (x,y) lies on the x axis, we have to have

$$S(x,0) = |x|,$$

and if the endpoint (x,y) lies on the y axis,

$$S(0,y) = |y|.$$

We see that no matter how we choose α and β in (5.3.9), S, as represented there, will *not* satisfy these boundary conditions.

Fortunately, our equation also possesses a *singular integral*,‡ namely, the envelope of (5.3.9), which we obtain by elimination of α§ from (5.3.9), and

$$S_\alpha = \frac{x}{2\sqrt{\alpha}} - \frac{y}{2\sqrt{1-\alpha}} = 0.$$

From the latter equation, we obtain

$$\frac{x^2}{x^2+y^2} = \alpha.$$

† *Ibid.*, p. 60.
‡ *Ibid.*
§ A. E. Taylor, "Advanced Calculus," pp. 399ff, Ginn and Company, Boston, 1955.

Substitution of this expression into (5.3.9) yields

$$S = \sqrt{x^2 + y^2} + \beta,$$

and we see that for $\beta = 0$, we have indeed

$$S(x,0) = |x|, \qquad S(0,y) = |y|.$$

So $S = \sqrt{x^2 + y^2}$ is the minimum of $I[y]$ between (0,0) and (x,y)—the euclidean distance—which, of course, we could have guessed from the outset.

In the next section, we shall generalize the ideas of this section to the problem in n unknown functions, and we shall also investigate the relation between the solution of the Hamilton-Jacobi equation and the extremals of the variational problem.

PROBLEMS 5.3

1. Suppose that $S = S(x,y)$ is defined in a simply connected domain relative to a fixed beginning point $P_a(a,y_a)$ and that the family of extremals $y = Y(x,c)$ defines a central field $(1,\phi(x,y))$ in Ω, with all the extremals emanating from P_a. Then, by necessity, the optimal trajectories that emanate from P_a—the existence of which was postulated in the definition of $S = S(x,y)$—coincide with the extremals $y = Y(x,c)$. In this setting, derive the relations (5.3.2) by utilizing Hilbert's invariant integral:

$$\begin{aligned} I[y] &= U[y] = U[\bar{y}] \\ &= \int_{(a,y_a)}^{(x,y)} [f(x,\bar{y},\phi(x,\bar{y})) + (\bar{y}' - \phi(x,\bar{y}))f_{y'}(x,\bar{y},\phi(x,\bar{y}))]\,dx \\ &= W(x,y) - W(a,y_a). \end{aligned}$$

2. With every point $(x,y) \in \Omega$, there is associated a unique slope y', namely, the slope of the optimal trajectory wherever the optimal trajectory is smooth: $y' = \phi(x,y)$. Assume that $\phi_x,\phi_y \in C(\Omega)$ and that $S = S(x,y) \in C^2(\Omega)$. Derive the Euler-Lagrange equation from the relations (5.3.2).
3. Assume that $S = S(x,y) \in C^1(\Omega)$ and derive the natural boundary conditions (2.7.4) from (5.3.2).
4. Assume that $S = S(x,y) \in C^1(\Omega)$ and derive the transversality conditions (2.8.6) from (5.3.2).
*5. Assume that $S = S(x,y) \in C^1(\Omega)$ and derive Weierstrass' necessary condition (3.9.6) for a strong relative minimum (Theorem 3.9.1) from Theorems 5.1 and 5.3.
6. Assume that $S = S(x,y) \in C^1(\Omega)$ and derive the Weierstrass-Erdmann corner conditions (2.9.1) and (2.9.4) from (5.3.2).
7. Assume that $f \in C^2(\mathfrak{R})$ and derive Legendre's necessary condition (3.9.7) for a weak relative minimum from Theorem 5.1.
8. Explain why and in what manner the validity of the natural boundary conditions, the transversality conditions, Weierstrass' necessary condition, the Weierstrass-Erdmann corner conditions, and the Legendre condition, as derived in problems 3 to 7, is more restricted by these derivations than by the derivations that were given in Chaps. 2 and 3.

5.4 SOLUTION OF THE HAMILTON-JACOBI EQUATION—JACOBI'S THEOREM

We shall now generalize the concept of Hamilton's characteristic function to the problem in n unknown functions,

$$I[\hat{y}] = \int_{(a,\hat{y}_a)}^{(x_o,\hat{y}_o)} f(x,\hat{y}(x),\hat{y}'(x))\,dx \to \text{minimum},$$

where the beginning point $(a,\hat{y}_a)$ is assumed to be fixed. [We remind the reader that $\hat{y} = (y_1, \ldots, y_n)$.]

We assume that there exists a simply connected domain in $x,\hat{y}$ space, $x > a$, so that every point $(x_o,\hat{y}_o)$ in that domain can be joined to $(a,\hat{y}_a)$ by a *uniquely determined* sectionally smooth optimal trajectory $\hat{y} = \hat{y}(x)$ which yields a minimum for $I[\hat{y}]$ between $(a,\hat{y}_a)$ and $(x_o,\hat{y}_o)$ and remains inside that domain for $x \in (a,x_o]$. As in Sec. 5.1 we call the union of this domain with the point $(a,\hat{y}_a)$ (which is a boundary point) an *admissible set of termination* Ω.

We call

$S(x_o,\hat{y}_o) =$ minimum value of $I[\hat{y}]$ between $(a,\hat{y}_a)$ and $(x_o,\hat{y}_o)$ as rendered by a uniquely determined sectionally smooth optimal trajectory that remains entirely in Ω (5.4.1)

Hamilton's characteristic function of the problem in n unknown functions.

If Ω is an admissible set of termination, then S is defined for all $(x,\hat{y}) \in \Omega$. If we assume that $S \in C^1(\Omega)$, and since Ω is simply connected, we can again write

$$S(x_o,\hat{y}_o) = \int_{(a,\hat{y}_a)}^{(x_o,\hat{y}_o)} dS\,(x,\hat{\hat{y}}(x))$$

$$= \int_{(a,\hat{y}_a)}^{(x_o,\hat{y}_o)} [S_x(x,\hat{\hat{y}}(x)) + S_{\hat{y}}(x,\hat{\hat{y}}(x))\hat{\hat{y}}'^T(x)]\,dx,\dagger$$

where $\hat{y} = \hat{\hat{y}}(x)$ is any sectionally smooth curve that joins $(a,\hat{y}_a)$ to $(x_o,\hat{y}_o)$ and remains entirely in Ω. As before, we have

$$I[\hat{\hat{y}}] - S(x_o,\hat{y}_o)$$

$$= \int_{(a,\hat{y}_a)}^{(x_o,\hat{y}_o)} [f(x,\hat{\hat{y}}(x),\hat{\hat{y}}'(x)) - S_x(x,\hat{\hat{y}}(x)) - S_{\hat{y}}(x,\hat{\hat{y}}(x))\hat{\hat{y}}'^T(x)]\,dx \geq 0$$

for all such curves $\hat{y} = \hat{\hat{y}}(x)$, and we obtain in the same manner as before that for any lineal element $(x,\hat{y},\hat{y}')$ of an optimal trajectory,

$$f(x,\hat{y},\hat{y}') - S_x(x,\hat{y}) - S_{\hat{y}}(x,\hat{y})\hat{y}'^T = 0, \qquad (5.4.2)$$

† Recall that T denotes the *transpose*, so that, in effect, $\hat{\hat{y}}'^T$ is a column vector.

while for all other lineal elements $(x,\hat{y},\hat{\bar{y}}')$, $(x,\hat{y}) \in \Omega$, we have

$$f(x,\hat{y},\hat{\bar{y}}') - S_x(x,\hat{y}) - S_{\hat{y}}(x,\hat{y})\hat{\bar{y}}'^T \geq 0. \tag{5.4.3}$$

(See Lemmas 5.1.1 and 5.1.2.)

Considering now the *characteristic surface*

$$\eta = f(x^o,y_1^o, \ldots,y_n^o,\xi_1, \ldots,\xi_n)$$

at a point $(x^o,y_1^o, \ldots,y_n^o)$ of an optimal trajectory and the plane

$$\eta = S_x(x^o,y_1^o, \ldots,y_n^o) + S_{\hat{y}}(x^o,y_1^o, \ldots,y_n^o)\hat{\xi}^T,$$

we come to realize, as in Sec. 5.3, that this plane has to be tangent to the characteristic surface for $\hat{\xi} = \hat{\xi}_o \equiv \hat{y}'(x^o)$, where $\hat{y}'(x^o)$ is the slope of the optimal trajectory at the point $(x^o,y_1^o, \ldots,y_n^o)$. We obtain that, by necessity,

$$\begin{aligned} S_x(x,\hat{y}) &= f(x,\hat{y},\hat{y}') - f_{\hat{y}'}(x,\hat{y},\hat{y}')\hat{y}'^T \\ S_{\hat{y}}(x,\hat{y}) &= f_{\hat{y}'}(x,\hat{y},\hat{y}') \end{aligned} \tag{5.4.4}$$

for every lineal element $(x,\hat{y},\hat{y}')$ of an optimal trajectory $\hat{y} = \hat{y}(x)$, except where $\hat{y}'$ has a jump discontinuity. As before, $\hat{y}'$ is to be viewed as a function of x and $\hat{y}$, being the uniquely defined slope of the optimal trajectory passing through $(x,\hat{y})$.

As in Prob. 5.3.5, we obtain from (5.4.2) to (5.4.4) for the optimal trajectory $\hat{y} = \hat{y}(x)$, the necessary condition

$$\mathcal{E}(x,\hat{y}(x),\hat{y}'(x),\hat{\bar{y}}') = f(x,\hat{y}(x),\hat{\bar{y}}') - f(x,\hat{y}(x),\hat{y}'(x)) + \sum_{i=1}^{n} (y_i'(x) - \bar{y}_i')f_{y'_i}(x,\hat{y}(x),\hat{y}'(x)) \geq 0 \tag{5.4.5}$$

for all $x \in [a,b]$ and all $-\infty < \bar{y}_i' < \infty$.

If we assume that $f \in C^2$ for all $(x,\hat{y}) \in \Omega$ and all $\hat{y}'$ and that the problem is regular in Ω and for all $\hat{y}'$, that is, $\det |f_{y'_iy'_k}| \neq 0$, we obtain by means of the Legendre transformation (2.12.9),

$$\begin{aligned} \hat{p} &= f_{\hat{y}'}(x,\hat{y},\hat{y}') \\ H(x,\hat{y},\hat{p}) &= f_{\hat{y}'}(x,\hat{y},\hat{y}')\hat{y}'^T - f(x,\hat{y},\hat{y}'), \end{aligned}$$

from (5.4.4)

$$\begin{aligned} S_x(x,\hat{y}) &= -H(x,\hat{y},\hat{p}) \\ S_{\hat{y}}(x,\hat{y}) &= \hat{p}. \end{aligned} \tag{5.4.6}$$

By elimination of $\hat{p}$, we obtain from (5.4.6) the *Hamilton-Jacobi equation* for the problem in n unknown functions:

$$S_x + H(x,\hat{y},S_{\hat{y}}) = 0. \tag{5.4.7}$$

Summarizing, we have:

Theorem 5.4.1 *If $f \in C^1$ for all $(x,\hat{y}) \in \Omega$ and for all $\hat{y}'$ and if $S \in C^1(\Omega)$, then for every lineal element $(x,\hat{y},\hat{y}')$ of an optimal trajectory $\hat{y} = \hat{y}(x)$, except where $\hat{y}'$ has a jump discontinuity, the equations*

$$S_x(x,\hat{y}) = f(x,\hat{y},\hat{y}') - f_{\hat{y}'}(x,\hat{y},\hat{y}')\hat{y}'^T$$
$$S_{\hat{y}}(x,\hat{y}) = f_{\hat{y}'}(x,\hat{y},\hat{y}')$$

have to hold, whereby $\hat{y}'$ is to be considered as a function of x, $\hat{y}$.

If $f \in C^2$ for all $(x,\hat{y}) \in \Omega$ and for all $\hat{y}'$ and if the problem is regular, then S has to satisfy the Hamilton-Jacobi equation

$$S_x + H(x,\hat{y},S_{\hat{y}}) = 0.$$

While an extremal is not necessarily an optimal trajectory, as we have seen in many instances, we realize, of course, that any smooth portion of an optimal trajectory has to be an extremal. The following theorem will show how extremals—solutions of the Euler-Lagrange equations—can be obtained from a solution of the Hamilton-Jacobi equation.

Theorem 5.4.2 Jacobi's theorem *If $S(x,y_1, \ldots,y_n,\alpha_1, \ldots,\alpha_n)$, where $\alpha_1, \ldots, \alpha_n$ are parameters, is a solution of the Hamilton-Jacobi equation (5.4.7) in a neighborhood $N(x_o,\hat{y}_o)$ of $(x_o,\hat{y}_o)$, for all $\hat{\alpha} = (\alpha_1, \ldots,\alpha_n)$ in a neighborhood $N(\hat{\alpha}_o)$ of some $\hat{\alpha}_o = (\alpha_1{}^o, \ldots,\alpha_n{}^o)$, if*

$$S_{\alpha_i},S_{\alpha_i x},S_{\alpha_i y_k},S_{y_i x},S_{y_i y_k} \in C(N(x_o,\hat{y}_o) \times N(\hat{\alpha}_o)),$$

and if $\qquad \det |S_{\alpha_i y_k}| \neq 0 \qquad$ *in $N(x_o,\hat{y}_o) \times N(\hat{\alpha}_o)$,*

then a two-parameter family of solutions $\hat{y} = \hat{y}(x,\hat{\alpha},\hat{\beta})$, $\hat{p} = \hat{p}(x,\hat{\alpha},\hat{\beta})$ of the Euler-Lagrange equations in canonical form (2.12.8), where $\hat{\beta} = (\beta_1, \ldots,\beta_n)$ are arbitrary constants, is obtained from

$$S_{\alpha_i}(x,\hat{y},\hat{\alpha}) = \beta_i$$
$$p_i = S_{y_i}(x,\hat{y},\hat{\alpha}),$$

$i = 1, 2, \ldots, n$, and this solution is valid in some neighborhood of $x = x_o$.

Proof: By the theorem on implicit functions we can solve

$$S_{\alpha_i}(x,\hat{y},\hat{\alpha}) = \beta_i$$

for arbitrary β_i, for y_i in some neighborhood of $(x_o,\hat{y}_o,\hat{\alpha}_o)$ and obtain

$$y_i = y_i(x,\alpha_1, \ldots,\alpha_n,\beta_1, \ldots,\beta_n).$$

We substitute these solutions back into $S_{\alpha_i} = \beta_i$ and differentiate the resulting identity

$$S_{\alpha_i}(x,\hat{y}(x,\hat{\alpha},\hat{\beta}),\hat{\alpha}) = \beta_i$$

totally with respect to x:

$$\frac{d}{dx} S_{\alpha_i}(x,\hat{y}(x,\hat{\alpha},\hat{\beta}),\hat{\alpha}) = 0.$$

If we carry out the differentiation on the left, we obtain

$$\frac{d}{dx}(S_{\alpha_i}) = S_{\alpha_i x} + \sum_{k=1}^{n} S_{\alpha_i y_k} y_k'.$$

From the Hamilton-Jacobi equation, we have $S_x = -H(x,\hat{y},S_{\hat{y}})$ and hence

$$S_{\alpha_i x} = -\sum_{k=1}^{n} H_{p_k} S_{y_k \alpha_i}.$$

Pulling these partial results together, we have

$$\frac{d}{dx}(S_{\alpha_i}) = \sum_{k=1}^{n} S_{y_k \alpha_i}(y_k' - H_{p_k}).$$

Since $(d/dx)(S_{\alpha_i}) = 0$ along $\hat{y} = \hat{y}(x,\hat{\alpha},\hat{\beta})$ and since $\det|S_{\alpha_i y_k}| \neq 0$, it follows that $\hat{y} = \hat{y}(x,\hat{\alpha},\hat{\beta})$ has to satisfy the first n of the canonical equations

$$y_k' = H_{p_k}, \qquad k = 1, 2, \ldots, n.$$

Next we substitute $\hat{y} = \hat{y}(x,\hat{\alpha},\hat{\beta})$ into

$$p_i = S_{y_i}(x,\hat{y}(x,\hat{\alpha},\hat{\beta}),\hat{\alpha}) = p_i(x,\hat{\alpha},\hat{\beta})$$

and check whether these functions satisfy the remaining equations of the canonical system. We have

$$\frac{dp_i}{dx} = \frac{d}{dx}(S_{y_i}) = S_{y_i x} + \sum_{k=1}^{n} S_{y_i y_k} y_k' = S_{y_i x} + \sum_{k=1}^{n} S_{y_i y_k} H_{p_k}.$$

From the Hamilton-Jacobi equation, we have

$$S_{x y_i} = -H_{y_i} + \sum_{k=1}^{n} H_{p_k} S_{y_k y_i}$$

and hence

$$\frac{dp_i}{dx} = -H_{y_i} + \sum_{k=1}^{n} S_{y_i y_k}(H_{p_k} - H_{p_k}) = -H_{y_i},$$

which shows that the remaining equations of the canonical system are also satisfied. The theorem is proved.

As a first illustration of this theorem, we consider the example of Sec. 5.3, with $I[y] = \int_a^b \sqrt{1 + y'^2(x)}\, dx$.

We obtain for the Hamilton-Jacobi equation (see page 267)

$$(S_x)^2 + (S_y)^2 = 1$$

and the solution

$$S = \sqrt{\alpha}x + \sqrt{1 - \alpha}y.$$

We see that

$$S_{y\alpha} = -\frac{1}{2\sqrt{1-\alpha}} \neq 0.$$

In accordance with Jacobi's theorem, we find $y = y(x,\alpha,\beta)$ from

$$S_\alpha = \beta,$$

which is, in our case,

$$\frac{x}{2\sqrt{\alpha}} - \frac{y}{2\sqrt{1-\alpha}} = \beta.$$

Hence

$$y = \frac{\sqrt{1-\alpha}}{\sqrt{\alpha}}x - 2\beta\sqrt{1-\alpha} \tag{5.4.8a}$$

and

$$p = S_y = \sqrt{1-\alpha}. \tag{5.4.8b}$$

Since $H = -\sqrt{1-p^2}$ (see page 266), the canonical system has the form

$$y' = \frac{p}{\sqrt{1-p^2}}$$

$$p' = 0,$$

and we see that y and p as given in (5.4.8a and b) are indeed solutions of this system. (Note that the trivial solution $y = 0$, $p = 0$ is obtained for $\alpha = 1$.)

We can also see that for any fixed $\alpha \in (0,1)$, the resulting one-parameter family of extremals (5.4.8a) defines a field on $-\infty < x < \infty$. (See also Prob. 5.4.1.)

As a second illustration, we consider the example from Sec. 3.4:

$$I[y] = \int_a^b (y'^2(x) - y^2(x))\, dx \to \text{minimum.}$$

Since $f_{y'} = 2y'$, we have $y' = p/2$ and $H = (p^2/4) + y^2$. Hence we have as the Hamilton-Jacobi equation

$$S_x + \tfrac{1}{4}(S_y)^2 + y^2 = 0,$$

which is again an equation of the type (5.3.8) and which we shall solve by the method that was outlined in Sec. 5.3: From $S_x = -\alpha$, we obtain

$$S = -\alpha x + A(y).$$

We solve $\frac{1}{4}(S_y)^2 + y^2 = \alpha$ for S_y and obtain $S_y = 2\sqrt{\alpha - y^2}$. Hence

$$S = y\sqrt{\alpha - y^2} + \alpha \arcsin\frac{y}{\sqrt{\alpha}} + B(x).$$

We match up the "constants" and have

$$S = -\alpha x + y\sqrt{\alpha - y^2} + \alpha \arcsin \frac{y}{\sqrt{\alpha}}.$$

We obtain for $S_\alpha = \beta$,

$$\arcsin \frac{y}{\sqrt{\alpha}} - x = \beta$$

and hence

$$y = \sqrt{\alpha} \sin (x + \beta). \tag{5.4.9}$$

Since $p = 2y'$, clearly, $p = 2\sqrt{\alpha} \cos (x + \beta)$, and this agrees with $p = S_y = 2\sqrt{\alpha - y^2}$ as obtained from the Hamilton-Jacobi theory.

For any given β, (5.4.9) generates a field for $-\beta < x < -\beta + \pi$. (Compare this with the result of Sec. 3.4 and Fig. 3.4.) If we take $S_y = -2\sqrt{\alpha - y^2}$ instead, we obtain $y = -\sqrt{\alpha} \sin (x + \beta)$.

For more examples, see Prob. 5.4.2*a* and *b*.

PROBLEMS 5.4

*1. It appears that in the example of this section, with $I[y] = \int_a^b \sqrt{1 + y'^2(x)}\, dx$, only points in the first quadrant can be joined with the origin by an extremal from $y = (\sqrt{1 - \alpha}/\sqrt{\alpha})x - 2\beta\sqrt{1 - \alpha}$. How do you obtain solutions from Jacobi's theorem that admit endpoints in the fourth quadrant?

2. Solve the Hamilton-Jacobi equation for the functionals

(*a*) $I[y] = \int_a^b y'^2(x)\, dx$

(*b*) $I[y] = \int_a^b y(x)\sqrt{1 + y'^2(x)}\, dx$

3. Explore the possibilities of obtaining fields from the solutions of the Hamilton-Jacobi equations in problem 2*a* and *b*.

5.5 THE HAMILTON-JACOBI EQUATION AND FIELD EXISTENCE

We saw from the two examples which we discussed at the end of the preceding section that fields can be obtained from solutions of the Hamilton-Jacobi equation.

The following theorem indicates when a field may be constructed from a solution of the Hamilton-Jacobi equation, and the subsequent proof will demonstrate how to go about it.

Theorem 5.5 *Let $S = S(x,\hat{y}) \in C^2(\mathcal{A})$ represent a solution of the Hamilton-Jacobi equation (5.4.7), where $\mathcal{A}$ represents a simply connected domain in $x,\hat{y}$ space. If $I[\hat{y}]$ is regular in $\mathcal{A}$ and for all $\hat{y}'$,* then $(1,\hat{\phi}(x,\hat{y}))$ *with*

$$\hat{\phi}(x,\hat{y}) = H_{\hat{p}}(x,\hat{y},S_{\hat{y}}(x,\hat{y}))$$

defines a Mayer field on that subset of $\mathcal{A}$ where $\hat{\phi},\hat{\phi}_x,\hat{\phi}_{\hat{y}} \in C$.

Proof: According to Definition 3.10.2, we have to show that the solutions of

$$\hat{y}' = \hat{\phi}(x,\hat{y})$$

are extremals of $I[\hat{y}]$ and that

$$(\partial/\partial y_i)f_{y'_k}(x,\hat{y},\hat{\phi}(x,\hat{y})) = (\partial/\partial y_k)f_{y'_i}(x,\hat{y},\hat{\phi}(x,\hat{y})).$$

(*a*) Suppose $\hat{y} = \hat{y}(x)$ is a solution of

$$\hat{y}' = H_{\hat{p}}(x,\hat{y},S_{\hat{y}}(x,\hat{y})), \tag{5.5.1}$$

that is, $\hat{y}'(x) = H_{\hat{p}}(x,\hat{y}(x),S_{\hat{y}}(x,\hat{y}(x)))$. If $S_{\hat{y}\hat{y}}$ denotes the matrix

$$S_{\hat{y}\hat{y}} = \begin{pmatrix} S_{y_1y_1} & \cdots & S_{y_1y_n} \\ \vdots & & \\ S_{y_ny_1} & \cdots & S_{y_ny_n} \end{pmatrix},$$

we have

$$\frac{d}{dx} S_{\hat{y}}(x,\hat{y}(x)) = S_{\hat{y}x}(x,\hat{y}(x)) + \hat{y}'(x) S_{\hat{y}\hat{y}}(x,\hat{y}(x)).$$

Since $S_{\hat{y}x} = S_{x\hat{y}}$, we have from the Hamilton-Jacobi equation,

$$S_{\hat{y}x} = -H_{\hat{y}}(x,\hat{y},S_{\hat{y}}(x,\hat{y})) - H_{\hat{p}}(x,\hat{y},S_{\hat{y}}(x,\hat{y})) S_{\hat{y}\hat{y}}(x,\hat{y}).$$

Hence,

$$\frac{d}{dx} S_{\hat{y}}(x,\hat{y}(x)) = -H_{\hat{p}}(x,\hat{y}(x),S_{\hat{y}}(x,\hat{y}(x))) S_{\hat{y}\hat{y}}(x,\hat{y}(x))$$
$$+ \hat{y}'(x) S_{\hat{y}\hat{y}}(x,\hat{y}(x)) - H_{\hat{y}}(x,\hat{y}(x),S_{\hat{y}}(x,\hat{y}(x))).$$

From (5.5.1),

$$\frac{d}{dx} S_{\hat{y}}(x,\hat{y}(x)) = -H_{\hat{y}}(x,\hat{y}(x),S_{\hat{y}}(x,\hat{y}(x))). \tag{5.5.2}$$

Hence, with $\hat{p}(x) = S_{\hat{y}}(x,\hat{y}(x))$, we have from (5.5.1) and (5.5.2),

$$\hat{y}'(x) = H_{\hat{p}}(x,\hat{y}(x),\hat{p}(x))$$
$$\hat{p}'(x) = -H_{\hat{y}}(x,\hat{y}(x),\hat{p}(x)),$$

which is the canonical form of the Euler-Lagrange equations. [See (2.12.8).] So we see that the solutions of $\hat{y}' = \hat{\phi}(x,\hat{y})$ are indeed extremals.

(*b*) By the involutary character of the Legendre transformation (see Lemma 2.12), we have from

$$\hat{\phi} = H_{\hat{p}}(x,\hat{y},S_{\hat{y}})$$

that

$$S_{\hat{y}} = f_{\hat{y}'}(x,\hat{y},\hat{\phi}).$$

Hence $\qquad \dfrac{\partial}{\partial y_i} f_{y'_k}(x,\hat{y},\hat{\phi}(x,\hat{y})) = S_{y_k y_i}, \qquad \dfrac{\partial}{\partial y_k} f_{y'_i}(x,\hat{y},\hat{\phi}(x,\hat{y})) = S_{y_i y_k},$

and it follows that condition 2 of Definition 3.10.2 is also satisfied.

For an illustration, we consider again the two examples of the preceding section.

First, we take $f = \sqrt{1 + y'^2}$. (See page 266.) We have seen that $H = -\sqrt{1 - p^2}$ and that $S = \sqrt{\alpha}x + \sqrt{1 - \alpha}y$. Hence $H_p = p/\sqrt{1 - p^2}$ and $S_y = \sqrt{1 - \alpha}$. By Theorem 5.5, $(1,\phi(x,y))$ with

$$\phi(x,y) = \frac{\sqrt{1 - \alpha}}{\sqrt{\alpha}}$$

defines a field for any fixed α in $0 < \alpha < 1$. It is obvious that this field is defined in the entire plane.

Another field is obtained if one takes the singular integral of the Hamilton-Jacobi equation, namely,

$$S = \sqrt{x^2 + y^2}$$

(see page 268). Then $(1,\phi(x,y))$ with

$$\phi(x,y) = \frac{y}{x}$$

defines a (central) field everywhere away from the origin.

Next we consider the example with $f = y'^2 - y^2$. We have found on page 273 that $H = \frac{1}{4}p^2 + y^2$ and $S = -\alpha x + y\sqrt{\alpha - y^2} + \alpha \arcsin(y/\sqrt{\alpha})$. Then $H_p = \frac{1}{2}p$ and $S_y = 2\sqrt{\alpha - y^2}$. Hence, by our theorem,

$$(1, \sqrt{\alpha - y^2})$$

defines a field for any fixed α but only on $|y| < \sqrt{\alpha}$ since $\phi_y = -y/\sqrt{\alpha - y^2}$ is only continuous in $|y| < \sqrt{\alpha}$. (See Fig. 5.5.)

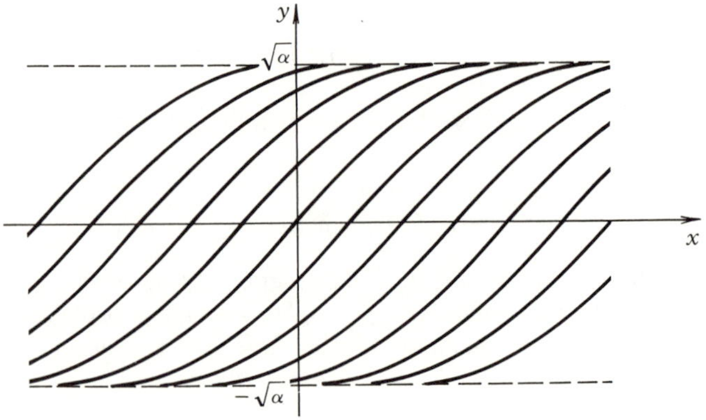

Figure 5.5

PROBLEMS 5.5

1. Solve the Hamilton-Jacobi equation for the variational problem

$$\int_a^b y'^2(x)\, dx \to \text{minimum},$$

and use the solution to construct a field when and where this is possible. (See also Prob. 5.4.2*a*.)

2. Same as in problem 1 for the variational problem

$$\int_a^b n(y(x)) \sqrt{1 + y'^2(x)}\, dx \to \text{minimum}.$$

[*Hint*: $S = \alpha^2 x + \int \sqrt{n^2(y) - \alpha^2}\, dy + \beta$.]

3. Same as in problem 1 for the variational problem

$$\int_a^b \sqrt{x^2 + y^2(x)} \sqrt{1 + y'^2(x)}\, dx \to \text{minimum}.$$

(*Hint*: $S = \alpha x^2 + \sqrt{1 - 4\alpha^2}\, xy - \alpha y^2$.)

5.6 A GENERAL MINIMUM-INTEGRAL CONTROL PROBLEM

We return again to the problem of Sec. 5.1, namely, to seek a sectionally smooth function $y = y(x)$ which satisfies certain boundary conditions $y(a) = y_a$, $y(b) = y_b$, for which $y'(x) \in U$ for all $x \in [a,b]$, and which is such that

$$I[y] = \int_a^b f(x,y(x),y'(x))\, dx \to \text{minimum}.$$

We shall formulate this problem in a different manner by introducing four new functions y_o, y_1, y_2, and u as follows:

$$\begin{aligned} y_o &= y_o(x) \equiv \int_a^x f(s,y(s),y'(s))\, ds \\ y_1 &= y_1(x) \equiv y(x) \\ y_2 &= y_2(x) \equiv x \\ u &= u(x) \equiv y'(x). \end{aligned}$$

These functions satisfy, in view of their definition, the following autonomous system of first-order differential equations:

$$\begin{aligned} y_o' &= f(y_2,y_1,u) \\ y_1' &= u \\ y_2' &= 1. \end{aligned} \qquad (5.6.1)$$

From the original boundary conditions and from

$$y_o(a) = \int_a^a f(s,y(s),y'(s))\,ds = 0,$$

we obtain the new initial conditions

$$y_o(a) = 0, \qquad y_1(a) = y_a, \qquad y_2(a) = a \tag{5.6.2a}$$

and the new terminal conditions

$$y_1(b) = y_b, \qquad y_2(b) = b. \tag{5.6.2b}$$

Since the integral which is to be minimized can now be written in terms of the new function y_o as,

$$\int_a^b f(s,y(s),y'(s))\,ds = y_o(b),$$

we can express the minimum condition simply as

$$y_o(b) \to \text{minimum}. \tag{5.6.3}$$

In view of all this, our problem can now be stated as follows: To be found are four functions $y_o = y_o(x)$, $y_1 = y_1(x)$, $y_2 = y_2(x)$, and $u = u(x)$, where $y_o,y_1,y_2 \in C[a,b]$, $u \in C_s[a,b]$, and $u(x) \in U$ for all $x \in [a,b]$, which satisfy the *underdetermined* autonomous system of first-order differential equations (5.6.1) and the boundary conditions (5.6.2a and b) and which are such that the terminal value $y_o(b)$ of y_o becomes a minimum.

Since a specific choice of u turns the underdetermined system (5.6.1) into a determined system and since different choices of u will lead, in general, to different solutions y_o,y_1,y_2 of (5.6.1) [and will, in this manner, control the solutions of (5.6.1)], one refers to u as a *control function* or, simply, a *control.*

Such controls for which (5.6.1) has solutions that satisfy the given boundary conditions (5.6.2a and b)—if there are such solutions—are called *admissible controls.*†

A curve $y_o = y_o(x)$, $y_1 = y_1(x)$, $y_2 = y_2(x)$ in (y_o,y_1,y_2) space that corresponds to an admissible control is called an *admissible trajectory.*

With this new terminology, our problem may be stated as follows: To find among all admissible controls the one for which $y_o(b)$ becomes a minimum. The control that leads to the minimum is called the *optimal control.* The trajectory that corresponds to the optimal control is called

† This terminology, which we find quite practical, is at variance with the standard terminology in the optimal-control-theory literature.

the *optimal trajectory*. A problem such as the one formulated in [(5.6.1) to (5.6.3)] is called an *optimal control problem.*

We shall now discuss an obvious generalization of this optimal control problem. As is customary, we shall designate the independent variable from now on by t.

We consider a physical system whose state at the time t is characterized by a point $(y_1(t), \ldots, y_n(t))$ of the n-dimensional $(y_1, \ldots, y_n)$ space. We assume that the behavior of the system can be influenced (controlled) at every instant of its evolution and that this influence can be characterized by the functions $u_1 = u_1(t), \ldots, u_m = u_m(t)$, the so-called *control functions* or, simply, *controls*.

For the sake of a mathematical analysis, as well as for practical reasons, it is important to restrict these controls to a certain specific class of functions, for example, C_s. For practical reasons, it is also of importance—as we pointed out in Sec. 5.1—to restrict the range of the controls to a certain subset U of the $(u_1, \ldots, u_m)$ space.

We assume that the behavior of the controlled system can be described by the following system of autonomous first-order differential equations:

$$\frac{dy_1}{dt} = f_1(y_1, \ldots, y_n, u_1, \ldots, u_m)$$
$$\vdots$$
$$\frac{dy_n}{dt} = f_n(y_1, \ldots, y_n, u_1, \ldots, u_m),$$

which is, as we note, a straightforward generalization of the system (5.6.1).

We shall study the problem of transferring the system from a given initial state $(y_1^a, \ldots, y_n^a)$ through suitable choice of the controls $u_i = u_i(t)$ to a given terminal state $(y_1^b, \ldots, y_n^b)$ in such a manner that a certain quantity that depends on the state variables and the controls and that can be expressed as an integral becomes a minimum:

$$I[y_1, \ldots, y_n, u_1, \ldots, u_m] = \int_{t_a}^{t_b} f_o(y_1(t), \ldots, y_n(t), u_1(t), \ldots, u_m(t))\, dt \to \text{minimum},$$

whereby the terminal time t_b is unspecified. (This integral may represent a business loss, temperature, fuel consumption, time, etc.) Since t_b is not specified and will depend on the choice of controls, our problem differs from the one we discussed in the beginning of this section, where b, the terminal value of the independent variable, was given. We shall see later (Sec. A5.9) that this difference is not significant.

As before, we introduce a new function

$$y_o = y_o(t) \equiv \int_{t_a}^{t} f_o(y_1(s), \ldots, y_n(s), u_1(s), \ldots, u_m(s))\, ds,$$

and we see that

$$y_o(t_a) = 0, \qquad y_o(t_b) = \int_{t_a}^{t_b} f_o(y_1(s), \ldots, y_n(s), u_1(s), \ldots, u_m(s))\, ds,$$

$$\frac{dy_o}{dt} = f_o(y_1, \ldots, y_n, u_1, \ldots, u_m).$$

In order to state this problem in a simple, uncluttered form, we agree on the following notation:

$\hat{y}, \hat{f}$ denote the $(n + 1)$-dimensional vectors

$$\hat{y} = (y_o, \ldots, y_n), \qquad \hat{f} = (f_o, \ldots, f_n).$$

y, f denote the n-dimensional vectors

$$y = (y_1, \ldots, y_n), \qquad f = (f_1, \ldots, f_n).$$

$\hat{u}$ denotes the m-dimensional vector

$$\hat{u} = (u_1, \ldots, u_m).$$

If U denotes a given subset of the $(u_1, \ldots, u_m)$ space, we can state our problem as follows: To be found is a control $\hat{u} = \hat{u}(t)$ with range in U, that is, $\hat{u}(t) \in U$ for all t, such that the solution $\hat{y} = \hat{y}(t)$ of the system of differential equations

$$\hat{y}' = \hat{f}(y, \hat{u}) \tag{5.6.4}$$

satisfies the initial conditions

$$\hat{y}(t_a) = \hat{y}_a, \qquad \text{where } \hat{y}_a = (0, y_1^a, \ldots, y_n^a), \tag{5.6.5a}$$

and the terminal conditions

$$\hat{y}(t_b) = \hat{y}_b, \qquad \text{where } \hat{y}_b = (y_o(t_b), y_1^b, \ldots, y_n^b), \tag{5.6.5b}$$

for some $t = t_b$ so that

$$y_o(t_b) \to \text{minimum}. \tag{5.6.6}$$

Such a problem is called a *terminal* (optimal) *control problem.* We now introduce the proper terminology for it:

Definition 5.6.1 *The space of all m-dimensional vector functions* $\hat{u} = \hat{u}(t) \in C_s[t_a, \infty)^m$ *with range in the* control region U *is called the* control space E.

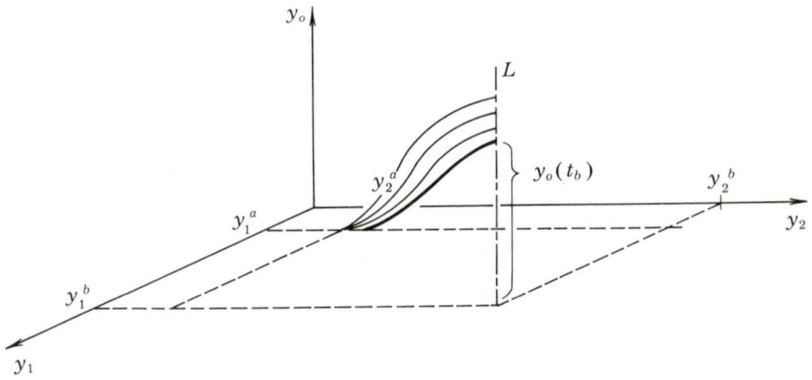

Figure 5.6

For example, U may be defined by $|u_i| \leq 1$, $i = 1, 2, \ldots, m$. Then the control space E is the space of all sectionally continuous vector functions for which $-1 \leq u_i(t) \leq 1$, $i = 1, 2, \ldots, n$.

For reasons of definiteness, we shall always assume that at any point $t = t^* \in [t_a, t_b)$ where at least one of the components of $\hat{u}$ has a jump discontinuity,

$$u_i(t^*) = u_i(t^* + 0)$$

and that

$$u_i(t_b) = u_i(t_b - 0).$$

Definition 5.6.2 *$\hat{u} \in E$ is called an* admissible control† *if the solution $\hat{y} = \hat{y}(t)$ of (5.6.4) for that particular control $\hat{u}$ exists and satisfies the initial conditions (5.6.5a) and the terminal conditions (5.6.5b) with some value of $y_0(t_b)$. The solution $\hat{y} = \hat{y}(t)$ that corresponds to an admissible control is called an* admissible trajectory.

Definition 5.6.3 *That admissible control $\hat{u} \in E$ for which $y_0(t_b)$ assumes a minimum value is called* the optimal control, *and the corresponding trajectory is called the* optimal trajectory.

We can easily interpret our problem geometrically: In a $(y_0, \ldots, y_n)$ space, there is given the initial point $(0, y_1^a, \ldots, y_n^a)$ and a line L, the terminal line, through the point $(0, y_1^b, \ldots, y_n^b)$ and parallel to the y_0 axis. (See Fig. 5.6.) The admissible trajectories are those curves $\hat{y} = \hat{y}(t)$ whose components satisfy (5.6.4) and which originate at $(0, y_1^a, \ldots, y_n^a)$ and terminate on L. The optimal trajectory—if it exists—is the one that intersects L at the lowest point.

† See footnote on page 278.

Let Y denote the $(y_1, \ldots, y_n)$ space. We shall assume for what is to follow here and in the later sections that

$$\hat{f}, \hat{f}_{y_k} \in C(Y \times U)^{n+1}. \tag{5.6.7}$$

We have:

Lemma 5.6 *If (5.6.7) is satisfied, then the admissible trajectories* $\hat{y} = \hat{y}(t)$ *of the terminal control problem [(5.6.4) to (5.6.6)] are sectionally smooth on* $[t_a, t_b]$, *provided any exist.*

Proof: Let $\tau_1 < \tau_2 < \cdots < \tau_r$, $t_a < \tau_1$, $\tau_r < t_b$, represent the points where at least one of the components of $\hat{u}$ has a jump discontinuity. If $\hat{y} = \hat{y}(t)$ exists, then it follows from (5.6.7) that $\hat{y} \in C^1[\tau_i, \tau_{i+1}]^{n+1}$. Hence,

$$\hat{y} = \hat{y}_a + \int_{t_a}^{t} \hat{f}(y(s), \hat{u}(s))\, ds \in C[t_a, t_b]^{n+1}$$

and

$$\hat{y}' = \hat{f}(y(t), \hat{u}(t)) \in C_s[t_a, t_b]^{n+1}$$

on account of (5.6.7) and $\hat{u} \in C_s[t_a, t_b]^m$.

PROBLEMS 5.6

1. Formulate the following variational problems as terminal control problems:

   (a) $\displaystyle\int_0^1 y'^2(x)\, dx \to \text{minimum}, \; y(0) = 0, \; y(1) = 1$

   (b) $\displaystyle\int_{\pi/4}^{\pi/2} (y'^2(x) - y^2(x))\, dx \to \text{minimum}, \; y(\pi/4) = 1, \; y(\pi/2) = -1$

   (c) $\displaystyle\int_a^b f(x, y_1(x), \ldots, y_n(x), y_1'(x), \ldots, y_n'(x))\, dx \to \text{minimum},$

   $\hat{y}(a) = \hat{y}_a, \; \hat{y}(b) = \hat{y}_b.$

2. Formulate the following optimal control problem as a variational problem:

$$y_0' = (1 - u^2)^2$$
$$y_1' = u$$
$$y_2' = 1,$$

   $y_0(0) = 0, \; y_1(0) = 0, \; y_2(0) = 0, \; y_1(b) = 1, \; y_2(b) = 1, \; y_0(b) \to \text{minimum}.$

3. Prove the following corollary to Lemma 5.6: If an admissible control $\hat{u} = \hat{u}(t)$ is continuous in $[t_a, t_b]^m$, then the corresponding admissible trajectory is smooth in $[t_a, t_b]^{n+1}$.

4. Formulate the following problem as a terminal control problem: Given the law of motion

$$\frac{d^2y}{dt^2} = u.$$

To be found is a force $u \in C_s$, with $|u(t)| \leq 1$ for all t, such that an object that is subjected to this law of motion is transferred from the state $(y(0),y'(0))$ to the state $(0,0)$ in the shortest possible time.

5. Show that any portion of an optimal trajectory is again an optimal trajectory.

5.7 THE MINIMUM PRINCIPLE OF PONTRYAGIN

In order to arrive at a necessary condition for a control to be optimal and, consequently, for a trajectory to be optimal, we shall proceed in a manner that was suggested by R. E. Bellman.† This will lead to a generalization of Carathéodory's lemma (Theorem 5.1), and this in turn will lead to the minimum principle of Pontryagin.‡

Although the subsequent argument is simple and compelling, the scope of its applicability is limited, as we shall point out at the proper time and place. Still, it is the easiest and quickest way to Pontryagin's minimum principle.

For easy reference, let us again state the terminal control problem of the preceding section: To be found is a control $\hat{u} \in E$ such that $\hat{y} = \hat{y}(t)$ satisfies the system of differential equations

$$\hat{y}' = \hat{f}(y,\hat{u}), \tag{5.7.1}$$

the initial conditions

$$\hat{y}(t_a) = \hat{y}_a, \qquad \text{where } \hat{y}_a = (0,y_1^a, \ldots,y_n^a), \tag{5.7.2a}$$

and the terminal conditions

$$\hat{y}(t_b) = \hat{y}_b, \qquad \text{where } \hat{y}_b = (y_o(t_b),y_1^b, \ldots,y_n^b), \tag{5.7.2b}$$

for some $t = t_b$ in such a manner that

$$y_o(t_b) \to \text{minimum}. \tag{5.7.3}$$

We have from Lemma 5.6 that an optimal trajectory is sectionally smooth, provided it exists.

Before stating the conditions under which we shall derive the minimum principle of Pontryagin, we must first generalize the concepts of admissible set of inception and Hamilton's characteristic function, which were first introduced in Sec. 5.1.

Definition 5.7.1 *A* point set Ω *which consists of a simply connected domain in* $(y_o, \ldots,y_n)$ *space and the line* L *is called an* admissible set of inception *if every point* $\hat{y}_o = (y_o^0, \ldots,y_n^0) \in \Omega$ *but not on* L *may be joined to the terminal*

† R. E. Bellman and S. E. Dreyfus, "Applied Dynamic Programming," p. 191, Princeton University Press, Princeton, N.J., 1962.

‡ L. S. Pontryagin et al., "The Mathematical Theory of Optimal Processes," Interscience Publishers, Inc., New York, 1962.

line L by a uniquely determined optimal trajectory $\hat{y} = \hat{y}(t)$ *so that* $\hat{y}(t) \in \Omega$ *for all* $t \in [\tau_a, \tau_b]$, *where* τ_a, τ_b *are the values of t for which* $\hat{y}(\tau_a) = \hat{y}_o$, $\hat{y}(\tau_b) = (y_o(\tau_b), y_b)$. (Note that all points in Ω except those that lie on L are interior points.)

Definition 5.7.2 *The function* $S = S(\hat{y}_o)$ *which denotes the minimum terminal value of* y_o *that is attainable by an optimal trajectory emanating from the point* $\hat{y}_o = (y_o^o, \ldots, y_1^o)$ *and terminating at* L *is called* Hamilton's characteristic *function for the terminal control problem* [(5.7.1) *to* (5.7.3)].

We observe that if Ω is an admissible set of inception, then $S = S(\hat{y})$ is defined for all $\hat{y} \in \Omega$.

We now make the following assumptions:

$$\hat{y}_a \in \Omega \tag{5.7.4}$$

$$S = S(\hat{y}) \in C^2(\Omega). \tag{5.7.5}$$

We assume that $\hat{u} = \hat{u}(t)$ is the optimal control and that $\hat{y} = \hat{y}(t)$ is the corresponding optimal trajectory of the problem [(5.7.1) to (5.7.3)]. By (5.7.4), $\hat{y} = \hat{y}(t)$ lies entirely in Ω.

We pick an arbitrary $t_1 \in [t_a, t_b)$ and replace the optimal control in the interval $[t_1, t_1 + \Delta t)$ by the constant control $\hat{v} = (v_1, \ldots, v_m) \in U$. Hereby, Δt is chosen sufficiently small so that $t_1 + \Delta t \in [t_a, t_b)$. Such a "variation" of the optimal control $\hat{u}$ is called a *needle-shaped variation.* (See Fig. 5.7.) This control will, in general, lead away from the optimal trajectory on a trajectory $\hat{y} = \hat{\bar{y}}(t)$ to a point whose coordinates can be found from (5.7.1) by integration:

$$\hat{\bar{y}}(t_1 + \Delta t) = \hat{y}(t_1) + \int_{t_1}^{t_1 + \Delta t} \hat{f}(\bar{y}(s), \hat{v})\, ds = \hat{y}(t_1) + \hat{f}(y(t_1), \hat{v})\,\Delta t + o(\Delta t), \tag{5.7.6}$$

where $\lim_{\Delta t \to 0} [o(\Delta t)/\Delta t] = 0$.

Since $\hat{y}(t_1) \in \Omega$, we have $\hat{\bar{y}}(t_1 + \Delta t) \in \Omega$, provided that $\Delta t > 0$ is sufficiently small. Hence there is an optimal trajectory that emanates from $\hat{\bar{y}}(t_1 + \Delta t)$ and leads to L, that is, $S(\hat{\bar{y}}(t_1 + \Delta t))$ is well defined. Clearly,

$$S(\hat{y}(t_1)) \le S(\hat{\bar{y}}(t_1 + \Delta t)). \tag{5.7.7}$$

Otherwise, the sectionally continuous function, which is composed of $\hat{u}$ in $[t_a, t_1]$, of $\hat{v}$ in $[t_1, t_1 + \Delta t)$, and of whatever may come after that in the remaining time, would lead to a smaller terminal value of y_o than does $\hat{u}$, and $\hat{u}$ would not be the optimal control, contrary to our hypothesis.

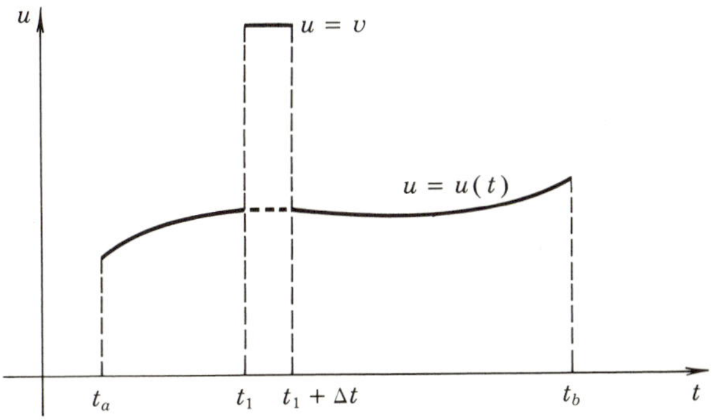

Figure 5.7

Since $\hat{f}_{v_k} \in C(\Omega)$, we can apply the mean value theorem to (5.7.7); using (5.7.6), we obtain

$$S_{\hat{y}}(\hat{\bar{y}}(\bar{t}))\hat{f}^T(\bar{y}(\bar{t}),\hat{v})\ \Delta t + o(\Delta t) \geq 0,$$

where $\bar{t} \in [t_1, t_1 + \Delta t)$ and where Δt is sufficiently small. If we divide by $\Delta t > 0$ and let $\Delta t \to 0$, we have

$$S_{\hat{y}}(\hat{y}(t_1))\hat{f}^T(y(t_1),\hat{v}) \geq 0.$$

This has to hold for all $t_1 \in [t_a,t_b)$ and for all $\hat{v} \in U$. Hence:

Lemma 5.7.1 *If the conditions (5.6.7), (5.7.4), and (5.7.5) are satisfied, then*

$$S_{\hat{y}}(\hat{y}(t))\hat{f}^T(y(t),\hat{v}) \geq 0 \tag{5.7.8}$$

for all $t \in [t_a,t_b)$ and for all $\hat{v} \in U$, where $\hat{y} = \hat{y}(t)$ is the optimal trajectory of the terminal control problem [(5.7.1) to (5.7.3)].

We also have

$$S(\hat{y}(t_a)) = S(\hat{y}(t)) \tag{5.7.9}$$

for all $t \in [t_a,t_b)$. Here, $S(\hat{y}(t))$ denotes the minimum that is attainable from the point $\hat{y}(t) \in \Omega$ on the optimal trajectory $\hat{y} = \hat{y}(t)$ for any given $t \in (t_a,t_b)$. Clearly, $S(\hat{y}(t_a)) < S(\hat{y}(t_1))$, $t_1 \in (t_a,t_b)$, is impossible because, by proceeding from $\hat{y}(t_1)$ on the optimal trajectory, the minimum $S(\hat{y}(t_a))$ is attained. Also, $S(\hat{y}(t_a)) > S(\hat{y}(t_1))$, $t_1 \in (t_a,t_b)$, is impossible because this would mean that there is an optimal trajectory $\hat{y} = \hat{\eta}(t)$ that emanates from $\hat{y}(t_1)$ and $\hat{y} \neq \hat{\eta}$ for $t > t_1$. But then,

$$\hat{y} = \hat{\bar{y}}(t) \equiv \begin{cases} \hat{y}(t) & \text{for } t_a \leq t \leq t_1 \\ \hat{\eta}(t) & \text{for } t_1 < t \end{cases},$$

$\hat{\hat{y}} \in C_s^1(t_a \leq t)$, will lead to a smaller value than $\hat{y} = \hat{y}(t)$ does, and it follows that $\hat{y} = \hat{y}(t)$ is not the optimal trajectory, contrary to our hypothesis.

We differentiate (5.7.9) with respect to t wherever $\hat{y}'$ exists, and we obtain

$$S_{\hat{y}}(\hat{y}(t))\hat{y}'^T(t) = 0. \tag{5.7.10}$$

For every fixed $t \in [t_a,t_b)$,

$$Q(\hat{y},\hat{v}) = S_{\hat{y}}(\hat{y})\,\hat{f}^T(y,\hat{v}) \tag{5.7.11}$$

may be viewed as a function of $(\hat{y},\hat{v})$. For a fixed point $\hat{y}_1 = \hat{y}(t_1)$ on the optimal trajectory $\hat{y} = \hat{y}(t)$, Q is a function of $\hat{v}$ only, and we have, in view of (5.7.8), that

$$Q(\hat{y},\hat{v}) \geq 0 \qquad \text{for all } \hat{v} \in U.$$

In view of (5.7.10) and because $\hat{y}' = \hat{f}(y,\hat{u})$ for every point on the optimal trajectory and the corresponding instantaneous value of the optimal control, we have

$$Q(\hat{y},\hat{u}) = 0.$$

Hence

$$Q(\hat{y},\hat{u}) = \min_{\hat{v}\in U} Q(\hat{y},\hat{v}) = 0,$$

and we have, if we replace Q according to (5.7.11):

Lemma 5.7.2 *Under the conditions of Lemma 5.7.1, we have for every point $\hat{y}(t)$ on the optimal trajectory $\hat{y} = \hat{y}(t)$ where $\hat{y}'$ is continuous that*

$$S_{\hat{y}}(\hat{y})\hat{f}^T(y,\hat{u}) = \min_{\hat{v}\in U} S_{\hat{y}}(\hat{y})\hat{f}^T(y,\hat{v}) = 0,$$

where $\hat{u}$ is the instantaneous value of the optimal control $\hat{u} = \hat{u}(t)$ at that point.

(This is a generalization of Carathéodory's lemma, as stated in (5.2.2), to terminal control problems of the type [(5.7.1) to (5.7.3)].)

Clearly, the principle stated in Lemma 5.7.2 is, for all practical purposes, useless since its application would require the knowledge of the partial derivatives of Hamilton's characteristic function along the optimal trajectory. As yet, we have no way of obtaining this information other than from a knowledge of $S(\hat{y})$ itself. But if we were to know $S(\hat{y})$, then the problem would be virtually solved and Lemma 5.7.2 would not make any further contribution to the solution.

The next step, however, will show how the values of $S_{\hat{y}}$ along the optimal trajectory $\hat{y} = \hat{y}(t)$ can be obtained as functions of t without any knowledge of S itself. (See also Probs. 5.1.3 and 5.1.4.)

If $\hat{y}_1 \notin L$, there exists a neighborhood $N(\hat{y}_1)$ of the point $\hat{y}_1$ such that $N(\hat{y}_1) \subset \Omega$ provided that $\hat{y}_1 \in \Omega$. Let $\hat{\eta} \in N(\hat{y}_1)$. Then there is an optimal trajectory that emanates from $\hat{\eta}$ and terminates on L, and we have from (5.7.8) and (5.7.11) that

$$Q(\hat{\eta},\hat{v}) \geq 0$$

for all $\hat{v} \in U$ and, in particular, for $\hat{v} = \hat{u}_1$, where $\hat{u}_1$ is the instantaneous value of the optimal control $\hat{u} = \hat{u}(t)$ that corresponds to the optimal trajectory $\hat{y} = \hat{y}(t)$ at the point $\hat{y}_1 = \hat{y}(t_1)$. Hence

$$Q(\hat{\eta},\hat{u}_1) \geq 0$$

for all $\hat{\eta} \in N(\hat{y}_1)$, and by Lemma 5.7.2,

$$Q(\hat{y}_1,\hat{u}_1) = 0.$$

This means that

$$Q(\hat{y}_1,\hat{u}_1) = \min_{\hat{\eta}\in N(\hat{y}_1)} Q(\hat{\eta},\hat{u}_1).$$

Therefore, it is necessary that

$$Q_{\hat{y}}(\hat{\eta},\hat{u}_1)\big|_{\hat{\eta}=\hat{y}_1} = 0.$$

[Note that $Q \in C^1(\Omega)$ since $S \in C^2(\Omega)$.]

We have from (5.7.11) that

$$Q_{\hat{y}} = \hat{f}S_{\hat{y}\hat{y}} + S_{\hat{y}}\hat{f}_{\hat{y}}, \tag{5.7.12}$$

where $S_{\hat{y}\hat{y}}$ denotes the matrix

$$S_{\hat{y}\hat{y}} = \begin{pmatrix} S_{y_oy_o} & \cdots & S_{y_oy_n} \\ \vdots & & \\ S_{y_ny_o} & \cdots & S_{y_ny_n} \end{pmatrix}$$

and where $\hat{f}_{\hat{y}}$ denotes the matrix

$$\hat{f}_{\hat{y}} = \begin{pmatrix} f_{oy_o} & \cdots & f_{oy_n} \\ \vdots & & \\ f_{ny_o} & \cdots & f_{ny_n} \end{pmatrix}. \tag{5.7.13}$$

(See also Prob. 5.7.8.)

On the optimal trajectory $\hat{y} = \hat{y}(t)$ that corresponds to the optimal control $\hat{u} = \hat{u}(t)$, we have $\hat{y}' = \hat{f}(y,\hat{u})$ and hence, if we also note that $S_{\hat{y}\hat{y}} = S_{\hat{y}\hat{y}}{}^T$,

$$\hat{f}(y(t),\hat{u}(t))S_{\hat{y}\hat{y}}(\hat{y}(t)) = \hat{y}'(t)S_{\hat{y}\hat{y}}(\hat{y}(t)) = \frac{d}{dt}S_{\hat{y}}(\hat{y}(t)).$$

Therefore, we obtain for $Q_{\hat{y}} = 0$, in view of (5.7.12),

$$\frac{d}{dt} S_{\hat{y}}(\hat{y}(t)) = -S_{\hat{y}}(\hat{y}(t))\hat{f}_{\hat{y}}(y(t),\hat{u}(t)),$$

which is a system of linear differential equations of first order for

$$\hat{p} = \hat{p}(t) \equiv S_{\hat{y}}(\hat{y}(t)), \qquad \hat{p} = (p_o, \ldots, p_n), \tag{5.7.14}$$

the value of $S_{\hat{y}}$ along the optimal trajectory $\hat{y} = \hat{y}(t)$. In terms of $\hat{p}$, this system can be written as

$$\hat{p}' = -\hat{p}\hat{f}_{\hat{y}}(y(t),\hat{u}(t)), \tag{5.7.15}$$

where $\hat{u} = \hat{u}(t)$ is the optimal control and $\hat{y} = \hat{y}(t)$ is the corresponding optimal trajectory and whereby $\hat{f}_{\hat{y}}$ is explained in (5.7.13). Equation (5.7.15) is called the *conjugate system* to (5.7.1).

Since $\hat{f}$ does not depend on y_o, we have $\hat{f}_{y_o} = 0$ and hence $p_o' = 0$, that is, p_o = constant. In view of this, we can write the conjugate system (5.7.15) as follows:

$$\begin{aligned} p_o' &= 0 \\ p_1' &= -\sum_{k=0}^{n} p_k f_{ky_1}(y(t),\hat{u}(t)) \\ &\vdots \\ p_n' &= -\sum_{k=0}^{n} p_k f_{ky_n}(y(t),\hat{u}(t)). \end{aligned}$$

Since $\hat{y} \in C_s^1[t_a,t_b]^{n+1}$ and $\hat{u} \in C_s[t_a,t_b]^m$, we can see that the solution $\hat{p} = \hat{p}(t)$ of (5.7.15) exists in the entire interval $[t_a,t_b]$ and that $\hat{p} \in C_s^1[t_a,t_b]^{n+1}$.

If $\hat{y} = \hat{\bar{y}}(t)$ is the optimal trajectory that emanates from a point $\hat{y}_1 = (y_o^1, \ldots, y_n^1) \in \Omega$ and if $\hat{u} = \hat{\bar{u}}(t)$ is the corresponding optimal control, then

$$S(\hat{y}_1) = y_o^1 + \int_{\tau_1}^{\tau_2} f_o(\bar{y}(s),\hat{\bar{u}}(s))\, ds,$$

and we see that $S_{y_o} = 1$. Hence, $p_o = 1$.

The function

$$H(\hat{p},y,\hat{u}) = \hat{p}\hat{f}^T(y,\hat{u}) \tag{5.7.16}$$

is called the *hamiltonian* of the terminal control problem [(5.7.1) to (5.7.3)]. In terms of the hamiltonian and with $\hat{p}$ as defined in (5.7.14), we can now state Lemma 5.7.2 as follows:

$$H(\hat{p},y,\hat{u}) = \min_{\hat{v}\in U} H(\hat{p},y,\hat{v}) = 0,$$

where $\hat{p} = \hat{p}(t)$ is a solution of the conjugate system (5.7.15) with $p_o = 1$. We summarize our result as a theorem:

Theorem 5.7.1 *If the conditions (5.6.7), (5.7.4), and (5.7.5) are satisfied and if* $\hat{u} = \hat{u}(t) \in C_s[t_a,t_b]^m$ *is the optimal control and* $\hat{y} = \hat{y}(t) \in C_s^1[t_a,t_b]^{n+1}$ *is the corresponding optimal trajectory of the terminal control problem [(5.7.1) to (5.7.3)], then it is necessary that there exist a solution* $\hat{p} = \hat{p}(t) \in C_s^1[t_a,t_b]^{n+1}$ *of the conjugate system (5.7.15) with* $p_o = 1$ *such that*

$$H(\hat{p}(t),y(t),\hat{u}(t)) = \min_{\hat{v}\epsilon U} H(\hat{p}(t),y(t),\hat{v}) = 0$$

at every $t \in [t_a,t_b]$ *where* $\hat{u}$ *is continuous.*

This theorem is a special case of the very general minimum principle of Pontryagin.† The theorem can be established without conditions (5.7.4) and (5.7.5) in the following form:

Theorem 5.7.2 Minimum principle of Pontryagin *If the condition (5.6.7) is satisfied and if* $\hat{u} = \hat{u}(t) \in C_s[t_a,t_b]^m$ *is the optimal control and* $\hat{y} = \hat{y}(t) \in C_s^1[t_a,t_b]^{n+1}$ *is the corresponding optimal trajectory of the terminal control problem [(5.7.1) to (5.7.3)], then it is necessary that there exist a solution* $\hat{p} = \hat{p}(t) \in C_s^1[t_a,t_b],^{n+1}$ $\hat{p} \neq 0$, *of the conjugate system (5.7.15) such that with* $p_o \geq 0$,

$$H(\hat{p}(t),y(t),\hat{u}(t)) = \min_{\hat{v}\epsilon U} H(\hat{p}(t),y(t),\hat{v}) = 0$$

at every $t \in [t_a,t_b]$ *where* $\hat{u}$ *is continuous.*‡

In order to find a control and trajectory that satisfy the conditions of Theorem 5.7.1 (or 5.7.2), one has to solve the two systems of differential equations

$$\hat{y}' = \hat{f}(y,\hat{u}(t)), \qquad \hat{p}' = -\hat{p}\hat{f}_{\hat{y}}(y,\hat{u}(t))$$

simultaneously in terms of $\hat{u} = \hat{u}(t)$. This will, in general, give rise to $2n + 2$ integration constants, of which only $2n + 1$ are essential since

† Originally, the principle was stated as a maximum principle, with $\hat{p} = -S_{\hat{y}}$ rather than $\hat{p} = S_{\hat{y}}$.

‡ For proof, see L. S. Pontryagin et al., "The Mathematical Theory of Optimal Processes," pp. 75–108, Interscience Publishers, Inc., New York, 1962; H. Halkin, On the Necessary Condition for the Optimal Control of Non-Linear Systems, *J. Analyse Math.*, vol. 12, pp. 1–82, 1964; M. Athans and P. L. Falb, "Optimal Control," pp. 304–347, McGraw-Hill Book Company, New York, 1966; E. B. Lee and L. Markus, "Foundations of Optimal Control Theory," pp. 308–340, John Wiley & Sons, Inc., New York, 1967, H. Halkin, Mathematical Foundations of System Optimization, in G. Leitman (ed.), "Topics in Optimization," pp. 197–262, Academic Press Inc., New York, 1967.

$\hat{p} = \hat{p}(t)$ need only be determined but for a multiplicative constant. To these parameters (constants) comes the unknown terminal time t_b, giving us altogether, in general, $2n + 2$ essential parameters to dispose of. These are balanced by $2n + 1$ boundary conditions (5.7.2*a* and *b*)—note that $y_o(t_b)$ in (5.7.2*b*) is free—and the condition $H(\hat{p},y,\hat{u}) = 0$. So we seem to have $2n + 2$ conditions for the $2n + 2$ essential parameters.

The assumptions (5.7.4) and (5.7.5) under which we derived Theorem 5.7.1 severely limit the scope of its applicability. Neither an admissible set of inception nor Hamilton's characteristic function that satisfies (5.7.5) need exist for an optimal control to exist. A case where an admissible set of inception does not exist will be discussed below, and a case where the partial derivatives of S are not continuous will be discussed in Sec. A5.8.

Let us consider the problem of finding a control $u \in E$ such that (y_o,y_1,y_2) satisfies the system of autonomous differential equations

$$y_o' = (1 - u^2)^2$$
$$y_1' = u$$
$$y_2' = 1,$$

the initial conditions

$$y_o(0) = 0, \qquad y_1(0) = 0, \qquad y_2(0) = 0,$$

and the terminal conditions

$$y_1(t_b) = 0, \qquad y_2(t_b) = 1$$

so that

$$y_o(t_b) \to \text{minimum}.$$

The control region U will be given by $|u| \leq \frac{1}{2}$.

Since $f_{iy_k} = 0$, $i, k = 0, 1, 2$, we obtain for the conjugate system (5.7.15)

$$p_o' = 0, \qquad p_1' = 0, \qquad p_2' = 0,$$

that is, p_o, p_1, p_2 are constants. $(p_o,p_1,p_2) = (1,0,-\frac{9}{16})$ is a nontrivial solution of the conjugate system, and we see that

$$H(1,0,-\tfrac{9}{16},y_1,y_2,\pm\tfrac{1}{2}) = \min_{|v|\leq 1/2} H(1,0,-\tfrac{9}{16},y_1,y_2,v)$$
$$= \min_{|v|\leq 1/2} [(1 - v^2)^2 - \tfrac{9}{16}] = 0,$$

that is, $u = \frac{1}{2}$ and $u = -\frac{1}{2}$ satisfy the conditions of the minimum principle. The corresponding trajectory's projection into the y_1,y_2 plane is a zigzag line with slope alternating between $\frac{1}{2}$ and $-\frac{1}{2}$ that emanates from (0,0) and terminates at (0,1). In between, there may be any number of corners (see Fig. 5.8).

In any case, we obtain

$$y_o(1) = \tfrac{9}{16}.$$

(Note that $t_b = 1$.)

If this problem has solutions at all, then, the solutions will be of the type as described above. We shall see in Sec. A5.10 that all these trajectories are indeed optimal.

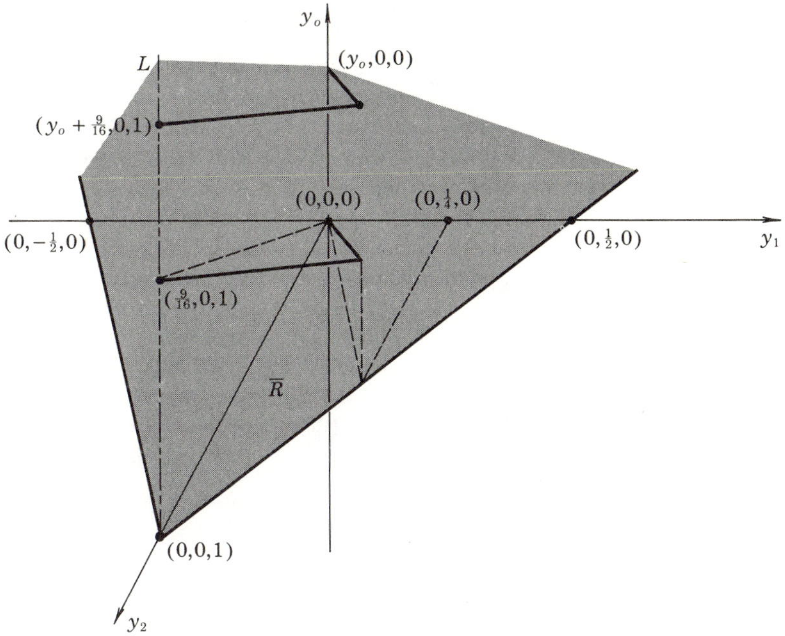

Figure 5.8

We see from Fig. 5.8 that any point (y_o,y_1,y_2) which lies outside the right prism that has the shaded region $\bar{\mathfrak{R}}$ as cross section with the y_1,y_2 plane cannot be joined to L by such a trajectory, while any point inside or on the wall of the prism can be joined to L by such a trajectory. At least part of any such trajectory will lie on the prism wall. Hence there is no admissible set of inception Ω which satisfies the requirements of Definition 5.7.1.

While the minimum principle in its general formulation of Theorem 5.7.2 is still applicable to this case, the derivation of Theorem 5.7.1 is not.

PROBLEMS 5.7

1. Find $S(y_o,y_1,y_2)$ for the problem that was discussed at the end of this section, and show that

$$S_{y_o} = 1, \qquad S_{y_1} = 0, \qquad S_{y_2} = -\tfrac{9}{16}$$

wherever these derivatives exist.

2. Show that (5.7.1), together with its conjugate system (5.7.15), can be written in terms of the hamiltonian H in the canonical form

$$\hat{y}' = H_{\hat{p}}, \qquad \hat{p}' = -H_{\hat{y}}.$$

*3. Prove: For an admissible control $\hat{u} = \hat{u}(t) \in C_s[t_a,t_b]^m$ and the corresponding admissible trajectory $\hat{y} = \hat{y}(t) \in C_s^1[t_a,t_b]^{n+1}$, we have $\hat{p} = \hat{p}(t) \in C_s^1[t_a,t_b]^{n+1}$, where $\hat{p}$ represents any solution of the conjugate system (5.7.15).

4. Prove: If $M(\hat{p}(t),y(t)) = \min_{\hat{v}\in U} H(\hat{p}(t),y(t),\hat{v})$ is differentiable at some $t = t_o$, then

$$\left.\frac{dM}{dt}\right|_{t=t_o} = 0.$$

5. Show that (5.7.7) leads to Bellman's functional equation (5.2.3) when applied to the problem

$$\int_a^b f(t,y(t),y'(t))\,dt \to \text{minimum},$$

$y(a) = y_1^a$, $y(b) = y_1^b$, $y' \in U$.

6. Formulate

$$\int_0^1 (1 - y'^2(x))^2\,dx \to \text{maximum}, \qquad y(0) = 0, \qquad y(1) = 0, \qquad |y'| \le \tfrac{1}{2}$$

as a terminal control problem and solve. (Replace in the minimum principle $p_o \ge 0$ by $p_o \le 0$. Why?)

7. Write (5.7.8) in components.

*8. Show that under the conditions (5.7.4) and (5.7.5), $Q_{\hat{y}}$ exists for all $\hat{y} \in \Omega$, where Q is defined in (5.7.11).

*9. Verify equation (5.7.12).

BRIEF SUMMARY

$S(x_o,y_o)$ denotes the minimum of $\int_{(a,y_a)}^{(x_o,y_o)} f(x,y(x),y'(x))\,dx$ as rendered by a sectionally smooth function $y = y(x)$ (optimal trajectory) for which $y'(x) \in U$ for all x, U denoting a given subset of the set of reals, and is called Hamilton's characteristic function.

If the problem admits an admissible set of inception Ω in the x,y plane, if $S \in C^1(\Omega)$, and if $f \in C$ for all $(x,y) \in \Omega$ and all $y' \in U$, then every lineal element (x_o,y_o,y_o') of an optimal trajectory has to satisfy the relation

$$f(x_o,y_o,y_o') - S_x(x_o,y_o) - S_y(x_o,y_o)y_o' = \min_{y'\in U}[f(x_o,y_o,y') - S_x(x_o,y_o) - S_y(x_o,y_o)y'] = 0$$

(Sec. 5.1).

If U represents the set of all reals, if $f \in C^2$ for all $(x,y) \in \Omega$ and all y', and if $f_{y'y'} \ne 0$ for all $(x,y) \in \Omega$ and all y', then S satisfies the Hamilton-Jacobi equation

$$S_x + H(x,y,S_y) = 0,$$

where H denotes the hamiltonian of the problem (Sec. 5.3).

This is equally true for the problem in n unknown functions where one obtains

$$S_x + H(x,y_1,\ldots,y_n,S_{y_1},\ldots,S_{y_n}) = 0,$$

where $S(x_o,y_1^o, \ldots, y_n^o)$ denotes the minimum of

$$\int_{(a,y_1^a, \ldots, y_n^a)}^{(x_o,y_1^o, \ldots, y_n^o)} f(x,y_1(x), \ldots, y_n(x), y_1'(x), \ldots, y_n'(x))\, dx$$

as a function of the coordinates of the endpoint $(x_o,y_1^o, \ldots, y_n^o)$ (Sec. 5.4).

If $S(x,y_1, \ldots, y_n, \alpha_1, \ldots, \alpha_n)$ is a solution of the Hamilton-Jacobi equation and $S \in C^2$, $\det |S_{\alpha_i y_k}| \neq 0$, then a $2n$-parameter family of solutions $\hat{y} = \hat{y}(x,\hat{\alpha},\hat{\beta})$, $\hat{p} = \hat{p}(x,\hat{\alpha},\hat{\beta})$ of the Euler-Lagrange equations in canonical form can be obtained from

$$S_{\alpha_i} = \beta_i, \qquad p_i = S_{y_i}, \qquad i = 1, 2, \ldots, n; \quad \beta_i \text{ arbitrary}$$

(Sec. 5.4).

If a solution $S = S(x,\hat{y})$ of the Hamilton-Jacobi equation is defined in a simply connected domain $\mathfrak{A}$ and if $S \in C^2(\mathfrak{A})$, then $(1,\hat{\phi}(x,\hat{y}))$, where

$$\hat{\phi}(x,\hat{y}) \equiv H_{\hat{p}}(x,\hat{y},S_{\hat{y}}(x,\hat{y})),$$

defines a Mayer field in some $\mathfrak{A}' \subset \mathfrak{A}$ (Sec. 5.5).

In a terminal control problem of not predetermined duration, where a control $\hat{u} = \hat{u}(t)$ with values $\hat{u}(t) \in U$ and a trajectory $\hat{y} = \hat{y}(t)$ are to be found such that

$$\hat{y}' = \hat{f}(y,\hat{u}),$$

$y = (y_1, \ldots, y_n)$, $\hat{y} = (y_o, \ldots, y_n)$, $\hat{u} = (u_1, \ldots, u_m)$, $\hat{f} = (f_o, \ldots, f_n)$, where $y_o(a) = 0$, $y(a) = y_a$, $y(b) = y_b$, and $y_o(b) \to$ minimum for some $b > a$, the minimum principle of Pontryagin is obtained as a necessary condition. By this principle, there has to exist a nontrivial solution $\hat{p} = \hat{p}(t)$ of the conjugate system

$$p_i' = -\sum_{k=0}^{n} f_{ky_i}(y(t),\hat{u}(t))p_k, \qquad i = 0, 1, \ldots, n,$$

such that at every point of the optimal trajectory with $p_o \geq 0$,

$$H(\hat{p}(t),y(t),\hat{u}(t)) \equiv \sum_{k=0}^{n} p_k(t)f_k(y(t),\hat{u}(t)) = \min_{\hat{v} \in U} H(\hat{p}(t),y(t),\hat{v}) = 0$$

(Secs. 5.6 and 5.7).

APPENDIX

A5.8 THE TIME-OPTIMAL CONTROL PROBLEM

We consider the terminal control problem [(5.7.1) to (5.7.3)] for the special case where

$$f_o(y,\hat{u}) = 1.$$

Then

$$y_o(t_b) = \int_{t_a}^{t_b} ds = t_b - t_a,$$

that is, the quantity to be minimized is the duration of the process of transferring the system from the initial state $y_a = (y_1{}^a, \ldots, y_n{}^a)$ into the terminal state $y_b = (y_1{}^b, \ldots, y_n{}^b)$. This problem is, for obvious reasons, called the *time-optimal control problem.*

We obtain for H as defined in (5.7.16) because of $f_o = 1$:

$$H(\hat{p},y,\hat{u}) = p_o + pf^T(y,\hat{u}),$$

where $p = (p_1, \ldots, p_n)$, or, with the notation

$$H^t(p,y,\hat{u}) = pf^T(y,\hat{u}), \tag{A5.8.1}$$
$$H(\hat{p},y,\hat{u}) = p_o + H^t(p,y,\hat{u}).$$

Here, $p_o' = 0$ and p is the solution of the conjugate system (5.7.15) as adapted to this problem with $f_o = 1$:

$$p' = -pf_y(y(t),\hat{u}(t)), \tag{A5.8.2}$$

where f_y represents the matrix

$$f_y = \begin{pmatrix} f_{1y_1} & \cdots & f_{1y_n} \\ \vdots & & \\ f_{ny_1} & \cdots & f_{ny_n} \end{pmatrix}.$$

By Theorem 5.7.2, we have to have

$$p_o + H^t(p,y,\hat{u}) = \min_{\hat{v}\epsilon U} [p_o + H^t(p,y,\hat{v})] = 0,$$

or as we may put it,

$$H^t(p,y,\hat{u}) = \min_{\hat{v}\epsilon U} H^t(p,y,\hat{v}) = -p_o \leq 0$$

since, by Theorem 5.7.2, $p_o \geq 0$.

If $p = 0$, then $H^t(p,y,\hat{u}) = 0$ and hence $p_o = 0$. Since the minimum principle as stated in Theorem 5.7.2 requires the existence of a nontrivial solution $\hat{p}$ of (5.7.15), we have to have $p \neq 0$, and we can state:

Theorem A5.8 *If* $\hat{u} = \hat{u}(t) \in C_s[t_a,t_b]^m$ *is the optimal control and*

$y = y(t) \in C_s^1[t_a,t_b]^n$ is the corresponding optimal trajectory of the time-optimal control problem

$$y_o' = 1, \qquad y' = f(y,\hat{u}),$$

$y(t_a) = y_a$, $y(t_b) = y_b$, and *$(t_b - t_a) \to$ minimum, where it is assumed that $f,f_y \in C(Y \times U)$, then it is necessary that there exist a solution $p = p(t) \in C_s^1[t_a,t_b]^n$, $p \neq 0$, of the conjugate system (A5.8.2) such that*

$$H^t(p(t),y(t),\hat{u}(t)) = \min_{\hat{v}\in U} H^t(p(t),y(t),\hat{v}) \leq 0$$

for every $t \in [t_a,t_b]$, where $\hat{u} = u(t)$ is continuous and whereby $\hat{u}(t)$ is the instantaneous value of the optimal control at that point.

[Note that on the optimal trajectory and for the optimal control, $H^t(p,y,\hat{u}) =$ constant since $p_o =$ constant.]

In order to find a control that satisfies the conditions of this theorem, we have to solve the two systems

$$y_o' = 1,\ y' = f(y,\hat{u}(t)) \qquad \text{and} \qquad p' = -pf_y(y,\hat{u}(t))$$

simultaneously in terms of $\hat{u} = \hat{u}(t)$. This will, in general, give rise to $2n + 1$ integration constants, $2n$ of which are essential since p need only be determined but for a multiplicative constant. To these constants comes the unknown terminal time t_b, thus giving us a total of $2n + 1$ constants, which are balanced by the $2n + 1$ boundary conditions $y_o(t_a) = 0$, $y(t_a) = y_a$, $y(t_b) = y_b$.

As an illustration of this theorem, we consider the following problem.

An object is assumed to obey the law of motion

$$y'' = u(t),$$

where the external force $u = u(t)$ is sectionally continuous and its values are restricted by $|u(t)| \leq 1$.

u is to be chosen such that an object that is located at the time $t = t_a$ at $y = y_1^a$ and possesses at that time the velocity $y' = y_2^a$ will be transferred to the location $y = 0$, where it shall arrive with velocity $y' = 0$ in the shortest possible time.

With $y_o = \int_{t_a}^{t_b} ds$, $y_1 = y$, $y_2 = y'$, we can formulate this problem as a time-optimal control problem as follows: To be found is a control $u \in E$ (that is $u \in C_s[t_a,t_b]$, $|u(t)| \leq 1$) such that the solution (y_o,y_1,y_2) of

$$y_0' = 1, \qquad y_1' = y_2, \qquad y_2' = u \tag{A5.8.3}$$

satisfies the initial conditions

$$y_o(t_a) = 0, \qquad y_1(t_a) = y_1^a, \qquad y_2(t_a) = y_2^a,$$

and the terminal conditions

$$y_1(t_b) = 0, \qquad y_2(t_b) = 0$$

and is such that

$$y_o(t_b) = t_b - t_a \rightarrow \text{minimum}.$$

Since $f_{1y_2} = 1$ and $f_{iy_k} = 0$ in all other cases, we obtain for the conjugate system (A5.8.2)

$$p_1' = 0, \qquad p_2' = -p_1$$

with the solutions

$$p_1 = \lambda_1, \qquad p_2 = -\lambda_1 t + \lambda_2,$$

where λ_1, λ_2 are constants.

We obtain for H^t,

$$H^t(p,y,u) = \lambda_1 y_2 + (\lambda_2 - \lambda_1 t)u,$$

and we have

$$H^t(p,y,1) = \min_{|v| \leq 1} H^t(p,y,v) \qquad \text{for } \lambda_2 - \lambda_1 t < 0$$

and

$$H^t(p,y,-1) = \min_{|v| \leq 1} H^t(p,y,v) \qquad \text{for } \lambda_2 - \lambda_1 t > 0.$$

Since there is at most one sign change in $\lambda_2 - \lambda_1 t$ as t increases, the control u that is defined by Theorem A5.8 is a step function that jumps from 1 to -1, or vice versa, if it jumps at all.

We shall solve the system (A5.8.3) for both cases and then investigate how the results are to be fitted together.

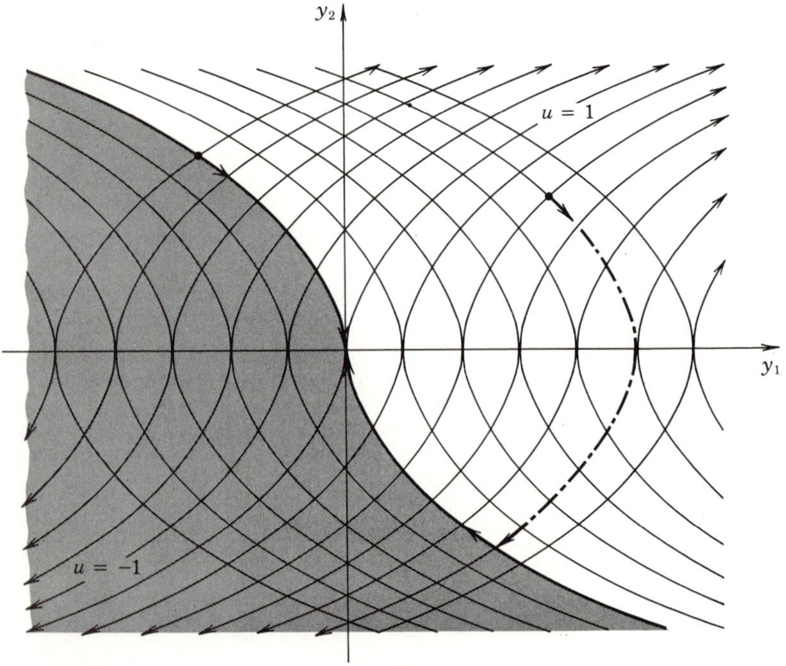

Figure 5.9

A. $u = 1$. Then (A5.8.3) becomes

$$y_o' = 1, \qquad y_1' = y_2, \qquad y_2' = 1$$

and hence

$$y_o = t + a, \qquad y_1 = \frac{t^2}{2} + bt + c, \qquad y_2 = t + b,$$

where a,b,c are some constants. Since these trajectories rise linearly above the y_1,y_2 plane as t increases, it suffices for our purpose to study their projection into the y_1,y_2 plane.

Elimination of t from y_1 and y_2 yields

$$y_1 = \frac{(y_2 - b)^2}{2} + b(y_2 - b) + c = \frac{y_2^2}{2} + \left(c - \frac{b^2}{2}\right).$$

These are parabolas that are open to the right and intersect the y_1 axis at $c - b^2/2$. (See Fig. 5.9.)

B. $u = -1$. Then (A5.8.3) becomes

$$y_o' = 1, \qquad y_1' = y_2, \qquad y_2' = -1$$

and hence

$$y_o = t + a, \qquad y_1 = -\frac{t^2}{2} + bt + c, \qquad y_2 = -t + b.$$

(These constants a, b, c are *not* the same as those in case *A*.)

Again, we eliminate t from y_1,y_2 and obtain

$$y_1 = -\frac{y_2^2}{2} + \left(\frac{b^2}{2} + c\right),$$

which are parabolas that are open to the left and intersect the y_1 axis at $b^2/2 + c$. (See Fig. 5.9.)

The original parameter representation of these parabolas determines their orientation, as indicated in Fig. 5.9 by arrows.

Since we are only allowed one switch from 1 to -1, or vice versa [remember that $(\lambda_2 - \lambda_1 t)$ has only one sign change as time progresses], it is now quite clear what we have to do. No matter what the initial position (y_1^a, y_2^a), this point will lie on two parabolas, one that corresponds to $u = 1$ and one that corresponds to $u = -1$. If one of these two parabolas leads to the origin of the coordinate system, then the control $u = 1$ (or $u = -1$) that corresponds to that parabola is to be applied. If neither parabola leads to the origin, then one of them will lead to a parabola that does lead to the origin and we shall have to switch the control from 1 to -1 (or vice versa) when we change from the one parabola to the other. We indicated two possibilities in Fig. 5.9. The initial point that lies in the second quadrant lies on a parabola that corresponds to the control $u = -1$ and leads to the origin. In the second case, where the initial point lies in the first quadrant, we have to start out with the control $u = -1$ and then switch to the control $u = 1$.

Inspection of Fig. 5.9 shows that if the initial point lies in the shaded region, the control $u = 1$ is to be chosen initially and then switched to $u = -1$ as soon as we arrive at the curve that separates the shaded and unshaded regions. If the initial point lies in the unshaded region, then $u = -1$ is to be chosen initially. If the initial point lies on the curve that separates the two regions, then when the point lies in the

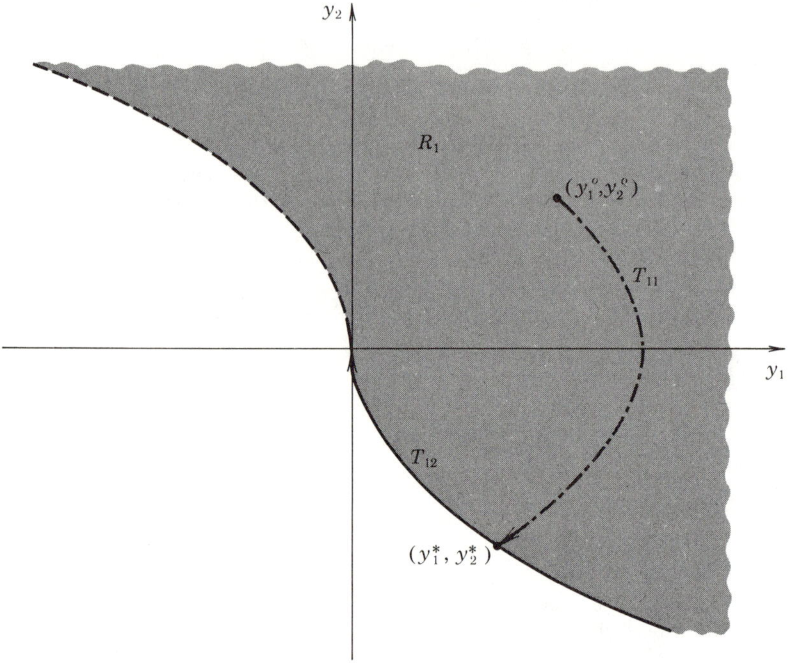

Figure 5.10

left half-plane, $u = -1$ will control the entire process, and when it lies in the right half-plane, $u = 1$ will control. The bold curve in Fig. 5.9 that separates the shaded and unshaded regions is called a *switching curve*—for obvious reasons.

For this problem, condition (5.7.4) is satisfied. As a matter of fact, Ω is the entire (y_o,y_1,y_2) space—assuming that the trajectories that were found are indeed optimal trajectories, which they are.† However, condition (5.7.5) is not met because S_{y_1} and S_{y_2} are not continuous along the switching curve, as we shall now show:

$S(y_o^o,y_1^o,y_2^o)$ represents the minimum duration of the process, given the initial state (y_o^o,y_1^o,y_2^o), where y_o^o represents the initial time. We may set $y_o^o = 0$ without loss of generality and study $S(0,y_1^o,y_2^o) = S(y_1^o,y_2^o)$ instead.

When evaluating $S(y_1^o,y_2^o)$, we have to distinguish between $(y_1^o,y_2^o) \in \mathcal{R}_1$ and $(y_1^o,y_2^o) \in \mathcal{R}_2$. (See Figs. 5.10 and 5.11.) In the first case, the optimal trajectory is put together from the parabolic arcs T_{11} ($u = -1$) and T_{12} ($u = 1$) (Fig. 5.10), and in the latter case, from the parabolic arcs T_{21} ($u = 1$) and T_{22} ($u = -1$) (Fig. 5.11).

The parabola of which T_{11} is a part has the equation

$$y_1 = -\frac{t^2}{2} + bt + c$$

$$y_2 = -t + b$$

(see page 297). This parabola corresponds to the control $u = -1$. We determine

† See L. S. Pontryagin et al., "The Mathematical Theory of Optimal Processes," p. 127, Interscience Publishers, Inc., New York, 1962.

b,c in such a manner that $y_1(0) = y_1^o$, $y_2(0) = y_2^o$, and we obtain

$$y_1 = -\frac{t^2}{2} + y_2^o t + y_1^o$$

$$y_2 = -t + y_2^o.$$

This curve reaches the lower portion of the parabola $y_1 = y_2^2/2$ (switching curve) at the time

$$t^* = y_2^o + \sqrt{\frac{y_2^{o2}}{2} + y_1^o}$$

and at the point

$$y_1^* = \frac{y_2^{o2}}{4} + \frac{y_1^o}{2}, \qquad y_2^* = -\sqrt{\frac{y_2^{o2}}{2} + y_1^o}.$$

Now we continue on T_{12} until we reach the origin. The parabola of which T_{12} is a part can be represented by

$$y_1 = \frac{t^2}{2}, \qquad y_2 = -t,$$

from which we obtain

$$t^{**} = -y_2^*$$

as the time required to travel from (y_1^*,y_2^*) to $(0,0)$.

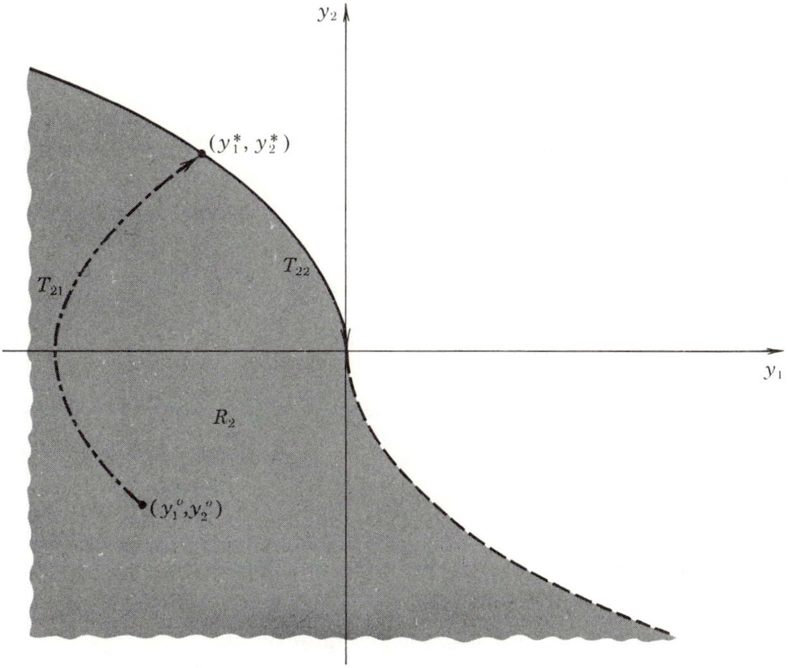

Figure 5.11

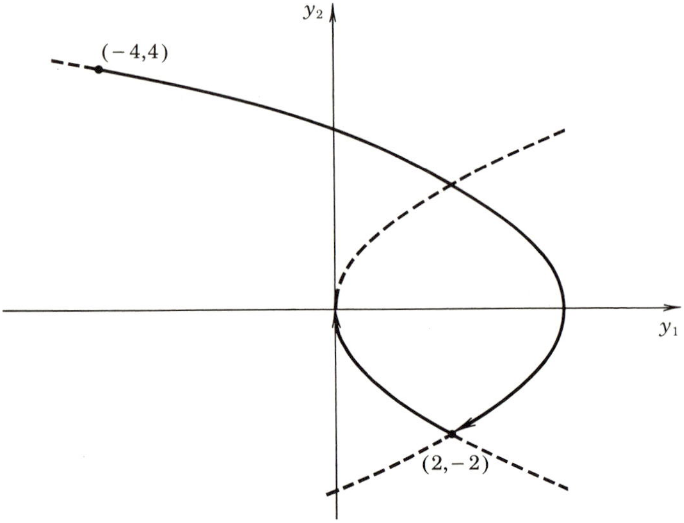

Figure 5.12

Hence $S(y_1^o,y_2^o) = t^* + t^{**}$, and we have

$$S(y_1^o,y_2^o) = y_2^o + 2\sqrt{\frac{y_2^{o2}}{2} + y_1^o}, \qquad (y_1^o,y_2^o) \in \mathcal{R}_1. \tag{A5.8.4}$$

Similarly, we obtain

$$S(y_1^o,y_2^o) = 2\sqrt{\frac{y_2^{o2}}{2} - y_1^o} - y_2^o, \qquad (y_1^o,y_2^o) \in \mathcal{R}_2. \tag{A5.8.5}$$

We obtain from (A5.8.4) and (A5.8.5)

$$S_{y_1}(y_1,y_2) = \begin{cases} \dfrac{1}{\sqrt{y_2^2/2 + y_1}}, & (y_1,y_2) \in \mathcal{R}_1 \\[2ex] -\dfrac{1}{\sqrt{y_2^2/2 - y_1}}, & (y_1,y_2) \in \mathcal{R}_2, \end{cases}$$

and we see that S_{y_1} is *not* continuous on the switching curve as $y_1 \to y_2^2/2$ (T_{12}) and $y_1 \to -y_2^2/2$ (T_{22}). (See also Prob. A5.8.4.)

In Prob. A5.8.5, the reader is asked to solve this time-optimal control problem for the specific initial conditions

$$y_1(0) = -4, \qquad y_2(0) = 4.$$

It will develop that one has to proceed initially with the control $u = -1$ along the parabola

$$y_1 = -\frac{t^2}{2} + 4t - 4, \qquad y_2 = -t + 4$$

from $t = 0$ to $t = 6$. Then the control has to be switched to $u = 1$, and we continue along the parabola

$$y_1 = \frac{t^2}{2} - 8t + 32, \qquad y_2 = t - 8$$

until we reach (0,0) at $t = 8$. (See Fig. 5.12.) With $\lambda_1 = 1$, $\lambda_2 = 6$, we obtain $\min H^t = -2$.

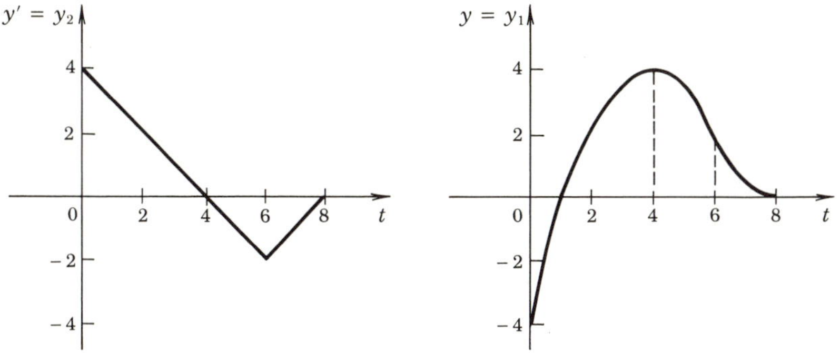

Figure 5.13

In Fig. 5.13, we have indicated the solution to the problem in t,y space and t,y' space.

PROBLEMS A5.8

1. Solve the time-optimal control problem

$$y_1' = y_2$$
$$y_2' = u - y_1,$$

$y_1(0) = y_1^o$, $y_2(0) = y_2^o$, $y_1(t_1) = 0$, $y_2(t_1) = 0$, $t_1 \to$ minimum, $|u| \leq 1$.

2. Same as in problem 1 for

$$y_1' = y_2 + u_1$$
$$y_2' = -y_1 + u_2,$$

$y_1(0) = y_1^o$, $y_2(0) = y_2^o$, $y_1(t_1) = 0$, $y_2(t_1) = 0$, $t_1 \to$ minimum, $|u_i| \leq 1$, $i = 1, 2$.

3. Give a physical interpretation of problems 1 and 2.

*4. Find $\lim (S_{y_2})$ as $y_1 \to -(y_2^2/2)$, $y_1 < 0$, $y_2 > 0$, and as $y_1 \to y_2^2/2$, $y_1 > 0$, $y_2 < 0$. [S is defined in (A5.8.4) and (A5.8.5).]

*5. Solve the time-optimal control problem that was discussed in this section for the initial conditions $y_1(0) = -4$, $y_2(0) = 4$.

A5.9 A NONAUTONOMOUS TERMINAL CONTROL PROBLEM OF PREDETERMINED DURATION

We shall now consider a terminal control problem of fixed duration that is associated with a nonautonomous system of differential equations.

We seek a control $\hat{u} = \hat{u}(t) \in E$ such that the corresponding trajectory $\hat{y} = \hat{y}(t)$ satisfies the differential equations

$$\hat{y}' = \hat{f}(y,t,\hat{u}), \tag{A5.9.1}$$

the initial conditions

$$\hat{y}(t_a) = \hat{y}_a, \tag{A5.9.2a}$$

and the terminal conditions

$$\hat{y}(t_b) = \hat{y}_b, \tag{A5.9.2b}$$

so that

$$y_o(t_b) \to \text{minimum} \tag{A5.9.3}$$

and where t_b *is given*. [The case where t_b is not given leads to a more complicated control problem, where the endpoint of the trajectory $\hat{y} = \hat{y}(t)$ is free to move in a plane that is parallel to the y_o axis and parallel to the t axis in t,y space.]†

If we introduce for t a new unknown function by

$$y_{n+1} = t,$$

then $y'_{n+1} = 1$, $y_{n+1}(t_a) = t_a$, $y_{n+1}(t_b) = t_b$, and we obtain a terminal control problem, as in Sec. 5.6, with $n + 2$ unknown functions (rather than $n + 1$) and $n + 2$ differential equations (rather than $n + 1$), namely,

$$\hat{y}' = \hat{f}(y,y_{n+1},\hat{u}), \qquad y'_{n+1} = 1$$

with the initial conditions

$$\hat{y}(t_a) = \hat{y}_a, \qquad y_{n+1}(t_a) = t_a,$$

and the terminal conditions

$$\hat{y}(t_b) = \hat{y}_b, \qquad y_{n+1}(t_b) = t_b,$$

whereby

$$y_o(t_b) \to \text{minimum.}$$

Here, t_b may be considered as unspecified. The fact that the last terminal condition together with $y'_{n+1} = 1$ forces t to assume the terminal value t_b is beside the point.

Since $f_{n+1} = 1$, we have $f_{n+1,y_k} = 0$ for all $k = 0, 1, \ldots, n + 1$, and we obtain for the conjugate system

$$\begin{aligned} p'_o &= 0 \\ p' &= -\hat{p}\hat{f}_y(y(t),y_{n+1}(t),\hat{u}(t)) \\ p'_{n+1} &= -\hat{p}\hat{f}_{y_{n+1}}(y(t),y_{n+1}(t),\hat{u}(t)), \end{aligned}$$

where we observe that p_{n+1} does not appear anywhere on the right side of these equations.

† *Ibid.*, pp. 55–66.

We obtain for the hamiltonian

$$H(\hat{p},p_{n+1},y,y_{n+1},\hat{u}) = \hat{p}\hat{f}^T(y,y_{n+1},\hat{u}) + p_{n+1},$$

or with

$$H^c(\hat{p},y,y_{n+1},\hat{u}) = \hat{p}\hat{f}^T(y,y_{n+1},\hat{u}), \tag{A5.9.4}$$

$$H(\hat{p},p_{n+1},y,y_{n+1},\hat{u}) = H^c(\hat{p},y,y_{n+1},\hat{u}) + p_{n+1}.$$

By Pontryagin's minimum principle (Theorem 5.7.2), we have to have

$$H^c(\hat{p},y,y_{n+1},\hat{u}) + p_{n+1} = \min_{\hat{v}\epsilon U} [H^c(\hat{p},y,y_{n+1},\hat{v}) + p_{n+1}] = 0, \tag{A5.9.5}$$

or as we may put it,

$$H^c(\hat{p},y,y_{n+1},\hat{u}) = \min_{\hat{v}\epsilon U} H^c(\hat{p},y,y_{n+1},\hat{v}) = -p_{n+1},$$

where $(\hat{p},p_{n+1}) \neq (0,\ldots,0)$ is a solution of the conjugate system and where $p_o \geq 0$.

Suppose for a moment that $\hat{p} = 0$. Then, by (A5.9.5), $p_{n+1} = 0$, contrary to the requirement by the minimum principle that $(\hat{p},p_{n+1}) \neq (0,0)$. Hence we may state:

Theorem A5.9 *For $\hat{u} = \hat{u}(t) \in C_s[t_a,t_b]^m$ to be the optimal control and for $\hat{y} = \hat{y}(t) \in C^1[t_a,t_b]^{n+1}$ to be the corresponding optimal trajectory of the terminal control problem of predetermined duration [(A5.9.1) to (A5.9.3)], it is necessary that there exist a nontrivial solution $\hat{p} = \hat{p}(t) \in C_s^1[t_a,t_b]^{n+1}$ of*

$$p_o' = 0, \qquad p' = -\hat{p}\hat{f}_y(y(t),t,\hat{u}(t))$$

such that with $p_o \geq 0$,

$$H^c(\hat{p}(t),y(t),t,\hat{u}(t)) = \min_{\hat{v}\epsilon U} H^c(\hat{p}(t),y(t),t,\hat{v})$$

for every $t \in [t_a,t_b]$, where $\hat{u} = \hat{u}(t)$ is continuous. Hereby, H^c is defined in (A5.9.4).

In order to find a control that satisfies the conditions of this theorem, we have to solve the two systems of differential equations

$$\hat{y}' = \hat{f}(y,t,\hat{u}(t))$$

and

$$p_o' = 0, \qquad p' = -\hat{p}\hat{f}_y(y,t,\hat{u}(t))$$

simultaneously in terms of $\hat{u} = \hat{u}(t)$. This gives rise to $2n + 2$ integration constants, one of which is redundant for the same reasons as before. The remaining constants are balanced by the $2n + 1$ boundary conditions (A5.9.2a and b)—recall that $y_o(t_b)$ is free.

PROBLEMS A5.9

1. Formulate and solve the following variational problem as a nonautonomous terminal control problem of predetermined duration:

$$\int_0^1 (1 - y'^2(x))^2\, dx \to \text{minimum}, \qquad y(0) = 0, \qquad y(1) = 0.$$

   (See also Prob. 2.9.2.)
2. Same as in problem 1 for

$$\int_0^1 (2 - y'^2(x))\, dx \to \text{minimum}, \qquad y(0) = 0, \qquad y(1) = 0, \qquad |y'| \le 1.$$

3. Formulate the following problem as a terminal control problem of fixed duration:

$$\int_a^b f(t,y_1(x), \ldots,y_n(x),y_1'(x), \ldots,y_n'(x))\, dt \to \text{minimum}$$

$$y_i(a) = y_i^a, \qquad y_i(b) = y_i^b, \qquad i = 1, 2, \ldots, n,$$

$$\varphi_1(t,y_1, \ldots,y_n,y_1', \ldots,y_n') = 0$$

$$\vdots$$

$$\varphi_\mu(t,y_1, \ldots,y_n,y_1', \ldots,y_n') = 0, \qquad \mu < n,$$

   where
$$\frac{\partial(\varphi_1, \ldots,\varphi_\mu)}{\partial(y_1', \ldots,y_\mu')} \neq 0.$$

A5.10 THE MINIMUM PRINCIPLE AS A SUFFICIENT CONDITION FOR LINEAR CONTROL PROBLEMS OF FIXED DURATION

The minimum principle as applied to various types of control problems and as stated in Theorems 5.7.2, A5.8, and A5.9 is, as we have pointed out repeatedly, strictly a necessary condition. A control $\hat{u} = \hat{u}(t)$ and the corresponding trajectory $\hat{y} = \hat{y}(t)$ which satisfy the minimum principle may or may not be optimal. (See Probs. A5.10.1 to A5.10.3.) Of course, if there is only one admissible control (see Definition 5.6.2) for a given problem to which there corresponds exactly one admissible trajectory, then the problem is to be considered as solved—within that particular control space, anyway. If there are more (finitely many or infinitely many) admissible controls, then it would be desirable to be in possession of a sufficient condition to obtain assurance that the control is indeed optimal.

Some sufficient conditions that are applicable to certain control problems have been developed.† We shall demonstrate here only that if

† M. Athans and P. L. Falb, "Optimal Controls," pp. 351–363, McGraw-Hill Book Company, New York, 1966; V. G. Boltyanskii, Sufficient Conditions for Optimality and the Justification of the Dynamic Programming Method, *SIAM Jour. on Control*, vol. 4, pp. 326–361, 1966; E. B. Lee and L. Markus, "Foundations of Optimal Control Theory," pp. 340–360, John Wiley & Sons, Inc., New York, 1967.

the system to be controlled is linear in the state variables and of predetermined duration, then the minimum principle as formulated in Theorem 5.7.1 constitutes not only a necessary condition but a sufficient condition as well. We consider the system of $n + 1$ nonautonomous differential equations that are linear in $y_1, \ldots, y_n$:

$$\hat{y}' = yA(t) + \hat{\varphi}(t,\hat{u}), \tag{A5.10.1}$$

where $A(t)$ represents the matrix

$$A(t) = \begin{pmatrix} a_{o1}(t) & \cdots & a_{n1}(t) \\ \vdots & & \\ a_{on}(t) & \cdots & a_{nn}(t) \end{pmatrix}$$

and where $\hat{\varphi}$ represents the row vector $\hat{\varphi} = (\varphi_o, \ldots, \varphi_n)$.

We assume that

$$a_{ik} \in C_s[t_a,t_b] \qquad \text{and} \qquad \hat{\varphi} \in C([t_a,t_b] \times U)^{n+1}.$$

As before, we consider the initial conditions

$$\hat{y}(t_a) = \hat{y}_a, \tag{A5.10.2a}$$

the terminal conditions

$$\hat{y}(t_b) = \hat{y}_b, \tag{A5.10.2b}$$

with

$$y_o(t_b) \to \text{minimum} \tag{A5.10.3}$$

and where t_b *is given.*

We assume that the admissible control $\hat{u} = \hat{u}(t)$ and its corresponding admissible trajectory $\hat{y} = \hat{y}(t)$ satisfy the conditions of the minimum principle as stated in Theorem 5.7.1, which, when modified to our specific control problem of fixed duration, reads as follows:

There has to exist a solution $\hat{p}$ of the conjugate system to (A5.10.1), namely,

$$p_o' = 0 \tag{A5.10.4}$$

$$p' = -\hat{p}A^T(t),$$

such that at every point of the trajectory with $p_o = 1$,

$$H^c(\hat{p},y,t,\hat{u}) = \min_{\hat{v}\in U} H^c(\hat{p},y,t,\hat{v}),$$

or as we may put it,

$$H^c(\hat{p},y,t,\hat{u}) \leq H^c(\hat{p},y,t,\hat{v}) \tag{A5.10.5}$$

for all $\hat{v} \in U$. Hereby,

$$H^c(\hat{p},y,t,\hat{u}) = \hat{p}(A^T(t)y^T + \hat{\varphi}^T(t,\hat{u})). \tag{A5.10.6}$$

Suppose now that $\hat{u} = \hat{\hat{u}}(t)$ is another admissible control and that

$\hat{y} = \hat{\bar{y}}(t)$ is the corresponding admissible trajectory. Then,

$$\hat{\bar{y}}'(t) = \bar{y}(t)A(t) + \varphi(t,\hat{\bar{u}}(t)) \quad \text{(A5.10.7)}$$

and
$$\bar{y}_o(t_a) = 0, \qquad \bar{y}(t_a) = y_a, \qquad \bar{y}(t_b) = y_b. \quad \text{(A5.10.8)}$$

For the same solution $\hat{p} = \hat{p}(t)$ of (A5.10.4) for which (A5.10.5) is satisfied, we consider

$$H^c(\hat{p},\bar{y},t,\hat{\bar{u}}) = \hat{p}(A^T(t)\bar{y}^T + \varphi^T(t,\hat{\bar{u}})). \quad \text{(A5.10.9)}$$

We obtain from (A5.10.6) and (A5.10.9) by shifting all terms to the left and integrating from t_a to t_b,

$$\int_{t_a}^{t_b} [H^c(\hat{p}(t),y(t),t,\hat{u}(t)) - \hat{p}(t)(A^T(t)y^T(t) + \varphi^T(t,\hat{u}(t)))]\,dt = 0 \quad \text{(A5.10.10)}$$

and
$$\int_{t_a}^{t_b} [H^c(\hat{p}(t),\bar{y}(t),t,\hat{\bar{u}}(t)) - \hat{p}(t)(A^T(t)\bar{y}^T(t) + \varphi^T(t,\hat{\bar{u}}(t)))]\,dt = 0. \quad \text{(A5.10.11)}$$

We subtract (A5.10.11) from (A5.10.10) and observe that we have from (A5.10.1) and (A5.10.7) by transposition,

$$\hat{y}'^T = A^T(t)y^T + \varphi^T(t,\hat{u}), \qquad \hat{\bar{y}}'^T = A^T(t)\bar{y}^T + \varphi^T(t,\hat{\bar{u}}).$$

Then,

$$\int_{t_a}^{t_b} [H^c(\hat{p}(t),y(t),t,\hat{u}(t)) - H^c(\hat{p}(t),\bar{y}(t),t,\hat{\bar{u}}(t)) + \hat{p}(t)(\hat{\bar{y}}'^T(t) - \hat{y}'^T(t))]\,dt = 0. \quad \text{(A5.10.12)}$$

We obtain from integration by parts,

$$\int_{t_a}^{t_b} \hat{p}(t)(\hat{\bar{y}}'^T(t) - \hat{y}'^T(t))\,dt = \hat{p}(t)(\hat{\bar{y}}^T(t) - \hat{y}^T(t))\Big|_{t_a}^{t_b} - \int_{t_a}^{t_b} \hat{p}'(t)(\hat{\bar{y}}^T(t) - \hat{y}^T(t))\,dt.$$

In view of (A5.10.2a and b) and (A5.10.8), and because of $p_o = 1$,

$$\hat{p}(t)(\hat{\bar{y}}^T(t) - \hat{y}^T(t))\Big|_{t_a}^{t_b} = \bar{y}_o(t_b) - y_o(t_b),$$

and we obtain for (A5.10.12), if we also note that $p_o' = 0$,

$$\int_{t_a}^{t_b} [H^c(\hat{p}(t),y(t),t,\hat{u}(t)) - H^c(\hat{p}(t),\bar{y}(t),t,\hat{\bar{u}}(t)) - p'(t)(\bar{y}^T(t) - y^T(t))]\,dt + \bar{y}_o(t_b) - y_o(t_b) = 0. \quad \text{(A5.10.13)}$$

From (A5.10.9),

$$\begin{aligned} H^c(\hat{p},\bar{y},t,\hat{\bar{u}}) &= \hat{p}(A^T(t)\bar{y}^T + \hat{\varphi}^T(t,\hat{\bar{u}})) \\ &= \hat{p}(A^T(t)y^T + \hat{\varphi}^T(t,\hat{\bar{u}})) + \hat{p}A^T(t)(\bar{y}^T - y^T) \\ &= H^c(\hat{p},y,t,\hat{\bar{u}}) + \hat{p}A^T(t)(\bar{y}^T - y^T). \end{aligned}$$

By (A5.10.4), $\hat{p}A^T(t) = -p'$, and we can write

$$H^c(\hat{p},\bar{y},t,\hat{\bar{u}}) = H^c(\hat{p},y,t,\hat{\bar{u}}) - p'(\bar{y}^T - y^T).$$

Hence we obtain for (A5.10.13),

$$\int_{t_a}^{t_b} [H^c(\hat{p}(t),y(t),t,\hat{u}(t)) - H^c(\hat{p}(t),y(t),t,\hat{\bar{u}}(t)) + p'(t)(\bar{y}^T(t) - y^T(t))$$
$$- p'(t)(\bar{y}^T(t) - y^T(t))]\,dt + \bar{y}_o(t_b) - y_o(t_b) = 0,$$

or as we may put it,

$$\int_{t_a}^{t_b} [H^c(\hat{p}(t),y(t),t,\hat{u}(t)) - H^c(\hat{p}(t),y(t),t,\hat{\bar{u}}(t))]\,dt = y_o(t_b) - \bar{y}_o(t_b).$$

Since (A5.10.5) is satisfied for all $\hat{v} \in U$, it will hold, in particular, for $\hat{v} = \hat{\bar{u}}$, and we have

$$y_o(t_b) - \bar{y}_o(t_b) \leq 0,$$

or

$$y_o(t_b) \leq \bar{y}_o(t_b),$$

that is, $y_o(t_b)$ is indeed the minimum.

Hence we can state:

Theorem A5.10 *If $\hat{u} = \hat{u}(t) \in C_s[t_a,t_b]^m$ is an admissible control and $\hat{y} = \hat{y}(t) \in C_s^1[t_a,t_b]^{n+1}$ is the corresponding admissible trajectory of the linear terminal control problem [(A.5.10.1) to (A5.10.3)] of predetermined duration, if there exists a solution $\hat{p} = \hat{p}(t) \in C_s^1[t_a,t_b]^{n+1}$ of the conjugate system (A5.10.4) with $p_o = 1$ and if for every $t \in [t_a,t_b]$ where $\hat{u} = \hat{u}(t)$ is continuous,*

$$H^c(\hat{p}(t),y(t),t,\hat{u}(t)) = \min_{\hat{v}\in U} H^c(\hat{p}(t),y(t),t,\hat{v}),$$

then $\hat{u} = \hat{u}(t)$ is the optimal control.

This theorem is still valid if one permits $\hat{\varphi}(t,\hat{u})$ to be sectionally continuous in t. (An analogous theorem pertaining to the linear control problem of not predetermined duration does not appear to hold.)

As an illustration, we consider the example from Sec. 5.7, which we shall now formulate as a fixed-duration control problem so that the theorem which we just proved becomes applicable. We shall also see from the discussion of this example that an optimal control is not necessarily uniquely determined. As a matter of fact, in this particular case there exist infinitely many optimal controls, all of which yield the same minimum value.

The problem, when formulated as a fixed-duration control problem, has the form

$$y_o' = [1 - u^2]^2$$
$$y_1' = u,$$

with the initial conditions $y_o(0) = 0$, $y_1(0) = 0$, the terminal condition $y_1(1) = 0$, and the minimum condition $y_o(1) \to$ minimum. The control region is given by $|u| \leq \frac{1}{2}$.

We have

$$p_o' = 0, \qquad p_1' = 0,$$

and we obtain with $(p_o,p_1) = (1,0)$,

$$H^c = (1 - u^2)^2 \geq \tfrac{9}{16} \qquad \text{for all } |u| \leq \tfrac{1}{2}.$$

The minimum is assumed for $u = \pm\frac{1}{2}$.

Since the system to be controlled is linear in y_1 (does not contain y_1 at all, as a matter of fact), it follows that any zigzag line with a slope that alternates between $\frac{1}{2}$ and $-\frac{1}{2}$, emanates from (0,0), and terminates at (1,0) is an optimal control. We obtain for all these optimal controls

$$y_o(1) = \tfrac{9}{16},$$

which is the minimum.

Finally, let us discuss a control problem the trajectories of which show a rather remarkable behavior. The system to be controlled is given by

$$\begin{aligned} y_o' &= u^2 \\ y_1' &= u \\ y_2' &= \phi(t)u, \end{aligned} \tag{A5.10.14}$$

where

$$\phi(t) = \begin{cases} 0 & \text{for } -\infty < t < 0 \\ \phi_0(t) & \text{for } 0 \leq t < \infty, \end{cases}$$

whereby $\phi_0 \in C_s[0,\infty)$. Then $\phi \in C_s(-\infty,\infty)$. (See the remark after Theorem A5.10.) This system is linear in y_1,y_2—it does not contain the y_i at all, as a matter of fact—and hence Theorem A5.10 is applicable. We impose the initial conditions

$$y_o(-1) = 0, \qquad y_1(-1) = 0, \qquad y_2(-1) = 0$$

and the terminal conditions

$$y_1(1) = 0, \qquad y_2(1) = 1.$$

$y_o(1)$ is to be minimized. The control region is given by $-\infty < u < \infty$.

Since the right side of (A5.10.14) is free of y_1, y_2, we obtain the simple conjugate system

$$p_o' = 0, \qquad p_1' = 0, \qquad p_2' = 0$$

and, consequently,

$$p_o = \lambda_o, \qquad p_1 = \lambda_1, \qquad p_2 = \lambda_2,$$

where λ_o, λ_1, λ_2 are constants. We choose $\lambda_o = 1$ and obtain for H^c

$$H^c(1,\lambda_1,\lambda_2,t,y_1,y_2,u) = u^2 + (\lambda_1 + \phi(t)\lambda_2)u.$$

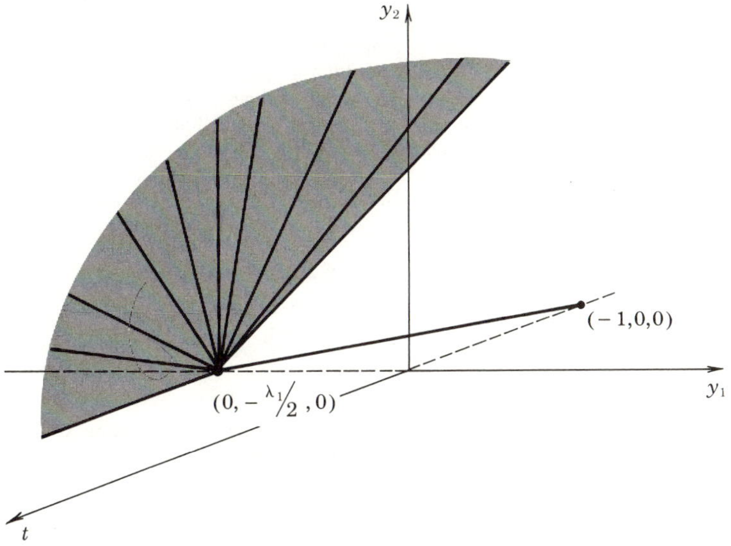

Figure 5.14

This function is minimized for each t by

$$u = -\tfrac{1}{2}(\lambda_1 + \phi(t)\lambda_2) = \begin{cases} -\tfrac{1}{2}\lambda_1 & \text{for } -1 \le t < 0 \\ -\tfrac{1}{2}(\lambda_1 + \phi_o(t)\lambda_2) & \text{for } 0 \le t \le 1. \end{cases} \tag{A5.10.15}$$

We consider the case where $\phi_o(t) = 1$, $0 \le t < \infty$. (For other cases, see Prob. A5.10.4.) Then

$$u = \begin{cases} -\tfrac{1}{2}\lambda_1 & \text{for } -1 \le t < 0 \\ -\tfrac{1}{2}(\lambda_1 + \lambda_2) & \text{for } 0 \le t \le 1. \end{cases}$$

If we integrate (A5.10.14), observing the initial conditions and taking care that the y_i remain continuous when we switch the control at $t = 0$, we obtain

$$y_o = \begin{cases} \dfrac{\lambda_1^2}{4} t + \dfrac{\lambda_1^2}{4} & \text{for } -1 \le t < 0 \\[2ex] \tfrac{1}{4}(\lambda_1 + \lambda_2)^2 t + \dfrac{\lambda_1^2}{4} & \text{for } 0 \le t \le 1 \end{cases}$$

$$y_1 = \begin{cases} -\dfrac{\lambda_1}{2} t - \dfrac{\lambda_1}{2} & \text{for } -1 \le t < 0 \\[2ex] -\tfrac{1}{2}(\lambda_1 + \lambda_2) t - \dfrac{\lambda_1}{2} & \text{for } 0 \le t \le 1 \end{cases}$$

$$y_2 = \begin{cases} 0 & \text{for } -1 \le t < 0 \\ -\tfrac{1}{2}(\lambda_1 + \lambda_2) t & \text{for } 0 \le t \le 1. \end{cases}$$

y_1, y_2 represent in a (t,y_1,y_2) space a two-parameter family of broken lines—broken at $t = 0$ on the y_1 axis—with the following remarkable property: Each line (for a given λ_1) that emanates from $(-1,0,0)$ remains in the t,y_1 plane until it reaches the y_1 axis. Then it fans out into a one-parameter (λ_2) family of lines all of which lie in the plane $y_1 = y_2 - (\lambda_1/2)$. Thus the totality of all such broken lines that emanate from $(-1,0,0)$ fill a two-dimensional portion of the (t,y_1,y_2) space for $-1 \leq t \leq 0$ and, from then on, a three-dimensional portion of the (t,y_1,y_2) space.† (See Fig. 5.14.)

The solution to our particular problem with the stated terminal conditions is obtained for $\lambda_1 = +2$, $\lambda_2 = -4$, namely,

$$y_o = t + 1, \qquad \text{for } -1 \leq t \leq 1,$$

$$y_1 = \begin{cases} -t - 1 & \text{for } -1 \leq t < 0 \\ t - 1 & \text{for } 0 \leq t \leq 1 \end{cases}$$

$$y_2 = \begin{cases} 0 & \text{for } -1 \leq t < 0 \\ t & \text{for } 0 \leq t \leq 1. \end{cases}$$

Since $p_o = 1$ and since the minimum condition of Pontryagin's minimum principle is satisfied, this is indeed the optimal trajectory, and we see that $y_o(1) = 2$ is the smallest possible value attainable for $y_o(1)$.

PROBLEMS A5.10

*1. Given the terminal control problem of predetermined duration

$$y_o' = u^2 - y_1^2, \qquad y_1' = u,$$

with the initial conditions $y_o(0) = 0$, $y_1(0) = 0$ and the terminal condition $y_1(3\pi/2) = 1$, where u is to be chosen so that $y_o(3\pi/2) \to$ minimum. The control region U is given by $-\infty < u < \infty$. Apply Theorem A5.9 and find a control and corresponding trajectory. Show that this control is *not* optimal.

*2. Show that the admissible control $u = \frac{1}{3}\cos(t/3)$ cannot be an optimal control in problem 1.

*3. Investigate whether or not $u = 2/3\pi$ is an admissible control for problem 1. If so, find $y_o(3\pi/2)$.

4. Solve the control problem

$$y_o' = u^2, \qquad y_1' = u, \qquad y_2' = \phi(t)u,$$

$y_o(-1) = 0$, $y_1(-1) = 0$, $y_2(-1) = 0$, $y_1(1) = 0$, $y_2(1) = 1$, $y_o(1) \to$ minimum, for:

(a) $$\phi(t) = \begin{cases} 0 & \text{for } -1 \leq t < 0 \\ t^n & \text{for } 0 \leq t \leq 1 \end{cases}$$

(b) $$\phi(t) = \begin{cases} 0 & \text{for } -1 \leq t \leq 0 \\ e^{-(1/t)} & \text{for } 0 < t \leq 1 \end{cases}$$

A5.11 BANG-BANG CONTROLS

It is one thing to establish a necessary condition for a minimum (maximum) of an integral in the calculus of variations, but it is quite another thing to

† See also C. Carathéodory, "Calculus of Variations and Partial Differential Equations of the First Order," vol. 2, pp. 355–360, Holden-Day, Inc., San Francisco, 1967.

actually find a function that satisfies such a necessary condition (Euler-Lagrange equation) and the given boundary conditions or the transversality conditions, whatever the case might be. But since the Euler-Lagrange equation is a differential equation (of second order), the problem of finding extremals is essentially shifted to the problem of solving boundary-value problems within the framework of the theory of differential equations. The existence of such a framework, in spite of all its built-in frustrations, facilitates matters considerably.

No such ready-made theory to fall back upon exists when the minimum principle in its various manifestations is applied. The development of methods for finding controls that satisfy the minimum principle is a matter of intensive study and great concern at the present time. Only for very simple cases—the simplest of which we shall discuss in this section—have successful solution techniques been developed.† In most cases, the finding of a control that satisfies the minimum principle is up to the cunning, ingenuity, and skill of the investigator. And even if such a control has been found, the question still remains as to whether or not it is optimal.

We shall discuss here the simplest terminal control problem, where the system to be controlled is autonomous and linear in the state variables as well as in the control variables and where the control region is a right parallelepiped in $(u_1, \ldots, u_m)$ space. For this case, the synthesis problem may be considered as completely solved.

Specifically, we consider the system

$$\hat{y}' = \hat{y}A + \hat{u}B, \tag{A5.11.1}$$

where $A = \begin{pmatrix} 0 & 0 & \cdots & 0 \\ a_{o1} & a_{11} & \cdots & a_{n1} \\ \vdots & & & \\ a_{on} & a_{1n} & \cdots & a_{nn} \end{pmatrix}, \qquad B = \begin{pmatrix} b_{o1} & b_{11} & \cdots & b_{n1} \\ \vdots & & & \\ b_{om} & b_{1m} & \cdots & b_{nm} \end{pmatrix}$

and where a_{ik} and b_{ij} are constants.

We impose the customary boundary conditions:

$$\hat{y}(t_a) = \hat{y}_a, \qquad \hat{y}(t_b) = \hat{y}_b, \tag{A5.11.2}$$

where

$$y_o(t_b) \to \text{minimum} \tag{A5.11.3}$$

and where we assume for the time being that t_b is *not given.*

We obtain for the conjugate system (5.7.15),

$$\hat{p}' = -\hat{p}A^T. \tag{A5.11.4}$$

† L. S. Pontryagin et al., "The Mathematical Theory of Optimal Processes," pp. 22–45, 135–140, 172–181, Interscience Publishers, Inc., New York, 1962. See also the comprehensive article by Bernard Paiewonsky, Optimal Control: A Review of Theory and Practice, *J. AIAA*, vol. 3, pp. 1985–2006. (This article also contains an extensive bibliography.)

The solution $\hat{p} = \hat{p}(t)$ of (A5.11.4) is independent of $\hat{u}$ and $\hat{y}$, and we obtain for any admissible control and corresponding admissible trajectory,

$$H(\hat{p},y,\hat{u}) = \hat{p}(A^T\hat{y}^T + B^T\hat{u}^T) = \hat{p}A^T\hat{y}^T + \hat{p}B^T\hat{u}^T. \quad \text{(A5.11.5)}$$

We see that the first term is independent of $\hat{u}$ and the second term is independent of $\hat{y}$.

We assume that the control region is given by

$$U = \{\hat{u} \mid |u_i| \leq U_i,\, i = 1, 2, \ldots, m\}$$

where the U_i are given constants.

The last term in (A5.11.5) can be written as

$$\hat{p}B^T\hat{u}^T = u_1 \sum_{k=0}^{n} b_{k1}p_k + \cdots + u_m \sum_{k=0}^{n} b_{km}p_k.$$

If the n vectors $(\hat{e}_jB, \hat{e}_jBA, \ldots, \hat{e}_jBA^{n-1})$, $j = 1, 2, 3, \ldots, m$, where $\hat{e}_j = (0, 0, \ldots, 0, 1, 0, \ldots, 0)$, are linearly independent (*general position condition*, cf. L. S. Pontryagin et al., *loc. cit.*, p. 118), none of the coefficients of the u_i can vanish identically and H will assume its minimum for any nontrivial solution $\hat{p} = \hat{p}(t)$ of (A5.11.4) for such values of u_i that lie in a vertex of U. Which vertex depends on the signs of the coefficients $\sum_{k=0}^{n} b_{ki}p_k$ of u_i in the above expression: $u_i = -\text{sign}\left(\sum_{k=0}^{n} b_{ki}p_k(t)\right)U_i$.

A switch from U_i to $-U_i$ (or vice versa) will occur whenever $\sum_{k=0}^{n} b_{ki}p_k(t)$ changes sign.

An upper bound for the number of sign changes (switches) can easily be obtained for the case where the eigenvalues of A are real. The general solution of the conjugate system (A5.11.4) is given by

$$\hat{p} = \hat{\lambda}e^{-A^Tt},$$

where $\hat{\lambda} = (\lambda_o, \ldots, \lambda_n)$ is an arbitrary constant vector.† If $\nu_1, \ldots, \nu_r$ are the distinct real eigenvalues of A, and hence of A^T, and if $\mu_1, \ldots, \mu_r$ are the multiplicities of these eigenvalues $\left(\sum_{j=1}^{r} \mu_j = n + 1\right)$, then the individual components of $\hat{p}$ will be of the form‡

$$p_k(t) = \pi_1(t)e^{\nu_1t} + \cdots + \pi_r(t)e^{\nu_rt}, \quad \text{(A5.11.6)}$$

where the $\pi_j(t)$ are polynomials of degree $\leq \mu_j - 1$. These polynomials

† E. A. Coddington and N. Levinson, "Theory of Ordinary Differential Equations," p. 76, McGraw-Hill Book Company, New York, 1955.
‡ *Ibid.*, p. 77.

also depend on k, of course, but marking this dependence with an additional subscript would confuse the formulas that are still to come.

Lemma A5.11 *If the distinct eigenvalues* $\nu_1, \ldots, \nu_r$ *with multiplicities* $\mu_1, \ldots, \mu_r$ *of* A *are all real, then* $\sum_{k=0}^{n} b_{ki}p_k(t)$ *has at most* $\sum_{i=1}^{r} \mu_i - 1 = n$ *real and distinct zeros.*

Proof: We have from (A5.11.6),

$$\sum_{k=0}^{n} b_{ki}p_k(t) = P_1(t)e^{\nu_1 t} + \cdots + P_r(t)e^{\nu_r t},$$

where the $P_j(t)$ are some polynomials of degree $\leq \mu_j - 1$, where μ_j is the multiplicity of ν_j.

We shall give a proof by induction:

If $r = 1$, then $P_1(t)e^{\nu_1 t}$ has at most $\mu_1 - 1$ real and distinct zeros because P_1 is a polynomial of degree $\leq \mu_1 - 1$.

We assume now that the lemma is true for all integers $< r$, and we consider

$$F(t) = P_1(t)e^{\nu_1 t} + \cdots + P_r(t)e^{\nu_r t}.$$

We want to prove that $F(t)$ has at most n real and distinct zeros. We assume that, on the contrary, $F(t)$ has at least $n + 1$ real and distinct zeros. Then

$$G(t) = F(t)e^{-\nu_1 t} = P_1(t) + P_2(t)e^{(\nu_2 - \nu_1)t} + \cdots + P_r(t)e^{(\nu_r - \nu_1)t}$$

also has at least $n + 1$ real and distinct zeros, and by the theorem of *Rolle*, $G'(t)$ has at least n real and distinct zeros, $G''(t)$ has at least $n - 1$ real and distinct zeros, $\ldots$, $G^{(\mu_1)}(t)$ has at least $n - \mu_1 + 1$ real and distinct zeros.

Since $P_1(t)$ is of degree $\leq \mu_1 - 1$, we have $P_1^{(\mu_1)}(t) = 0$, and hence we obtain

$$G^{(\mu_1)}(t) = Q_2(t)e^{(\nu_2 - \nu_1)t} + \cdots + Q_r(t)e^{(\nu_r - \nu_1)t},$$

where the $Q_j(t)$ are some polynomials of degree $\leq \mu_j - 1$.

Since the $(r - 1)$ exponents $(\nu_2 - \nu_1), \ldots, (\nu_r - \nu_1)$ are all distinct, it follows from our induction hypothesis that $G^{(\mu_1)}(t)$ has at most $\mu_2 + \cdots + \mu_r - 1 = n - \mu_1$ real and distinct zeros and *not* at least $n - \mu_1 + 1$ real and distinct zeros. This leads to a contradiction, and our lemma is proved.

Collecting the results obtained thus far, we can state:

Theorem A5.11.1 Bang-bang principle *The minimum principle (Theorem 5.7.2) when applied to the linear terminal control problem [(A5.11.1) to*

[(A5.11.3)] of not predetermined duration, with a control region that is given by $|u_i| \le U_i$*, defines for any nontrivial solution* $\hat{p} = \hat{p}(t)$ *of the conjugate system (A5.11.4) a unique control* $\hat{u} = \hat{u}(t)$*, where* $u_i = U_i$ *or* $u_i = -U_i$*.*

If the eigenvalues of the coefficient matrix A in (A5.11.1) are all real, then each component of $\hat{u}$ *is switched from* U_i *to* $-U_i$ *(or vice versa) at most n times.*

A control such as this is called a bang-bang *control.*†

Note that the fact that a control is uniquely defined by the minimum principle does not necessarily imply that this control is an optimal control of the given problem. As a matter of fact, not even an admissible control may exist (recall Definition 5.6.2) for the given problem. However, one can show that if there exists at least one admissible control for such a linear problem, then there exists an optimal control.‡

If we consider instead of (A5.11.1), the linear time-optimal control problem (see also Sec. A5.8)

$$y_0' = 1 \qquad \text{(A5.11.7)}$$

$$y' = yA_1 + \hat{u}B_1,$$

where A_1 is obtained from A by omission of the first row and the first column and B_1 from B by omission of the first column, with the boundary conditions (A5.11.2) and the minimum condition (A5.11.3), then, by Theorem A5.8, we have to minimize H^t.

Since

$$H^t(p,y,\hat{u}) = pA_1{}^T y^T + pB_1{}^T \hat{u}^T,$$

where p is a solution of the conjugate system

$$p' = -pA_1{}^T, \qquad \text{(A5.11.8)}$$

we see that we have exactly the same problem as before, only reduced by one dimension, and we can state, assuming that the general position condition (see p. 312) is satisfied:

Corollary to Theorem A5.11.1 *The minimum principle (Theorem A5.8) when applied to the linear time-optimal control problem [(A5.11.7), (A5.11.2), (A5.11.3)], with a control region that is given by* $|u_i| < U_i$*, defines for every nontrivial solution of the conjugate system (A5.11.8) a unique control* $\hat{u}$*, where* $u_i = U_i$ *or* $u_i = -U_i$*.*

† A generalization of this theorem to a control region that is a closed and bounded convex polyhedron can be found in L. S. Pontryagin et al., "The Mathematical Theory of Optimal Processes," pp. 115–119, Interscience Publishers, Inc., New York, 1962.
‡ *Ibid.*, pp. 127–135.

If the eigenvalues of A_1 are all real, then the components of $\hat{u}$ are switched from U_i to $-U_i$ (or vice versa) at most $n-1$ times.

The time-optimal control problem which we discussed at great length in Sec. A5.8 and which is associated with the system

$$y_o' = 1$$
$$y_1' = y_2$$
$$y_2' = u,$$

where the control region is given by $|u| \leq 1$, may serve as an illustration.

$$A_1 = \begin{pmatrix} 0 & 0 \\ 1 & 0 \end{pmatrix}$$

has only real eigenvalues ($\nu_{1,2} = 0$), and hence we can deduce from the above corollary that at most one switch ($n = 2$) from -1 to 1, or vice versa, is possible. (We also obtained this result in Sec. A5.8 from a direct analysis of the problem.)

Finally, let us consider a terminal control problem of fixed duration that is associated with a system such as (A5.11.1). We add the equation

$$y_{n+1}' = 1$$

and the boundary conditions $y_{n+1}(t_a) = t_a$, $y_{n+1}(t_b) = t_b$, and we apply Theorem A5.9, according to which we have to minimize

$$H^c(\hat{p},y,t,\hat{u}) = \hat{p}A^T\hat{y}^T + \hat{p}B^T\hat{u}^T.$$

But this is exactly the same expression as that which we considered in the case of the terminal control problem of not predetermined duration discussed in the beginning of this section, and Theorem A5.11.1 becomes applicable.

Keeping also in mind that the minimum principle when applied in the strengthened form with $p_o = 1$ to a terminal control problem of fixed duration that is linear in $\hat{y}$ constitutes a sufficient condition as well (Theorem A5.10), we can state:

Theorem A5.11.2 *If there exists an admissible control $\hat{u} = \hat{u}(t) \in C_s[t_a,t_b]^m$ of the terminal control problem of fixed duration*

$$\hat{y}' = \hat{y}A + \hat{u}B,$$

$\hat{y}(t_a) = \hat{y}_a$, $\hat{y}(t_b) = \hat{y}_b$, $y_o(t_b) \to$ minimum, where t_b is given and where the control region U is given by $|u_i| \leq U_i$, and if this control satisfies the conditions of the minimum principle as stated in Theorem A5.9, with $p_o = 1$, then this control is optimal and its components are step functions that assume the values U_i and $-U_i$ only. If all eigenvalues of A are real, then u_i is switched from U_i to $-U_i$ (and vice versa) at most n times.

To illustrate this theorem, we consider the problem

$$\int_0^b (\alpha y + \beta y')\, dt \to \text{minimum},$$

$$y(0) = 0, \qquad y(b) = y_b.$$

As a variational problem without restriction on y', this problem does not make sense at all: If $\alpha \neq 0$, the integral can be made as small as one pleases, and if $\alpha = 0$, the integral is independent of the path and hence has the same value no matter what sectionally smooth curve one might choose to join $(0,0)$ to (b,y_b). (See Prob. A5.11.1.)

However, if we impose an inequality constraint on y', for example, $|y'| \leq 1$, then we obtain a control problem of exactly the type that is covered by the above theorem, namely,

$$y_o' = \alpha y_1 + \beta u$$
$$y_1' = u$$
$$y_2' = 1,$$

with the boundary conditions

$$y_o(0) = 0, \qquad y_1(0) = 0, \qquad y_2(0) = 0, \qquad y_1(b) = y_b, \qquad y_2(b) = b,$$

the minimum condition $y_o(b) \to$ minimum, and the control region $|u| \leq 1$.

We know from Theorem A5.11.2 that if there is an optimal control, then it is a step function that assumes the values 1 and -1 and switches from one to the other at most once ($n = 1$). We can see from $|u| \leq 1$ that only endpoints with $|y_b| \leq |b|$ can be reached from $(0,0,0)$ by an admissible trajectory, and these points can indeed be reached by an optimal trajectory. If $|y_b| = |b|$, then no switch is required, and if $|y_b| < |b|$, then one switch is required. (See also Prob. A5.11.2.)

PROBLEMS A5.11

*1. Given $I[y] = \int_0^b (\alpha y(x) + \beta y'(x))\, dt$, $y(0) = 0$, $y(b) = y_b$. Show that $I[y]$ can be made arbitrarily small through proper choice of $y \in C^1[0,b]$, provided that $\alpha \neq 0$ and that $I[y]$ depends only on (b,y_b) if $\alpha = 0$.

*2. Consider the terminal control problem of fixed duration that was discussed at the end of this section. Investigate when a bang-bang control yields a minimum and when it yields a maximum. (*Hint*: Sufficient for maximum is that $H^c \to$ minimum, $p_o = -1$. Why?)

A5.12 A PROBLEM OF LAGRANGE AS AN OPTIMAL CONTROL PROBLEM

Lagrange considered the problem of finding a vector function $y = y(x) \equiv (y_1(x), \ldots, y_n(x))$ which minimizes the integral

$$I[y] = \int_a^b f(x,y(x),y'(x))\, dx, \tag{A5.12.1}$$

satisfies the initial conditions and terminal conditions

$$y(a) = y_a, \qquad y(b) = y_b, \tag{A5.12.2}$$

where $y_a = (y_1{}^a, \ldots, y_n{}^a)$ and $y_b = (y_1{}^b, \ldots, y_n{}^b)$, and also satisfies the constraining differential equations

$$\varphi_\rho(x,y,y') = 0, \qquad \rho = 1,2, \ldots, \mu < n.$$

Such a problem is called a *problem of Lagrange,* or simply, a *Lagrange problem.*

The next chapter will be devoted to a detailed analysis of this problem. Necessary and sufficient conditions will be derived by generalizing the ideas of Chaps. 2 and 3. Here we shall discuss only a special case of this problem which may be formulated as a terminal control problem of predetermined duration and may, as such, be subjected to the minimum principle as stated in Theorem A5.9.

The case we are going to discuss is the one where the constraining differential equations can be solved for a certain subset $y'_{k_1}, \ldots, y'_{k_\mu}$ of $y'_1, \ldots, y'_n$, for example, $y'_1, \ldots, y'_\mu$. Then the constraining differential equations can be written as

$$y'_\rho = f_\rho(x,y,y'_{\mu+1}, \ldots, y'_n), \qquad \rho = 1, 2, \ldots, \mu. \tag{A5.12.3}$$

If we also introduce the new functions

$$y_o = y_o(x) \equiv \int_a^x f(s,y(s),y'(s))\, ds, \qquad y_o(a) = 0,$$

$$u_j = y'_{\mu+j}, \qquad j = 1, 2, \ldots, m,\ m = n - \mu, \tag{A5.12.4}$$

and denote

$$f(x,y,f_1(x,y,\hat{u}), \ldots, f_\mu(x,y,\hat{u}),\hat{u}) = f_o(x,y,\hat{u}), \tag{A5.12.5}$$

where $\hat{u} = (u_1, \ldots, u_m)$, then we can formulate our problem as follows:

To be found is a control $\hat{u} = \hat{u}(x)$ and a corresponding trajectory $\hat{y} = \hat{y}(x) \equiv (y_o(x), \ldots, y_n(x))$ such that

$$\begin{aligned} y'_\rho &= f_\rho(x,y,\hat{u}), & \rho &= 0, 1, \ldots, \mu, \\ y'_{\mu+j} &= u_j, & j &= 1, 2, \ldots, m, \end{aligned} \tag{A5.12.6}$$

where $\hat{y} = \hat{y}(x)$ satisfies the initial conditions

$$\hat{y}(a) = \hat{y}_a, \qquad \hat{y}_a = (0,y_1{}^a, \ldots, y_n{}^a), \tag{A5.12.7a}$$

and the terminal conditions

$$\hat{y}(b) = \hat{y}_b, \qquad \hat{y}_b = (y_o(b),y_1{}^b, \ldots, y_n{}^b), \tag{A5.12.7b}$$

whereby $\qquad y_o(b) \to \text{minimum}. \qquad$ (A5.12.8)

This is a special case of a terminal control problem of predetermined duration of the type [(A5.9.1) to (A5.9.3)] with t, t_a, t_b being replaced by x, a, b. It is a special case on two counts: First, the control region U is now the entire $(u_1, \ldots, u_m)$ space, and second, the functions $f_{\mu+j}$, $j = 1, 2, \ldots, m$, have the special form $f_{\mu+j}(x,y,\hat{u}) \equiv u_j$.

In order to apply the minimum principle as formulated in Theorem A5.9, we need the conjugate system to (A5.12.6) and the hamiltonian H^c. Toward this end, we make the following observations:

$$f_{ky_o} = 0, \qquad k = 0, 1, \ldots, n, \tag{A5.12.9}$$

$$f_{oy_i} = f_{y_i} + \sum_{\rho=1}^{\mu} f_{y'_\rho} f_{\rho y_i}, \qquad i = 1, 2, \ldots, n. \tag{A5.12.10}$$

The latter relation follows from the definition of f_o in (A5.12.5).

$$f_{\mu+j,y_i} = 0, \qquad j = 1,2, \ldots,m, \; i = 1,2, \ldots,n, \tag{A5.12.11}$$

since $f_{\mu+j} = u_j$, independent of y_i.

Hence we have, in view of (A5.12.9) to (A5.12.11), with the notation that was introduced in (A5.7.13),

$$\hat{f}_{\hat{y}} = \begin{pmatrix} 0 & f_{y_1} + \sum_{\rho=1}^{\mu} f_{y'_\rho} f_{\rho y_1} & \cdots & f_{y_n} + \sum_{\rho=1}^{\mu} f_{y'_\rho} f_{\rho y_n} \\ 0 & f_{1y_1} & \cdots & f_{1y_n} \\ \vdots & & & \\ 0 & f_{\mu y_1} & \cdots & f_{\mu y_n} \\ 0 & 0 & \cdots & 0 \\ \vdots & & & \\ 0 & 0 & \cdots & 0 \end{pmatrix}$$

and, consequently,

$$\hat{p}\hat{f}_{\hat{y}} = \Big(0, p_o\big(f_{y_1} + \sum_{\rho=1}^{\mu} f_{y'_\rho} f_{\rho y_1}\big) + \sum_{\rho=1}^{\mu} f_{\rho y_1} p_\rho,$$
$$\ldots, p_o\big(f_{y_n} + \sum_{\rho=1}^{\mu} f_{y'_\rho} f_{\rho y_n}\big) + \sum_{\rho=1}^{\mu} f_{\rho y_n} p_\rho\Big).$$

Hence the conjugate system to (A5.12.6) (see Theorem A5.9) can be written as

$$p'_o = 0$$

$$p'_k = -p_o\big(f_{y_k} + \sum_{\rho=1}^{\mu} f_{y'_\rho} f_{\rho y_k}\big) - \sum_{\rho=1}^{\mu} f_{\rho y_k} p_\rho, \qquad k = 1, 2, \ldots, n. \tag{A5.12.12}$$

We obtain for H^c, as defined in (A5.9.4),

$$H^c(\hat{p},y,x,\hat{u}) = \hat{p}\hat{f}^T = p_o f(x,y,f_1(x,y,\hat{u}),\ldots,f_\mu(x,y,\hat{u}),\hat{u}) + \sum_{\rho=1}^{\mu} f_\rho(x,y,\hat{u})p_\rho + \sum_{j=1}^{m} u_j p_{\mu+j}.$$

By Theorem A5.9, it is necessary, in order for $\hat{u} = \hat{u}(x) \in C_s[a,b]^m$ to be the optimal control and for $\hat{y} = \hat{y}(x) \in C_s^1[a,b]^{n+1}$ to be the corresponding optimal trajectory of the terminal control problem of predetermined duration [(A5.12.6) to (A5.12.8)], that there exist a nontrivial solution $\hat{p} = \hat{p}(x) \in C_s^1[a,b]^{n+1}$ of the conjugate system (A5.12.12) such that, with $p_o \geq 0$,

$$H^c(\hat{p}(x),y(x),x,\hat{u}(x)) = \min_{(\hat{v})} H^c(\hat{p}(x),y(x),x,\hat{v}),$$

whereby $-\infty < v_j < \infty, j = 1, 2, \ldots, m$, since U is the entire $(u_1, \ldots, u_m)$ space.

Since $H^c(\hat{p},y,x,\hat{v})$ has to assume its minimum for $\hat{v} = \hat{u}$, we have, by necessity, that

$$H^c_{u_j}(\hat{p},y,x,\hat{v})|_{\hat{v}=\hat{u}} = 0, \qquad j = 1, 2, \ldots, m, \qquad \text{(A5.12.13)}$$

if we also impose suitable differentiability conditions on f,f_ρ.

By (A5.12.5),

$$f_{ou_j} = f_{u_j} + \sum_{\rho=1}^{\mu} f_{y'_\rho} f_{\rho u_j}, \qquad j = 1, 2, \ldots, m.$$

Hence,

$$H^c_{u_j} = p_o(f_{u_j} + \sum_{\rho=1}^{\mu} f_{y'_\rho} f_{\rho u_j}) + \sum_{\rho=1}^{\mu} f_{\rho u_j} p_\rho + p_{\mu+j}, \qquad j = 1, 2, \ldots, m,$$

and (A5.10.13) can be written as

$$p_{\mu+j} = -p_o f_{u_j} - \sum_{\rho=1}^{\mu} f_{\rho u_j}(f_{y'_\rho} p_o + p_\rho), \qquad j = 1, 2, \ldots, m.$$

If we introduce the abbreviating notation

$$\lambda_\rho(x) = f_{y'_\rho}(x,y(x),y'(x))p_o + p_\rho(x), \qquad \rho = 1, 2, \ldots, \mu, \qquad \text{(A5.12.14)}$$

we can write the above equation as

$$p_{\mu+j} = -p_o f_{u_j} - \sum_{\rho=1}^{\mu} \lambda_\rho f_{\rho u_j}, \qquad j = 1, 2, \ldots, m. \qquad \text{(A5.12.15)}$$

If we convert the last m equations of the conjugate system (A5.12.12) to equivalent integral equations, we obtain, with the new notation

(A5.12.14),

$$p_{\mu+j}(x) = p_{\mu+j}(a) - \int_a^x [p_o f_{y_{\mu+j}}(s,y(s),y'(s)) + \sum_{\rho=1}^{\mu} \lambda_\rho(s) f_{\rho,y_{\mu+j}}(s,y(s),\hat{u}(s))]\, ds. \quad \text{(A5.12.16)}$$

Equating the right sides of (A5.12.15) and (A5.12.16) and noting that the partial derivatives of f with respect to u_j are really partial derivatives with respect to $y'_{\mu+j}$ in the original notation, we have

$$p_o f_{y'_{\mu+j}} + \sum_{\rho=1}^{\mu} \lambda_\rho f_{\rho,y'_{\mu+j}} = \int_a^x [p_o f_{y_{\mu+j}}(s,y(s),y'(s)) + \sum_{\rho=1}^{\mu} \lambda_\rho(s) f_{\rho,y_{\mu+j}}(s,y(s),\hat{u}(s))]\, ds - p_{\mu+j}(a).$$

With the notation

$$h(x,y,y',\hat{\lambda}) = -p_o f(x,y,y') + \sum_{\rho=1}^{\mu} \lambda_\rho (y'_\rho - f_\rho(x,y,y'_{\mu+1}, \ldots, y'_n)), \quad \text{(A5.12.17)}$$

where $\hat{\lambda} = (p_o, \lambda_1, \ldots, \lambda_\mu)$, we have, in view of

$$h_{y_{\mu+j}} = -p_o f_{y_{\mu+j}} - \sum_{\rho=1}^{\mu} \lambda_\rho f_{\rho,y_{\mu+j}},$$

$$h_{y'_{\mu+j}} = -p_o f_{y'_{\mu+j}} - \sum_{\rho=1}^{\mu} \lambda_\rho f_{\rho,y'_{\mu+j}}, \qquad j = 1, 2, \ldots, m,$$

for the above equation,

$$h_{y'_{\mu+j}}(x,y(x),y'(x),\hat{\lambda}(x)) = \int_a^x h_{y_{\mu+j}}(s,y(s),y'(s),\hat{\lambda}(s))\, ds + p_{\mu+j}(a), \qquad j = 1, 2, \ldots, m, \quad \text{(A5.12.18)}$$

which is already part of the result we are aiming for.

We have not yet utilized the first $\mu + 1$ equations of the conjugate system (A5.12.12). These may be written in terms of the new notation that was introduced in (A5.12.14) and after integration from a to x as

$$p_o = \text{constant},$$

$$p_\nu(x) = -\int_a^x [p_o f_{y_\nu}(s,y(s),y'(s)) + \sum_{\rho=1}^{\mu} f_{\rho y_\nu}(s,y(s),\hat{u}(s))\lambda_\rho(s)]\, ds + p_\nu(a)$$

($\nu = 1, 2, \ldots, \mu$), where $\hat{u}(s)$ stands for $(y'_{\mu+1}(s), \ldots, y'_n(s))$, or in terms of h,

$$p_o = \text{constant},$$

$$p_\nu(x) = \int_a^x h_{y_\nu}(s,y(s),y'(s),\hat{\lambda}(s))\, ds + p_\nu(a), \qquad \nu = 1, 2, \ldots, \mu,$$

since $h_{y_\nu} = -p_o f_{y_\nu} - \sum_{\rho=0}^{\mu} \lambda_\rho f_{\rho y_\nu}, \qquad \nu = 1, 2, \ldots, \mu.$

From (A5.12.14) and (A5.12.17),

$$p_\nu = \lambda_\nu - p_o f_{y'_\nu} = h_{y'_\nu}, \qquad \nu = 1, 2, \ldots, \mu,$$

and we obtain, in addition to the equations in (A5.12.18),

$$h_{y'_\nu}(x,y(x),y'(x),\hat{\lambda}(x)) = \int_a^x h_{y_\nu}(s,y(s),y'(s),\hat{\lambda}(s))\, ds + p_\nu(a), \qquad \nu = 1, 2, \ldots, \mu. \quad \text{(A5.12.19)}$$

Suppose for a moment that $p_o = \lambda_1 = \cdots = \lambda_\mu = 0$. Then, by (A5.12.14), $p_1 = \cdots = p_\mu = 0$, and by (A5.12.15), also $p_{\mu+1} = \cdots = p_n = 0$, contrary to the minimum principle. Hence we have to have

$$\hat{\lambda} = (p_o, \lambda_1, \ldots, \lambda_\mu) \neq (0,0, \ldots, 0).$$

If we assume that $f, f_{y_k}, f_{y'_k}, f_{\rho y_k}, f_{\rho y'_{\mu+j}} (k = 1, 2, \ldots, n; \rho = 1, 2, \ldots, m)$ are continuous for all x, y, y', then the conditions (A5.6.7) under which Theorem A5.7.2 and hence Theorem A5.9 are valid are met, and we can state:

Theorem A5.12 *For $y = y(x) \in C_s^1[a,b]^n$ to be a solution of the Lagrange problem [(A5.12.1) to (A5.12.3)], it is necessary that there exist a vector function $\hat{\lambda} = (p_o, \lambda_1, \ldots, \lambda_\mu) \in C_s[a,b]^{n+1}$, $\hat{\lambda} \neq 0$, such that*

$$h_{y'_k}(x,y(x),y'(x),\hat{\lambda}(x)) = \int_a^x h_{y_k}(s,y(s),y'(s),\hat{\lambda}(s))\, ds + C_k, \quad \text{(A5.12.20)}$$

$k = 1,2, \ldots, n$, for some constants C_k, whereby

$$h(x,y,y',\hat{\lambda}) = -p_o f(x,y,y') + \sum_{\rho=1}^{\mu} \lambda_\rho (y'_\rho - f_\rho(x,y,y'_{\mu+1}, \ldots, y'_n)).$$

Any smooth portion of $y = y(x)$ has to satisfy the differential equations

$$h_{y_k}(x,y,y',\hat{\lambda}) - \frac{d}{dx} h_{y'_k}(x,y,y',\hat{\lambda}) = 0, \qquad k = 1, 2, \ldots, n. \quad \text{(A5.12.21)}$$

Equation (A5.12.20) is called the *Mayer equation in integrated form* of the

Lagrange problem, and (A5.12.21) is called the *Mayer equation of the Lagrange problem.* (See also Sec. 6.3.)

Note that (A5.12.21) follows from (A5.12.20) by the same argument by which the Euler-Lagrange equation (2.3.8) follows from the Euler-Lagrange equation in integrated form (2.3.7).

The condition stated in Theorem A5.12—the so-called *Lagrange multiplier rule*—appears to be a fairly straightforward generalization of the multiplier rule for extreme-value problems of functions of more than one variable with constraints:

$$F(x_1, \ldots, x_n) \to \text{minimum}$$

$$G_\rho(x_1, \ldots, x_n) = 0, \qquad \rho = 1, 2, \ldots, \mu < n, \text{ rank} \left(\frac{\partial G_\rho}{\partial x_k}\right) = \mu.$$

For $(x_1, \ldots, x_n)$ to yield a relative minimum of $F(x_1, \ldots, x_n)$ under the given constraints, it is necessary that there are $\mu + 1$ constants $(l_o, \ldots, l_\mu) \neq (0, \ldots, 0)$ such that $H_{x_i} = 0$, $i = 1, 2, \ldots, n$, where $H = l_oF + l_1G_1 + \cdots + l_\mu G_\mu$.†

The Lagrange multiplier rule, as we have derived it here, may be considered as proved, provided that we base the derivation on Theorem 5.7.1, which we have proved. In this case, we have obtained a necessary condition for what we may call a *strong relative minimum* in an obvious generalization of this concept as it was defined in Chap. 2. This is quite clear if we recall that an optimal trajectory, by definition, renders a minimum, as compared with the value that is obtained for all other admissible trajectories that lie in a weak neighborhood—the admissible set of inception—of the optimal trajectory. This remains true if one bases the derivation of the multiplier rule on Theorem 5.7.2.

The fact that the minimum principle, as stated in Theorem 5.7.1 as well as in Theorem 5.7.2, constitutes a necessary condition for a strong relative minimum suggests that it can also be employed to arrive at a generalization of the Weierstrass condition in Theorem 3.9.1 as a necessary condition for a strong relative minimum. This is indeed the case, as we shall now show.

We assume that $\hat{u} = \hat{u}(x) \equiv (y'_{\mu+1}(x), \ldots, y'_n(x))$ is an optimal control and that $y = y(x) \equiv (y_1(x), \ldots, y_n(x))$ is the corresponding optimal trajectory. For a nontrivial solution $\hat{p} = \hat{p}(x)$ of the conjugate system (A5.12.12), we have from the minimum principle that

$$H^c(\hat{p}, y, x, \bar{y}'_{\mu+1}, \ldots, \bar{y}'_n) - H^c(\hat{p}, y, x, y'_{\mu+1}, \ldots, y'_n) \geq 0 \quad \text{(A5.12.22)}$$

† T. M. Apostol, "Mathematical Analysis," pp. 152ff, Addison-Wesley Publishing Company, Inc., Reading, Mass., 1957.

for *all* $(\bar{y}'_{\mu+1}, \ldots, \bar{y}'_n)$ and, in particular, for those for which the lineal element $(x,y,\bar{y}')$ satisfies the constraining equations (A5.12.3).

We have

$$H^c(\hat{p},y,x,\bar{y}'_{\mu+1}, \ldots, \bar{y}'_n) - H^c(\hat{p},y,x,y'_{\mu+1}, \ldots, y'_n)$$
$$= p_o(f(x,y,\bar{y}') - f(x,y,y')) + \sum_{\rho=1}^{\mu} p_\rho[f_\rho(x,y,\bar{y}'_{\mu+1}, \ldots, \bar{y}'_n)$$
$$- f_\rho(x,y,y'_{\mu+1}, \ldots, y'_n)] + \sum_{j=1}^{m} (\bar{y}'_{\mu+j} - y'_{\mu+j})p_{\mu+j}.$$

We have from (A5.12.14) and (A5.12.15),

$$p_\rho = \lambda_\rho - f_{y'_\rho}p_o, \qquad \rho = 1, 2, \ldots, \mu,$$
$$p_{\mu+j} = -p_o f_{y'_{\mu+j}} - \sum_{\rho=1}^{\mu} \lambda_\rho f_{\rho,y'_{\mu+j}}, \qquad j = 1, 2, \ldots, m.$$

If we also note that

$$f_\rho(x,y,y'_{\mu+1}, \ldots, y'_n) = y'_\rho, \qquad f_\rho(x,y,\bar{y}'_{\mu+1}, \ldots, \bar{y}'_n) = \bar{y}'_\rho,$$

we obtain

$$H^c(\hat{p},y,x,\bar{y}'_{\mu+1}, \ldots, \bar{y}'_n) - H^c(\hat{p},y,x,y'_{\mu+1}, \ldots, y'_n)$$
$$= p_o(f(x,y,\bar{y}') - f(x,y,y')) + \sum_{\rho=1}^{\mu} (\lambda_\rho - f_{y_\rho'}p_o)(\bar{y}'_\rho - y'_\rho)$$
$$- \sum_{j=1}^{m} (p_o f_{y'_{\mu+j}} + \sum_{\rho=1}^{\mu} \lambda_\rho f_{\rho,y'_{\mu+j}})(\bar{y}'_{\mu+j} - y'_{\mu+j})$$
$$= p_o(f(x,y,\bar{y}') - f(x,y,y')) + \sum_{k=1}^{n} h_{y'_k}(x,y,y',\hat{\lambda})(\bar{y}'_k - y'_k).$$

If we define an excess function as

$$\mathcal{E}(x,y,y',\bar{y}',\hat{\lambda}) = p_o(f(x,y,\bar{y}') - f(x,y,y')) + \sum_{k=1}^{n} h_{y'_k}(x,y,y',\hat{\lambda})(\bar{y}'_k - y'_k),$$

we see that

$$\mathcal{E}(x,y,y',\bar{y}',\hat{\lambda}) \geq 0$$

for all lineal elements (x,y,y') of $y = y(x)$ and the corresponding $\hat{\lambda} = \hat{\lambda}(x)$ and for all lineal elements $(x,y,\bar{y}')$ that satisfy the constraining equations (A5.12.3) is a necessary condition for $y = y(x)$ to yield a strong relative minimum for the Lagrange problem [(A5.12.1) to (A5.12.3)]. (See also Sec. 6.10, and beware of the discrepancy in the definitions of the excess functions here and there.)

PROBLEMS A5.12

1. Derive corner conditions for the solutions of the Lagrange problem [(A5.12.1) to (A5.12.3)].
2. Formulate the *isoperimetric* problem

$$\int_a^b f_0(x,y_1(x), \ldots, y_m(x), y_1'(x), \ldots, y_m'(x))\, dx \to \text{minimum},$$

$$y_i(a) = y_i^a,\ y_i(b) = y_i^b,\ i = 1, 2, \ldots, m,$$

$$\int_a^b f_\rho(x,y_1(x), \ldots, y_m(x), y_1'(x), \ldots, y_m'(x))\, dx = l_\rho, \qquad \rho = 1, 2, \ldots, \mu,$$

as a Lagrange problem, apply the multiplier rule, and show that all the multipliers p_0, λ_ρ have to be constants.

CHAPTER 6

THE PROBLEM OF LAGRANGE AND THE ISOPERIMETRIC PROBLEM

APPENDIX

6.1 VARIATIONAL PROBLEMS WITH CONSTRAINTS

In 1744, L. Euler formulated the general *isoperimetric problem*—of which problem C, Sec. 1.1, is a special case—as follows:

To be found is a vector function $(y_1, \ldots, y_m)$ which minimizes the functional

$$I[y_1, \ldots, y_m] = \int_a^b f(x, y_1(x), \ldots, y_m(x), y_1'(x), \ldots, y_m'(x))\, dx, \qquad (6.1.1)$$

satisfies the initial conditions

$$y_i(a) = y_i^a, \qquad i = 1, 2, \ldots, m, \qquad (6.1.2a)$$

and the terminal conditions

$$y_i(b) = y_i^b, \qquad i = 1, 2, \ldots, m, \qquad (6.1.2b)$$

and renders

$$\begin{aligned} \int_a^b f_1(x, y_1(x), \ldots, y_m(x), y_1'(x), \ldots, y_m'(x))\, dx &= l_1, \\ &\vdots \\ \int_a^b f_\mu(x, y_1(x), \ldots, y_m(x), y_1'(x), \ldots, y_m'(x))\, dx &= l_\mu, \end{aligned} \qquad (6.1.3)$$

where $l_1, \ldots, l_\mu$ are given numbers.

This problem may be viewed as a special case of the more general problem that was formulated by Lagrange and has ever since been associated with his name. The *Lagrange problem* consists of finding a vector function $(y_1, \ldots, y_n)$ which minimizes the functional

$$I[y_1, \ldots, y_n] = \int_a^b f(x, y_1(x), \ldots, y_n(x), y_1'(x), \ldots, y_n'(x))\, dx \qquad (6.1.4)$$

and satisfies the initial conditions

$$y_i(a) = y_i^a, \qquad i = 1, 2, \ldots, n, \qquad (6.1.5a)$$

the terminal conditions

$$y_i(b) = y_i^b, \qquad i = 1, 2, \ldots, n, \qquad (6.1.5b)$$

and the underdetermined system of constraining differential equations

$$\begin{aligned} \varphi_1(x, y_1, \ldots, y_n, y_1', \ldots, y_n') &= 0 \\ &\vdots \\ \varphi_\mu(x, y_1, \ldots, y_n, y_1', \ldots, y_n') &= 0, \qquad \mu < n. \end{aligned} \qquad (6.1.6)$$

We shall first demonstrate how the isoperimetric problem can be formulated as a Lagrange problem and then proceed to an even more general formulation of such a problem.

We introduce in the isoperimetric problem [(6.1.1) to (6.1.3)] μ new functions:

$$y_{m+1} = \int_a^x f_1(t, y_1(t), \ldots, y_m(t), y_1'(t), \ldots, y_m'(t))\, dt$$
$$\vdots$$
$$y_{m+\mu} = \int_a^x f_\mu(t, y_1(t), \ldots, y_m(t), y_1'(t), \ldots, y_m'(t))\, dt.$$

Then

$$y_{m+i}' = f_i(x, y_1, \ldots, y_m, y_1', \ldots, y_m'), \qquad i = 1, 2, \ldots, \mu,$$

and

$$y_{m+i}(a) = 0, \qquad y_{m+i}(b) = l_i, \qquad i = 1, 2, \ldots, \mu.$$

If we rearrange the enumeration of the functions y_k by denoting y_i as $y_{\mu+i}$ for $i = 1, 2, \ldots, m$, and y_{m+j} as y_j for $j = 1, 2, \ldots, \mu$, and if we let $\mu + m = n$, then we may formulate the isoperimetric problem as a Lagrange problem in the following manner:

To be found is a vector function $(y_1, \ldots, y_n)$ which minimizes the functional

$$I[y_{\mu+1}, \ldots, y_n] = \int_a^b f(x, y_{\mu+1}(x), \ldots, y_n(x), y_{\mu+1}'(x), \ldots, y_n'(x))\, dx \quad (6.1.7)$$

and satisfies the initial conditions

$$y_1(a) = 0, \ldots, y_\mu(a) = 0, y_{\mu+1}(a) = y_1^a, \ldots, y_n(a) = y_m^a, \qquad (6.1.8a)$$

the terminal conditions

$$y_1(b) = l_1, \ldots, y_\mu(b) = l_\mu,\ y_{\mu+1}(b) = y_1^b, \ldots, y_n(b) = y_m^b, \quad (6.1.8b)$$

and the constraining differential equations

$$\begin{aligned} y_1' &= f_1(x, y_{\mu+1}, \ldots, y_n, y_{\mu+1}', \ldots, y_n') \\ &\vdots \\ y_\mu' &= f_\mu(x, y_{\mu+1}, \ldots, y_n, y_{\mu+1}', \ldots, y_n'). \end{aligned} \qquad (6.1.9)$$

Thus we see that the isoperimetric problem is indeed a special case of a Lagrange problem with the constraining equations explicitly solved for $y_1', \ldots, y_\mu'$. (As such, it is also a special case of the terminal control problem of predetermined duration which we discussed in Secs. A.5.9 and A5.12.)

We shall now reformulate the Lagrange problem [(6.1.4) to (6.1.6)] by introducing—as we have done repeatedly when dealing with terminal control problems in Chap. 5—the new function

$$y_o(x) = \int_a^x f(t,y_1(t), \ldots ,y_n(t),y_1'(t), \ldots ,y_n'(t))\,dt.$$

In view of

$$y_o' = f(x,y_1, \ldots ,y_n,y_1', \ldots ,y_n'), \qquad y_o(a) = 0,$$

$$y_o(b) = \int_a^b f(x,y_1(x), \ldots ,y_n(x),y_1'(x), \ldots ,y_n'(x))\,dx,$$

we can now formulate the Lagrange problem as follows:

To be found is an $(n + 1)$-dimensional vector function

$$\hat{y} = (y_0(x), \ldots ,y_n(x)),$$

where $\hat{y} \equiv (y_o, \ldots ,y_n)$, which is a solution of the underdetermined system of constraining differential equations

$$\begin{aligned} y_o' - f(x,y_1, \ldots ,y_n,y_1', \ldots ,y_n') &= 0 \\ \varphi_1(x,y_1, \ldots ,y_n,y_1', \ldots ,y_n') &= 0 \\ &\vdots \\ \varphi_\mu(x,y_1, \ldots ,y_n,y_1', \ldots ,y_n') &= 0, \qquad \mu < n, \end{aligned}$$

which satisfies the initial conditions

$$y_o(a) = 0, \qquad y_i(a) = y_i^a, \qquad i = 1, 2, \ldots, n,$$

and the terminal conditions

$$y_i(b) = y_i^b, \qquad i = 1, 2, \ldots, n,$$

and which is such that

$$y_o(b) \rightarrow \text{minimum}.$$

Uniformity of notation and greater generality are achieved by replacing the first constraining equation by a differential equation of the type

$$\varphi_o(x,y_1, \ldots ,y_n,y_o', \ldots ,y_n') = 0.$$

To further simplify the notation,we shall introduce vector notation. We stipulate that throughout this chaper, Greek letters will denote $(\mu + 1)$-dimensional vectors, as in

$$\hat{\varphi} = (\varphi_o,\varphi_1, \ldots ,\varphi_\mu), \tag{6.1.10}$$

and Roman letters will denote $(n + 1)$-dimensional vectors, as in

$$\hat{y} = (y_o,y_1, \ldots ,y_n), \qquad \hat{y}_a = (0,y_1^a, \ldots ,y_n^a), \qquad \hat{y}_b = (y_o(b),y_1^b, \ldots ,y_n^b). \tag{6.1.11}$$

(No notational convention can be rigidly enforced, and there will be some exceptions.)

With this notation, we can formulate our problem as follows:

To be found is a vector function $\hat{y} = \hat{y}(x)$ which is a solution of the constraining differential equations

$$\hat{\varphi}(x,\hat{y},\hat{y}') = 0$$

and which satisfies the initial conditions

$$\hat{y}(a) = \hat{y}_a$$

and the terminal conditions

$$\hat{y}(b) = \hat{y}_b$$

so that

$$y_o(b) \to \text{minimum}.$$

The problem as formulated here is generally referred to as the *problem of Mayer*.†

6.2 THE PROBLEM OF MAYER AND A FUNDAMENTAL THEOREM ON UNDERDETERMINED SYSTEMS

We consider the Mayer problem, as formulated at the end of the preceding section, of finding a vector function $\hat{y} = \hat{y}(x)$ which satisfies the underdetermined system of constraining differential equations

$$\hat{\varphi}(x,\hat{y},\hat{y}') = 0, \tag{6.2.1}$$

the initial conditions

$$\hat{y}(a) = \hat{y}_a, \tag{6.2.2a}$$

and the terminal conditions

$$\hat{y}(b) = \hat{y}_b, \tag{6.2.2b}$$

whereby

$$y_o(b) \to \text{minimum}. \tag{6.2.3}$$

[We remind the reader of the notational convention which we adopted in (6.1.10) and (6.1.11).]

We may give the following geometric interpretation of this problem: To be found is a curve $\hat{y} = \hat{y}(x)$ in $x,\hat{y}$ space which satisfies (6.2.1), emanates from the point $(a,\hat{y}(a)) = (a,0,y_1{}^a, \ldots,y_n{}^a)$ and terminates with the smallest possible y_o coordinate on a line L through $(b,0,y_1{}^b, \ldots,y_n{}^b)$ that is parallel to the y_o axis. (See Fig. 6.1.)

† See G. A. Bliss, "Lectures on the Calculus of Variations," pp. 187ff, The University of Chicago Press, Chicago, 1946. See also Sec. 6.8.

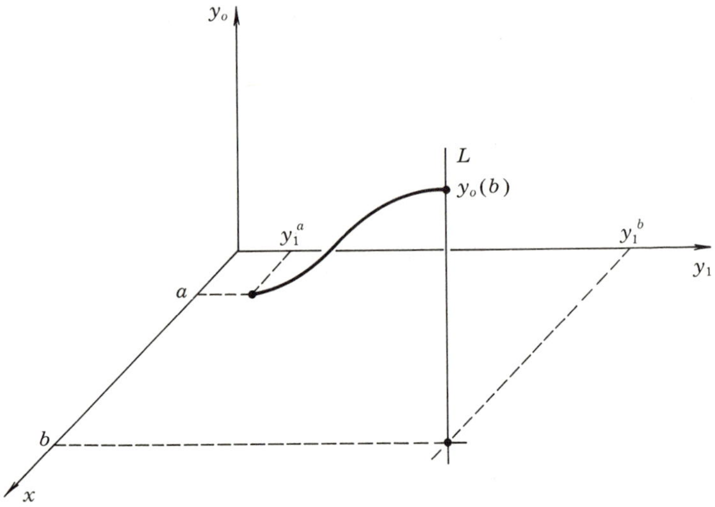

Figure 6.1

Definition 6.2 *The set* $\mathfrak{A}$ *of all points* $(b, y_o, \ldots, y_n)$ *in* $(x, \hat{y})$ *space that can be reached by a solution* $\hat{y} = \hat{y}(x)$ *of (6.2.1) which satisfies the initial conditions (6.2.2a) is called the* attainable set *of the initial-value problem [(6.2.1), (6.2.2a)] in the interval* $[a,b]$.

Suppose now that a solution $\hat{y} = \hat{y}(x)$ of [(6.2.1), (6.2.2a)] terminates at the point $(b, y_o(b), y_1^b, \ldots, y_n^b) \in \mathfrak{A}$ for some $y_o(b)$ and that $(b, y_o(b), y_1^b \ldots, y_n^b)$ is an *interior point* of the attainable set $\mathfrak{A}$. Then for some $\delta > 0$, the points $(b, y_0(b) + \Delta, y_1^b, \ldots, y_n^b)$, $|\Delta| < \delta$, are also in $\mathfrak{A}$; that is, these points can be reached by a solution of [(6.2.1), (6.2.2a)]. We may choose $\Delta < 0$ as long as $|\Delta| < \delta$, and we see that $y_o(b)$ cannot be the minimum because there is another solution that terminates at $(b, y_o(b) + \Delta, y_1^b, \ldots, y_n^b)$ and $y_o(b) + \Delta < y_o(b)$. Consequently, $\hat{y} = \hat{y}(x)$ cannot be a solution of the Mayer problem [(6.2.1) to (6.2.3)].

We state this intermediate result as a lemma:

Lemma 6.2.1 *For* $\hat{y} = \hat{y}(x)$ *to be a solution of the Mayer problem [(6.2.1) to (6.2.3)] with* $y_o(b) = y_o^b$, *it is necessary that* $(b, y_o^b, y_1^b, \ldots, y_n^b)$ *be a boundary point of the attainable set* $\mathfrak{A}$.

(Note that if the attainable set is open, then the problem cannot have a solution.)

If we can establish a necessary condition for $(b, y_o^b, y_1^b, \ldots, y_n^b)$ to be a boundary point of the attainable set $\mathfrak{A}$, we than have a necessary condition for $\hat{y} = \hat{y}(x)$, $\hat{y}(b) = (y_0{}^b, y_1{}^b, \ldots, y_n{}^b)$ to be a solution of the Mayer problem. By specialization, we shall then obtain necessary conditions for a solution of the Lagrange problem and the isoperimetric problem.

In order to state such a necessary condition, we make the following assumptions:

If $\mathfrak{D}$ denotes the $(2n + 3)$-dimensional point set

$$\mathfrak{D} = \{(x,\hat{y},\hat{y}') \mid a \leq x \leq b, -\infty < y_i < \infty, \\ -\infty < y_i' < \infty, i = 0, 1, \ldots, n\}, \quad (6.2.4)$$

we assume that

$$\hat{\varphi} = \hat{\varphi}(x,\hat{y},\hat{y}') \in C^1(\mathfrak{D}) \quad (6.2.5)$$

and that for the solution $\hat{y} = \hat{y}(x)$ of the Mayer problem [(6.2.1) to (6.2.3)]

$$\text{rank}\left(\frac{\partial \varphi_\nu}{\partial y_k'}\right) = \mu + 1 \qquad \text{for all } x \in [a,b]. \quad (6.2.6)$$

Here, $\nu = 0, 1, \ldots, \mu$; $k = 0, 1, \ldots, n$. We shall agree, in accord with our previous convention on vector notation, that Greek subscripts will run from 0 to μ (or from 1 to μ on occasions) and that Roman subscripts will run from 0 to n (or from 1 to n on occasions).

We are now ready to state a fundamental theorem that will lead us to a necessary condition for a solution of the Mayer problem, the Lagrange problem, and the isoperimetric problem.

Lemma 6.2.2 Fundamental theorem on underdetermined systems *For a solution $\hat{y} = \hat{y}(x) \in C_s{}^1[a,b]^{n+1}$ of [(6.2.1), (6.2.2a)] for which (6.2.6) is satisfied to terminate at a boundary point $(b,\hat{y}(b))$ of the attainable set $\mathfrak{A}$, it is necessary that there exist a $(\mu + 1)$-dimensional vector function $\hat{\lambda} = (\lambda_0(x), \ldots, \lambda_\mu(x)) \neq (0, \ldots, 0)$, $\hat{\lambda} \in C_s[a,b]^{\mu+1}$, such that $\hat{\lambda} = \hat{\lambda}(x)$ together with $\hat{y} = \hat{y}(x)$ satisfies the $n + 1$ equations*

$$h_{y'_k}(x,\hat{y}(x),\hat{y}'(x),\hat{\lambda}(x)) = \int_a^x h_{y_k}(s,\hat{y}(s),\hat{y}'(s),\hat{\lambda}(s))\,ds + C_k, \\ k = 0, 1, \ldots, n, \quad (6.2.7)$$

for some constants $C_0, \ldots, C_n$, everywhere except where $\hat{y}' = \hat{y}'(x)$ has a jump discontinuity. Hereby

$$h(x,\hat{y},\hat{y}',\hat{\lambda}) = \sum_{\nu=0}^{\mu} \lambda_\nu \varphi_\nu(x,\hat{y},\hat{y}'). \quad (6.2.8)$$

The discontinuities of $\hat{\lambda}$ can only occur at points where $\hat{y}'$ has a discontinuity. (A proof of this lemma will be given in Sec. 6.7.)

Lemmas 6.2.1 and 6.2.2 yield:

Theorem 6.2 Multiplier rule for the Mayer problem *For* $\hat{y} = \hat{y}(x) \in C^1_{sp}[a,b]^{n+1}$ *which satisfies (6.2.6) to be a solution of the Mayer problem [(6.2.1) to (6.2.3)], it is necessary that there exist a* $(\mu + 1)$*-dimensional vector function*

$$\hat{\lambda} = (\lambda_0(x), \ldots, \lambda_\mu(x)) \not\equiv (0, \ldots, 0), \qquad \hat{\lambda} \in C_{sp}[a,b],$$

such that $\hat{\lambda} = \hat{\lambda}(x)$ *together with* $\hat{y} = \hat{y}(x)$ *satisfies, in addition to the constraining equations (6.2.1), the* $n + 1$ *equations (6.2.7) for some constants* $C_0, \ldots, C_n$ *everywhere except where* $\hat{y}' = \hat{y}'(x)$ *has a jump discontinuity.*

Every smooth portion of $\hat{y} = \hat{y}(x)$ *has to satisfy the differential equations*

$$h_{y_k}(x,\hat{y},\hat{y}',\hat{\lambda}) - \frac{d}{dx} h_{y'_k}(x,\hat{y},\hat{y}',\hat{\lambda}) = 0, \qquad k = 0, 1, \ldots, n, \tag{6.2.9}$$

and at every point $x = c$ *where* $\hat{y}' = \hat{y}'(x)$ *has a jump discontinuity, the* corner conditions of the Mayer problem

$$h_{y'_k}(c, \hat{y}(c), \hat{y}'(c - 0), \hat{\lambda}(c - 0))$$
$$= h_{y'_k}(c, \hat{y}(c), \hat{y}'(c + 0), \hat{\lambda}(c + 0)), \tag{6.2.10}$$

$k = 0, 1, \ldots, n,$ *have to be satisfied.*

Equations (6.2.7) are called the *Mayer equations in integrated form*, and (6.2.9) are called the *Mayer equations*. We note that the Mayer equations have the same structure as the Euler-Lagrange equations of variational problems without constraints.

A solution $\hat{y} = \hat{y}(x) \in C^1[a,b]^{n+1}$ of the Mayer equations (6.2.9) and (6.2.1) is called an *extremal* of the Mayer problem. The components of $\hat{\lambda} = \hat{\lambda}(x)$ are called the *multipliers* of the Mayer problem.

Observe that the Mayer equations (6.2.9) and the corner conditions (6.2.10) follow from (6.2.7) by the same reasoning that was employed in Secs 2.3 and 2.9.

The next four sections will be devoted to the application of Theorem 6.2 to the Lagrange problem and the isoperimetric problem.

6.3 THE LAGRANGE MULTIPLIER RULE

We shall now apply Theorem 6.2 to the Lagrange problem which we formulated in Sec. 6.1 and which we restate here for easy reference:

To be found is a vector function $(y_1, \ldots, y_n)$ which satisfies the constraining differential equations

$$\begin{aligned} \varphi_1(x,y_1, \ldots, y_n, y_1', \ldots, y_n') &= 0 \\ &\vdots \\ \varphi_\mu(x,y_1, \ldots, y_n, y_1', \ldots, y_n') &= 0, \qquad \mu < n, \end{aligned} \tag{6.3.1}$$

the initial conditions

$$y_i(a) = y_i^a, \qquad i = 1, 2, \ldots, n, \tag{6.3.2a}$$

and the terminal conditions

$$y_i(b) = y_i^b, \qquad i = 1, 2, \ldots, n, \tag{6.3.2b}$$

and is such that

$$\int_a^b f(x,y_1(x), \ldots, y_n(x), y_1'(x), \ldots, y_n'(x))\, dx \to \text{minimum}. \tag{6.3.3}$$

As we saw in Sec. 6.1, we can formulate this problem as a Mayer problem with

$$\varphi_o(x,y_1, \ldots, y_n, y_o', \ldots, y_n') = y_o' - f(x,y_1, \ldots, y_n, y_1', \ldots, y_n'). \tag{6.3.4}$$

Since

$$\begin{aligned} \varphi_{oy'_o} = 1, \qquad \varphi_{oy'_k} = -f_{y'_k}, \qquad k = 1, 2, \ldots, n; \\ \varphi_{\nu y'_o} = 0, \qquad \nu = 1, 2, \ldots, \mu, \end{aligned} \tag{6.3.5}$$

we see that

$$\left(\frac{\partial \varphi_\nu}{\partial y_k'}\right) = \begin{pmatrix} 1 & -f_{y'_1} & \cdots & -f_{y'_n} \\ 0 & \varphi_{1y'_1} & \cdots & \varphi_{1y'_n} \\ \vdots & & & \\ 0 & \varphi_{\mu y'_1} & \cdots & \varphi_{\mu y'_n} \end{pmatrix}$$

Hence condition (6.2.6) can be replaced by

$$\operatorname{rank}\left(\frac{\partial \varphi_\nu}{\partial y_k'}\right)_{\substack{\nu=1,\ldots,\mu \\ k=1,\ldots,n}} = \mu. \tag{6.3.6}$$

Instead of (6.2.5), we now have

$$f, \varphi_1, \ldots, \varphi_\mu \in C^1(\mathfrak{D}_1), \tag{6.3.7}$$

where

$$\mathfrak{D}_1 = \{(x,y_1, \ldots, y_n, y_1', \ldots, y_n') \mid a \leq x \leq b, -\infty < y_i < \infty, \\ -\infty < y_i' < \infty, i = 1, 2, \ldots, n\}.$$

h, as defined in (6.2.7), becomes

$$h = \lambda_o(y_o' - f) + \sum_{\nu=1}^{\mu} \lambda_\nu \varphi_\nu.$$

Since neither the φ_ν nor f depend on y_o explicitly, we have

$$h_{y_k} = \begin{cases} 0 & \text{for } k = 0 \\ -\lambda_o f_{y_k} + \sum_{\nu=1}^{\mu} \lambda_\nu \varphi_{\nu y_k}, & k = 1, 2, \ldots, n. \end{cases}$$

In view of (6.3.5)

$$h_{y'_k} = \begin{cases} \lambda_o & \text{for } k = 0 \\ -\lambda_o f_{y'_k} + \sum_{\nu=1}^{\mu} \lambda_\nu \varphi_{\nu y'_k}, & k = 1, 2, \ldots, n. \end{cases}$$

Hence we have, for the first equation in (6.2.7),

$$\lambda_o = \text{constant},$$

while the remaining equations can be written in the form

$$h^L_{y'_k} = \int_a^x h^L_{y_k}\, dt + C_k, \qquad k = 1, 2, \ldots, n,$$

where
$$h^L = -\lambda_o f + \sum_{\nu=1}^{\mu} \lambda_\nu \varphi_\nu.$$

In view of all this, we state Theorem 6.2, as applied to the Lagrange problem, as follows:

Theorem 6.3 Lagrange's multiplier rule *If* $(y_1, \ldots, y_n) \in C^1_{sP}[a,b]^n$ *is a solution of the Lagrange problem* [*(6.3.1) to (6.3.3)*] *and if the conditions (6.3.6) and (6.3.7) are satisfied, then it is necessary that there exist a vector function* $\hat{\lambda} = (\lambda_o, \ldots, \lambda_\mu) \neq (0,0, \ldots, 0)$, $\hat{\lambda} \in C_{sP}[a,b]^{\mu+1}$, $\lambda_o =$ *constant, such that, in addition to the constraining equations* (6.3.1), *the equations*

$$h^L_{y'_k}(x, y_1(x), \ldots, y_n(x), y'_1(x), \ldots, y'_n(x), \hat{\lambda}(x))$$
$$= \int_a^x h^L_{y_k}(t, y_1(t), \ldots, y_n(t), y'_1(t), \ldots, y'_n(t), \hat{\lambda}(t))\, dt + C_k \qquad (6.3.8)$$

$(k = 1, 2, \ldots, n)$ *are satisfied for some constants* C_k, *whereby*

$$h^L(x, y_1, \ldots, y_n, y'_1, \ldots, y'_n, \hat{\lambda}) = -\lambda_o f(x, y_1, \ldots, y_n, y'_1, \ldots, y'_n)$$
$$+ \sum_{\nu=1}^{\mu} \lambda_\nu \varphi_\nu(x, y_1, \ldots, y_n, y'_1, \ldots, y'_n). \qquad (6.3.9)$$

Every smooth portion of $(y_1, \ldots, y_n)$ *has to satisfy the differential equations*

$$h^L_{y_k}(x,y_1, \ldots,y_n,y'_1, \ldots,y'_n,\hat{\lambda}) - \frac{d}{dx} h^L_{y'_k}(x,y_1, \ldots,y_n,y'_1, \ldots,y'_n,\hat{\lambda}) = 0 \tag{6.3.10}$$

$(k = 1, 2, \ldots, n)$, *and at every point* $x = c$, *where* $(y'_1, \ldots,y'_n)$ *has a jump discontinuity, the conditions*

$$h^L_{y'_k}(c, y_1(c), \ldots, y_n(c), y'_1(c-0), \ldots, y'_n(c-0), \hat{\lambda}(c-0))$$
$$= h^L_{y'_k}(c, y_1(c), \ldots, y_n(c), y'_1(c+0), \ldots, y'_n(c+0), \hat{\lambda}(c+0)) \tag{6.3.11}$$

$(k = 1, 2, \ldots n)$ *have to be satisfied.*

We call (6.3.8) the Mayer equations in integrated form of the Lagrange problem, (6.3.10) the Mayer equations of the Lagrange problem, and (6.3.11) the corner conditions of the Lagrange problem.

PROBLEMS 6.3

1. Apply Lagrange's multiplier rule to the problem

$$\int_0^{\pi/2} w^2(x)\, dx \to \text{minimum,}$$
$$w' + y - (y-z)^2 y = 0$$
$$y' - w = 0$$

with the boundary conditions $y(0) = 0$, $z(0) = 0$, $w(0) = 1$, $y(\pi/2) = 1$, $z(\pi/2) = 1$, $w(\pi/2) = 0$, and solve for y, z, w, λ_0, λ_1, λ_2.

2. Same as in problem 1 for

$$\int_0^1 y'^2(x)\, dx \to \text{minimum,}$$
$$w' + y' - z' = 0$$
$$w' + 2z' = 0$$

with the boundary conditions $y(0) = 0$, $z(0) = 0$, $w(0) = 0$, $y(1) = 3$, $z(1) = 1$, $w(1) = -2$.

3. Same as in problem 1 for

$$\int_0^{\pi/2} (y'^2(x) - y^2(x))\, dx \to \text{minimum,}$$
$$w' + y' - z' = 0$$
$$w' + 2z' = 0$$

with the boundary conditions $y(0) = 0$, $z(0) = 0$, $w(0) = 0$, $y(\pi/2) = 3\pi/2$, $z(\pi/2) = \pi/2$, $w(\pi/2) = -\pi$.

6.4 DISCUSSION OF THE LAGRANGE MULTIPLIER RULE

Theorem 6.3 states a necessary condition which a vector function

$$(y_1, \ldots, y_n) \in C_s^1[a,b]^n$$

has to satisfy in order to be a solution of the Lagrange problem. While it is not true, in general, that a vector function that does satisfy this condition is indeed the solution of the Lagrange problem, we know at least that if the problem has a solution in $C_s^1[a,b]^n$ at all, then it is among those vector functions that satisfy the condition of Theorem 6.3.

We shall now discuss the more formal aspects of finding such a vector function for the simpler case where $(y_1, \ldots, y_n) \in C^1[a,b]^n$.

By Theorem 6.3 we have to find $(y_1, \ldots, y_n) \in C^1[a,b]^n$ such that (6.3.10), namely,

$$h^L_{y_k} - \frac{d}{dx} h^L_{y'_k} = 0, \qquad \text{where } h^L = -\lambda_o f + \sum_{\nu=1}^{\mu} \lambda_\nu \varphi_\nu,$$

is satisfied with $\lambda_o =$ constant.

These are n differential equations (in general, nonlinear) of second order in $y_1, \ldots, y_n$ and are linear and of first order in $\lambda_1, \ldots, \lambda_\mu$. Along with these equations we have the constraining equations (6.3.1), namely,

$$\varphi_\nu(x, y_1, \ldots, y_n, y'_1, \ldots, y'_n) = 0, \qquad \nu = 1, 2, \ldots, \mu,$$

which are μ (in general, nonlinear) first-order differential equations for $y_1, \ldots, y_n$.

Since $(y_1, \ldots, y_n)$ has to satisfy n initial conditions (6.3.2*a*) and n terminal conditions (6.3.2*b*), we hope that the solutions $y_i = y_i(x)$, $i = 1, 2, \ldots, n$, will contain a total of $2n$ integration constants. (Note that this is, in general, *not* sufficient to satisfy the initial and terminal conditions, but it is the best we can hope for.)

In order to investigate this matter, we shall proceed as in Sec. 2.12 and introduce *canonical* variables by

$$p_i = h^L_{y'_i}(x, y_1, \ldots, y_n, y'_1, \ldots, y'_n, \lambda_o, \ldots, \lambda_\mu), \qquad i = 1, 2, \ldots, n. \tag{6.4.1}$$

We consider these transformations, together with the constraining equations (6.3.1), that is, we consider the following system:

$$h^L_{y'_i}(x, y_1, \ldots, y_n, y'_1, \ldots, y'_n, \lambda_o, \ldots, \lambda_\mu) - p_i = 0 \qquad i = 1, 2, \ldots, n,$$
$$\varphi_\nu(x, y_1, \ldots, y_n, y'_1, \ldots, y'_n) = 0, \qquad \nu = 1, 2, \ldots, \mu. \tag{6.4.2}$$

We shall try to solve these $n + \mu$ equations for $y'_1, \ldots, y'_n, \lambda_1, \ldots, \lambda_\mu$.

Definition 6.4 $(y_1, \ldots, y_n) \in C^1[a,b]^n$ *is called a* regular extremal *of the Lagrange problem if*

$$\left.\frac{\partial(h^L_{y'_1}, \ldots, h^L_{y'_n}, \varphi_1, \ldots, \varphi_\mu)}{\partial(y'_1, \ldots, y'_n, \lambda_1, \ldots, \lambda_\mu)}\right|_{\substack{y_i=y_i(x)\\ \lambda_\nu=\lambda_\nu(x)}} \neq 0 \qquad \textit{for all } x \in [a,b].$$

An extremal for which this jacobian is zero is called singular. (Compare this definition with Definition 2.5.2 of a regular extremal.)

If $(y_1, \ldots, y_n)$ is a regular extremal in the sense of this definition, then we can solve (6.4.2) in a strong neighborhood of $(y_1, \ldots, y_n)$ and a weak neighborhood of $(\lambda_1, \ldots, \lambda_\mu)$ for $y'_1, \ldots, y'_n, \lambda_1, \ldots, \lambda_\mu$ and obtain

$$\begin{aligned} y'_i &= \Phi_i(x, y_1, \ldots, y_n, p_1, \ldots, p_n, \lambda_o), \qquad i = 1, 2, \ldots, n. \\ \lambda_\nu &= \Psi_\nu(x, y_1, \ldots, y_n, p_1, \ldots, p_n, \lambda_o), \qquad \nu = 1, 2, \ldots, \mu. \end{aligned} \tag{6.4.3}$$

As in Sec. 2.12, we introduce the *hamiltonian:*

$$H(x, y_1, \ldots, y_n, p_1, \ldots, p_n, \lambda_o) = \left[\sum_{i=1}^n h^L_{y'_i} y'_i - h^L\right]_{\substack{y'_i=\Phi_i(x,y,p,\lambda_o)\\ \lambda_\nu=\Psi_\nu(x,y,p,\lambda_o)}} \tag{6.4.4}$$

Then $\quad dH = \sum_{i=1}^n h^L_{y'_i}\, dy'_i + \sum_{i=1}^n y'_i\, dh^L_{y'_i} - dh^L$

where $\quad dh^L = h^L_x\, dx + \sum_{i=1}^n h^L_{y_i}\, dy_i + \sum_{i=1}^n h^L_{y'_i}\, dy'_i + \sum_{\nu=1}^\mu h^L_{\lambda_\nu}\, d\lambda_\nu.$

(Note that λ_o = constant.) From the definition of h, we have

$$h^L_{\lambda_\nu} = \varphi_\nu, \qquad \nu = 1, 2, \ldots, \mu,$$

and from (6.4.2),

$$h^L_{y'_i} = p_i, \qquad dh^L_{y'_i} = dp_i, \qquad i = 1, 2, \ldots, n.$$

From the Mayer equations (6.3.10),

$$h^L_{y_i} = \frac{d}{dx} h^L_{y'_i} = \frac{dp_i}{dx}, \qquad i = 1, 2, \ldots, n,$$

and from the restraining equations,

$$\varphi_\nu = 0, \qquad \nu = 1, 2, \ldots, \mu.$$

In view of all this, we have

$$dH = \cancel{\sum_{i=1}^{n} p_i\,dy'_i} + \sum_{i=1}^{n} y'_i\,dp_i - h_x^L\,dx - \sum_{i=1}^{n} p'_i\,dy_i - \cancel{\sum_{i=1}^{n} p_i\,dy'_i}$$

$$= \sum_{i=1}^{n} y'_i\,dp_i - h_x^L\,dx - \sum_{i=1}^{n} p'_i\,dy_i,$$

and we obtain, as in (2.12.8),

$$\begin{aligned} y'_i &= H_{p_i} \\ p'_i &= -H_{y_i} \end{aligned} \tag{6.4.5}$$

for $i = 1, 2, \ldots, n$. These are $2n$ first-order differential equations for $y_1, \ldots, y_n, p_1, \ldots, p_n$, the solutions of which will, in general, contain $2n$ arbitrary integration constants. (See also Prob. 6.4.8.)

Equations (6.4.5) still contain λ_o. From the Euler-Lagrange equations (6.3.10), it is clear that $\lambda_o, \lambda_1, \ldots, \lambda_\mu$ are only determined but for a common constant multiple, and it appears that we may choose λ_o, which is a constant, freely. This is only permissible, however, if $\lambda_o \neq 0$. Then and only then may we choose $\lambda_o = 1$ without loss of generality.

However, if $\lambda_o = 0$, then the equations (6.3.10) become

$$\frac{\partial}{\partial y_i}\sum_{\nu=1}^{\mu} \lambda_\nu \varphi_\nu - \frac{d}{dx}\frac{\partial}{\partial y'_i}\sum_{\nu=1}^{\mu} \lambda_\nu \varphi_\nu = 0, \qquad i = 1, 2, \ldots, n, \tag{6.4.6}$$

and f does not appear anymore. We easily recognize (6.4.6) as a necessary condition that $(b, y_1(b), \ldots, y_n(b))$ is a boundary point of the attainable set of the initial value problem

$$\varphi_\nu(x, y_1, \ldots, y_n, y'_1, \ldots, y'_n) = 0, \qquad \nu = 1, 2, \ldots, \mu,$$

$y_i(a) = y_i^a$, $i = 1, 2, \ldots, n$. (See Lemma 6.2.2.)

Suppose for a moment that the attainable set of this initial value problem consists of the point $(b, y_1(b), \ldots, y_n(b))$ only, i.e., the solution of (6.3.1) together with the boundary conditions (6.3.2a and b) is uniquely determined. Then, (6.4.6) will be satisfied, by necessity, and (6.3.10) becomes

$$\lambda_o\left(f_{y_i} - \frac{d}{dx} f_{y'_i}\right) = 0, \qquad i = 1, 2, \ldots, n.$$

Hence, unless $(y_1, \ldots, y_n)$ is an extremal of the Euler-Lagrange equations of the variational problem $\int_a^b f(x, y_1, \ldots, y_n, y'_1, \ldots, y'_n)\,dx \to \text{minimum}$ without regard to the constraining equations [for example, if $(y_1, \ldots, y_n)$ minimizes the integral $\int_a^b f(x, y_1(x), \ldots, y_n(x), y'_1(x), \ldots, y'_n(x))\,dx$], we have to have $\lambda_o = 0$. A solution $(y_1, y_2, \ldots, y_n)$ for which (6.4.6.) has a nontrivial solution is called an *anormal*—as opposed to a *normal*—solution. (See also Probs. 6.4.5 and 6.4.9.)

PROBLEMS 6.4

1. Show that $h^L_{\lambda_\rho} = 0$ for the solution of the Lagrange problem.
2. Consider the extremals of the Lagrange problem in Prob. 6.3.1. Show that these are singular extremals in the sense of Definition 6.4.
3. Consider the extremals of Probs. 6.3.2 and 6.3.3, for which $\lambda_o = 0$. Show that these are singular in the sense of Definition 6.4.
4. Find the hamiltonian $H(x,y,z,w,p,q,r,\lambda_o)$ for the Lagrange problem in Probs. 6.3.2 and 6.3.3. What about $\lim_{\lambda_0 \to 0} H(x,y,z,w,p,q,r,\lambda_o)$?
5. Investigate the solutions of Prob. 6.3.1, with $\lambda_o = 0$, with respect to anormality.
6. Show that the extremals $y = 3x$ and $y = (3\pi/2)\sin x$, respectively, in Probs. 6.3.2 and 6.3.3 are extremals of the variational problems that are obtained from the stated Lagrange problems by omitting the constraining equations.
7. Solve the Lagrange problem

$$\int_0^{\pi/2} w(x)\,dx \to \text{minimum},$$

$$w' + y - (y - z')^2 y = 0$$

$$y' - w = 0,$$

$y(0) = 0$, $z(0) = -1$, $w(0) = 1$, $y(\pi/2) = 1$, $z(\pi/2) = 0$, $w(\pi/2) = 0$. Check whether or not the extremal is singular.

*8. Show: If

$$\frac{\partial(h^L_{y'_1}, \ldots, h^L_{y'_n}, \varphi_1, \ldots, \varphi_\mu)}{\partial(y'_1, \ldots, y'_n, \lambda_1, \ldots, \lambda_\mu)} > 0$$

(or < 0) for all $x \in [a,b]$ when taken for $y_i = y_i(x)$, $\lambda_\rho = \lambda_\rho(x)$, the solutions of the Mayer equations and the constraining equations of the Lagrange problem, and if $f, \varphi_i \in C^2$, then $y_i, \lambda_\rho \in C^1[a,b]$.

9. Find an extremal of

$$\int_0^1 (y'^2(x) - y^2(x))\,dx \to \text{minimum},$$

$y(0) = 0$, $y(1) = 1$, $z(0) = 0$, $z(1) = \sqrt{2}$ under the constraint

$$z' - \sqrt{1 + y'^2} = 0$$

and show that it is anormal.

6.5 THE ISOPERIMETRIC PROBLEM

We have already seen, in Sec. 6.1, that the *isoperimetric problem* of finding an m-dimensional vector function $(y_1, y_2, \ldots y_m) \in C^1[a,b]^m$ which minimizes the functional

$$I[y_1, \ldots, y_m] = \int_a^b f(x, y_1(x), \ldots, y_m(x), y_1'(x), \ldots, y_m'(x))\, dx, \tag{6.5.1}$$

satisfies the initial conditions

$$y_i(a) = y_i^a, \qquad i = 1, 2, \ldots, m, \tag{6.5.2a}$$

and the terminal conditions

$$y_i(b) = y_i^b, \qquad i = 1, 2, \ldots, m, \tag{6.5.2b}$$

and is subject to the μ constraints

$$\int_a^b f_\rho(x, y_1, \ldots, y_m, y_1', \ldots, y_m')\, dx = l_\rho, \qquad \rho = 1, 2, \ldots, \mu, \tag{6.5.3}$$

where l_ρ are given numbers, is equivalent to a Lagrange problem of finding an n-dimensional vector function $(y_1, y_2, \ldots, y_n) \in C^1[a,b]^n$ that minimizes the functional

$$I[y_{\mu+1}, \ldots, y_n] = \int_a^b f(x, y_{\mu+1}(x), \ldots, y_n(x), y_{\mu+1}'(x), \ldots, y_n'(x))\, dx,$$

satisfies the initial conditions

$$y_\rho(a) = 0, \qquad \rho = 1, 2, \ldots, \mu, \qquad y_{\mu+i}(a) = y_i^a, \qquad i = 1, 2, \ldots, n - \mu,$$

and the terminal conditions

$$y_\rho(b) = l_\rho, \qquad \rho = 1, 2, \ldots, \mu, \qquad y_{\mu+i}(b) = y_i^b, \qquad i = 1, 2, \ldots, n - \mu,$$

and is subject to the constraints

$$y_\rho' - f_\rho(x, y_{\mu+1}, \ldots, y_n, y_{\mu+1}', \ldots, y_n') = 0, \qquad \rho = 1, 2, \ldots, \mu.$$

In the transition from one problem to the other, we denoted $m + \mu = n$ and changed the enumeration of the y_i as follows:

$$\begin{aligned} y_{m+\rho} \text{ is now } y_\rho, \qquad & \rho = 1, 2, \ldots, \mu; \\ y_i \text{ is now } y_{\mu+i}, \qquad & i = 1, 2, \ldots, m. \end{aligned} \tag{6.5.4}$$

We point out that there is no restriction on μ relative to m; that is, both $\mu < m$ and $\mu \geq m$ are possible.

It is clear, of course, that with $\varphi_\rho = y_\rho' - f_\rho$,

$$\operatorname{rank}\left(\frac{\partial \varphi_\rho}{\partial y_k'}\right)\Bigg|_{\substack{\rho=1,2,\ldots,\mu \\ k=1,2,\ldots,n}} = \mu$$

since the constraining equations are *already solved* for $y_1', y_2', \ldots, y_\mu'$.

In order to investigate the Mayer equation of Theorem 6.3 as it applies to this particular case, we note that

$$\frac{\partial \varphi_\rho}{\partial y_k} = \begin{cases} 0 & \text{for } k \leq \mu;\ \rho = 1, \ldots, \mu \\ -\dfrac{\partial f_\rho}{\partial y_k} & \text{for } k = \mu + 1, \mu + 2, \ldots, n;\ \rho = 1, \ldots, \mu \end{cases}$$

and

$$\frac{\partial \varphi_\rho}{\partial y'_\alpha} = \begin{cases} 1 & \text{for } \rho = \alpha,\ \alpha = 1, 2, \ldots, \mu \\ 0 & \text{for } \rho \neq \alpha,\ \alpha = 1, 2, \ldots, \mu;\ \rho = 1, 2, \ldots, \mu \\ -\dfrac{\partial f_\rho}{\partial y'_\alpha} & \text{for } \alpha = \mu + 1, \mu + 2, \ldots, n;\ \rho = 1, 2, \ldots, \mu. \end{cases}$$

Hence, with

$$h^L = -\lambda_o f + \sum_{\rho=1}^{\mu} \lambda_\rho \varphi_\rho,$$

we have

$$h^L_{y_k} = -\lambda_o \frac{\partial f}{\partial y_k} + \sum_{\rho=1}^{\mu} \lambda_\rho \frac{\partial \varphi_\rho}{\partial y_k} = \begin{cases} 0 & \text{for } k \leq \mu \\ -\lambda_o \dfrac{\partial f}{\partial y_k} - \displaystyle\sum_{\rho=1}^{\mu} \lambda_\rho \frac{\partial f_\rho}{\partial y_k} & \text{for } k > \mu \end{cases}$$

and

$$h^L_{y'_k} = -\lambda_o \frac{\partial f}{\partial y'_k} + \sum_{\rho=1}^{\mu} \lambda_\rho \frac{\partial \varphi_\rho}{\partial y'_k} = \begin{cases} \lambda_k & \text{for } k \leq \mu \\ -\lambda_o \dfrac{\partial f}{\partial y'_k} - \displaystyle\sum_{\rho=1}^{\mu} \lambda_\rho \frac{\partial f_\rho}{\partial y'_k} & \text{for } k > \mu. \end{cases}$$

Thus we obtain for the first μ equations (6.3.10),

$$\frac{d\lambda_\rho}{dx} = 0, \qquad \rho = 1, 2, \ldots, \mu,$$

that is, $\lambda_1, \lambda_2, \ldots, \lambda_\mu$ *are constants.*

The remaining $n - \mu$ $(=m)$ equations can be written in the form

$$h^I_{y_k} - \frac{d}{dx} h^I_{y'_k} = 0, \qquad k = \mu + 1, \mu + 2, \ldots, n,$$

with

$$h^I = \lambda_o f + \sum_{\rho=1}^{\mu} \lambda_\rho f_\rho,$$

If we return to our original notation—see (6.5.4)—and change the signs of all the multipliers λ_ρ, we can state:

Theorem 6.5 *For* $(y_1, \ldots, y_m) \in C^1[a,b]^m$ *to be a solution of the isoperimetric problem* [(6.5.1) *to* (6.5.3)], *it is necessary that there exist* $\mu + 1$ *constants* $(\lambda_o, \ldots, \lambda_\mu) \neq (0, \ldots, 0)$ *such that*

$$h^I_{y_k}(x,y_1, \ldots, y_m, y'_1, \ldots, y'_m, \hat{\lambda}) - \frac{d}{dx} h^I_{y'_k}(x,y_1, \ldots, y_m, y'_1, \ldots, y'_m, \hat{\lambda}) = 0 \tag{6.5.5}$$

$(k = 1, 2, \ldots, m)$, where

$$h^I(x,y_1, \ldots, y_m, y'_1, \ldots, y'_m, \hat{\lambda}) = \lambda_o f(x,y_1, \ldots, y_m, y'_1, \ldots, y'_m) + \sum_{\rho=1}^{\mu} \lambda_\rho f_\rho(x,y_1, \ldots, y_m, y'_1, \ldots, y'_m).$$

Equation (6.5.5) is called the *Mayer equation of the isoperimetric problem*.†

PROBLEMS 6.5

1. Solve the isoperimetric problem

$$\int_0^1 y(x)\, dx \to \text{maximum}$$

$$\int_0^1 \sqrt{1 + y'^2(x)}\, dx = \frac{\pi}{2}, \qquad y(0) = 0, \qquad y(1) = 0.$$

2. Solve the isoperimetric problem

$$\int_0^1 \sqrt{1 + y'^2(x)}\, dx \to \text{minimum}$$

$$\int_0^1 y(x)\, dx = \frac{\pi}{8}, \qquad y(0) = 0, \qquad y(1) = 0.$$

3. Find that curve of constant mass density with a given length that joins the points (0,0), (1,0) and is such that the y coordinate of its centroid is a minimum. (Determine the solution but for a multiplicative constant.)
4. State Theorem 6.5 for the case where $(y_1, \ldots, y_m) \in C_s^1[a,b]^m$.

† For a derivation of the multiplier rule for the isoperimetric problem without reference to the more general Lagrange problem and in a functional analytical setting, see L. A. Liusternik and V. J. Sobolev, "Elements of Functional Analysis," pp. 208–210, Frederick Ungar Publishing Co., New York, 1961.

6.6 DISCUSSION OF THE ISOPERIMETRIC PROBLEM

Before we discuss the solution of the isoperimetric problem and some of the pitfalls, let us examine such a hazard in a particular case:

The following problem is obviously an isoperimetric problem:

$$\int_0^1 y'^2(x)\,dx \to \text{minimum}$$

$$\int_0^1 \sqrt{1+y'^2(x)}\,dx = 2, \qquad y(0) = 0, \qquad y(1) = 1.$$

If $y = y(x) \in C^1[0,1]$ is a solution of this problem, then by Theorem 6.5 it is necessary that there be two constants $(\lambda_o,\lambda_1) \neq (0,0)$ such that

$$\frac{\partial}{\partial y}\left(\lambda_o y'^2 + \lambda_1\sqrt{1+y'^2}\right) - \frac{d}{dx}\frac{\partial}{\partial y'}\left(\lambda_o y'^2 + \lambda_1\sqrt{1+y'^2}\right) = 0,$$

or after the differentiations have been carried out and after a few simplifications:

$$\frac{d}{dx}\,y'\left(2\lambda_o + \frac{\lambda_1}{\sqrt{1+y'^2}}\right) = 0.$$

Hence

$$y' = \text{constant}$$

where this constant contains λ_o, λ_1 and one integration constant.

Thus

$$y = ax + b,$$

and after having determined the constants a,b such that the boundary conditions are satisfied, we are left with

$$y = x.$$

We see that

$$\int_0^1 \sqrt{1+y'^2}\,dx = \int_0^1 \sqrt{2}\,dx = \sqrt{2} \neq 2,$$

that is, that it is *impossible* to satisfy the constraint.

We realize, of course, what went wrong. The function $y = x$ that appears as the solution not only is the solution of the Mayer equation of the isoperimetric problem but also is an extremal of $\int_0^1 \sqrt{1+y'^2(x)}\,dx$, and hence we cannot, in general, expect the constraining condition to be satisfied—unless the constraint is $\int_0^1 \sqrt{1+y'^2(x)}\,dx = \sqrt{2}$.

Let us investigate this matter in an orderly fashion. In general, the solutions of the Mayer equations of the isoperimetric problem will contain $2m+\mu$ essential constants:

$$y_i = y_i(x,c_1,\ldots,c_{2m},\lambda_1,\ldots,\lambda_\mu), \qquad i = 1, 2, \ldots, m.$$

Suppose we utilize the constants $c_1, \ldots, c_{2m}$ to satisfy the initial and terminal conditions (which is not always possible either—see Sec. 2.6). Then

$$y_i(a,\lambda_1, \ldots,\lambda_\mu) = y_i^a, \qquad y_i(b,\lambda_1, \ldots,\lambda_\mu) = y_i^b \qquad \text{for all } (\lambda_1, \ldots,\lambda_\mu),$$

and hence

$$\left.\frac{\partial y_i}{\partial \lambda_\nu}\right|_{x=a} = 0, \qquad \left.\frac{\partial y_i}{\partial \lambda_\nu}\right|_{x=b} = 0, \qquad i = 1, 2, \ldots, m; \nu = 1, 2, \ldots, \mu. \tag{6.6.1}$$

In order for the constraining conditions to be satisfied, we have to determine the remaining μ constants $\lambda_1, \ldots, \lambda_\mu$ such that

$$\int_a^b f_\rho(x,y_1(x,\lambda), \ldots,y_m(x,\lambda),y_1'(x,\lambda), \ldots,y_m'(x,\lambda))\, dx = l_\rho \quad \rho = 1, 2, \ldots, \mu,$$

where we have written λ instead of $\lambda_1, \ldots, \lambda_\mu$ in order to simplify the notation.

We can guarantee that these constraints can be satisfied for any $\{l_1,l_2, \ldots,l_\mu\}$ if

$$D = \frac{\partial\left(\int_a^b f_1\, dx, \ldots, \int_a^b f_\mu\, dx\right)}{\partial(\lambda_1, \ldots,\lambda_\mu)} \neq 0.$$

If we make suitable assumptions to guarantee the interchangeability of integration and differentiation with respect to λ_ρ, and the following integration-by-parts process (see Prob. 6.6.1), we obtain

$$\begin{aligned} \frac{\partial \int_a^b f_\nu\, dx}{\partial \lambda_\rho} &= \int_a^b \sum_{k=1}^m \left(f_{\nu y_k} \frac{\partial y_k}{\partial \lambda_\rho} + f_{\nu y'_k} \frac{\partial y_k'}{\partial \lambda_\rho}\right) dx \\ &= \int_a^b \sum_{k=1}^m \frac{\partial y_k}{\partial \lambda_\rho}\left(f_{\nu y_k} - \frac{d}{dx} f_{\nu y'_k}\right) dx + \left.\sum_{k=1}^m \frac{\partial y_k}{\partial \lambda_\rho} f_{\nu y'_k}\right|_a^b . \end{aligned} \tag{6.6.2}$$

By (6.6.1), the second term vanishes, and we have

$$\frac{\partial \int_a^b f_\nu\, dx}{\partial \lambda_\rho} = \int_a^b \sum_{k=1}^m \frac{\partial y_k}{\partial \lambda_\rho}\left(f_{\nu y_k} - \frac{d}{dx} f_{\nu y'_k}\right) dx.$$

Suppose now that the extremal of the isoperimetric problem is also an extremal of the integral of the αth constraint, that is,

$$f_{\alpha y_k} - \frac{d}{dx} f_{\alpha y'_k} = 0, \qquad k = 1, 2, \ldots, m.$$

Then an entire row in the jacobian D vanishes, and $D = 0$.

Thus we can say: *If the extremal of the isoperimetric problem is also an extremal of at least one of the constraining integrals, then it is not possible to guarantee the existence of a solution of the Mayer equations of the isoperimetric problem that also satisfies the boundary conditions and the integral constraints.*

As an example of an isoperimetric problem, we consider an *elastic rod* of length l which is clamped at (a,y_a) and (b,y_b) at angles τ_a and τ_b with the positive x axis. Hereby, $\sqrt{(b-a)^2 + (y_b - y_a)^2} < l$.

The equilibrium position of the rod is characterized by the minimum of the potential energy, which in turn is proportional to $\int_0^l \kappa^2\, ds$, where κ represents the curvature of the rod and where s represents the arc length on the rod.

The projections of the rod onto the x axis and the y axis are given by

$$\int_0^l \cos\tau\, ds = \int_a^b dx = b - a$$

$$\int_0^l \sin\tau\, ds = \int_a^b dy = y_b - y_a,$$

where τ is the angle of the tangent line to the rod with the positive x axis.

Thus we have the following isoperimetric problem, which we formulate for reasons of convenience in natural coordinates τ and s (see also Prob. 6.6.2):

$$\int_0^l \left(\frac{d\tau}{ds}\right)^2 ds \to \text{minimum}$$

$$\int_0^l \cos\tau\, ds = b - a$$

$$\int_0^l \sin\tau\, ds = y_b - y_a$$

with the initial condition $\tau(0) = \tau_a$ and the terminal condition $\tau(l) = \tau_b$. (Remember that $\kappa = d\tau/ds$.)

We have

$$h^I = \lambda_o \left(\frac{d\tau}{ds}\right)^2 + \lambda_1 \cos\tau + \lambda_2 \sin\tau,$$

and hence

$$h^I_\tau = -\lambda_1 \sin\tau + \lambda_2 \cos\tau, \qquad h^I_{\tau'} = 2\lambda_o \frac{d\tau}{ds}.$$

Thus we obtain for the Mayer equation of this isoperimetric problem,

$$-\lambda_1 \sin\tau + \lambda_2 \cos\tau - 2\lambda_o \frac{d^2\tau}{ds^2} = 0.$$

Integration with respect to s yields

$$-\lambda_1 y + \lambda_2 x - 2\lambda_o \frac{d\tau}{ds} + C = 0,$$

or
$$\frac{d\tau}{ds} = \mu_1 x + \mu_2 y + \gamma,$$

where μ_1, μ_2, and γ are suitably chosen in terms of λ_o, λ_1, λ_2, and C.

This equation tells us already that *all points of inflection of the elastic rod have to lie on a straight line* $\mu_1 x + \mu_2 y + \gamma = 0$.

We rotate and translate the coordinate system so that $\mu_1 x + \mu_2 y + \gamma = 0$ becomes the new $\bar{y}$ axis, and we change the scale on the new $\bar{x}$ axis so that we may write in the new $\bar{x},\bar{y}$ coordinate system

$$\frac{d\tau}{ds} = \bar{x}.$$

Since
$$\frac{d\tau}{ds} = \frac{1}{\rho} = \frac{d}{dx}\left(\frac{\bar{y}'}{\sqrt{1 + \bar{y}'^2}}\right),$$

we have
$$\frac{\bar{y}'}{\sqrt{1 + \bar{y}'^2}} = \frac{\bar{x}^2}{2} + a,$$

and hence
$$\bar{y} = \bar{y}(\bar{x}) \equiv \int \frac{(\bar{x}^2/2) + a}{\sqrt{1 - [(\bar{x}^2/2) + a]^2}}\, d\bar{x} + b.$$

This representation of $\bar{y} = \bar{y}(\bar{x})$ contains two arbitrary integration constants a, b. Three more constants have been swallowed up by the rotation and scale change and may be recovered by reverting the process. Thus we have, altogether, five arbitrary constants, which are also needed to satisfy the boundary conditions $y(a) = y_a$, $y(b) = y_b$, $y'(a) = \tan \tau_a$, $y'(b) = \tan \tau_b$ and to make sure that the length of the solution curve is l.

PROBLEMS 6.6

*1. Formulate suitable conditions to ensure the validity of the operations in (6.6.2).

*2. Give an interpretation of weak and strong relative minima when the variational problem is formulated in natural coordinates such as the problem at the end of this section.

3. Solve the isoperimetric problem

$$\int_0^1 (y'^2(x) - y^2(x))\, dx \to \text{minimum},$$

$$\int_0^1 \sqrt{1 + y'^2(x)}\, dx = \sqrt{2},$$

with the boundary conditions $y(0) = 0$, $y(1) = 1$. (See also problem 6.4.9.)

6.7 PROOF OF THE FUNDAMENTAL THEOREM ON UNDERDETERMINED SYSTEMS

The proof of Lemma 6.2.2 that we are going to present here follows, in outline, a paper by J. Radon.†

† J. Radon, Zum Problem von Lagrange, *Ab. Math. Seminar Hamburg. Univ.*, vol. 6, pp. 273–299, 1928.

Suppose that the statement A represents a sufficient condition for the point $(b,\hat{y}(b))$ to be an interior point of the attainable set $\mathcal{A}$ of [(6.2.1), (6.2.2a)] in the interval $[a,b]$ (see Definition 6.2) and that B represents the statement "$(b,\hat{y}(b))$ is an interior point of $\mathcal{A}$." Since $A \Rightarrow B$ is equivalent to its contrapositive $\bar{B} \Rightarrow \bar{A}$, we see that $\bar{A}$ ("*not* A") represents a necessary condition for $(b,\hat{y}(b))$ to be a *boundary point* of $\mathcal{A}$.

In the proof that is to follow, we shall first establish a sufficient condition for $(b,\hat{y}(b))$ to be an interior point of $\mathcal{A}$ and then, by negation, convert this condition into a necessary condition for $(b,\hat{y}(b))$ to be a boundary point of $\mathcal{A}$.

For easy reference, let us restate the Mayer problem:

$$\hat{\varphi}(x,\hat{y},\hat{y}') = 0 \tag{6.7.1}$$

$$\hat{y}(a) = \hat{y}_a \tag{6.7.2a}$$

$$\hat{y}(b) = \hat{y}_b \tag{6.7.2b}$$

$$y_0(b) \rightarrow \text{minimum.} \tag{6.7.3}$$

We remind the reader of our notational convention whereby Greek letters denote $(\mu + 1)$-dimensional vectors and Roman letters denote $(n + 1)$-dimensional vectors. Greek subscripts range from 0 to μ, and Roman subscripts range from 0 to n.

We shall present the proof of Lemma 6.2.2 for the case where it is assumed that $\hat{y} = \hat{y}(x) \in C^1[a,b]^{n+1}$ and then point out how to adapt this proof so that it becomes applicable to the case where $\hat{y}'$ has finitely many jump discontinuities in $[a,b]^{n+1}$.

Suppose that $\hat{y} = \hat{y}(x) \in C^1[a,b]^{n+1}$ is a uniquely defined solution of [(6.7.1), (6.7.2a and b)] with $y_o(b) = y_o^b$ and satisfies (6.2.6). Then

$$\varphi_{\nu y'_k}(x,\hat{y}(x),\hat{y}'(x)) \equiv a_{\nu k}(x) \in C[a,b],$$

and by (6.2.6),

$$\text{rank } (a_{\nu k}(x)) = \mu + 1 \qquad \text{for all } x \in [a,b].$$

Hence we can find $(n - \mu)(n + 1)$ functions

$$a_{\mu+j,k}(x) \in C[a,b], \qquad j = 1, 2, \ldots, n - \mu,$$
$$k = 0, \ldots, n,$$

such that $\quad \text{rank } (a_{ik}(x)) = n + 1 \qquad \text{for all } x \in [a,b].$

(See Theorem A6.11.)

Let

$$\varphi_{\mu+j}(x,\hat{y},\hat{y}') = a_{\mu+j,o}(x)y'_o + \cdots + a_{\mu+j,n}(x)y'_n, \qquad j = 1, 2, \ldots, n - \mu,$$

and let

$$\hat{f} = (f_o, \ldots, f_n) = (\varphi_o, \ldots, \varphi_\mu, \varphi_{\mu+1}, \ldots, \varphi_n).$$

Then $\hat{f}, \hat{f}_{\hat{y}}, \hat{f}_{\hat{y}'}$ are continuous for $x \in [a,b]$ and for all $-\infty < y_i < \infty$, $-\infty < y_i' < \infty$ because of (6.2.5) and the above definition of $\varphi_{\mu+j}$, and we have

$$\operatorname{rank}\left(\frac{\partial f_i}{\partial y_k'}\right)_{\hat{y}=\hat{y}(x)} = n + 1. \tag{6.7.4}$$

If we define $n - \mu$ functions $w_{\mu+j}(x)$ by

$$w_{\mu+j}(x) = \varphi_{\mu+j}(x,\hat{y}(x),\hat{y}'(x)), \qquad j = 1, 2, \ldots, n - \mu,$$

we see that $\hat{y} = \hat{y}(x)$ satisfies the following system of $n + 1$ first-order differential equations:

$$\hat{f}(x,\hat{y},\hat{y}') = \hat{w}(x) \tag{6.7.5}$$

where $$\hat{w}(x) = (0,0, \ldots,0,w_{\mu+1}(x), \ldots,w_n(x)).$$

We call (6.7.5) the *augmented system.*

Obviously, the ideas of Chap. 1, which were based on the existence of a linear space of admissible variations and led to a first necessary condition for a minimum for variational problems without constraints, are of no relevance here. For a linear space of admissible variations $\mathcal{H}$ for the Mayer problem to exist, the following would have to be true: If $\hat{y} = \hat{y}(x)$ is a solution of the Mayer problem, then $\hat{y} = \hat{y}(x) + \hat{h}(x)$ would have to satisfy the constraining equations and the boundary conditions for $\hat{h} \in \mathcal{H}$ since the space of competing functions is made up of functions that satisfy the constraining equations and the boundary conditions. Even if there were such a function $\hat{h}$, $\hat{y} = \hat{y}(x) + t\hat{h}(x)$ for all $t \in R$ would, in general, not be a solution of the constraining equations.

The variations which we shall construct in the sequel are, in a sense, of a more specialized nature than the variations that were considered in Chaps. 1 and 2, but they will suffice for our purpose. We shall utilize the augmented system (6.7.5) to generate such variations in the following manner:

Let $Z = Z(x)$ denote the $(n + 1)\times(n+1)$ matrix function

$$Z \equiv \begin{pmatrix} 0 & 0 & \cdots & 0 \\ \vdots & & & \\ 0 & 0 & \cdots & 0 \\ & & & \\ z_{\mu+1}^{o} & z_{\mu+1}^{1} & \cdots & z_{\mu+1}^{n} \\ \vdots & & & \\ z_n^{o} & z_n^{1} & \cdots & z_n^{n} \end{pmatrix}, \qquad z_k^{i} \in C[a,b], \tag{6.7.6}$$

where the functions $z_k^{i} = z_k^{i}(x)$ are chosen quite arbitrarily, and let $\hat{e}$ denote the row vector

$$\hat{e} = (e_o, \ldots, e_n),$$

where the e_i are real numbers.

We shall now disturb the augmented system (6.7.5) by changing the right side in the following manner:

$$\hat{f}(x,\hat{y},\hat{y}') = \hat{w}(x) + \hat{e}Z(x)^T, \tag{6.7.7}$$

where the superscript T denotes the transpose. (See Prob. 6.7.1.)

In view of (6.7.4), we can solve the *disturbed augmented system* (6.7.7) for $\hat{y}'$ in a domain of the $(x,\hat{y},\hat{y}')$ space that contains all lineal elements of $\hat{y} = \hat{y}(x)$. For $\hat{e} = 0$, the system has, by hypothesis, the unique solution $\hat{y} = \hat{y}(x)$, which satisfies the initial conditions (6.7.2*a*). If the $|e_i|$ are sufficiently small, then (6.7.7) possesses unique solutions

$$\hat{y} = \hat{y}(x,\hat{e}) \qquad \text{where } \hat{y}(x,0) = \hat{y}(x) \tag{6.7.8}$$

which satisfy the initial conditions (6.7.2*a*)

$$\hat{y}(a,\hat{e}) = \hat{y}_a. \tag{6.7.9}$$

We have from a theorem of *Gronwall*† that $\hat{y} = \hat{y}(x, \hat{e})$, $\hat{y}' = \hat{y}'(x,\hat{e})$, $\hat{y}_{\hat{e}} = \hat{y}_{\hat{e}}(x,\hat{e})$, $\hat{y}'_{\hat{e}} = \hat{y}'_{\hat{e}}(x,\hat{e})$ are continuous for all $x \in [a,b]$ and for all $\hat{e}$, provided that $|e_i| < \delta$ for some $\delta > 0$. Since $\hat{y}(x,0) = \hat{y}(x)$, we know that $\hat{y}(b,0) = \hat{y}_b$ with $y_o(b) = y_o{}^b$.

Suppose that for some choice of the matrix Z,

$$\left.\frac{\partial(y_o(x,\hat{e}), \ldots, y_n(x,\hat{e}))}{\partial(e_o, \ldots, e_n)}\right|_{\substack{\hat{e}=0\\ x=b}} \neq 0. \tag{6.7.10}$$

Then, by the theorem on implicit functions,‡ we can find for any given $\hat{\Delta} = (\Delta_o, \ldots, \Delta_n)$ a vector $\hat{e}$ such that

$$\hat{y}(b,\hat{e}) = \hat{y}_b + \hat{\Delta},$$

provided that $\max\limits_{(i)} |\Delta_i|$ is sufficiently small.

This means that $(b,\hat{y}_b)$ is an interior point of $\mathfrak{A}$ in the sense of Definition 6.2. Hence:

Lemma 6.7.1 *If* $\hat{y} = \hat{y}(x) \in C^1[a,b]^{n+1}$ *is a solution of* [(6.7.1), (6.7.2a

† T. H. Gronwall, Note on the Derivative with Respect to a Parameter of the Solutions of a System of Differential Equations, *Ann. Math.*, ser. 2, vol. 20, pp. 292–296, 1919.

‡ T. M. Apostol, "Mathematical Analysis," p. 147, Addison-Wesley Publishing Company, Inc., Reading, Mass., 1957.

and b)] for which (6.7.4) is satisfied and if for some choice of $Z = Z(x)$,

$$\left.\frac{\partial(y_o(x,\hat{e}), \ldots, y_n(x,\hat{e}))}{\partial(e_o, \ldots, e_n)}\right|_{\substack{\hat{e}=0 \\ x=b}} \neq 0,$$

where $\hat{y} = \hat{y}(x,\hat{e})$ *is the solution of the disturbed augmented system (6.7.7), satisfies the initial condition (6.7.2a), and contains* $\hat{y} = \hat{y}(x)$ *for* $\hat{e} = 0$, *then* $(b,\hat{y}(b))$ *is an interior point of the attainable set* $\mathfrak{A}$.

From this lemma we obtain immediately:

Corollary to Lemma 6.7.1 *For* $(b,\hat{y}(b))$ *to be a boundary point of the attainable set* $\mathfrak{A}$, *it is necessary that*

$$\left.\frac{\partial(y_o(x,\hat{e}), \ldots, y_n(x,\hat{e}))}{\partial(e_o, \ldots, e_n)}\right|_{\substack{\hat{e}=0 \\ x=b}} = 0 \tag{6.7.11}$$

for all possible choices of $Z = Z(x)$.

["All possible choices of $Z = Z(x)$" means that all possible continuous functions in all possible combinations are chosen as entries in the last $n - \mu$ rows of the matrix Z that is defined in (6.7.6).]

The necessary condition for a solution of the Mayer problem which is stated in this corollary is not very practical because it makes reference to the arbitrary matrix Z and to the inaccessible functions $\hat{y} = \hat{y}(x,\hat{e})$. It is the goal of the subsequent investigation to eliminate these extraneous references.

We denote by V the *variation matrix*

$$V = V(x) \equiv \begin{pmatrix} \dfrac{\partial y_o(x,\hat{e})}{\partial e_o} & \cdots & \dfrac{\partial y_o(x,\hat{e})}{\partial e_n} \\ \vdots & & \\ \dfrac{\partial y_n(x,\hat{e})}{\partial e_o} & \cdots & \dfrac{\partial y_n(x,\hat{e})}{\partial e_n} \end{pmatrix}_{\hat{e}=0}$$

$$\equiv \begin{pmatrix} v_o{}^o(x) & \cdots & v_o{}^n(x) \\ \vdots & & \\ v_n{}^o(x) & \cdots & v_n{}^n(x) \end{pmatrix}. \tag{6.7.12}$$

Then we can write (6.7.11) in the simple form

$$\det |V(b)| = 0. \tag{6.7.13}$$

We can obtain direct access to the matrix V without having to resort to the (unknown) family of solutions $\hat{y} = \hat{y}(x,\hat{e})$ of (6.7.7) by substituting

$\hat{y} = \hat{y}(x,\hat{e})$ into (6.7.7) and noting that the resulting equation is an identity in $\hat{e}$:

$$\hat{f}(x,\hat{y}(x,\hat{e}),\hat{y}'(x,\hat{e})) = \hat{w}(x) + \hat{e}Z(x)^T. \tag{6.7.14}$$

Consequently,
$$\frac{d}{de_m}\hat{f}(x,\hat{y}(x,\hat{e}),\hat{y}'(x,\hat{e})) = \frac{d}{de_m}(\hat{e}Z(x)^T). \tag{6.7.15}$$

We obtain

$$\frac{d}{de_m}(\hat{e}Z(x)^T) = (0, \ldots, 0, z_{\mu+1}^m, \ldots, z_n^m)$$

(see Prob. 6.7.2), and

$$\frac{d}{de_m}\hat{f}(x,\hat{y}(x,\hat{e}),\hat{y}'(x,\hat{e})) = \sum_{k=0}^{n}\left[\hat{f}_{y_k}\frac{\partial y_k}{\partial e_m} + \hat{f}_{y'_k}\frac{\partial y'_k}{\partial e_m}\right].$$

Hence, we can write (6.7.15) for $\hat{e} = 0$ as

$$F_y V + F_{y'} V' = Z \tag{6.7.16}$$

(see Probs. 6.7.3 and 6.7.4), whereby

$$F_y = \begin{pmatrix} \frac{\partial f_o}{\partial y_o} & \cdots & \frac{\partial f_o}{\partial y_n} \\ \vdots & & \\ \frac{\partial f_n}{\partial y_o} & \cdots & \frac{\partial f_n}{\partial y_n} \end{pmatrix}_{\hat{e}=0}, \qquad F_{y'} = \begin{pmatrix} \frac{\partial f_o}{\partial y'_o} & \cdots & \frac{\partial f_o}{\partial y'_n} \\ \vdots & & \\ \frac{\partial f_n}{\partial y'_o} & \cdots & \frac{\partial f_n}{\partial y'_n} \end{pmatrix}_{\hat{e}=0} \tag{6.7.17}$$

and where, in accord with (6.7.12),

$$V' = \begin{pmatrix} v_o^{o\prime} & \cdots & v_o^{n\prime} \\ \vdots & & \\ v_n^{o\prime} & \cdots & v_n^{n\prime} \end{pmatrix}.$$

(The matrix equation (6.7.16)—the so-called *variation equation*—reads, when written out in components,

$$\sum_{k=0}^{n}[f_{iy_k}v_k^m + f_{iy'_k}v_k^{m\prime}] = z_i^m, \qquad m = 0, 1, \ldots, n; \qquad i = 0, 1, \ldots, n,$$

whereby $z_\nu^m = 0$ for $\nu = 0, 1, \ldots, \mu$.)

By (6.7.4), rank $(F_{y'}) = n + 1$, that is, $F_{y'}$ is nonsingular and possesses an inverse $F_{y'}^{-1}$. Pre-multiplication of (6.7.16) by $F_{y'}^{-1}$ yields

$$V' = F_{y'}^{-1}Z - F_{y'}^{-1}F_yV.$$

This is a linear matrix differential equation for the variation matrix V in normal form ($n + 1$ systems of $n + 1$ linear differential equations each) with continuous coefficients in $[a,b]$. Hence the solution $V = V(x)$ is uniquely determined by some initial conditions $V(a) = V_a$ and exists throughout the interval $[a,b]$.

In order to find the initial values V_a, we recall that $\hat{y} = y(x,\hat{e})$ satisfies the initial conditions $\hat{y}(a,\hat{e}) = \hat{y}_a$. Hence $[\partial y_k(a,\hat{e})/\partial e_m]|_{\hat{e}=0} = 0$, and we have $V_a = 0$. This gives us the initial conditions

$$V(a) = 0 \tag{6.7.18}$$

for the solutions $V = V(x)$ of the variation equation (6.7.16).

Let us consider the following linear vector differential equation of first order, the adjoint equation to (6.7.16),

$$\frac{d}{dx}(\hat{\lambda}F_{y'}) - \hat{\lambda}F_y = 0 \tag{6.7.19}$$

for $\hat{\lambda} = \hat{\lambda}(x) = (\lambda_o(x), \ldots, \lambda_n(x))$. [We make here an exception to our notational convention and use the Greek letter $\hat{\lambda}$ to denote this particular $n + 1$ vector. There are two reasons: First, it is customary to denote this particular vector by $\hat{\lambda}$, and second, it will turn out later that the solutions of (6.7.19) we are interested in are such that their last $n - \mu$ components are zero and hence are, in effect, $\mu + 1$ vectors.]

Equation (6.7.19) has $n + 1$ linearly independent solutions $\hat{\lambda}^o, \ldots, \hat{\lambda}^n$ which exist throughout the interval $[a,b]$ and $\hat{\lambda}^k \in C[a,b]^{n+1}$. (See Prob. 6.7.11.)

Let us define the $(n + 1)\times(n + 1)$ matrix

$$\Lambda = \begin{pmatrix} \hat{\lambda}^o \\ \vdots \\ \hat{\lambda}^n \end{pmatrix}. \tag{6.7.20}$$

We pre-multiply (6.7.16) by Λ and obtain

$$\Lambda F_y V + \Lambda F_{y'} V' = \Lambda Z.$$

From (6.7.19),

$$\Lambda F_y = \frac{d}{dx}(\Lambda F_{y'}).$$

Since
$$\frac{d}{dx}(\Lambda F_{y'} V) = \frac{d}{dx}(\Lambda F_{y'}) V + \Lambda F_{y'} V'$$

(see Prob. 6.7.5), we finally obtain

$$\frac{d}{dx}(\Lambda F_{y'} V) = \Lambda Z.$$

Integration from a to b yields, in view of the initial conditions (6.7.18),

$$\Lambda(b)F_{y'}(b)V(b) = \int_a^b \Lambda(x)Z(x)\,dx. \tag{6.7.21}$$

By hypothesis, $\det|F_{y'}(b)| \neq 0$ since rank $(F_{y'}) = n+1$ for all $x \in [a,b]$. $\det|\Lambda(b)| \neq 0$ since the rows of Λ are linearly independent solutions in $[a,b]$ of (6.7.19). Hence $\det|V(b)| = 0$ if and only if $\det|\int_a^b \Lambda(x)Z(x)\,dx| = 0$.

We state this intermediate result as a lemma:

Lemma 6.7.2 *$\det|V(b)| = 0$ for all possible choices of $Z = Z(x)$ if and only if*

$$\det\left|\int_a^b \Lambda(x)Z(x)\,dx\right| = 0 \tag{6.7.22}$$

for all possible choices of $Z = Z(x)$.

By means of this intermediate result, we have shifted the investigation from condition (6.7.13) to (6.7.22).

In the next step, we shall achieve a considerable simplification of condition (6.7.22). We shall show:

Lemma 6.7.3 *If $\det|\int_a^b \Lambda(x)Z(x)\,dx| = 0$ for all possible choices of $Z = Z(x)$, then there exists a nontrivial solution $\hat{\lambda} = \hat{\lambda}(x)$ of (6.7.19) such that*

$$\int_a^b \hat{\lambda}(x)\hat{z}^T(x)\,dx = 0 \tag{6.7.23}$$

for any vector $\hat{z} = \hat{z}(x) = (0,\ldots,0,z_{\mu+1}(x),\ldots,z_n(x))$, $z_k \in C[a,b]$.

Proof: We have from (6.7.22) that

$$\hat{a}\int_a^b \Lambda(x)\,Z(x)\,dx = 0$$

has a nontrivial solution $\hat{a} = (a_o,\ldots,a_n)$ that is determined by at most n of these $n+1$ equations. At least one, say the jth, is superfluous. This means, in turn, that the entries in the jth column of Z are not utilized in the determination of $\hat{a}$ and can be replaced by any column $\hat{z}^T$ with $\hat{z} = (0,\ldots,0, z_{\mu+1}(x),\ldots,z_n(x))$. Since

$$\hat{\lambda} = \hat{a}\Lambda$$

is a nontrivial solution of (6.7.19), we have

$$\hat{a}\int_a^b \Lambda(x)\hat{z}^T(x)\,dx = \int_a^b \hat{\lambda}(x)\,\hat{z}^T(x)\,dx = 0$$

for any such vector $\hat{z} = (0, \ldots, 0, z_{\mu+1}(x), \ldots, z_n(x))$.

Let us choose

$$\hat{z} = (0, \ldots, 0, z(x), 0, \ldots, 0),$$

where $z = z(x)$ stands in the $(\mu + 1)$th place and is any continuous function in $[a,b]$. Then, by Lemma 6.7.3,

$$\int_a^b \lambda_{\mu+1}(x)z(x)\,dx = 0$$

for all continuous functions $z = z(x)$, and we obtain from the fundamental lemma of the calculus of variations (see Prob. 2.4.1) that, by necessity,

$$\lambda_{\mu+1}(x) = 0 \qquad \text{for all } x \in [a,b].$$

Similarly, we can see that

$$\lambda_{\mu+2}(x) = \cdots = \lambda_n(x) = 0 \qquad \text{for all } x \in [a,b].$$

(See Prob. 6.7.6.)

However, $(\lambda_o, \ldots, \lambda_\mu) \neq (0, \ldots, 0)$ since $\hat{\lambda} = (\lambda_o, \ldots, \lambda_\mu, 0, \ldots, 0)$ is one of the linearly independent solutions of (6.7.19) or a linear combination thereof.

Thus we can state that if $(b, \hat{y}(b))$ is a boundary point of the attainable set $\mathcal{A}$, then, by necessity, there exists a nontrivial solution $\hat{\lambda} = (\lambda_o, \ldots, \lambda_\mu, 0, \ldots, 0) \in C[a,b]^{n+1}$ of

$$\frac{d}{dx}(\hat{\lambda}F_{y'}) - \hat{\lambda}F_y = 0.$$

We observe that

$$\hat{\lambda}F_{y'} = (\lambda_o, \ldots, \lambda_\mu, 0, \ldots, 0)\begin{pmatrix} f_{oy'_o} & \cdots & f_{oy'_n} \\ \vdots & & \\ f_{ny'_o} & \cdots & f_{ny'_n} \end{pmatrix}$$

$$= (\lambda_o, \ldots, \lambda_\mu)\begin{pmatrix} \varphi_{oy'_o} & \cdots & \varphi_{oy'_n} \\ \vdots & & \\ \varphi_{\mu y'_o} & \cdots & \varphi_{\mu y'_n} \end{pmatrix}.$$

Hence, (6.7.19) becomes (6.2.9)—the Mayer equation—and the lemma is proved for the case where $\hat{y} = \hat{y}(x) \in C^1[a,b]^{n+1}$.

Let us now consider how this proof can be adapted to apply to the case where $\hat{y} = \hat{y}(x) \in C_{sP}{}^1[a,b]^{n+1}$, where $P = \{x_1, \ldots, x_r\}$, $x_i \in (a,b)$. [We may assume without loss of generality that $\hat{y}'(a) = \hat{y}'(a+0)$, $\hat{y}'(b) = \hat{y}'(b-0)$.]

In this case,

$$\varphi_{\nu y'_k}(x,\hat{y}(x),\hat{y}'(x)) = a_{\nu k}(x) \in C_{sP}[a,b],$$

and we can augment the matrix $(a_{\nu k})$ by elements $a_{\mu+j,k} \in C_{sP}[a,b]$ so that

$$\text{rank } (a_{ik}) = n+1 \qquad \text{for all } x \in [a,b] \text{ and } a_{ik} \in C_{sP}[a,b].$$

Then $\hat{w} = \hat{w}(x) \in C_{sP}[a,b]^{n+1}$ [see (6.7.5)], and the augmented system $f(x,\hat{y},\hat{y}') = \hat{w}(x)$ is satisfied by sections. For the entries of Z, we again use continuous functions, but the solutions $\hat{y} = \hat{y}(x,\hat{e})$ of the disturbed augmented system (6.7.7) will now be sectionally smooth as functions of x, rather than smooth. We also have for the elements of V that $v_i{}^k \in C_{sP}{}^1[a,b]$, that is, that the variation equation (6.7.16) is satisfied by sections. Since the entries of $F_{y'}$ and F_y are in $C_{sP}[a,b]$, we have to solve (6.7.19) by sections and piece the solutions together in such a manner that $\hat{\lambda}F_{y'}$ remains continuous throughout $[a,b]$. (See Prob. 6.7.12.) Then

$$\hat{\lambda}F_{y'} = \int_a^x \hat{\lambda}(t)F_y(t)\,dt + \hat{C} \tag{6.7.24}$$

for some constant vector $\hat{C}$. Hence

$$\Lambda F_{y'} = \int_a^x \Lambda(t)F_y(t)\,dt + C$$

where the rows of Λ are linearly independent vectors in each subinterval $[x_i, x_{i+1}]$ and where C is a constant matrix.

With the notation

$$\int_a^x \Lambda(t)F_y(t)\,dt = \Phi(x),$$

we have

$$\int_a^b \Lambda(x)[F_y(x)V(x) + F_{y'}(x)V'(x)]\,dx = \Phi(x)V(x)\Big|_a^b$$
$$+ \int_a^b [\Lambda(x)F_{y'}(x) - \Phi(x)]V'(x)\,dx = [\Phi(b) + C]V(b).$$

(See Prob. 6.7.8.)

Since $\Phi(b) = \int_a^b \Lambda(x) F_y(x)\, dx = \Lambda(b) F_{y'}(b) - C$, we have

$$\int_a^b \Lambda(x)[F_y(x) V(x) + F_{y'}(x) V'(x)]\, dx = \Lambda(b) F_{y'}(b) V(b).$$

On the other hand, we obtain, after pre-multiplication of (6.7.16) by Λ and integration from a to b:

$$\int_a^b \Lambda(x)[F_y(x) V(x) + F_{y'}(x) V'(x)]\, dx = \int_a^b \Lambda(x) Z(x)\, dx.$$

Hence we again have

$$\Lambda(b) F_{y'}(b) V(b) = \int_a^b \Lambda(x) Z(x)\, dx,$$

which is (6.7.21). We continue the argument as before, with (6.7.24) taking the place of (6.7.19), and obtain $\lambda_{\mu+j}(x) = 0$, $j = 1, 2, \ldots, n-\mu$, everywhere except on P.

PROBLEMS 6.7

*1. Write the disturbed augmented system (6.7.7) in components.

*2. Show that

$$\frac{d}{de_m}(\hat{e} Z(x)^T) = (0, \ldots, 0, z_{\mu+1}^m, \ldots, z_n^m).$$

*3. Show that

$$\frac{d}{de_m}\hat{f}(x, \hat{y}(x,\hat{e}), \hat{y}'(x,\hat{e})) = \sum_{k=0}^{n}\left[\hat{f}_{y_k}\frac{\partial y_k}{\partial e_m} + \hat{f}_{y'_k}\frac{\partial y'_k}{\partial e_m}\right].$$

*4. (*a*) Show that the solutions V of (6.7.16) define a linear space.

(*b*) Impose suitable differentiability conditions and show that

$$\hat{y}(x,\hat{e}) = \hat{y}(x) + \hat{e} V^T + \cdots.$$

(*c*) Explain the name "variation matrix" in the light of the results in parts *a* and *b*.

*5. (*a*) Let $A(x)$, $B(x)$ denote square matrices of the same order, the elements of which are differentiable functions of x. Show that

$$\frac{d}{dx}[A(x) B(x)] = A'(x) B(x) + A(x) B'(x).$$

(*b*) Generalize this result to $(d/dx)\,[A_1(x) \cdots A_r(x)]$, where the A_i are square matrices of the same order and have differentiable functions of x as elements.

(*c*) Show that

$$\frac{d}{dx}(\Lambda(x) F_{y'}(x) V(x)) = \frac{d}{dx}(\Lambda(x) F_{y'}(x)) V(x) + \Lambda(x) F_{y'}(x) V'(x)$$

where V, $F_{y'}$, Λ are defined in (6.7.12), (6.7.17), (6.7.20).

*6. Show that $\lambda_{\mu+2}(x) = \cdots = \lambda_n(x) = 0$ for all $x \in [a,b]$.

7. Show that $\hat{C} = \hat{\lambda}(a)F_{y'}(a)$ in $\hat{\lambda}(x)F_{y'}(x) = \int_a^x \hat{\lambda}(t)F_y(t)\,dt + \hat{C}$.

*8. Show that

$$\int_a^b \Lambda(x)F_y(x)V(x)\,dx = \Phi(x)V(x)\Big|_a^b - \int_a^b \Phi(x)V'(x)\,dx,$$

where $\Phi(x) = \int_a^x \Lambda(t)F_y(t)\,dt$. (See also problem 5.)

9. Apply the theorem on implicit functions to (6.7.5) to obtain

$$\hat{y}' = \hat{F}(x,\hat{y},\hat{w}),$$

and show that $\hat{F}_x$, $\hat{F}_{y_k}$, $\hat{F}_{w_j}$ are continuous functions in a suitable chosen domain of $(x,y_o, \ldots,y_n,w_{\mu+1}, \ldots,w_n)$ space.

10. Given the constraining equation

$$\varphi \equiv \alpha(x)y' + \beta(x)z' = 1,$$

where $\alpha(x) = \begin{cases} 0 & \text{for } x < 0 \\ 1 & \text{for } x \geq 0, \end{cases} \qquad \beta(x) = \begin{cases} 1 & \text{for } x < 0 \\ 0 & \text{for } x \geq 0. \end{cases}$

Find rank $(\partial\varphi/\partial y', \partial\varphi/\partial z')$.

*11. The elements of F_y and $F_{y'}$ are continuous functions of x in $[a,b]$ and det $|F_{y'}| \neq 0$ for all $x \in [a,b]$. Show that (6.7.19) has $n + 1$ linearly independent solution vectors $\hat{\lambda}^o, \ldots, \hat{\lambda}^n$ which are continuous in $[a,b]^{n+1}$. If the elements of F_y and $F_{y'}$ have a continuous derivative in $[a,b]^{n+1}$ (for example, if $\hat{f} \in C^2$ and $\hat{y} = \hat{y}(x) \in C^2[a,b]^{n+1}$), show that $\hat{\lambda}'$ is continuous in $[a,b]^{n+1}$. (*Hint*: consider $(d/dx)\,\hat{\mu} - \hat{\mu}C = 0$, where $\hat{\mu} = \hat{\lambda}F_{y'}$ and $C = F_{y'}^{-1}F_{y}$.)

*12. If the elements of F_y and $F_{y'}$ are in $C_{sp}[a,b]$, show that there are $n + 1$ vectors $\hat{\lambda}^o, \ldots, \hat{\lambda}^n \in C_{sp}[a,b]^{n+1}$ which satisfy (6.7.19) by sections, are linearly independent in each subinterval $[x_k,x_{k+1}]$, and are so that $\hat{\lambda}^iF_{y'} \in C[a,b]^{n+1}$ for $i = 0, \ldots, n$.

13. Suppose that the constraining differential equations $\hat{\varphi}(x,\hat{y},\hat{y}') = 0$ can be solved for $y_1', \ldots, y_\mu'$ in the entire interval $[a,b]$. Show that even if $\hat{y} = \hat{y}(x) \in C_s^1[a,b]^{n+1}$, the multipliers $\lambda_o, \ldots, \lambda_\mu$ are continuous in $[a,b]$.

6.8 THE MAYER PROBLEM WITH A VARIABLE ENDPOINT

We shall now consider the Mayer problem and allow the endpoint to lie on a given manifold rather than prescribe the coordinates of the endpoint, as we did before.

Let such a *terminal manifold* be given by

$$\begin{aligned} \psi_1(x,y_1, \ldots,y_n) &= 0 \\ &\vdots \\ \psi_k(x,y_1, \ldots,y_n) &= 0, \qquad k \leq n. \end{aligned}$$

If ψ_{jx}, ψ_{jy_i} are continuous for all $(x,y_1, \ldots,y_n)$ and if

$$\operatorname{rank}\left(\frac{\partial\psi_j}{\partial y_i}\right)_{\substack{j=1,\ldots,k \\ i=1,\ldots,n}} = k,$$

then the above equations are said to define an $(n + 1 - k)$ dimensional manifold in $(x,y_1, \ldots,y_n)$ space.

At no extra cost, we obtain a notational advantage and achieve greater generality if we now require that instead of $y_o(b) \to$ minimum, a given function $\psi_o(b,\hat{y}(b))$ be minimized. [The case of Sec. 6.2 is contained in this more general case for $\psi_o(b,\hat{y}(b)) \equiv y_o(b)$.] We shall also allow the ψ_j to be functions of y_o in addition to $x, y_1, \ldots, y_n$.

Then we can formulate the Mayer problem with a variable endpoint as follows: To be found is an $(n + 1)$-dimensional vector function $\hat{y} = \hat{y}(x)$ such that it satisfies the constraining differential equations

$$\hat{\varphi}(x,\hat{y},\hat{y}') = 0, \tag{6.8.1}$$

the initial conditions

$$\hat{y}(a) = \hat{y}_a, \qquad \hat{y}_a = (0,y_1{}^a, \ldots,y_n{}^a), \tag{6.8.2}$$

and the terminal conditions

$$\begin{aligned} &\psi_o(b,\hat{y}(b)) \to \text{minimum} \\ &\psi_j(b,\hat{y}(b)) = 0, \qquad j = 1, 2, \ldots, k \leq n, \end{aligned} \tag{6.8.3}$$

for some $x = b$.

We remind the reader of the notational convention that we adopted in Sec. 6.2. In addition to this, we shall agree that from now on, $\hat{\psi}$, $\hat{\Psi}$, $\hat{\nu}$ will denote $(k + 1)$-dimensional vectors $\hat{\psi} = (\psi_o, \ldots,\psi_k)$, etc., and that the subscript j will range from 0 to k (or from 1 to k). We shall reserve the subscript ρ for the case $\rho = 1, 2, \ldots, n - r$.

We retain the hypotheses (6.2.5) and (6.2.6), under which we proved Lemma 6.2.2, with the additional provision that the interval $[a,b]$ be replaced by the half-line $x \geq a$ or $x \leq a$ and the hypothesis that

$$\text{rank}\left(\frac{\partial \psi_j}{\partial y_i}\right)_{\substack{j=1,\ldots,k \\ i=1,\ldots,n}} = k.$$

We add two more hypotheses, namely,

$$\hat{\psi} \in C^1\{(x,\hat{y}) \mid x,y_i \in (-\infty,\infty), i = 0, 1, \ldots, n\} \tag{6.8.4}$$

and

$$\text{rank}\left(\frac{\partial \psi_j}{\partial y_i}\right) = k + 1, \qquad j = 0, 1, \ldots, k; i = 0, 1, \ldots, n. \tag{6.8.5}$$

We assume that $\hat{y} = \hat{y}(x) \in C^1[a,b]^{n+1}$ is the solution of the Mayer problem with a variable endpoint [(6.8.1) to (6.8.3)]. Then $\hat{y} = \hat{y}(x)$

will emanate from the point $(a,\hat{y}_a)$ and terminate at some point $(b,\hat{y}_b)$, so that $\psi_j(b,\hat{y}_b) = 0$, $j = 1, 2, \ldots, k$, and so that $\psi_o(b,\hat{y}_b)$ is a relative minimum.

We define now the attainable set $\mathfrak{A}_T$ on the terminal manifold $\psi_j(x,\hat{y}) = 0$, $j = 1, 2, \ldots, k$, as the set of all points $(x,y_o, \ldots,y_n)$ for which $\psi_j(x,\hat{y}) = 0$, $j = 1, 2, \ldots, k$, that can be reached by a solution $\hat{y} = \hat{y}(x)$ of the initial-value problem [(6.8.1), (6.8.2)].

Suppose that $(b,\hat{y}_b)$ is an interior point of $\mathfrak{A}_T$. Then all points $(x,\hat{y})$ on the terminal manifold for which $(x - b)^2 + \sum_{i=0}^{n} (y_i - y_i^b)^2 < \delta$ lie also in $\mathfrak{A}_T$ provided that $\delta > 0$ is sufficiently small.

Since $\psi_o(x,\hat{y})$ has to assume a minimum at $(b,\hat{y}_b)$ under the constraints $\psi_j(x,\hat{y}) = 0$, $j = 1, 2, \ldots, k$, and since $(b,\hat{y}_b)$ is an interior point of $\mathfrak{A}_T$, we have, from the Lagrange multiplier rule for extreme-value problems of functions of more than one variable with constraints, that there exist k constants $(l_1, \ldots,l_k)$ such that at $x = b$, $\hat{y} = \hat{y}_b$,

$$\frac{\partial\psi_o}{\partial x} + l_1\frac{\partial\psi_1}{\partial x} + \cdots + l_k\frac{\partial\psi_k}{\partial x} = 0$$

$$\frac{\partial\psi_o}{\partial y_i} + l_1\frac{\partial\psi_1}{\partial y_i} + \cdots + l_k\frac{\partial\psi_k}{\partial y_i} = 0, \qquad i = 0, 1, \ldots, n.$$

But then

$$\operatorname{rank}\left(\frac{\partial\psi_j}{\partial y_i}\right)_{\substack{j=0,1,\ldots,k\\ i=0,1,\ldots,n}} < k + 1,$$

contrary to our hypothesis (6.8.5).

Hence, if $\hat{y} = \hat{y}(x)$ is to be a solution of the Mayer problem with variable endpoints [(6.8.1) to (6.8.3)], then, by necessity, $(b,\hat{y}(b)) = (b,\hat{y}_b)$ is to be a boundary point of the attainable set $\mathfrak{A}_T$.

As in the preceding section, we consider a family of solutions $\hat{y} = \hat{y}(x,\hat{e})$ of the disturbed augmented system (6.7.7) which satisfies the initial conditions (6.8.2). If for some choice of Z,

$$\left.\frac{\partial(y_o(x,\hat{e}), \ldots,y_n(x,\hat{e}))}{\partial(e_o, \ldots,e_n)}\right|_{\substack{\hat{e}=0\\ x=b}} \neq 0,$$

or equivalently,

$$\left.\frac{\partial(x,y_o(x,\hat{e}), \ldots,y_n(x,\hat{e}))}{\partial(x,e_o, \ldots,e_n)}\right|_{\substack{\hat{e}=0\\ x=b}} \neq 0,$$

then there is a $\delta > 0$ such that we can solve

$$x = b + \Delta$$
$$\hat{y}(x,\hat{e}) = \hat{y}_b + \hat{\Delta}, \qquad \hat{\Delta} = (\Delta_o, \ldots, \Delta_n), \tag{6.8.6}$$

for x, $\hat{e}$ provided that $|\Delta| < \delta$, $\max |\Delta_i| < \delta$.

We intersect the open cube with center at $(b,\hat{y}_b)$ and side length 2δ with the terminal manifold $\psi_j(x,\hat{y}) = 0, j = 1, 2, \ldots, k$, and see that $(b,\hat{y}_b)$ is an interior point of the attainable set $\mathfrak{A}_T$. We may therefore conclude that the condition (6.7.11), which is stated in the corollary to Lemma 6.7.1, is necessary if $\hat{y} = \hat{y}(x)$ is to be a solution of our problem.

From there, we may proceed, as in the preceding section, and obtain Theorem 6.2 as a necessary condition for a solution of the Mayer problem with a variable endpoint:

Lemma 6.8.1 *If $\hat{y} = \hat{y}(x) \in C^1[a,b]^{n+1}$ is the solution of the Mayer problem with a variable endpoint [(6.8.1 to 6.8.3)], where it is assumed that the conditions (6.2.5), (6.2.6), (6.8.4), (6.8.5) are satisfied, then it is necessary that $\hat{y} = \hat{y}(x)$ satisfy the condition of Theorem 6.2, namely, that there exist a vector function $\hat{\lambda} = \hat{\lambda}(x) \in C[a,b]^{\mu+1}$, $\hat{\lambda} \neq 0$, such that the Mayer equation*

$$h_{y_k}(x,\hat{y},\hat{y}',\hat{\lambda}) - \frac{d}{dx} h_{y'_k}(x,\hat{y},\hat{y}',\hat{\lambda}) = 0, \qquad k = 0, 1, \ldots, n,$$

is satisfied, where $h = \sum_{\alpha=0}^{\mu} \lambda_\alpha \varphi_\alpha$.

An additional necessary condition—a generalization of the transversality condition of Chap. 2—can be obtained, as we shall demonstrate in the sequel.

We consider again a solution of $\hat{y} = \hat{y}(x,\hat{e})$ of the disturbed augmented system (6.7.7) for some choice of Z which satisfies the initial condition $\hat{y}(a,\hat{e}) = \hat{y}_a$ and which contains $\hat{y} = \hat{y}(x)$ for $\hat{e} = 0$: $\hat{y}(x,0) = \hat{y}(x)$. Then

$$\hat{\psi}(x,\hat{y}(x,\hat{e})) = \hat{M},$$

where $\hat{M} = (\psi_o(b,\hat{y}(b)),0, \ldots,0)$, is satisfied for $x = b$, $\hat{e} = 0$. Hereby, $\psi_o(b,\hat{y}(b))$ is the relative minimum value that can be achieved by a solution of the constraining equations (6.8.1) that emanates from $(a,\hat{y}_a)$.

With the notation

$$\hat{\Psi}(x,\hat{e}) = \hat{\psi}(x,\hat{y}(x,\hat{e})), \tag{6.8.7}$$

we can restate this as follows:

$$\hat{\Psi}(x,\hat{e}) = \hat{M} \tag{6.8.8}$$

is satisfied for $x = b$, $\hat{e} = 0$.

Suppose for a moment that

$$\operatorname{rank}\begin{pmatrix}\dfrac{\partial\hat{\Psi}}{\partial x}\\[2ex] \dfrac{\partial\hat{\Psi}}{\partial e_o}\\ \vdots\\ \dfrac{\partial\hat{\Psi}}{\partial e_n}\end{pmatrix}_{\substack{\hat{e}=0\\ x=b}} = k+1.$$

Then it is possible to solve

$$\hat{\Psi}(x,\hat{e}) = \hat{M} + \hat{\Delta}, \qquad \hat{\Delta} = (\Delta,0,\ldots,0),$$

(in more than one way) for x, $\hat{e}$, provided that $|\Delta|$ is sufficiently small. This means that we can obtain for a suitable choice of $\hat{e}$ a vector function $\hat{y} = \hat{y}(x,\hat{e})$ that satisfies the constraining equations (6.8.1) together with the initial conditions (6.8.2) and that terminates at the manifold $\psi_j(x,\hat{y}) = 0$ for some $x = b^*$ such that

$$\psi_o(b^*,\hat{y}(b^*,\hat{e})) = \psi_o(b,\hat{y}(b)) + \Delta.$$

We may choose $\Delta < 0$ as long as $|\Delta|$ is sufficiently small, and we see that $\hat{y} = \hat{y}(x)$ was not the solution of the Mayer problem with a variable endpoint after all. Hence:

Lemma 6.8.2 *For $\hat{y} = \hat{y}(x) \in C^1[a,b]^{n+1}$ to be a solution of the Mayer problem with a variable endpoint [(6.8.1) to (6.8.3)] under the conditions that are stated in Lemma 6.8.1, it is necessary that*

$$\operatorname{rank}\begin{pmatrix}\dfrac{\partial\hat{\Psi}}{\partial x}\\[2ex] \dfrac{\partial\hat{\Psi}}{\partial e_0}\\ \vdots\\ \dfrac{\partial\hat{\Psi}}{\partial e_n}\end{pmatrix}_{\substack{\hat{e}=0\\ x=b}} < k+1 \qquad (6.8.9)$$

for all possible choices of $Z = Z(x)$. [$\hat{\Psi}$ is defined in (6.8.7).]

If (6.8.9) is true, then there exists a constant vector $\hat{\nu} = (\nu_o, \ldots, \nu_k) \neq (0, \ldots, 0)$ such that

$$\hat{\nu} \begin{pmatrix} \dfrac{\partial \hat{\Psi}}{dx} \\ \\ \dfrac{\partial \hat{\Psi}}{\partial e_m} \end{pmatrix}^T_{\substack{\hat{e}=0 \\ x=b}} = 0. \tag{6.8.10}$$

We have from (6.8.7) and (6.7.12) that

$$\begin{pmatrix} \dfrac{\partial \hat{\Psi}}{\partial x} \\ \\ \dfrac{\partial \hat{\Psi}}{\partial e_m} \end{pmatrix}^T_{\substack{\hat{e}=0 \\ x=b}} = \left([\hat{\psi}_x{}^T + \sum_{i=0}^{n} \hat{\psi}_{y_i}{}^T y_i']_{\substack{\hat{e}=0, \\ x=b}} \Psi_y V(b)\right),$$

where

$$\Psi_y = \begin{pmatrix} \psi_{oy_o} & \cdots & \psi_{oy_n} \\ \vdots & & \\ \psi_{ky_o} & \cdots & \psi_{ky_n} \end{pmatrix}_{\substack{\hat{e}=0. \\ x=b}}$$

Hence we can write (6.8.10) as follows:

$$\hat{\nu}\left([\hat{\psi}_x{}^T + \sum_{i=0}^{n} \hat{\psi}_{y_i}{}^T y_i']_{\substack{\hat{e}=0, \\ x=b}} \Psi_y V(b)\right) = 0. \tag{6.8.11}$$

We shall first discuss the last $n + 1$ of these equations, namely,

$$\hat{\nu}\Psi_y V(b) = 0.$$

Let

$$\hat{c} = \hat{\nu}\Psi_y. \tag{6.8.12}$$

Clearly, $\hat{c} \neq 0$ since rank $(\Psi_y) = k + 1$ [see (6.8.5)] and $\hat{\nu} \neq 0$. Hence we have to have

$$\hat{c}V(b) = 0 \tag{6.8.13}$$

for any choice of Z.

We know that, by necessity, det $|V(b)| = 0$ [see (6.7.13)]. Hence rank $(V(b)) < n + 1$. Let us choose that matrix Z for which $V(b)$ assumes maximum rank $r + 1$, $r < n$. Then (6.8.13) may be viewed as a system of $n + 1$ linear homogeneous equations for $n + 1$ unknowns $\hat{c} = (c_o, \ldots, c_n)$ with a coefficient matrix of rank $r + 1$. Hence there are exactly $n - r$ linearly independent solutions $\hat{c} = \hat{c}^\rho$, $\rho = 1, 2, \ldots, n - r$. On the other hand, if $\hat{\lambda} = (\lambda_o, \ldots, \lambda_\mu, 0, \ldots, 0)$ together with $\hat{y} = \hat{y}(x)$ is a solution of the Mayer equation (6.2.9), we know from (6.7.21) that

$$\hat{\lambda}(b)F_{y'}(b)V(b) = 0$$

if we also note that there are only zeros in the first $\mu + 1$ rows of the matrix Z. We may view this as a system of $n + 1$ linear homogeneous equations for $[\hat{\lambda}(b)F_{y'}(b)]$ with a coefficient matrix of rank $r + 1$. Again, there are precisely $n - r$ linearly independent solutions $[\hat{\lambda}^{\rho}(b)F_{y'}(b)]$, $\rho = 1, 2, \ldots n - r$. [Note that $F_{y'}(b)$ is fixed.] Hence $\hat{c}$ in (6.8.13) has to be of the form

$$\hat{c} = \sum_{\rho=1}^{n-r} \beta_{\rho}\hat{\lambda}^{\rho}(b)F_{y'}(b).$$

If $\hat{\lambda} = \hat{\lambda}^{\rho}(x)$ are those solutions of the Mayer equations that assume the terminal values $\hat{\lambda}^{\rho}(b)$, then $\hat{\lambda} = \sum_{\rho=1}^{n-r} \beta_{\rho}\hat{\lambda}^{\rho}(x)$ is a solution of the Mayer equations that assumes the terminal value $\hat{\lambda}(b) = \sum_{\rho=1}^{n-r} \beta_{\rho}\hat{\lambda}^{\rho}(b)$, and we have

$$\hat{c} = \hat{\lambda}(b)F_{y'}(b). \tag{6.8.14}$$

(6.8.12) and (6.8.14) yield the condition

$$p\Psi_y = \hat{\lambda}(b)F_{y'}(b). \tag{6.8.15}$$

The first equation in (6.8.11), namely,

$$p[\hat{\psi}_x{}^T + \sum_{i=0}^{n} \hat{\psi}_{y_i}{}^T y_i']_{\hat{e}=0,\, x=b} = 0,$$

can be written as

$$[p\hat{\psi}_x{}^T = -p \sum_{i=0}^{n} \hat{\psi}_{y_i}{}^T y_i']_{\hat{e}=0,\, x=b}.$$

Since $(\hat{\psi}_{y_i}{}^T)_{i=0,1,\ldots,n} = \Psi_y$, we have, in view of (6.8.15),

$$p\hat{\psi}_x{}^T(b,\hat{y}(b)) = -\hat{\lambda}(b)F_{y'}(b)\hat{y}'(b)^T. \tag{6.8.16}$$

We have seen in the preceding section that

$$\hat{\lambda}F_{y'} = (\lambda_o, \ldots, \lambda_\mu)\begin{pmatrix} \varphi_{oy'_o} & \cdots & \varphi_{oy'_n} \\ \vdots & & \\ \varphi_{\mu y'_o} & \cdots & \varphi_{\mu y'_n} \end{pmatrix}.$$

Hence, with $h = \sum_{\alpha=0}^{\mu} \lambda_\alpha \varphi_\alpha$,

$$\hat{\lambda}F_{y'} = (h_{y'_o}, \ldots, h_{y'_n}).$$

Therefore the conditions (6.8.15) and (6.8.16), in reverse order, can be written as

$$\sum_{j=0}^{k} \nu_j \psi_{jx}(b,\hat{y}(b)) = -\sum_{i=0}^{n} h_{y'_i}(b,\hat{y}(b),\hat{y}'(b),\hat{\lambda}(b))y'_i(b), \quad (6.8.17)$$

$$\sum_{j=0}^{k} \nu_j \psi_{jy_i}(b,\hat{y}(b)) = h_{y'_i}(b,\hat{y}(b),\hat{y}'(b),\hat{\lambda}(b)),$$

$$i = 0, 1, \ldots, n. \quad (6.8.18)$$

These two (really $n + 2$) conditions are called the *traversality conditions* of the Mayer problem with a variable endpoint.

We formulate our result as a theorem:

Theorem 6.8 *If $\hat{y} = \hat{y}(x) \in C^1[a,b]^{n+1}$ is to be the solution of the Mayer problem with a variable endpoint [(6.8.1) to (6.8.3)], where it is assumed that the conditions (6.2.5), (6.2.6), (6.8.4), (6.8.5) are satisfied, then it is necessary that there exist a vector function $\hat{\lambda} = (\lambda_0, \ldots, \lambda_\mu) \neq (0, \ldots, 0)$, $\hat{\lambda} \in C[a,b]^{\mu+1}$, such that the Mayer equations (6.2.9) are satisfied, and it is also necessary that there exist a constant vector $\hat{\nu} = (\nu_0, \ldots, \nu_k) \neq (0, \ldots, 0)$ such that the transversality conditions (6.8.17) and (6.8.18) are satisfied.*

PROBLEMS 6.8

1. Make suitable assumptions and generalize Theorem 6.8 to a Mayer problem with a variable beginning point on $\psi_i^{(1)}(x,\hat{y}) = 0$, $i = 1, 2, \ldots, l$, and a variable endpoint on $\psi_j^{(2)}(x,\hat{y}) = 0$, $j = 1, 2, \ldots, k$, so that $\psi_0(a,b,\hat{y}(a),\hat{y}(b)) \to$ minimum.
2. State a condition that will guarantee that $(\nu_0, \ldots, \nu_k)$ can be eliminated from the system (6.8.17), (6.8.18) of transversality conditions for the Mayer problem.
3. Formulate the problem $\int_0^1 (1 - y'^2(x))^2\,dx \to$ minimum, $|y'| \leq \frac{1}{2}$, $y(0) = 0$, $y(1) = 0$ as a Mayer problem with variable endpoints. Check if the conditions (6.2.5), (6.2.6), (6.8.4), and (6.8.5) are satisfied and in which domains and solve. (*Hint*: Replace the inequality constraint $|y'| \leq \frac{1}{2}$ by the differential equation constraint $y'^2 + z'^2 - \frac{1}{4} = 0$, where z is a new unknown function.)
4. Same as in problem 3 for the problem of reaching the terminal state $y = 0$, $y' = 0$ from an initial state $y(0) = y_1^0$, $y'(0) = y_2^0$ by appropriate choice of $u = u(x)$, $|u| \leq 1$, in $y'' = u$, in the shortest possible time, where x denotes the time.

6.9 TRANSVERSALITY CONDITIONS FOR THE LAGRANGE PROBLEM WITH A VARIABLE ENDPOINT

We shall now apply Theorem 6.8 to the Lagrange problem of Sec. 6.3 with a variable endpoint on an $(n + 1 - k)$-dimensional manifold in $(x,y_1, \ldots, y_n)$ space.

We formulate the problem as follows: To be found is a vector function $(y_1, \ldots, y_n)$ which satisfies the constraining equations

$$\begin{aligned} &\varphi_1(x,y_1, \ldots,y_n,y_1', \ldots,y_n') = 0 \\ &\vdots \\ &\varphi_\mu(x,y_1, \ldots,y_n,y_1', \ldots,y_n') = 0, \qquad \mu < n, \end{aligned} \tag{6.9.1}$$

the initial conditions

$$y_i(a) = y_i^a, \qquad i = 1, 2, \ldots, n, \tag{6.9.2a}$$

and the terminal conditions

$$\begin{aligned} &\psi_1(b,y_1(b), \ldots,y_n(b)) = 0 \\ &\vdots \\ &\psi_k(b,y_1(b), \ldots,y_n(b)) = 0, \qquad k \le n, \end{aligned} \tag{6.9.2b}$$

for some b and is such that

$$\int_a^b f(x,y_1(x), \ldots,y_n(x),y_1'(x), \ldots,y_n'(x))\, dx \to \text{minimum.} \tag{6.9.3}$$

In addition to the assumptions under which we pronounced the Lagrange multiplier rule (Theorem 6.3), we shall now assume that

$$\psi_j \in C^1\{(x,y_1, \ldots,y_n) \mid x,y_i \in (-\infty,\infty), i = 1, 2, \ldots, n\} \tag{6.9.4}$$

for $j = 1, 2, \ldots, k$, and

$$\operatorname{rank}\left(\frac{\partial\psi_j}{\partial y_i}\right) = k, \qquad j = 1, 2, \ldots, k; i = 1, 2, \ldots, n. \tag{6.9.5}$$

Then, with

$$\varphi_o = y_o' - f(x,y_1, \ldots,y_n,y_1', \ldots,y_n')$$

and

$$\psi_o = y_o,$$

the conditions (6.8.4) and (6.8.5) are fulfilled, and Theorem 6.8 becomes applicable.

We note that

$$\begin{aligned} &\varphi_{\alpha y_o} = 0 && && \text{for } \alpha = 0, 1, \ldots, \mu, \\ &\varphi_{oy'_o} = 1, && \varphi_{\alpha y'_o} = 0 && \text{for } \alpha = 1, 2, \ldots, \mu, \\ &\psi_{oy_o} = 1, && \psi_{jy_o} = 0 && \text{for } j = 1, 2, \ldots, k, \\ &\psi_{ox} = 0, && \psi_{oy_i} = 0 && \text{for } i = 1, 2, \ldots, n. \end{aligned}$$

In view of all this, we obtain from the first equation in (6.8.18) (for $i = 0$)

$$\nu_o = \lambda_o,$$

and we can write the remaining equations in (6.8.17) and (6.8.18) with

$$h^L = -\lambda_o f + \sum_{\alpha=1}^{\mu} \lambda_\alpha \varphi_\alpha$$

as follows:

$$\sum_{j=1}^{k} \nu_j \psi_{jx}(b,y(b)) = -\lambda_o f(b,y(b),y'(b))$$

$$- \sum_{i=1}^{n} h^L_{y'_i}(b,y(b),y'(b),\hat{\lambda}(b))y'_i(b) \qquad (6.9.6)$$

$$\sum_{j=1}^{k} \nu_j \psi_{jy_i}(b,y(b)) = h^L_{y'_i}(b,y(b),y'(b),\hat{\lambda}(b)), \qquad i = 1, 2, \ldots, n, \qquad (6.9.7)$$

where we have written y and y' for $(y_1, \ldots, y_n)$ and $(y'_1, \ldots, y'_n)$.

These are the *transversality conditions* for the Lagrange problem with a variable endpoint. Hence Theorem 6.8 as applied to this problem reads as follows:

Theorem 6.9 *If* $(y_1, \ldots, y_n) \in C^1[a,b]^n$ *is the solution of the Lagrange problem with a variable endpoint* [*(6.9.1) to (6.9.3)*], *where it is assumed that the conditions (6.3.6), (6.3.7), (6.9.4), and (6.9.5) are satisfied, then it is necessary that there exist a* $(\mu + 1)$*-dimensional vector function* $\hat{\lambda} = \hat{\lambda}(x) \in C[a,b]^{n+1}$, $\hat{\lambda} \neq 0$, $\lambda_o =$ constant, *which satisfies the Mayer equations of the Lagrange problem (6.3.10), and that there exist a constant vector*

$$\hat{\nu} = (\nu_o, \ldots, \nu_k) \neq (0, \ldots, 0),$$

where $\nu_o = \lambda_o$, *such that the transversality conditions (6.9.6) and (6.9.7) are satisfied.*

We shall illustrate this theorem with a discussion of the problem of finding the shortest distance from the point (2,0,0) in (x,y_1,y_2) space to the surface of the circular cylinder $y_1^2 + x^2 - 1 = 0$, under the restriction that the path that renders the shortest distance lies in the plane $x - y_2 - 2 = 0$. (See Fig. 6.2.) Let us formulate this problem as a Lagrange problem with a variable endpoint.

Although the constraining equation $x - y_2 - 2 = 0$ is not a differential equation, we can easily obtain a differential-equation constraint if we note that the solution $y_1 = y_1(x)$, $y_2 = y_2(x)$ will have to satisfy the constraint $x - y_2(x) - 2 = 0$ for all x. (See also Sec. A6.12.) Hence, we obtain after differentiation with respect to x, the differential-equation constraint

$$\varphi_1 \equiv 1 - y'_2 = 0.$$

We have the initial conditions

$$y_1(2) = 0, \qquad y_2(2) = 0$$

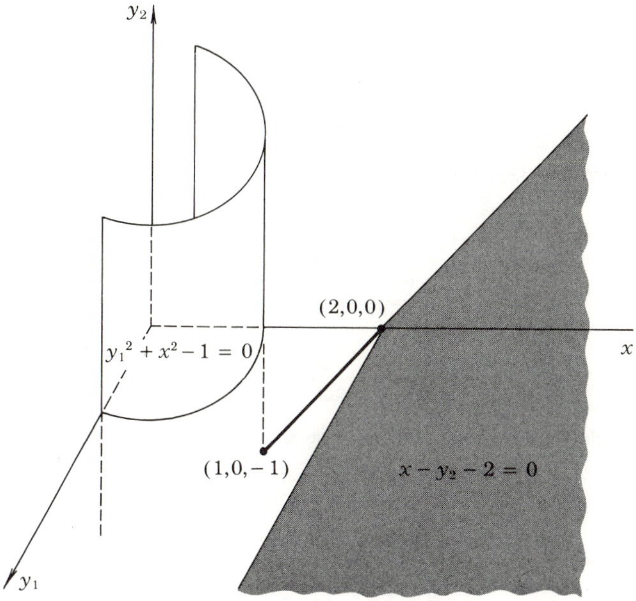

Figure 6.2

and the terminal condition

$$\psi_1 \equiv y_1^2(b) + b^2 - 1 = 0.$$

Finally, we have

$$\int_2^b \sqrt{1 + y_1'^2(x) + y_2'^2(x)}\, dx \to \text{minimum.}$$

We integrate the Mayer equations (6.3.10) and obtain, with $\lambda_o = 1$,

$$-\frac{y_1'}{\sqrt{1 + y_1'^2 + y_2'^2}} = \alpha, \qquad -\frac{y_2'}{\sqrt{1 + y_1'^2 + y_2'^2}} - \lambda = \beta,$$

where α, β are integration constants. These, together with the constraining equation, have to be solved for y_1, y_2, λ. We obtain $y_2' = 1$, and hence

$$-\frac{y_1'}{\sqrt{2 + y_1'^2}} = \alpha, \qquad -\frac{1}{\sqrt{2 + y_1'^2}} - \lambda = \beta.$$

These equations yield

$$y_1 = Ax + B, \qquad y_2 = x + C, \qquad \lambda = -\frac{1}{\sqrt{2 + A^2}} - \beta,$$

where A, B, C are arbitrary constants.

We obtain from the initial conditions that $B = -2A$, $C = -2$. Next, we have to satisfy, with our intermediate result,

$$y_1 = Ax - 2A, \qquad y_2 = x - 2, \qquad \lambda = -\frac{1}{\sqrt{2 + A^2}} - \beta,$$

the terminal condition

$$y_1^2(b) + b^2 - 1 = 0,$$

and the transversality conditions

$$2\nu b = -\sqrt{2 + A^2} + \frac{A^2}{\sqrt{2 + A^2}} + \frac{1}{\sqrt{2 + A^2}} + \lambda(b),$$

$$2\nu y_1(b) = -\frac{A}{\sqrt{2 + A^2}},$$

$$0 = -\frac{1}{\sqrt{2 + A^2}} - \lambda(b).$$

In view of the third transversality condition and because of $y_1(b) = Ab - 2A$, we have

$$2\nu(Ab - 2A) = -\frac{A}{\sqrt{2 + A^2}}, \qquad 2\nu b = -\sqrt{2 + A^2} + \frac{A^2}{\sqrt{2 + A^2}}.$$

These two equations, together with the terminal condition have the only real solution

$$A = 0, \qquad b = \pm 1, \qquad \nu = \mp\frac{\sqrt{2}}{2}.$$

Hence our solution is given by

$$y_1 = 0, \qquad y_2 = x - 2,$$

which penetrates the terminal manifold at $b = 1$, $y_1 = 0$, $y_2 = -1$ and at $b = -1$, $y_1 = 0$, $y_2 = -3$. Obviously, the first penetration point is the one that is closest to the beginning point.

Before closing this section, let us consider a few special cases of the tranversality conditions (6.9.6) and (6.9.7).

In Sec. 2.8, we derived a transversality condition for the simplest variational problem

$$I[y] = \int_a^b f(x,y(x),y'(x))\,dx \to \text{minimum}$$

with a fixed beginning point

$$y(a) = y_a$$

and a variable endpoint on the curve $y = \varphi(x)$, which we may write as

$$\psi \equiv y - \varphi(x) = 0.$$

There are no constraining equations, that is, $\varphi_\alpha \equiv 0$ for all $\alpha = 1, 2, \ldots, \mu$, and we have, with $\lambda_o = 1$,

$$h^L = -f(x,y,y').$$

(See also Prob. 6.9.1.)

If φ' is continuous for all x and if $f \in C^1$ in a suitably chosen domain of the (x,y,y') space, all conditions of Theorem 6.9 are met, and we obtain from (6.9.6) and (6.9.7), because of $h^L_{y'} = -f_{y'}$ and since $\nu_o = \lambda_o = 1$,

$$\begin{aligned} -\nu_1\varphi'(b) &= -f(b,y(b),y'(b)) + f_{y'}(b,y(b),y'(b))y'(b), \\ \nu_1 &= -f_{y'}(b,y(b),y'(b)). \end{aligned}$$

Elimination of ν_1 leads to the transversality condition (2.8.5):

$$f_{y'}(b,y(b),y'(b))(\varphi'(b)-y'(b)) + f(b,y(b),y'(b)) = 0.$$

Next we consider the variational problem in n unknown functions

$$I[y_1, \ldots, y_n] = \int_a^b f(x,y_1(x), \ldots, y_n(x), y_1'(x), \ldots, y_n'(x))\,dx \to \text{minimum} \tag{6.9.8}$$

with a fixed beginning point

$$y_i(a) = y_i^a, \qquad i = 1, 2, \ldots, n. \tag{6.9.9a}$$

If the terminal manifold is an n-dimensional surface $x = u(y_1, \ldots, y_n)$, we can write the terminal condition as

$$\psi \equiv b - u(y_1(b), \ldots, y_n(b)) = 0. \tag{6.9.9b}$$

In the absence of constraining equations and with $\lambda_o = 1$, we have

$$h^L = -f(x,y_1, \ldots, y_n, y_1', \ldots, y_n'), \qquad h^L_{y'_i} = -f_{y'_i}.$$

(See also Prob. 6.9.3.) Further,

$$\psi_x = 1, \qquad \psi_{y_i} = -u_{y_i}.$$

Hence we obtain from (6.9.6) and (6.9.7)

$$\begin{aligned} \nu_1 &= -f(b,y(b),y'(b)) + \sum_{i=1}^n f_{y'_i}(b,y(b),y'(b))y_i'(b) \\ \nu_1 u_{y_i}(y(b)) &= f_{y'_i}(b,y(b),y'(b)), \qquad i = 1, 2, \ldots, n, \end{aligned}$$

where we again wrote y and y' for $(y_1, \ldots, y_n)$ and $(y_1', \ldots, y_n')$. Elimination of ν_1 leads to the transversality conditions

$$\begin{aligned} \Big[f(b,y(b),y'(b)) - \sum_{i=1}^n f_{y'_i}(b,y(b),y'(b))y_i'(b)\Big]u_{y_k}(y(b)) \\ + f_{y'_k}(b,y(b),y'(b)) = 0, \qquad k = 1, 2, \ldots, n, \end{aligned} \tag{6.9.10}$$

of the variational problem [(6.9.8), (6.9.9*a* and *b*)]. For $n = 2$, we obtain a system of two conditions that is equivalent to (A2.13.5). (See Probs. A2.13.1 and 6.9.4.)

In Chap. 2, we avoided dealing with the case where the variable endpoint is permitted to lie on a surface of dimension less than n. In fact, we could not have dealt with this case with the tools that had been developed in Chaps. 1 and 2. Now we can take care of this case very easily.

We consider the terminal conditions (6.9.2*b*). Then, again assuming no other constraints, we obtain from (6.9.6) and (6.9.7) as transversality conditions for the problem [(6.9.8), (6.9.9*a*), (6.9.2*b*)],

$$\sum_{j=1}^{k} \nu_j \psi_{jx}(b,y(b)) = -f(b,y(b),y'(b)) + \sum_{i=1}^{n} f_{y'_i}(b,y(b),y'(b))y'_i(b),$$

$$\sum_{i=1}^{k} \nu_j \psi_{jy_i}(b,y(b)) = -f_{y'_i}(b,y(b),y'(b)), \qquad i = 1, 2, \ldots, n, \tag{6.9.11}$$

for some constants $\nu_1, \ldots, \nu_k$. (See also Prob. 6.9.5.)

PROBLEMS 6.9

1. Show that Theorem 6.9 is indeed applicable to the variational problem $\int_a^b f(x,y(x),y'(x))\,dx \to \text{minimum}$, $y(a) = y_a$, $y(b) = \varphi(b)$, where $y = \varphi(x) \in C^1(-\infty,\infty)$. In particular, demonstrate that one may choose $\lambda_o = 1$ and that one may choose, in place of the solutions of the nonexistent disturbed augmented system, the family $y = y(x,e) = y(x) + eh(x)$, $h(a) = 0$, $h(b) \neq 0$, $h \in C^1(a \leq x)$, whereby $y = y(x)$ represents the solution of the problem.
2. Show that the transversality conditions that were derived in this section for the problem in 1 also apply to the case where $\varphi'(b) = y'(b)$.
*3. Show that Theorem 6.9 applies to the variational problem

$$\int_a^b f(x,y_1(x), \ldots, y_n(x), y'_1(x), \ldots, y'_n(x))\,dx \to \text{minimum},$$

$y_i(a) = y_i{}^a$, $b = u(y_1(b), \ldots, y_n(b))$, the transversality conditions of which are listed in (6.9.10).
*4. Write down the transversality conditions (6.9.10) for $n = 2$ and show that they are indeed identical to the ones listed in Prob. A2.13.1.
*5. Show that Theorem 6.9 is applicable to the variational problem [(6.9.8), (6.9.9*a*), (6.9.2*b*)].
6. Consider an isoperimetric problem as in Sec. 6.5 with a variable endpoint on the manifold (6.9.2*b*) for $n = m$ and state the transversality conditions.
7. Let $x = X(t)$, $y_i = Y_i(t)$, $i = 1, 2, \ldots, n$, represent a smooth curve that lies on the terminal surface (6.9.2*b*) and passes through the terminal point $(b,y_1(b), \ldots, y_n(b))$. Show that the transversality conditions (6.9.6), (6.9.7) imply

$$\sum_{i=1}^{n} h^L_{y'_i}(b,y(b),y'(b),\hat{\lambda}(b))\left(\frac{dY_i}{dt} - y'_i(b)\frac{dX}{dt}\right) - \nu_o f(b,y(b),y'(b))\frac{dX}{dt} = 0,$$

where $(dX/dt, dY_i/dt)$ are to be evaluated at the point $(b,y(b))$ and where y denotes the vector $(y_1, \ldots, y_n)$.

6.10 A SUFFICIENT CONDITION FOR THE LAGRANGE PROBLEM

Since we shall deal in this section exclusively with the Lagrange problem [(6.3.1) to (6.3.3)], we shall adopt the following notational convention, which we have also used to some extent in Chap. 5: From now on, $y = (y_1, \ldots, y_n)$, $\phi = (\phi_1, \ldots, \phi_n)$, and Roman subscripts will range from 1 to n. λ will denote the μ vector $(\lambda_1, \ldots, \lambda_\mu)$, and Greek subscripts will range from 1 to μ.

In order to establish a sufficient condition, patterned after the theory developed in Chap. 3, for a vector function $y = y(x)$ to be a solution of a Lagrange problem (with fixed endpoints), we shall have to generalize the concepts of field, invariant integral, and excess function so that they will fit our present needs.

Our goal will be to transform

$$\Delta I = \int_a^b f(x,\bar{y}(x),\bar{y}'(x))\,dx - \int_a^b f(x,y_o(x),y_o'(x))\,dx$$

into an expression that is easily accessible to investigation. Here, $y = y_o(x)$ denotes the solution that is to be investigated and $y = \bar{y}(x)$ denotes any smooth vector function that joins the same two endpoints and also satisfies the constraining equations

$$\varphi_\alpha(x,\bar{y},\bar{y}') = 0, \qquad \alpha = 1, 2, \ldots, \mu.$$

In particular, we shall have to define a field of lineal elements $(x,y,\phi(x,y))$, each of which satisfies the constraining equations, in such a manner that exactly one lineal element is defined for every point (x,y) in a weak neighborhood of the extremal $y = y_o(x)$ and so that $y = y_o(x)$ appears as a solution of $y' = \phi(x,y)$.

Further, we shall need a suitable function $W(x,y)$ in terms of which we can express the invariant integral as

$$U[y] = \int_a^b \left(W_x + \sum_{i=1}^n W_{y_i} y_i'\right) dx$$

and an excess function

$$\mathcal{E} = f(x,y,\bar{y}') - W_x(x,y) - \sum_{i=1}^n W_{y_i}(x,y)\bar{y}_i' \tag{6.10.1}$$

with the property that for every fixed point (x,y), $\mathcal{E} \geq 0$ for all lineal elements $(x,y,\bar{y}')$ that satisfy the constraining equations $\varphi_\alpha(x,y,\bar{y}') = 0$, $\alpha = 1, 2, \ldots, \mu$, and so that $\mathcal{E} = 0$ if $\bar{y}' = \phi(x,y)$, which is the slope assigned to the point (x,y) by the field-defining function $\phi(x,y)$. In other words, we seek a function $\mathcal{E}$ which, for every fixed point (x,y), possesses the minimum value 0 at $\bar{y}' = \phi(x,y)$ under the constraints $\varphi_\alpha(x,y,\bar{y}') = 0$.

This is an ordinary minimum-value problem with constraints in the variables $y_1', \ldots, y_n'$, and we have as a necessary condition that

$$\Big(\mathcal{E} + \sum_{\alpha=1}^{\mu} \lambda_\alpha \varphi_\alpha\Big)_{y'_k} = 0, \qquad k = 1, 2, \ldots, n, \tag{6.10.2}$$

for some "constants" $\lambda_\alpha = \lambda_\alpha(x,y)$.

With the notation

$$g(x,y,\bar{y}',\lambda) = f(x,y,\bar{y}') + \sum_{\alpha=1}^{\mu} \lambda_\alpha \varphi_\alpha(x,y,\bar{y}'),$$

we can write, in view of (6.10.1),

$$\mathcal{E} + \sum_{\alpha=1}^{\mu} \lambda_\alpha \varphi_\alpha = g(x,y,\bar{y}',\lambda) - W_x(x,y) - \sum_{i=1}^{n} W_{y_i}(x,y)\bar{y}_i'$$

and obtain from (6.10.2) at $\bar{y}' = \phi(x,y)$

$$\Big(\mathcal{E} + \sum_{\alpha=1}^{\mu} \lambda_\alpha \varphi_\alpha\Big)_{y'_k} = \Big(g - W_x - \sum_{i=1}^{n} W_{y_i}\bar{y}_i'\Big)_{y'_k} = g_{y'_k} - W_{y_k} = 0,$$

that is, $\qquad W_{y_k}(x,y) = g_{y'_k}(x,y,\phi(x,y),\lambda(x,y)). \qquad$ (6.10.3)

We remind the reader of our differentiation convention whereby a variable as a subscript indicates a partial derivative with respect to the position that is normally occupied by that variable, as in $g_{y_i}(x,y,y',\lambda)$. A partial differentiation symbol denotes a partial derivative with respect to the indicated variable wherever this variable occurs. For example, if ϕ and λ are functions of x and y, then $(\partial/\partial x)\, g(x,y,\phi(x,y),\lambda(x,y)) = g_x + \sum_{i=1}^{n} g_{y'_i}\phi_{ix} + \sum_{\rho=1}^{\mu} g_{\lambda_\rho}\lambda_{\rho x}$. d/dx is used when $y = y(x)$ and means a derivative with respect to x wherever x occurs, such as in

$$(d/dx)\, g(x,y(x),\phi(x,y(x)),\lambda(x,y(x))).$$

Since we want $\mathcal{E} = 0$ at $\bar{y}' = \phi(x,y)$, we also have, from (6.10.1) and (6.10.3),

$$f(x,y,\phi(x,y)) - W_x(x,y) - \sum_{i=1}^{n} g_{y'_i}(x,y,\phi(x,y),\lambda(x,y))\phi_i(x,y) = 0,$$

and hence

$$W_x(x,y) = g(x,y,\phi(x,y),\lambda(x,y)) - \sum_{i=1}^{n} g_{y'_i}(x,y,\phi(x,y),\lambda(x,y))\phi_i(x,y) \tag{6.10.4}$$

if we also note that f may be replaced by g in view of $\varphi_\alpha = 0$.

For a function $W(x,y)$ which satisfies (6.10.3) and (6.10.4) to exist, it is necessary and sufficient that†

$$\frac{\partial}{\partial y_k} W_x = \frac{\partial}{\partial x} W_{y_k}, \qquad \frac{\partial}{\partial y_k} W_{y_i} = \frac{\partial}{\partial y_i} W_{y_k}, \qquad i, k = 1, 2, \ldots, n. \quad (6.10.5)$$

From (6.10.4),

$$\frac{\partial}{\partial y_k} W_x = g_{y_k} + \sum_{i=1}^{n} g_{y'_i}\phi_{iy_k} - \sum_{i=1}^{n} g_{y'_i}\phi_{iy_k} - \sum_{i=1}^{n} \frac{\partial}{\partial y_k}(g_{y_i'})\phi_i,$$

and from (6.10.3),

$$\frac{\partial}{\partial x} W_{y_k} = \frac{\partial}{\partial x} g_{y'_k}, \qquad \frac{\partial}{\partial y_i} W_{y_k} = \frac{\partial}{\partial y_i} g_{y'_k}.$$

Hence, the integrability conditions (6.10.5) appear as

$$g_{y_k} - \sum_{i=1}^{n} \frac{\partial g_{y'_i}}{\partial y_k}\phi_i - \frac{\partial g_{y'_k}}{\partial x} = 0, \qquad k = 1, 2, \ldots, n, \quad (6.10.6)$$

and

$$\frac{\partial g_{y'_k}}{\partial y_i} = \frac{\partial g_{y'_i}}{\partial y_k}, \qquad i, k = 1, 2, \ldots, n. \quad (6.10.7)$$

In view of (6.10.7) and because of

$$\frac{\partial g_{y'_k}}{\partial x} + \sum_{i=1}^{n} \frac{\partial g_{y'_k}}{\partial y_i}\phi_i = \frac{d}{dx} g_{y'_k},$$

we can write (6.10.6) as

$$g_{y_k} - \frac{d}{dx} g_{y'_k} = 0, \qquad k = 1, 2, \ldots, n, \quad (6.10.8)$$

which are the Mayer equations (6.3.10) for

$$g(x,y,y',\lambda) = f(x,y,y') + \sum_{\alpha=1}^{\mu} \lambda_\alpha \varphi_\alpha(x,y,y').$$

In view of all this, we assume that $\varphi \in C^2$ in a weak neighborhood of the extremal and for all y' and give the following definition:

Definition 6.10 *The extremal $y = y_o(x) \in C^1[a,b]^n$ of the Lagrange problem with fixed endpoints [(6.3.1) to (6.3.3)] is embeddable in a Mayer field if there exists in a weak neighborhood of $y = y_o(x)$ a vector function $\phi = \phi(x,y) \in C^1$ such that $y_o' = \phi(x,y_o)$ and a vector function $\lambda = \lambda(x,y) \in C$ such that*

† T. M. Apostol, "Mathematical Analysis," pp. 292ff, Addison-Wesley Publishing Company, Inc., Reading, Mass., 1957.

the solutions of $y' = \phi(x,y)$ *satisfy*

$$g_{y_k} - \frac{d}{dx} g_{y'_k} = 0, \qquad k = 1, 2, \ldots, n,$$

for $g(x,y,\phi(x,y),\lambda(x,y)) = f(x,y,\phi(x,y)) + \sum_{\alpha=1}^{\mu} \lambda_\alpha(x,y)\varphi_\alpha(x,y,\phi(x,y))$,

$$\frac{\partial}{\partial y_i} g_{y'_k} = \frac{\partial}{\partial y_k} g_{y'_i}, \qquad i, k = 1, 2, \ldots, n,$$

and $\varphi_\alpha(x,y,\phi(x,y)) = 0$, $\alpha = 1, 2, \ldots, \mu$.

Note that for $y = y_o(x)$, we have

$$\begin{aligned} g(x,y_o(x),\phi(x,y_o(x)),\lambda(x,y_o(x))) &= f(x,y_o(x),y_o'(x)) \\ &\quad + \sum_{\alpha=1}^{\mu} \lambda_\alpha(x,y_o(x))\varphi_\alpha(x,y_o(x),y_o'(x)) \\ &= h^L(x,y_o(x),y_o'(x),\lambda(x)) \end{aligned} \tag{6.10.9}$$

with
$$h^L_{y_k} - \frac{d}{dx} h^L_{y'_k} = 0, \qquad k = 1, 2, \ldots, n,$$

where $\lambda_o = -1$, $\lambda_\alpha(x) = \lambda_\alpha(x,y_o(x))$ are a set of Lagrange multipliers, as called for by Theorem 6.3.

Now, if $\phi = \phi(x,y)$ and $\lambda = \lambda(x,y)$ define a field in the sense of this definition, then there exists a function $W = W(x,y)$ such that (6.10.3) and (6.10.4) are satisfied, and we can write the invariant integral as

$$\begin{aligned} U[y] = \int_a^b [g(x,y,\phi(x,y),\lambda(x,y)) \\ + \sum_{i=1}^{n} (y_i' - \phi_i(x,y))g_{y'_i}(x,y,\phi(x,y),\lambda(x,y))]\,dx, \end{aligned}$$

where $y = y(x)$.

We have, in view of (6.10.9), $\varphi_\alpha(x,y,\phi(x,y)) = 0$, and $y_o' = \phi(x,y_o)$, that

$$U[y_o] = \int_a^b f(x,y_o(x),y_o'(x))\,dx = I[y_o].$$

Hence, we obtain for any smooth curve $y = \bar{y}(x)$ which joins the same two endpoints that $y = y_o(x)$ does and satisfies the constraining equations:

$$\Delta I = I[\bar{y}] - I[y_o] = I[\bar{y}] - U[y_o] = I[\bar{y}] - U[\bar{y}]$$

$$= \int_a^b [f(x,\bar{y},\bar{y}') - g(x,\bar{y},\phi(x,\bar{y}),\lambda(x,\bar{y}))$$

$$+ \sum_{i=1}^{n} (\phi_i(x,\bar{y}) - \bar{y}'_i) g_{y'_i}(x,\bar{y},\phi(x,\bar{y}),\lambda(x,\bar{y}))]\, dx,$$

where $\bar{y} = \bar{y}(x)$, $\bar{y}' = \bar{y}'(x)$.

In view of $\varphi_\alpha(x,\bar{y},\bar{y}') = 0$, we can replace $f(x,\bar{y},\bar{y}')$ by $g(x,\bar{y},\bar{y}',\lambda)$ and write instead

$$\Delta I = \int_a^b [g(x,\bar{y},\bar{y}',\lambda(x,\bar{y})) - g(x,\bar{y},\phi(x,\bar{y}),\lambda(x,\bar{y}))$$

$$+ \sum_{i=1}^{n} (\phi_i(x,\bar{y}) - \bar{y}'_i) g_{y'_i}(x,\bar{y},\phi(x,\bar{y}),\lambda(x,\bar{y}))]\, dx.$$

Accordingly, we define

$$\mathcal{E}(x,y,y',\bar{y}',\lambda) = g(x,y,\bar{y}',\lambda) - g(x,y,y',\lambda) + \sum_{i=1}^{n} (y'_i - \bar{y}'_i) g_{y'_i}(x,y,y',\lambda). \tag{6.10.10}$$

[Note that the excess function as defined here differs from the one defined in Sec. A5.12. This is due to the fact that we now have $h^L = f + \sum_{\alpha=1}^{\mu} \lambda_\alpha \varphi_\alpha$, while we had in Sec. A5.12, $h = -p_o f + \sum_{\alpha=1}^{\mu} \lambda_\alpha (y'_\alpha - f_\alpha)$ with $p_o \geq 0$.]

In terms of this generalized excess function, we can now state the following sufficient condition for a strong relative minimum of the Lagrange problem:

Theorem 6.10 *If $y = y_o(x)$ is embeddable in a field in the sense of Definition 6.10 and if*

$$\mathcal{E}(x,y,\phi(x,y),\bar{y}',\lambda(x,y)) \geq 0$$

for all points (x,y) in a weak neighborhood of $y = y_o(x)$ and all $\bar{y}'$ for which $(x,y,\bar{y}')$ satisfies the constraining equations $\varphi_\alpha(x,y,\bar{y}') = 0$, then $y = y_o(x)$ yields a strong relative minimum for the Lagrange problem with fixed endpoints [(6.3.1) to (6.3.3)].

The reader is now in possession of all the tools and techniques that are required to investigate the conditions under which the existence of a Mayer field can be guaranteed and also to discuss ways of constructing such fields. So, rather than dwell on this problem in general terms, we shall discuss an example in some detail.

We consider the Lagrange problem

$$\frac{1}{2}\int_0^1 (y'^2(x) + z'^2(x))\, dx \to \text{minimum}$$

$$yz' - zy' - 1 = 0$$

with the boundary conditions

$$y(0) = 1, \qquad z(0) = 0, \qquad y(1) = \cos 1, \qquad z(1) = \sin 1.$$

One obtains some simplifications if one introduces cylindrical coordinates by

$$y = r\cos\theta, \qquad z = r\sin\theta,$$

where $r = r(x)$, $\theta = \theta(x)$.

Then the problem can be formulated as follows:

$$\frac{1}{2}\int_0^1 (r'^2 + r^2\theta'^2)\, dx \to \text{minimum},$$

$$r^2\theta' = 1$$

with the boundary conditions

$$r(0) = 1, \qquad \theta(0) = 0, \qquad r(1) = 1, \qquad \theta(1) = 1.$$

(The reader is challenged to justify what is to follow in view of this transformation.)

First, we apply the multiplier rule to obtain an extremal: We have

$$h = \tfrac{1}{2}(r'^2 + r^2\theta'^2) + \lambda(r^2\theta' - 1),$$

and hence

$$h_r = r\theta'^2 + 2\lambda r\theta', \qquad h_{r'} = r'$$

$$h_\theta = 0, \qquad h_{\theta'} = r^2\theta' + \lambda r^2,$$

which leads to the Mayer equations

$$r'' = r\theta'^2 + 2\lambda r\theta',$$

$$\frac{d}{dx}(r^2\theta' + \lambda r^2) = 0.$$

For the second equation, we obtain immediately a first integral,

$$r^2\theta' + \lambda r^2 = C_1. \tag{6.10.11}$$

Since h does not contain x explicitly, the hamiltonian H will not contain x explicitly either, and hence H = constant appears as a first integral (see Prob. 6.10.1). Since

$$H = r'h_{r'} + \theta' h_{\theta'} - h,$$

we obtain immediately

$$\tfrac{1}{2}(r'^2 + r^2\theta'^2) + \lambda = C_2. \tag{6.10.12}$$

Equation (6.10.11) together with the constraining equation $r^2\theta' = 1$ yields

$$\lambda = \frac{C_3}{r^2},$$

and we obtain from (6.10.12)

$$\frac{r\,dr}{\sqrt{C_4 r^2 + C_5}} = dx$$

and hence

$$r^2 = Ax^2 + Bx + C. \tag{6.10.13}$$

Then

$$\theta = \int_0^x \frac{dt}{At^2 + Bt + C} + D, \tag{6.10.14}$$

and

$$\lambda = \frac{C_3}{Ax^2 + Bx + C}.$$

Thus far, we have been quite liberal with the various constants and have ignored their interdependence. We can take care of this now by substitution into (6.10.12) and comparison of coefficients. We obtain

$$C_3 = \frac{AC - 1 - (B^2/4)}{2}.$$

Hence

$$\lambda = \frac{AC - 1 - (B^2/4)}{2(Ax^2 + Bx + C)}. \tag{6.10.15}$$

From the first three boundary conditions, we obtain

$$A = -B, \qquad C = 1, \quad D = 0.$$

The last boundary condition leads to

$$1 = \int_0^1 \frac{dx}{Ax^2 - Ax + 1}.$$

One can show, either by direct computation or, better still, by a simple geometric consideration, that $A = 0$.

Hence the extremal and the corresponding multipliers are given by

$$\begin{aligned} r &= 1, \qquad \theta = x \\ \lambda_o &= -1, \qquad \lambda = -\tfrac{1}{2}. \end{aligned} \tag{6.10.16}$$

We recognize $r = 1$, $\theta = x$ as a *helix* that winds itself around a circular cylinder of radius 1 with the x axis as axis, originates from (0,1,0) and climbs at the rate $dx/d\theta = 1$ to terminate at the point (1, cos 1, sin 1).

We shall now try to embed this extremal in a Mayer field, and we shall apply Theorem 6.10 in order to check whether or not a strong relative minimum is ensured.

We let $A = B = 0$ in (6.10.13) and (6.10.14), and we are led to a two-parameter family of solutions of the Euler-Lagrange equations of the Lagrange problem, namely,

$$\begin{aligned} r^2 &= C \\ \theta &= \frac{x}{C} + D \end{aligned} \tag{6.10.17}$$

Equations (6.10.17) represent a two-parameter family of helices that wind around circular cylinders with the x axis as axis and radius $\sqrt{C}$ and that climb at the rate $dx/d\theta = C$. (See Fig. 6.3.)

Along that helix, which corresponds to the parameter pair (C,D), we have

$$r' = 0$$
$$\theta' = \frac{1}{C}.$$

From (6.10.17), $C = r^2$, and hence

$$\begin{aligned} r' &= \phi_1(x,r,\theta) \equiv 0 \\ \theta' &= \phi_2(x,r,\theta) \equiv \frac{1}{r^2}. \end{aligned} \tag{6.10.18}$$

From (6.10.15), we have

$$\lambda(x,r,\theta) = -\frac{1}{2r^2}, \tag{6.10.19}$$

which, on the extremal $r = 1$, $\theta = x$, becomes the Lagrange multiplier $\lambda = -\frac{1}{2}$.

We shall now show that (6.10.18), (6.10.19) define a Mayer field for all $r > 0$. Since

$$g(x,r,\theta,r',\theta',\lambda) = \tfrac{1}{2}(r'^2 + r^2\theta'^2) + \lambda(r^2\theta' - 1)$$

and

$$g_r = r\theta'^2 + 2\lambda r\theta', \qquad g_\theta = 0,$$
$$g_{r'} = r', \qquad g_{\theta'} = r^2\theta' + \lambda r^2,$$

we obtain for $r' = 0$, $\theta' = 1/r^2$, $\lambda = -(1/2r^2)$,

$$g(x,r,\theta,\phi_1(x,r,\theta),\phi_2(x,r,\theta),\lambda(x,r,\theta)) = \frac{1}{2r^2}$$
$$g_r(x,r,\theta,\phi_1(x,r,\theta),\phi_2(x,r,\theta),\lambda(x,r,\theta)) = 0$$
$$g_\theta(x,r,\theta,\phi_1(x,r,\theta),\phi_2(x,r,\theta),\lambda(x,r,\theta)) = 0$$
$$g_{r'}(x,r,\theta,\phi_1(x,r,\theta),\phi_2(x,r,\theta),\lambda(x,r,\theta)) = 0$$
$$g_{\theta'}(x,r,\theta,\phi_1(x,r,\theta),\phi_2(x,r,\theta),\lambda(x,r,\theta)) = \tfrac{1}{2},$$

and hence

$$g_r - \frac{d}{dx}g_{r'} = 0 - 0 = 0,$$

$$g_\theta - \frac{d}{dx}g_{\theta'} = 0 - \frac{d}{dx}(\tfrac{1}{2}) = 0,$$

$$\frac{\partial g_{r'}}{\partial \theta} = 0, \qquad \frac{\partial g_{\theta'}}{\partial r} = 0.$$

Hence

$$\frac{\partial g_{r'}}{\partial \theta} = \frac{\partial g_{\theta'}}{\partial r}.$$

Thus we see that (6.10.18) together with $\lambda = -(1/2r^2)$ does indeed define a Mayer field according to Definition 6.10 and that the extremal (6.10.16) appears embedded in this Mayer field.

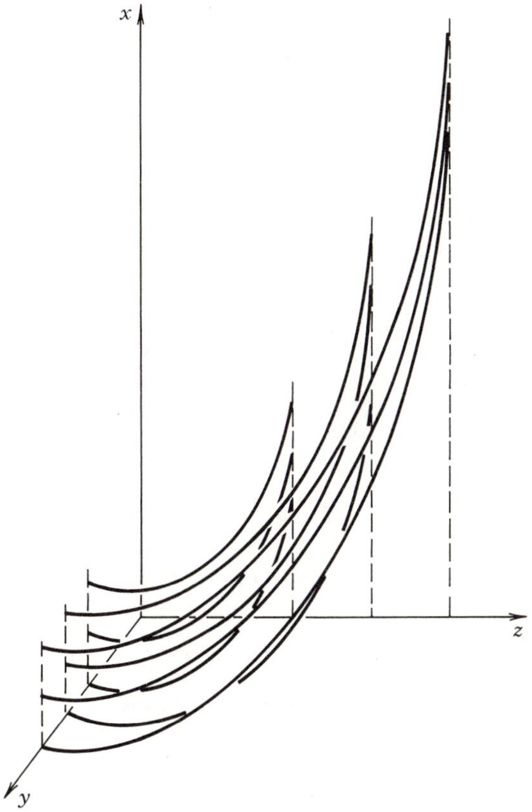

Figure 6.3

As a final step, we have to evaluate and discuss $\mathcal{E}$. From (6.10.10), we have

$$\mathcal{E}\,(x,r,\theta,\phi_1(x,r,\theta),\phi_2(x,r,\theta),\bar{r}',\bar{\theta}',\lambda(x,r,\theta)) = \tfrac{1}{2}\bar{r}'^2 > 0$$

for all $\bar{r}' \neq \phi_1(x,r,\theta) \equiv 0$. (Note that in view of the constraining equation, we have to have $\bar{\theta}' = \theta' = 1/r^2$.)

So we see that the extremal in (6.10.16) does indeed yield a strong relative minimum.

In this problem, no matter what the boundary conditions, the extremal, provided that it exists, is always a helix that winds around a surface of revolution with the x axis as axis of rotation, namely,

$$r^2 = Ax^2 + Bx + C.$$

Specifically, one obtains for $A > 0$, $CA - B^2 > 0$ a hyperboloid of one sheet; for $A > 0$, $CA - B^2 < 0$ a hyperboloid of two sheets; for $A > 0$, $CA - B^2 = 0$ a cone; for $A < 0$, $CA - B^2 < 0$ an ellipsoid; for $A = 0$, $B \neq 0$ a paraboloid; and for $A = B = 0$, $C > 0$ a cylinder. No other combinations will yield real surfaces. (See also Probs. 6.10.2 to 6.10.4.)

PROBLEMS 6.10

*1. Show that if $H = H(y,p,\lambda_o)$ does not explicitly depend on x, then H = constant in an integral of the Mayer equations of the Lagrange problem.

2. Consider the example that was discussed in this section, namely,

$$\frac{1}{2}\int_0^1 (y'^2(x) + z'^2(x))\,dx \to \text{minimum}, \qquad yz' - zy' - 1 = 0,$$

under the boundary conditions $y(0) = 1$, $z(0) = 0$, $y(1) = 1$, $z(1) = 1$. Show that the extremal is a straight line in (x,y,z) space and lies on a hyperboloid of one sheet.

3. Show that in a Lagrange problem, such as the one in problem 2, straight lines $y = cx + d$, $z = ax + b$ can only appear as extremals if $ad - cb = 1$.

4. Investigate the possibilities of constructing Mayer fields for a Lagrange problem such as the one in problem 2 with suitable boundary conditions if the extremal lies on:

   (*a*) A hyperboloid of one sheet
   (*b*) A hyperboloid of two sheets
   (*c*) A cone
   (*d*) An ellipsoid
   (*e*) A paraboloid

5. Given a Mayer field by $\phi = \phi(x,y)$ and $\lambda = \lambda(x,y)$. Show that the n-dimensional hypersurface, $W(x,y)$ = constant, is transversal to the integral curves of the field; i.e., every integral curve of the field satisfies the transversality condition on W = constant.

6. Impose suitable conditions and show that an n-parameter family of extremals of the Lagrange problem and the corresponding multipliers that satisfy the transversality conditions on a given n-dimensional hypersurface $\psi(x,y) = 0$ can be used to construct a Mayer field for the Lagrange problem in a neighborhood of that surface.

7. Show that

$$\mathcal{E}\,(x,y_o(x),y_o'(x),\bar{y}',\lambda(x)) \geq 0 \qquad \text{for all } x \in [a,b]$$

and for all $\bar{y}'$ for which $\varphi_\alpha(x,y_o(x),\bar{y}') = 0$ is a necessary condition for $y = y_o(x)$ to yield a strong relative minimum for the Lagrange problem. Hereby, $\lambda = \lambda(x)$ are the Lagrange multipliers that correspond to $y = y_o(x)$.

BRIEF SUMMARY

With the notation $\hat{y} = (y_o, \ldots, y_n)$, $\hat{\varphi} = (\varphi_o, \ldots, \varphi_\mu)$, $\mu < n$, the Mayer problem with a variable endpoint may be stated as follows: To be found is a vector function $\hat{y} = \hat{y}(x)$ such that

$$\hat{\varphi}(x,\hat{y},\hat{y}') = 0, \qquad \hat{y}(a) = \hat{y}_a,$$

so that for some b,

$$\psi_o(b,\hat{y}(b)) \to \text{minimum},$$

while

$$\psi_j(b,\hat{y}(b)) = 0, \qquad j = 1, 2, \ldots, k \leq n$$

(Secs. 6.2 and 6.8).

For $\hat{y} = \hat{y}(x) \in C^1[a,b]^n$ to be a solution of this problem, it is necessary that there exist a vector function $\hat{\lambda} = (\lambda_o, \ldots, \lambda_\mu) \in C[a,b]^{\mu+1}$, $\hat{\lambda} \neq 0$, such that the Mayer equations

$$h_{y_i}(x,\hat{y},\hat{y}',\hat{\lambda}) - \frac{d}{dx} h_{y'_i}(x,\hat{y},\hat{y}',\hat{\lambda}) = 0, \qquad i = 0, 1, \ldots, n,$$

are satisfied where

$$h = \sum_{\alpha=0}^{\mu} \lambda_\alpha \varphi_\alpha$$

(Secs. 6.2 and 6.7). In addition, there has to exist a constant vector $\hat{\nu} = (\nu_o, \ldots, \nu_k) \neq (0, \ldots, 0)$ such that the transversality conditions

$$\sum_{j=0}^{k} \nu_j \psi_{jx}(b,\hat{y}(b)) = -\sum_{i=0}^{n} h_{y'_i}(b,\hat{y}(b),\hat{y}'(b),\hat{\lambda}(b)) y'_i(b)$$

$$\sum_{j=0}^{k} \nu_j \psi_{jy_i}(b,\hat{y}(b)) = h_{y'_i}(b,\hat{y}(b),\hat{y}'(b),\hat{\lambda}(b)), \qquad i = 0, 1, \ldots, n,$$

are satisfied (Sec. 6.8).

When applied to the Lagrange problem of finding a vector function $y = (y_1, \ldots, y_n)$ such that, with $\varphi = (\varphi_1, \ldots, \varphi_\mu)$, $\mu < n$,

$$\varphi(x,y,y') = 0, \qquad y(a) = y_a,$$

$$\int_a^b f(x,y(x),y'(x))\, dx \to \text{minimum},$$

while

$$\psi_j(b,y(b)) = 0, \qquad j = 1, 2, \ldots, k \leq n,$$

the Mayer equations become

$$\lambda_o = \text{constant},$$

$$h^L_{y_i}(x,y,y',\hat{\lambda}) - \frac{d}{dx} h^L_{y'_i}(x,y,y',\hat{\lambda}) = 0, \qquad i = 1, 2, \ldots, n,$$

where $h^L = -\lambda_o f + \sum_{\alpha=1}^{\mu} \lambda_\alpha \varphi_\alpha$, (Sec. 6.3), and the transversality conditions become

$$\nu_o = \lambda_o$$

$$\sum_{j=1}^{n} \nu_j \psi_{jx}(b,y(b)) = -\lambda_o f(b,y(b),y'(b)) - \sum_{i=1}^{n} h^L_{y'_i}(b,y(b),y'(b),\hat{\lambda}(b)) y'_i(b)$$

$$\sum_{j=1}^{n} \nu_j \psi_{jy_i}(b,y(b)) = h^L_{y'_i}(b,y(b),y'(b),\hat{\lambda}(b)), \qquad i = 1, 2, \ldots n$$

(Sec. 6.9).

For the isoperimetric problem of finding a vector function $y = (y_1, \ldots, y_m)$ such that

$$\int_a^b f(x,y(x),y'(x))\,dx \to \text{minimum}, \qquad y(a) = y_a, \qquad y(b) = y_b,$$

and

$$\int_a^b f_\rho(x,y(x),y'(x))\,dx = l_\rho, \qquad \rho = 1,2, \ldots, \mu,$$

where $l_1, \ldots, l_\mu$ are given numbers, the Mayer equations assume the form

$$h^I_{y_i}(x,y,y',\hat{\lambda}) - \frac{d}{dx} h^I_{y'_i}(x,y,y',\hat{\lambda}) = 0, \qquad i = 1, 2, \ldots, m,$$

where $$h^I(x,y,y',\hat{\lambda}) = \lambda_o f(x,y,y') + \sum_{\rho=1}^{\mu} \lambda_\rho f_\rho(x,y,y')$$

and $(\lambda_o, \ldots,\lambda_\mu) \neq (0, \ldots,0)$, $\lambda_\rho =$ constant for all $\rho = 0, 1, \ldots, \mu$ (Secs. 6.5, 6.6).

If the solution $y = y_o(x)$ of the Lagrange problem with fixed endpoints, $\varphi(x,y,y') = 0$, $y(a) = y_a$, $y(b) = y_b$, $\int_a^b f(x,y(x),y'(x))\,dx \to$ minimum, is embeddable in a field, i.e., if there exists in a weak neighborhood of $y = y_o(x)$ two vector functions $\phi = \phi(x,y)$, $\lambda = \lambda(x,y)$ such that $y'_o = \phi(x,y_o)$, $\varphi(x,y,\phi(x,y)) = 0$, and for the solutions of $y' = \phi(x,y)$, $g_{y_k} - (d/dx)\, g_{y'_k} = 0$, $k = 1, 2, \ldots, n$, $(\partial/\partial y_i)\, g_{y'_k} = (\partial/\partial y_k)\, g_{y'_i}$, $i, k = 1, 2, \ldots, n$, where

$$g(x,y,\phi(x,y),\lambda(x,y)) = f(x,y,\phi(x,y)) + \sum_{\alpha=1}^{\mu} \lambda_\alpha(x,y)\varphi_\alpha(x,y,\phi(x,y)),$$

and if

$$\mathcal{E}(x,y,\phi(x,y),\bar{y}',\lambda(x,y)) \geq 0$$

for all (x,y) in a weak neighborhood of $y = y_o(x)$ and all $\bar{y}'$ for which $\varphi(x,y,\bar{y}') = 0$, then $y = y_o(x)$ yields a strong relative minimum. Hereby,

$$\mathcal{E}(x,y,y',\bar{y}',\lambda) = g(x,y,\bar{y}',\lambda) - g(x,y,y',\lambda) + \sum_{i=1}^{n} (y'_i - \bar{y}'_i) g_{y'_i}(x,y,y',\lambda)$$

(Sec. 6.10).

APPENDIX

A6.11 ON THE AUGMENTATION OF A MATRIX

This section is devoted exclusively to the proof of a theorem that was used in the augmentation of the system of constraining equation (6.7.1) in the proof of Lemma 6.2.2:

Theorem A6.11 *Let* $a_{\nu k} = a_{\nu k}(x) \in C[a,b]$, $\nu = 1, 2, \ldots, \mu < n$, $k = 1, 2, \ldots, n$, *represent the elements of the* $\mu \times n$ *matrix*

$$A = \begin{pmatrix} a_{11}(x) & \cdots & a_{1n}(x) \\ \vdots & & \\ a_{\mu 1}(x) & \cdots & a_{\mu n}(x) \end{pmatrix},$$

and let

$$\text{rank } A = \mu \qquad \textit{for all } x \in [a,b].$$

Then one can find $(n - \mu) \cdot n$ *functions* $a^*_{\mu+j,k} \in C[a,b]$, $j = 1, 2, \ldots, n - \mu$, $k = 1, 2, \ldots, n$, *such that*

$$\textit{rank } A^* = n \qquad \textit{for all } x \in [a,b],$$

where

$$A^* = \begin{pmatrix} a_{11}(x) & \cdots & a_{1n}(x) \\ \vdots & & \\ a_{\mu 1}(x) & \cdots & a_{\mu n}(x) \\ a^*_{\mu+1,1}(x) & \cdots & a^*_{\mu+1,n}(x) \\ \vdots & & \\ a^*_{n1}(x) & \cdots & a^*_{nn}(x) \end{pmatrix}.$$

Proof: If $x_o \in [a,b]$, then, by hypothesis, there exists a nonvanishing major in A, say

$$\begin{vmatrix} a_{1k_1}(x_o) & \cdots & a_{1k_\mu}(x_o) \\ \vdots & & \\ a_{\mu k_1}(x_o) & \cdots & a_{\mu k_\mu}(x_o) \end{vmatrix} \neq 0,$$

where $\{k_1, k_2, \ldots, k_\mu\} \subset \{1, 2, \ldots, n\}$.

Now, we choose constants $a'_{\mu+j,k}$, $j = 1, 2, \ldots, n - \mu$, $k = 1, 2, \ldots, n$, such that

$$\det |A'| = \begin{vmatrix} a_{1k_1}(x_o) & \cdots & a_{1k_\mu}(x_o) & a_{1k_{\mu+1}}(x_o) & \cdots & a_{1k_n}(x_o) \\ \vdots & & & & & \\ a_{\mu k_1}(x_o) & \cdots & a_{\mu k_\mu}(x_o) & a_{\mu k_{\mu+1}}(x_o) & \cdots & a_{\mu k_n}(x_o) \\ a'_{\mu+1,k_1} & \cdots & a'_{\mu+1,k_\mu} & a'_{\mu+1,k_{\mu+1}} & \cdots & a'_{\mu+1,k_n} \\ \vdots & & & & & \\ a'_{nk_1} & \cdots & a'_{nk_\mu} & a'_{nk_{\mu+1}} & \cdots & a'_{nk_n} \end{vmatrix} \neq 0,$$

This is always possible. All we do is choose $a'_{\mu+j,k_{\mu+i}}$, $i = 1, 2, \ldots, n - \mu$,

$j = 1, 2, \ldots, n - \mu$, such that

$$\begin{vmatrix} a'_{\mu+1,k_{\mu+1}} & \cdots & a'_{\mu+1,k_n} \\ \vdots & & \\ a'_{nk_{\mu+1}} & \cdots & a'_{nk_n} \end{vmatrix} \neq 0$$

and let the remaining $a'_{\mu+j,k_\rho} = 0$. Then, by Laplace's theorem,

$$\det |A'| = \left|\begin{array}{ccc:ccc} a_{1k_1}(x_o) & \cdots & a_{1k_\mu}(x_o) & a_{1k_{\mu+1}}(x_o) & \cdots & a_{1k_n}(x_o) \\ \vdots & & & & & \\ a_{\mu k_1}(x_o) & \cdots & a_{\mu k_\mu}(x_o) & a_{\mu k_{\mu+1}}(x_o) & \cdots & a_{\mu k_n}(x_o) \\ \hdashline 0 & \cdots & 0 & a'_{\mu+1,k_{\mu+1}} & \cdots & a'_{\mu+1,k_n} \\ \vdots & & & & & \\ 0 & \cdots & 0 & a'_{nk_{\mu+1}} & \cdots & a'_{nk_n} \end{array}\right|$$

$$= \pm \begin{vmatrix} a_{1k_1}(x_o) & \cdots & a_{1k_\mu}(x_o) \\ \vdots & & \\ a_{\mu k_1}(x_o) & \cdots & a_{\mu k_\mu}(x_o) \end{vmatrix} \cdot \begin{vmatrix} a'_{\mu+1,k_{\mu+1}} & \cdots & a'_{\mu+1,k_n} \\ \vdots & & \\ a'_{nk_{\mu+1}} & \cdots & a'_{nk_n} \end{vmatrix} \neq 0.$$

Such an augmentation of the matrix A is possible at each point of the interval $[a,b]$, but the augmented constants may, of course, differ from point to point.

Since $a_{ik} \in C[a,b]$, it follows that for each such point x_o, there is a neighborhood where the determinant of the augmented matrix does not vanish either. By the Heine-Borel theorem, we can select a finite number, say r, such neighborhoods so that this finite collection will cover the entire interval $[a,b]$. In the overlapping portions of these neighborhoods, we select points $x_1, x_2, \ldots, x_{r-1}$. Then we can say that in each closed interval $[x_k,x_{k+1}]$, $k = 0,1, \ldots . r - 1$, $a = x_o$, $b = x_r$, the matrix A may be augmented by constants as outlined above in such a manner that the augmented matrix A' has rank n in the entire interval $[x_k,x_{k+1}]$, that is $\det |A'| \neq 0$.

We take the first interval $[a,x_1] = [x_o,x_1]$ and augment the matrix A by constants in the indicated manner so that $\det |A'| \neq 0$ in $[x_o,x_1]$. We call this augmented matrix A'_1, and we know that the elements of A'_1 are continuous functions of x and $\det |A'| \neq 0$ in $[x_o,x_1]$.

We repeat this procedure for the first h intervals, and we obtain matrices $A'_1, A'_2, \ldots, A'_h$ such that $\det |A'_i| \neq 0$ in $[x_{i-1},x_i]$, $i = 1, 2, \ldots, h$. We assume that these matrices $A'_1, A'_2, \ldots, A'_h$ can be replaced by matrices $A^*_1, A^*_2, \ldots, A^*_h$ such that $\det |A^*_i| \neq 0$ on $[x_{i-1},x_i]$, $i = 1, 2, \ldots, h$, and such that the elements of A^*_i are continuous extensions of the elements of A^*_{i-1} on $[x_{i-2},x_{i-1}]$ into $[x_{i-1},x_i]$, $i = 1, 2, \ldots, h$.

Now we shall show *by induction* that this can be done for all $i = 1, 2, \ldots, r$ by demonstrating that A'_{h+1} can be replaced by a matrix A^*_{h+1} such that $\det |A^*_{h+1}| \neq 0$ on $[x_h, x_{h+1}]$ and that the elements of A^*_{h+1} are continuous extensions of the elements of A_h^*. (Observe that during the entire process, the elements of the first μ rows remain unchanged!)

We have

$$A_h^* = \begin{pmatrix} a_{11}(x) & \cdots & a_{1n}(x) \\ \vdots & & \\ a_{\mu 1}(x) & \cdots & a_{\mu n}(x) \\ a_{\mu+1,1}(x) & \cdots & a_{\mu+1,n}(x) \\ \vdots & & \\ a_{n1}(x) & \cdots & a_{nn}(x) \end{pmatrix}$$

and

$$A'_{h+1} = \begin{pmatrix} a_{11}(x) & \cdots & a_{1n}(x) \\ \vdots & & \\ a_{\mu 1}(x) & \cdots & a_{\mu n}(x) \\ a'_{\mu+1,1} & \cdots & a'_{\mu+1,n} \\ \vdots & & \\ a'_{n1} & \cdots & a'_{nn} \end{pmatrix}$$

We define

$$A_{h+1}^* = CA'_{h+1}$$

where

$$C = \begin{pmatrix} 1 & 0 & \cdots & 0 & 0 & \cdots & 0 \\ 0 & 1 & \cdots & 0 & 0 & \cdots & 0 \\ \vdots & & & & & & \\ 0 & 0 & \cdots & 1 & 0 & \cdots & 0 \\ c_{\mu+1,1} & c_{\mu+1,2} & \cdots & c_{\mu+1,\mu} & c_{\mu+1,\mu+1} & \cdots & c_{\mu+1,n} \\ \vdots & & & & & & \\ c_{n1} & c_{n2} & \cdots & c_{n\mu} & c_{n\mu+1} & \cdots & c_{nn} \end{pmatrix}$$

and where the c_{ik} are constants.

Then

$$A_{h+1}^* = \begin{pmatrix} a_{11}(x) & \cdots & a_{1n}(x) \\ \vdots & & \\ a_{\mu 1}(x) & \cdots & a_{\mu n}(x) \\ \bar{a}_{\mu+1,1}(x) & \cdots & \bar{a}_{\mu+1,n}(x) \\ \vdots & & \\ \bar{a}_{n1}(x) & \cdots & \bar{a}_{nn}(x) \end{pmatrix},$$

where

$$\bar{a}_{ik}(x) = c_{i1}a_{1k}(x) + \cdots + c_{i\mu}a_{\mu k}(x) + c_{i,\mu+1}a'_{\mu+1,k} + \cdots + c_{in}a'_{nk},$$

$i = \mu + 1, \ldots, n;\ k = 1, 2, \ldots, n$. We try to determine the c_{ij} in such a manner that

$$\bar{a}_{\mu+j,k}(x_h) = a_{\mu+j,k}(x_h), \qquad j = 1, 2, \ldots, n - \mu;\ k = 1, 2, \ldots, n,$$

where the $a_{\mu+j,k}(x)$ are the elements of the preceding matrix A_h^*. This can be done because we have for each set of unknowns c_{ij}, i fixed, $j = 1, 2, \ldots, n$, the following system of nonhomogeneous linear equations:

$$c_{i1}a_{1k}(x_h) + \cdots + c_{i\mu}a_{\mu k}(x_h) + c_{i,\mu+1}a'_{\mu+1,k} + \cdots + c_{in}a'_{nk} = a_{ik}(x_h),$$

$k = 1, 2, \ldots, n$, which has $\det |A'_{h+1}(x_h)|$ as coefficient determinant, and this determinant is, by hypothesis, different from zero. Hence there exists a unique solution $c_{i1}, c_{i2}, \ldots, c_{in}$ for each $i = \mu + 1, \mu + 2, \ldots, n$.

All that is left to be shown is that $\det |A^*_{h+1}| \neq 0$ on $[x_h, x_{h+1}]$. By construction,

$$\det |A^*_{h+1}| = \det |C| \det |A'_{h+1}|.$$

At $x = x_h$, we have $A^*_{h+1} = A^*_h$ and, by hypothesis, $\det |A^*_h| \neq 0$ at x_h. Hence

$$\det |C| \det |A'_{h+1}(x_h)| = \det |A^*_{h+1}(x_h)| = \det |A^*_h(x_h)| \neq 0,$$

and it follows that $\det |C| \neq 0$. Thus,

$$\det |A^*_{h+1}(x)| = \det |C| \det |A'_{h+1}(x)| \neq 0 \quad \text{on} \quad [x_h, x_{h+1}]. \quad \text{Q.E.D.}$$

A6.12 A LAGRANGE PROBLEM WITH FINITE CONSTRAINTS

We shall now formulate a variational problem with fixed endpoints where the constraining equations are not differential equations but ordinary equations. With the notation of Sec. 6.10, we can state the problem as follows:

To be found is a vector function $y = y(x)$ which satisfies the equations

$$\varphi^*(x,y) = 0, \qquad \varphi^* = (\varphi_1, \ldots, \varphi_\mu), \tag{A6.12.1}$$

and the boundary conditions

$$y(a) = y_a, \qquad y(b) = y_b, \tag{A6.12.2}$$

and is such that

$$\int_a^b f(x,y(x),y'(x))\, dx \to \text{minimum}. \tag{A6.12.3}$$

[We remind the reader that $y = (y_1, \ldots, y_n)$,] We assume that the boundary conditions are compatible with the constraining equations (A6.12.1), that is,

$$\varphi^*(a,y_a) = 0, \qquad \varphi^*(b,y_b) = 0.$$

Such a problem is called a *Lagrange problem with finite constraints.*

The multiplier rule (Theorem 6.3) is not immediately applicable because one condition that is essential for the validity of the multiplier rule, namely, (6.3.6) rank $(\partial\varphi_\alpha^*/\partial y_k') = \mu$, is *not* satisfied. As a matter of fact, $(\partial\varphi_\alpha^*/\partial y_k') = 0$ for all α, k, and we have rank $(\partial\varphi_\alpha^*/\partial y_k') = 0$.

One could deal with this problem as follows: Whenever possible, solve the constraining equations for $y_{k_1}, \ldots, y_{k_\mu}$ for some $\{k_1, \ldots, k_\mu\} \subset \{1,2, \ldots, n\}$ in terms of x and the remaining y's, substitute the result into (A6.12.3), and arrive in this manner at an ordinary variational problem without constraints.

We shall use a different approach. We shall change the constraints to differential-equation constraints and then apply the Lagrange multiplier rule.

If $y = y(x)$ is a solution of the Lagrange problem with finite constraints, then

$$\varphi^*(x,y(x)) = 0$$

for all $x \in [a,b]$ and hence, by necessity,

$$\frac{d}{dx}\varphi^*(x,y(x)) = 0$$

for all $x \in [a,b]$.

Since

$$\frac{d}{dx}\varphi^*(x,y(x)) = \varphi_x^*(x,y(x)) + \sum_{k=1}^{n} \varphi_{y_k}^*(x,y(x))y_k'(x),$$

it follows that the solution will have to satisfy, by necessity, the constraining differential equations

$$\varphi(x,y,y') \equiv \varphi_x^*(x,y) + \sum_{k=1}^{n} \varphi_{y_k}^*(x,y)y_k' = 0.$$

We find that

$$\varphi_{y'_k} = \varphi_{y_k}^*$$

and hence
$$\operatorname{rank}\left(\frac{\partial\varphi_\alpha}{\partial y_k'}\right) = \operatorname{rank}\left(\frac{\partial\varphi_\alpha^*}{\partial y_k}\right).$$

Thus, if we require that

$$\operatorname{rank}\left(\frac{\partial\varphi_\alpha^*}{\partial y_k}\right) = \mu \qquad \text{for all } x \in [a,b] \tag{A6.12.4}$$

and
$$\varphi^* \in C^2([a,b] \times \{y \mid -\infty < y_i < \infty, i = 1, 2, \ldots, n\}) \tag{A6.12.5}$$

then we can apply Theorem 6.3:

It is necessary that there exist $\mu + 1$ continuous functions

$(\lambda_o, \ldots, \lambda_\mu) \neq (0, \ldots, 0)$, λ_o = constant, such that

$$h^L_{y_i} - \frac{d}{dx} h^L_{y'_i} = 0, \qquad i = 1, 2, \ldots, n.$$

where $h^L = -\lambda_o f + \sum_{\alpha=1}^{\mu} \lambda_\alpha (\varphi^*_{\alpha x} + \sum_{k=1}^{n} \varphi^*_{\alpha y_k} y'_k)$.

A simplification of this rule can be obtained if λ' is continuous, e.g., if $\varphi^* \in C^3$ and $y \in C^2[a,b]^n$.

We have in that case

$$h^L_{y_i} = -\lambda_o f_{y_i} + \sum_{\alpha=1}^{\mu} \lambda_\alpha (\varphi^*_{\alpha x y_i} + \sum_{k=1}^{n} \varphi^*_{\alpha y_k y_i} y'_k)$$

and $$h^L_{y'_i} = -\lambda_o f_{y'_i} + \sum_{\alpha=1}^{\mu} \lambda_\alpha \varphi^*_{\alpha y_i},$$

$$\frac{d}{dx} h^L_{y'_i} = -\lambda_o \frac{d}{dx} f_{y'_i} + \sum_{\alpha=1}^{\mu} \lambda'_\alpha \varphi^*_{\alpha y_i} + \sum_{\alpha=1}^{\mu} \lambda_\alpha (\varphi^*_{\alpha y_i x} + \sum_{k=1}^{n} \varphi^*_{\alpha y_i y_k} y'_k).$$

Hence the Mayer equation appears, after appropriate cancellations, as

$$-\lambda_o f_{y_i} + \lambda_o \frac{d}{dx} f_{y'_i} - \sum_{\alpha=1}^{\mu} \lambda'_\alpha \varphi^*_{\alpha y_i} = 0.$$

If we let $\lambda'_\alpha = -\lambda^*_\alpha$ and $h^* = -\lambda_o f + \sum_{\alpha=1}^{\mu} \lambda^*_\alpha \varphi^*_\alpha$, we see that

$$h^*_{y_i} = -\lambda_o f_{y_i} + \sum_{\alpha=1}^{\mu} \lambda^*_\alpha \varphi^*_{\alpha y_i},$$

$$h^*_{y'_i} = -\lambda_o f_{y'_i},$$

and we can write the Mayer equation in the form

$$h^*_{y_i} - \frac{d}{dx} h^*_{y'_i} = 0, \qquad i = 1, 2, \ldots, n.$$

Thus we can state:

Theorem A6.12 *If $y = y(x) \in C^2[a,b]^n$ is the solution of the Lagrange problem with finite constraints [(A6.12.1) to (A6.12.3)], where it is assumed that the conditions (A6.12.4) and $\varphi^* \in C^3$ are fulfilled, then it is necessary that there exist μ continuous functions $\lambda^* = (\lambda^*_1, \ldots, \lambda^*_\mu)$ such that*

$$h^*_{y_i}(x,y,y',\lambda^*) - \frac{d}{dx} h^*_{y'_i}(x,y,y',\lambda^*) = 0, \qquad i = 1, 2, \ldots, n,$$

where $$h^*(x,y,y',\lambda^*) = -\lambda_o f(x,y,y') + \sum_{\alpha=1}^{\mu} \lambda_\alpha^* \varphi_\alpha^*(x,y)$$

and where λ_o = *constant.*

[Note that $(\lambda_o,\lambda_1^*, \ldots,\lambda_\mu^*) \neq (0, \ldots,0)$ is not necessary anymore since $\lambda_\alpha^* = -\lambda'_\alpha$, where λ_α are the Lagrange multipliers of Theorem 6.3.]

PROBLEM A6.12

Impose suitable conditions and derive a multiplier rule for the Lagrange problem with mixed constraints:

$$\int_a^b f(x,y_1(x), \ldots,y_n(x),y_1'(x), \ldots,y_n'(x))\, dx \to \text{minimum},$$

$y_i(a) = y_i^a$, $y_i(b) = y_i^b$,

$$\begin{aligned}
&\varphi_1(x,y_1, \ldots,y_n,y_1', \ldots,y_n') = 0\\
&\vdots\\
&\varphi_\rho(x,y_1, \ldots,y_n,y_1', \ldots,y_n') = 0\\
&\psi_{\rho+1}(x,y_1, \ldots,y_n) = 0\\
&\vdots\\
&\psi_\mu(x,y_1, \ldots,y_n) = 0, \qquad \mu < n.
\end{aligned}$$

CHAPTER 7

THE THEORY OF THE SECOND VARIATION

APPENDIX

7.1 NECESSARY AND SUFFICIENT CONDITIONS FOR A WEAK MINIMUM

The results that will be obtained in this chapter are, with one exception (Jacobi's necessary condition), in one form or another embodied in Chap. 3, where they appeared on the fringes of the theory of fields. We shall now achieve these results by an entirely different process, namely, by a study of the second variation of a functional. The Jacobi equation will again appear quite naturally and will thus establish a link between the investigations of this chapter and those of Chap. 3.

For the convenience of the reader, we shall briefly recapitulate some of the concepts and results from Chap. 1.

We shall assume that the functional $I[y]$ is defined on an open subset Y of a normed linear space $\mathcal{S}$ and that it possesses a second Gâteaux variation which satisfies the conditions of Theorem 1.8.1. That is, for all $y_o \in Y$,

$$I[y_o + h] - I[y_o] = \delta I[h] + \tfrac{1}{2}\delta^2 I[h] + \alpha[h]$$

for all $h \in \mathcal{S}$ for which $||h|| < \delta$ for some $\delta > 0$ and $\lim_{t\to 0} (\alpha[th]/t^2) = 0$. We called $\delta^2 I[h]$ the second variation of $I[y]$ at $y = y_o$.

We saw in Chap. 1 that if $I[y]$ possesses a relative minimum at $y = y_o \in Y$ relative to Σ and Σ admits a linear space $\mathcal{H}$ of admissible variations, then it is necessary that

$$\delta I[h] = 0 \qquad \text{for all } h \in \mathcal{H}, \tag{7.1.1}$$

and

$$\delta^2 I[h] \geq 0 \qquad \text{for all } h \in \mathcal{H}. \tag{7.1.2}$$

(See Theorem 1.8.2.)

We saw also that if $I[y]$ possesses a second Fréchet differential $B_f[h,h]$ at $y = y_o$ (see Prob. 1.8.6), then

$$B_f[h,h] \geq \mu ||h||^2 \tag{7.1.3}$$

for fixed $\mu > 0$ and all $h \in \{h \mid h \in \mathcal{S}, h = y - y_o, \forall y \in \Sigma\}$ is sufficient for a relative minimum at $y = y_o$ (see Prob. 1.8.7). (If $B_f[h,h]$ satisfies (7.1.3), then the second variation is called *strongly positive.*)

In all the preceding chapters and, in particular, in Chap. 2, we were preoccupied with the interpretation of condition (7.1.1) for various variational problems and we were led one way or another to an Euler-Lagrange equation, or Euler-Lagrange equations. In this chapter, we shall concern ourselves chiefly with condition (7.1.2). Condition (7.1.3) is, as we shall see in Sec. 7.4, not very practical because it is too strong. Still, we shall see that a practical sufficient condition for a weak relative minimum can be found by a study of the second variation (Sec. 7.5).

Let us state at the outset, and then not mention it again, that the necessary conditions that will be derived in this chapter for the fixed-beginning-point–fixed-endpoint problem hold as well for the variable-beginning-point and/or variable-endpoint problem. As we have mentioned repeatedly in Chap. 2, a curve which yields a relative minimum for the latter problem, by necessity yields a relative minimum for a problem with fixed beginning point and fixed endpoint.

We saw in Sec. 1.8 that if

$$f = f(x,y,y') \in C^2(\mathcal{R}),$$

where $\mathcal{R}$ is a domain in (x,y,y') space that contains all lineal elements of $y = y_o(x) \in C^1[a,b]$, then

$$I[y] = \int_a^b f(x,y(x),y'(x))\,dx, \qquad y(a) = y_a, \qquad y(b) = y_b,$$

possesses a second Gâteaux variation at $y = y_o(x)$ which satisfies the conditions of Theorem 1.8.1, and we obtained for the second variation

$$\begin{aligned}\delta^2 I[h] = \int_a^b [&f_{yy}(x,y_o(x),y_o'(x))h^2(x) \\ &+ 2f_{yy'}(x,y_o(x),y_o'(x))h(x)h'(x) \\ &+ f_{y'y'}(x,y_o(x),y_o'(x))h'^2(x)]\,dx\end{aligned} \tag{7.1.4}$$

[See (1.8.3).] This formula will serve as the point of departure in our investigation of the second variation. Our primary goal will be to express the condition (7.1.2) in terms of f and y_o without reference to the space of admissible variations.

PROBLEMS 7.1

1. Find $\delta^2 I[h]$ for the following functionals:

(a) $I[y] = \int_0^{\pi/2} (y'^2(x) - y^2(x))\,dx$

(b) $I[y] = \int_0^1 y'^3(x)\,dx$

(c) $I[y] = \int_0^1 (1 - y'^2(x))^2\,dx$

(d) $I[y] = \int_{-1}^1 y^2(x)(1 - y'(x))^2\,dx$

(e) $I[y] = \int_0^1 \sqrt{1 + y'^2(x)}\,dx$

(f) $I[y] = \int_a^b y(x)\sqrt{1 + y'^2(x)}\,dx$

(g) $I[y] = \int_a^b \frac{\sqrt{1 + y'^2(x)}\,dx}{\sqrt{y(x)}}$

2. Given the homogeneous problem $I[\gamma] = \int_{Pa}^{Pb} F(x,y,\dot{x},\dot{y})\,dt \to$ minimum. Assume that the extremal arc $x = x_o(t)$, $y = y_o(t)$ can be represented by $y = y_o(x) \in C^1[a,b]$ and find $\delta^2 I[\gamma]$. (See also Sec. 4.1, page 202.)

7.2 LEGENDRE'S NECESSARY CONDITION

We consider the variational problem

$$I[y] = \int_a^b f(x,y(x),y'(x))\,dx \to \text{minimum}, \qquad y(a) = y_a, \qquad y(b) = y_b, \tag{7.2.1}$$

where we now assume that $f \in C^2(\mathcal{R})$, $\mathcal{R}$ being a domain in (x,y,y') space. By condition (7.1.2) and in view of (7.1.4), for $y = y_o(x) \in C^1[a,b]$, where $(x,y_o(x),y_o'(x)) \in \mathcal{R}$ for all $x \in [a,b]$, to yield a relative minimum for $I[y]$, it is necessary that

$$\delta^2 I[h] = \int_a^b [(f_{yy})_o h^2(x) + 2(f_{yy'})_o h(x)h'(x) + (f_{y'y'})_o h'^2(x)]\,dx \geq 0 \tag{7.2.2}$$

for all $h \in \mathcal{H}$, where

$$\mathcal{H} = \{h \mid h \in \mathcal{C}^1[a,b],\, h(a) = h(b) = 0\}.$$

In (7.2.2), $(\quad)_o$ means "to be taken at $(x,y_o(x),y_o'(x))$."

In order to achieve some simplifications of (7.2.2), we shall assume for the moment that $f \in C^3(\mathcal{R})$ and $y_o \in C^2[a,b]$. Then we can apply integration by parts to the middle term of the expression for $\delta^2 I[h]$ in (7.2.2) and obtain

$$2\int_a^b (f_{yy'})_o hh'\,dx = (f_{yy'})_o h^2 \Big|_a^b - \int_a^b h^2 \frac{d}{dx}(f_{yy'})_o\,dx.$$

Since $h(a) = h(b) = 0$, the first term vanishes, and we obtain

$$\delta^2 I[h] = \int_a^b \left[\left(f_{yy} - \frac{d}{dx} f_{yy'}\right)_o h^2 + (f_{y'y'})_o h'^2\right] dx,$$

where f_{yy}, $f_{yy'}$, $f_{y'y'}$ are to be taken at $(x,y_o(x),y_o'(x))$ and are to be considered as known functions of x.

We introduce the abbreviating notation

$$\begin{aligned} \alpha(x) &= f_{y'y'}(x,y_o(x),y_o'(x)), \\ \beta(x) &= f_{yy}(x,y_o(x),y_o'(x)) - \frac{d}{dx} f_{yy'}(x,y_o(x),y_o'(x)) \end{aligned} \tag{7.2.3}$$

Then we can write

$$\delta^2 I[h] = \int_a^b [\alpha(x)h'^2(x) + \beta(x)h^2(x)]\,dx. \tag{7.2.4}$$

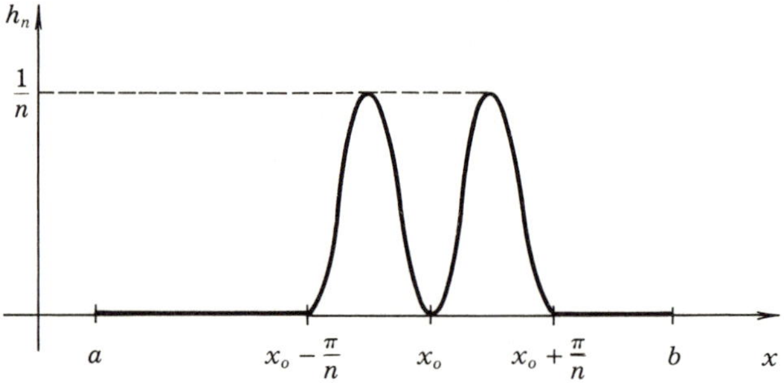

Figure 7.1

Since (7.2.2) has to hold for all $h \in \mathcal{H}$, it has to hold, in particular, for the functions

$$h_n = h_n(x) \equiv \begin{cases} \frac{1}{n} \sin^2 n(x - x_o) & \text{for } |x - x_o| \leq \frac{\pi}{n} \\ 0 & \text{for } |x - x_o| > \frac{\pi}{n} \end{cases} \tag{7.2.5}$$

(see Fig. 7.1), where $x_o \in (a,b)$ and where we choose n sufficiently large so that $(x_o - \pi/n, x_o + \pi/n) \subset [a,b]$. Clearly, $h_n \in \mathcal{H}$.

For $h = h_n(x)$, we obtain

$$\int_a^b \alpha(x) h_n'^2(x)\, dx = 4 \int_{x_o - \pi/n}^{x_o + \pi/n} \alpha(x) \sin^2 n(x - x_o) \cos^2 n(x - x_o)\, dx$$

$$= 4\alpha\left(x_o + \Theta \frac{\pi}{n}\right) \int_{x_o - \pi/n}^{x_o + \pi/n} \sin^2 n(x - x_o) \cos^2 n(x - x_o)\, dx$$

$$= \alpha\left(x_o + \Theta \frac{\pi}{n}\right) \frac{\pi}{n},$$

where $|\Theta| \leq 1$. We also have

$$\left| \int_a^b h_n^2(x) \beta(x)\, dx \right| = \frac{1}{n^2} \left| \int_{x_o - \pi/n}^{x_o + \pi/n} \beta(x) \sin^4 n(x - x_o)\, dx \right| \leq \frac{2M\pi}{n^3}$$

since $|\beta(x)| \leq M$ on $[a,b]$ for some M in view of $f \in C^3(\mathcal{R})$. Thus we can say that, in view of (7.2.4),

$$\delta^2 I[h_n] = \alpha\left(x_o + \Theta \frac{\pi}{n}\right) \frac{\pi}{n} + \int_a^b \beta(x) h_n^2(x)\, dx, \tag{7.2.6}$$

where

$$\left|\int_a^b \beta(x)h_n^2(x)\,dx\right| \le \frac{2M\pi}{n^3}.$$

Suppose now that $\alpha(x_o) < 0$ for some $x_o \in (a,b)$. Then there exists a $\delta > 0$ such that

$$\alpha(x) \le -m^2 \qquad \text{for all } |x - x_o| < \delta \text{ and for some } m.$$

Now we choose n sufficiently large so that $\pi/n < \delta$ and $(2M/n^2) - m^2 < 0$. Then, by (7.2.6),

$$\delta^2 I[h_n] \le \frac{\pi}{n}\left(-m^2 + \frac{2M}{n^2}\right) < 0$$

and not ≥ 0, as it should be.

We note that a similar estimate can be obtained without the transformation of the middle term (see Prob. 7.2.3) and that we can free ourselves from the restricting assumptions $f \in C^3(\mathcal{R})$, $y_o \in C^2[a,b]$. (See also Sec. 7.7.)

Thus we can state:

Theorem 7.2 Legendre's condition *For $y = y_o(x) \in C^1[a,b]$ to yield a relative minimum (maximum) for $I[y]$, it is necessary that*

$$f_{y'y'}(x,y_o(x),y_o'(x)) \ge 0 \qquad (\le 0)$$

for all $x \in [a,b]$.

Proof: We have seen above that if for some $x_o \in (a,b)$ the condition is violated, then it is possible to find an admissible h_n such that the second variation becomes negative, as opposed to (7.2.2). Hence $\alpha(x) \ge 0$ for all $x \in (a,b)$. That the condition also has to hold at the endpoints of the interval follows from the continuity of $\alpha(x)$ in $[a,b]$.

Note that the same condition was obtained in Theorem 3.9.2. (See also Prob. 3.9.5.)

PROBLEMS 7.2

1. Check whether or not the Legendre condition is satisfied for the variational problems in Prob. 7.1.1. When of relevance, state the range of y, y' where the Legendre condition is satisfied.
2. Derive the Legendre condition for the homogeneous problem under the assumption stated in Prob. 7.1.2.

*3. Derive the Legendre condition directly, without transforming the second variation into the form (7.2.4).

7.3 BLISS' SECONDARY VARIATIONAL PROBLEM AND JACOBI'S NECESSARY CONDITION

If $y = y_o(x)$ is a regular extremal of $I[y]$, then by Definition 2.5.2,

$$f_{y'y'}(x,y_o(x),y_o'(x)) \neq 0 \qquad \text{for all } x \in [a,b].$$

But then, for $y = y_o(x)$ to yield a relative minimum for $I[y]$, it is necessary according to Theorem 7.2 that

$$f_{y'y'}(x,y_o(x),y_o'(x)) > 0 \qquad \text{for all } x \in [a,b]. \tag{7.3.1}$$

We call this the *strengthened Legendre condition.*

We shall show in this section that, in this case, the open interval (a,b) cannot contain a point a^* conjugate to a, and thus we shall arrive at a new necessary condition for a minimum. For this purpose, we shall assume that $y_o \in C^2[a,b]$ (see Theorem 2.5.2) and $f \in C^3(\mathfrak{R})$.

We shall use the abbreviating notation

$$\begin{aligned} P(x) &= f_{yy}(x,y_o(x),y_o'(x)), \\ Q(x) &= f_{yy'}(x,y_o(x),y_o'(x)), \\ R(x) &= f_{y'y'}(x,y_o(x),y_o'(x)), \end{aligned}$$

and we consider, as suggested by *Gilbert A. Bliss* (1876–1951),

$$\delta^2 I[h] = \int_a^b [P(x)h^2(x) + 2Q(x)h(x)h'(x) + R(x)h'^2(x)]\,dx \to \text{minimum}$$

$$h(a) = h(b) = 0, \tag{7.3.2}$$

as a variational problem with $\mathcal{H}$ as space of competing functions—the so-called *secondary variational problem.*

For $h = h(x)$ to yield a minimum, it is necessary that h satisfy the Euler-Lagrange equation of the secondary variational problem, which is, in view of

$$f_h = 2Ph + 2Qh', \qquad f_{h'} = 2Qh + 2Rh',$$

the *Jacobi equation* of Chap. 3 [see (3.5.4), where $\alpha = R$, $\beta = P - Q'$]:

$$\frac{d}{dx}(Rh') + (Q' - P)h = 0. \tag{7.3.3}$$

Thus we can characterize the Jacobi equation as the Euler-Lagrange equation of the secondary variational problem.

Suppose now that $a < a^* < b$, that is, there exists a nontrivial solution $h = h^*(x)$ of the Jacobi equation (7.3.3) such that

$$h^*(a) = h^*(a^*) = 0, \qquad \text{but } h^*(x) \neq 0 \text{ for } x \in (a,a^*).$$

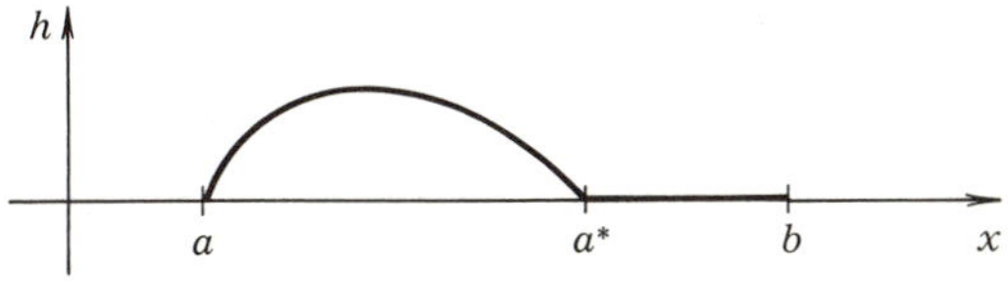

Figure 7.2

Since $h = h(x) \equiv 0$ is also a solution of the Jacobi equation, the function

$$h = \bar{h}(x) \equiv \begin{cases} h^*(x) & \text{for } a \leq x \leq a^* \\ 0 & \text{for } a^* \leq x \leq b \end{cases}$$

is a sectionally smooth solution of the Euler-Lagrange equation of the secondary variational problem. (See Fig. 7.2.)

We have

$$\delta^2 I[\bar{h}] = \int_a^{a^*} (Ph^{*2} + 2Qh^*h^{*\prime} + Rh^{*\prime 2})\, dx$$

$$= \int_a^{a^*} [(P - Q')h^{*2} + Rh^{*\prime 2}]\, dx$$

$$= Rh^*h^{*\prime}\Big|_a^{a^*} - \int_a^{a^*} h^*\left[\frac{d}{dx}(Rh^{*\prime}) + (Q' - P)h^*\right] dx = 0.$$

Suppose now that 0 is the minimum of $\delta^2 I[h]$. Then it is necessary that $h = \bar{h}(x)$, which yields the value 0 (which we assume is the minimum), satisfy the Weierstrass-Erdmann corner condition (2.9.1) at $x = a^*$:

$$g_{h'}\big|_{a^*-0} = g_{h'}\big|_{a^*+0},$$

where $g = Rh'^2 + (P - Q')h^2$. Since $g_{h'} = 2Rh'$, we have for $h = \bar{h}(x)$,

$$g_{h'}\big|_{a^*-0} = 2R(a^*)h^{*\prime}(a^*)$$

and

$$g_{h'}\big|_{a^*+0} = 0.$$

By hypothesis, $R(x) > 0$ for all $x \in [a,b]$. Hence the corner condition can only be satisfied if $h^{*\prime}(a^*) = 0$. Since we also have $h^*(a^*) = 0$, there follows from the uniqueness theorem that $h^*(x) = 0$ for all $x \in [a,a^*]$, as opposed to our assumption that $h = h^*(x)$ is a nontrivial solution of the Jacobi equation.

Since $h = \bar{h}(x)$ does not satisfy the necessary condition for a minimum, $\delta^2 I[\bar{h}] = 0$ cannot be the minimum of the second variation relative to functions with a corner at $x = a^*$. Hence there exists a function $h = \bar{\bar{h}}(x)$ with a corner at $x = a^*$ such that $\delta^2 I[\bar{\bar{h}}] < 0$. Then, by Corollary 1 to Theorem 2.9 (fairing theorem), there exists a function $h \in \mathcal{C}^1[a,b]$, $h(a) = h(b) = 0$, such that $\delta^2 I[h] < 0$.

Thus we can state:

Lemma 7.3 *If $R(x) > 0$ for all $x \in [a,b]$ and if (a,b) contains a conjugate point to a, then it is possible to find a function $h \in \mathcal{H}$ such that*

$$\delta^2 I[h] < 0.$$

Lemma 7.3 leads immediately to:

Theorem 7.3: Jacobi's necessary condition *For the regular extremal $y = y_0(x) \in C^2[a,b]$ to yield a weak relative minimum for $I[y]$, it is necessary that (a,b) contain no point that is conjugate to a.*

Proof: By (7.1.2) for $y = y_0(x)$ to yield a weak minimum for $I[y]$ it is necessary that

$$\delta^2 I[h] \geq 0 \qquad \text{for all } h \in \mathcal{H}.$$

If (a,b) contains a conjugate point to a, then by Lemma 7.3, it is possible to find an admissible h so that $\delta^2 I[h] < 0$. Hence, by necessity, (a,b) cannot contain a conjugate point to a.

To illustrate Lemma 7.3, let us consider the example of Sec. 3.4:

$$I[y] = \int_0^b (y'^2(x) - y^2(x))\,dx \to \text{minimum}$$

$$y(0) = 0, \qquad y(b) = 1.$$

We have $R(x) = f_{y'y'} = 2 > 0$, and we obtain the Jacobi equation

$$h'' + h = 0$$

with the general solution

$$h = c \sin(x - \Delta).$$

We see that $0^* = \pi$, that is, π is the conjugate point to 0. Since $f_{yy'} = 0$, $f_{yy} = -2$, we have for the second variation,

$$\delta^2 I[h] = 2 \int_0^b (h'^2(x) - h^2(x))\,dx.$$

If $b = 3\pi/2$, then $(0,b)$ contains the conjugate point π, and we should be able to make the second variation negative by a suitable choice of h.

Take

$$h = \frac{1}{2} \sin \frac{2x}{3}.$$

Then $h(0) = h(3\pi/2) = 0$, and we obtain

$$\delta^2 I[h] = 2 \int_0^{3\pi/2} (h'^2 - h^2)\,dx = \frac{1}{2} \int_0^{3\pi/2} \left(\frac{13}{9} \cos^2 \frac{2x}{3} - 1\right) dx = -\frac{5\pi}{24} < 0.$$

When we discussed this problem in Sec. 3.4, we saw that for $b > \pi$, an extremal does not yield a relative minimum. Theorem 7.3 leads us to the same result by an entirely different method.

PROBLEMS 7.3

1. Solve the Jacobi equation for the variational problems in Prob. 7.1.1b and c and check whether or not the Jacobi condition (Theorem 7.3) is satisfied.
2. Same as in problem 1 for the variational problem

$$I[y] = \int_0^1 y^2(x)\,(1 - y'^2(x))^2\,dx \to \text{minimum}, \qquad y(0) = 0, \qquad y(1) = 1.$$

3. Same as in problem 1 for the variational problem

$$I[y] = \int_0^b y(x)\sqrt{1 + y'^2(x)}\,dx \to \text{minimum}, \qquad y(0) = 1, \qquad y(b) = \cosh b.$$

7.4 LEGENDRE'S TRANSFORMATION OF THE SECOND VARIATION

In an attempt to establish a sufficient condition for at least a weak (relative) minimum, Legendre subjected the integrand of the second variation to a certain transformation that might be vaguely described as a generalized "completing the square" process. This transformation will be the subject treated in this section.

We assume that $y = y_o(x) \in C^1[a,b]$ is a solution of the Euler-Lagrange equation and satisfies the boundary conditions $y(a) = y_a$, $y(b) = y_b$. With the notation of Sec. 7.3, we write the second variation as

$$\delta^2 I[h] = \int_a^b [P(x)h^2(x) + 2Q(x)h(x)h'(x) + R(x)h'^2(x)]\,dx.$$

We shall now add to the integrand an expression, the integral of which vanishes. This expression will ultimately be chosen in such a manner that the integrand of $\delta^2 I[h]$ may be written as a complete square.

We have for any $u \in C^1[a,b]$

$$\int_a^b \frac{d}{dx}[u(x)h^2(x)]\,dx = u(x)h^2(x)\Big|_a^b = 0 \tag{7.4.1}$$

because $h(a) = h(b) = 0$. Hence we do not change $\delta^2 I[h]$ if we augment it by (7.4.1). We obtain then, if we also note that $(uh^2)' = u'h^2 + 2uhh'$,

$$\delta^2 I[h] = \int_a^b \left[Ph^2 + 2Qhh' + Rh'^2 + \frac{d}{dx}(uh^2)\right]dx$$

$$= \int_a^b [(P + u')h^2 + 2(Q + u)hh' + Rh'^2]\,dx.$$

We shall assume from now on that $y = y_o(x)$ is a *regular extremal*, that is,

$$R(x) = f_{y'y'}(x,y_o(x),y_o'(x)) \neq 0 \qquad \text{for all } x \in [a,b].$$

Then we can transform the integrand as follows:

$$(P + u')h^2 + 2(Q + u)hh' + Rh'^2$$
$$= R\left(h' + \frac{Q+u}{R}h\right)^2 + \left(P + u' - \frac{(Q+u)^2}{R}\right)h^2.$$

As yet, we have not imposed any restrictions on u other than demanding that $u \in C^1[a,b]$. We shall now try to determine u in such a manner that

$$P + u' - \frac{(Q+u)^2}{R} = 0, \tag{7.4.2}$$

so that the second term in the integrand of the second variation drops out and we are left with

$$\delta^2 I[h] = \int_a^b R(x)\left[h'(x) + \frac{Q(x) + u(x)}{R(x)}h(x)\right]^2 dx. \tag{7.4.3}$$

The process that we just described is *Legendre's transformation of the second variation.*

Let us now investigate whether it is at all possible to determine a function $u = u(x)$ that satisfies (7.4.2). We write (7.4.2) as

$$u' = -P + \frac{(Q+u)^2}{R} \tag{7.4.4}$$

and recognize this as a *Riccati equation* for u. So it seems that in order to carry out the Legendre transformation, one needs a solution of the Riccati equation (7.4.4) for the entire interval $[a,b]$.

This appears to be rather difficult since a solution of a Riccati equation cannot, in general, be obtained by elementary methods, unless one already knows a particular solution. Also, since the equation is nonlinear, the general existence theorems will only guarantee a solution locally and not in the entire interval.

Fortunately, these difficulties can be circumvented by a transformation of the Riccati equation into a linear second-order equation, as Jacobi has demonstrated. Jacobi proposed to introduce, instead of $u = u(x)$, a new function $\eta = \eta(x)$ by means of the substitution

$$u(x) = -R(x)\frac{\eta'(x)}{\eta(x)} - Q(x), \tag{7.4.5}$$

where we have to assume that $\eta(x) \neq 0$ in $[a,b]$. Then

$$u' = -R'\frac{\eta'}{\eta} - R\frac{\eta''\eta - \eta'^2}{\eta^2} - Q',$$

and we obtain, instead of the Riccati equation (7.4.4),

$$\frac{d}{dx}(R\eta') + (Q' - P)\eta = 0, \tag{7.4.6}$$

which is again the *Jacobi equation* (7.3.3).

We recall from the definition of the norm that $||h|| = 0$ if and only if $h = 0$. With this in mind, we give the following definition:

Definition 7.4 *$\delta^2 I[h]$ is called* positive definite *if, with $h \in \mathcal{H}$,*

$$\delta^2 I[h]\begin{cases} > 0 & \text{for all } ||h|| \neq 0 \\ = 0 & \text{if } ||h|| = 0. \end{cases}$$

We can now state the following preliminary result:

Lemma 7.4.1 *If the Jacobi equation (7.4.6) has a solution $\eta = \eta(x) \neq 0$ for all $x \in [a,b]$ and if $R(x) > 0$ for all $x \in [a,b]$, then $\delta^2 I[h]$ is positive definite.*

Proof: First we note that if the Jacobi equation has a solution that does not vanish in the entire interval $[a,b]$, then the Legendre transformation can be carried out, and we can write $\delta^2 I[h]$ in the form (7.4.3).

Then we note that, since $R(x) > 0$ for all $x \in [a,b]$, the only way $\delta^2 I[h]$ can vanish is that

$$h' + \frac{Q+u}{R}h = 0 \qquad \text{for all } x \in [a,b],$$

$h \in \mathcal{H}$, that is, $h(a) = h(b) = 0$, $h \in C^1[a,b]$. This is a linear first-order differential equation for h, and its solution is uniquely determined by its initial value $h(a) = 0$. But $h(x) \equiv 0$ is that solution and hence the only one. Therefore, if $h \in \mathcal{H}$ and $h(x) \not\equiv 0$, then $h' + [(Q+u)/R]h \neq 0$, and hence $\delta^2 I[h] > 0$.

This lemma has a generalization, the significance of which is not apparent at the moment but will become so in Sec. 7.5, where we shall derive a sufficient condition for a weak minimum. Here is the generalization:

Lemma 7.4.2 *If the Jacobi equation (7.4.6) has a solution $\eta = \eta(x) \neq 0$ in*

$[a,b]$ and if $R(x) > 0$ for all $x \in [a,b]$, then there exists a $\lambda_o > 0$ such that

$$\delta^2 I_\lambda[h] = \int_a^b [Ph^2 + 2Qhh' + (R - \lambda^2)h'^2]\,dx$$

is positive definite for all $|\lambda| < \lambda_o$.

Proof: Since $R(x) > 0$ for all $x \in [a,b]$, there exists a λ_1 such that $R(x) - \lambda^2 > 0$ for all $|\lambda| < \lambda_1$.

Let us now subject the integrand of $\delta^2 I_\lambda[h]$ to the Legendre transformation. We realize that all we have to do is replace $R(x)$ by $R(x) - \lambda^2$ in (7.4.5) and (7.4.6) to come up with the Jacobi equation

$$\frac{d}{dx}[(R - \lambda^2)\eta'(x,\lambda)] + (Q' - P)\eta(x,\lambda) = 0,$$

the solutions $\eta = \eta(x,\lambda)$ of which will be uniformly continuous for $x \in [a,b]$ and $|\lambda| \le \lambda_2$ for some λ_2. For definiteness, we consider that solution $\eta = \eta(x,\lambda)$ for which $\eta(a,\lambda) = \eta(a)$, $\eta'(a,\lambda) = \eta'(a)$. Then $\eta(x,0) \equiv \eta(x)$.

By hypothesis,

$$\eta(x) = \eta(x,0) \neq 0 \qquad \text{for all } x \in [a,b].$$

Hence there is a $\lambda_3 > 0$ such that $\eta(x,\lambda) \neq 0$ for all $x \in [a,b]$ and all $|\lambda| < \lambda_3$, and we can write

$$\delta^2 I_\lambda[h] = \int_a^b (R - \lambda^2)\left(h' + \frac{Q + u(x,\lambda)}{R - \lambda^2}\,h\right)^2 dx.$$

It follows, as in the proof of Lemma 7.4.1, that $\delta^2 I_\lambda[h]$ is positive definite for all $|\lambda| < \lambda_o = \min(\lambda_1,\lambda_2,\lambda_3)$.

Corollary to Lemma 7.4.2 *If $R(x) > 0$ for all $x \in [a,b]$ and if the interval $(a,b]$ does not contain a conjugate point to a, then $\delta^2 I_\lambda[h]$ is positive definite for sufficiently small $|\lambda|$.*

Proof: By Definition 3.6, the hypothesis that $(a,b]$ contains no conjugate point to a means that the Jacobi equation (7.4.6) has a solution that vanishes at a but does not vanish again in $(a,b]$. Then, by Theorem 3.6.2, there exists an interval $(a - \Delta, b]$, $\Delta > 0$, that does not contain a conjugate point to $a - \Delta$ either; that is, there is a solution of the Jacobi equation that does not vanish in $[a,b]$. The corollary follows now from Lemma 7.4.2.

In order to achieve a sufficient condition for a weak (relative) minimum, one might be tempted to try to establish that, under the conditions of Lemma 7.4.1, the second variation is not only *positive definite* but also *strongly positive.* This is quite impossible, as a simple example will show.

Suppose we want to demonstrate that if $R(x) > 0$ in $[a,b]$ and if $(a,b]$ does not contain a conjugate point to a, then there exists a constant $\mu > 0$ such that

$$B_f[h,h] \geq \mu||h||^2 \qquad \text{for all } h \in \mathcal{H}.$$

In particular, this must be true for the example

$$I[y] = \frac{1}{2}\int_0^1 y'^2(x)\, dx \to \text{minimum}, \qquad y(0) = 0, \qquad y(1) = 0.$$

Clearly, $y = 0$ yields the absolute and strong minimum of this functional and hence, also, a weak and relative minimum.

We have $R(x) = f_{y'y'} = 1 > 0$, and we obtain for the Jacobi equation

$$\eta'' = 0.$$

The solution $\eta = x$, which vanishes at $x = 0$, never vanishes again. Hence $(0,1]$ does not contain a conjugate point to 0.

We note that $B_f[h,h] = \delta^2 I[h]$ provided that $B_f[h,h]$ exists, and we also note that $B_f[h,h]$ exists for the above functional. (See Probs. 7.4.3 and 7.4.4.) We have

$$B_f[h,h] = \delta^2 I[h] = \int_0^1 h'^2(x)\, dx,$$

and we shall show that no matter how small we choose $\mu > 0$, there is always an admissible $h \in \{h \mid h \in C^1[a,b],\ h = y - 0,\ \forall y \in \Sigma\}$ such that $\delta^2 I[h] < \mu||h||^2$.

First, we consider a sequence of functions $\bar{h}_n$ which are not smooth, namely,

$$\bar{h}_n = \bar{h}_n(x) \equiv \begin{cases} nx & \text{for } 0 \leq x \leq \dfrac{1}{n} \\ 1 & \text{for } \dfrac{1}{n} < x < 1 - \dfrac{1}{n} \\ -nx + n & \text{for } 1 - \dfrac{1}{n} \leq x \leq 1. \end{cases}$$

We shall take care of the corners at $x = 1/n$ and $x = 1 - (1/n)$ later by invoking the fairing theorem (Corollary 1 to Theorem 2.10). We see that

$$||\bar{h}_n|| = \max_{[0,1]} |\bar{h}_n(x)| + \sup_{[0,1]} |\bar{h}'_n(x)| = 1 + n$$

and

$$\delta^2 I[\bar{h}_n] = \int_0^{1/n} n^2\, dx + \int_{1-1/n}^1 n^2\, dx = 2n$$

Since $[2n/(1 + n)^2] \to 0$ as $n \to \infty$, we have for any $\mu > 0$, no matter how small, an n such that $[2n/(1 + n)^2] < \mu$, and hence

$$\delta^2 I[\bar{h}_n] = 2n < \mu(1 + n)^2 = \mu||\bar{h}_n||^2.$$

When we round off the corners of $\bar{h}_n$ at $x = 1/n$ and $x = 1 - (1/n)$ and, by this process, replace $\bar{h}_n$ by a smooth function h_n, we note that $||\bar{h}_n|| = ||h_n||$. Hence the upper bound in the above inequality will not change, and we have from the fairing theorem that the corners can be rounded off in such a manner that the integral $\int_0^1 h_n'^2(x)\,dx$ remains smaller than $\mu||h_n||^2$.

Thus we have seen that for any $\mu > 0$, no matter how small, there is a smooth function h_n with $h_n(0) = h_n(1) = 0$ such that

$$\delta^2 I[h_n] < \mu||h_n||^2.$$

We shall see in Sec. 7.5 that it is nevertheless possible to obtain a practical sufficient condition for a weak relative minimum under the hypotheses of Lemma 7.4.1, but not in terms of (7.1.3).

The following theorem will serve to round out the discussions of this and the preceding section:

Theorem 7.4 *If $R(x) = f_{y'y'}(x,y_o(x),y_o'(x)) > 0$ for all $x \in [a,b]$, then $\delta^2 I[h]$ is positive definite if and only if $(a,b]$ contains no conjugate point to a.*

Proof: If $(a,b]$ does not contain a conjugate point to a, then, by the corollary to Lemma 7.4.2, $\delta^2 I[h]$ is positive definite. That the converse is also true may be seen as follows:

If $\delta^2 I[h]$ is positive definite, then $\delta^2 I[h] \geq 0$ for all admissible functions h. Then, by Lemma 7.3, it follows that (a,b) cannot contain a conjugate point to a.

So the only case still to be taken care of is the case $a^* = b$. Now, if $a^* = b$, then there exists a nontrivial solution $h = h(x)$ of the Jacobi equation such that $h(a) = h(b) = 0$, and we have

$$0 = \int_a^b \left[\frac{d}{dx}(Rh') - (P - Q')h\right]h\,dx = Rhh'\Big|_a^b$$

$$- \int_a^b [Rh'^2 + (P - Q')h^2]\,dx = -\delta^2 I[h],$$

that is, $\delta^2 I[h] = 0$ for some $||h|| \neq 0$, which contradicts our assumption that $\delta^2 I[h]$ is positive definite. Hence b cannot be a conjugate point to a either.

PROBLEMS 7.4

1. Carry out the Legendre transformation of $\delta^2 I[h]$ for the variational problem

$$I[y] = \int_0^b (y'^2(x) - y^2(x))\,dx \to \text{minimum}, \qquad 0 < b < \pi.$$

2. Same as in problem 1 for

$$I[y] = \int_0^b y(x)\sqrt{1 + y'^2(x)}\,dx \to \text{minimum},$$

$y(0) = 1$, $y(b) = \cosh b$, $0 < b < 1$.

*3. Show that if $B_f[h,h]$ exists for a given functional $I[y]$ at $y = y_o$, then $B_f[h,h] = \delta^2 I[h]$, where $\delta^2 I[h]$ is the second Gâteaux variation at $y = y_o$. (See also Probs. 1.8.6 and 1.8.7.)

*4. Given $I[y] = \frac{1}{2}\int_0^1 y'^2(x)\,dx$. Find $B_f[h,h]$.

7.5 A SUFFICIENT CONDITION FOR A WEAK RELATIVE MINIMUM

In Chap. 3, we studied the embeddability of an extremal $y = y_o(x) \in C^1[a,b]$ into a field and the Weierstrass excess function. In the course of this study, we were led to a sufficient condition for a weak relative minimum, namely, that $y = y_o(x)$ be embeddable in a field and that

$$f_{y'y'}(x,y_o(x),y_o'(x)) > 0 \qquad \text{for all } x \in [a,b]$$

(Theorem 3.8.1).

In this section, we shall obtain the same condition but without reference to fields, embeddability, and excess function. Our argument will be strictly on the basis of the corollary to Lemma 7.4.2.

First we note that if $f \in C^2(\mathfrak{R})$ and $(x,y_o(x),y_o'(x)) \in \mathfrak{R}$ for all $x \in [a,b]$, then

$$\begin{aligned}\Delta I &= I[y_o + h] - I[y_o] \\ &= \int_a^b [f(x,y_o(x) + h(x),y_o'(x) + h'(x)) - f(x,y_o(x),y_o'(x))]\,dx \\ &= \delta I[h] + \frac{1}{2}\int_a^b [\bar{f}_{yy}h^2(x) + 2\bar{f}_{yy'}h(x)h'(x) + \bar{f}_{y'y'}h'^2(x)]\,dx,\end{aligned}$$

where the quantities designated by a bar are to be taken at

$$(x,\, y_o(x) + \Theta_1(x)h(x),\, y_o'(x) + \Theta_2(x)h'(x)),$$

whereby $|\Theta_i(x)| \leq 1$ for all $x \in [a,b]$.

If we introduce the functions

$$\begin{aligned}\varepsilon_1(x) &= f_{yy}(x,\, y_o(x) + \Theta_1(x)h(x),\, y_o'(x) + \Theta_2(x)h'(x)) \\ &\quad - f_{yy}(x,y_o(x),y_o'(x)) \\ \varepsilon_2(x) &= f_{yy'}(x,\, y_o(x) + \Theta_1(x)h(x),\, y_o'(x) + \Theta_2(x)h'(x)) \\ &\quad - f_{yy'}(x,y_o(x),y_o'(x)) \\ \varepsilon_3(x) &= f_{y'y'}(x,\, y_o(x) + \Theta_1(x)h(x),\, y_o'(x) + \Theta_2(x)h'(x)) \\ &\quad - f_{y'y'}(x,y_o(x),y_o'(x))\end{aligned}$$

and assume that $y = y_o(x)$ is an extremal of $I[y]$, that is, $\delta I[h] = 0$, we can write the total variation as follows:

$$\begin{aligned}\Delta I &= \frac{1}{2}\int_a^b [f_{yy}(x,y_o(x),y_o'(x))h^2(x) + 2f_{yy'}(x,y_o(x),y_o'(x))h(x)h'(x)\\ &\quad + f_{y'y'}(x,y_o(x),y_o'(x))h'^2(x)]\,dx\\ &\quad + \frac{1}{2}\int_a^b [\varepsilon_1(x)h^2(x) + 2\varepsilon_2(x)h(x)h'(x) + \varepsilon_3(x)h'^2(x)]\,dx\\ &= \frac{1}{2}\,\delta^2 I[h] + \frac{1}{2}\int_a^b [\varepsilon_1(x)h^2(x) + 2\varepsilon_2(x)h(x)h'(x) + \varepsilon_3(x)h'^2(x)]\,dx.\end{aligned} \tag{7.5.1}$$

By the *Cauchy-Buniakovskii-Schwarz inequality*,

$$\begin{aligned}h^2(x) = \left(\int_a^x h'(x)\,dx\right)^2 &\le \int_a^x dx \int_a^x h'^2(x)\,dx\\ &\le (x-a)\int_a^b h'^2(x)\,dx,\end{aligned}$$

and hence
$$\int_a^b h^2(x)\,dx \le \frac{(b-a)^2}{2}\int_a^b h'^2(x)\,dx. \tag{7.5.2}$$

Also,

$$\left|\int_a^b h(x)h'(x)\,dx\right| \le \sqrt{\int_a^b h^2(x)\,dx}\sqrt{\int_a^b h'^2(x)\,dx} \le \frac{b-a}{\sqrt{2}}\int_a^b h'^2(x)\,dx,$$

that is,

$$\left|\int_a^b h(x)h'(x)\,dx\right| \le \frac{b-a}{\sqrt{2}}\int_a^b h'^2(x)\,dx. \tag{7.5.3}$$

We also note that $\lim_{||h||\to 0} \varepsilon_i(x) = 0$ uniformly in x (see Prob. 7.5.1) and hence that there is for any $\varepsilon > 0$, a $\delta_\varepsilon > 0$ independent of x such that

$$|\varepsilon_i(x)| < \varepsilon \qquad \text{provided that } ||h|| < \delta_\varepsilon. \tag{7.5.4}$$

Thus, if $||h|| < \delta_\varepsilon$, then we have, in view of (7.5.2) to (7.5.4),

$$\begin{aligned}\left|\int_a^b (\varepsilon_1 h^2 + 2\varepsilon_2 hh' + \varepsilon_3 h'^2)\,dx\right| &\le \varepsilon\left[\int_a^b h^2\,dx + 2\left|\int_a^b hh'\,dx\right| + \int_a^b h'^2\,dx\right]\\ &\le \varepsilon\left[\frac{(b-a)^2}{2} + 2\,\frac{b-a}{\sqrt{2}} + 1\right]\int_a^b h'^2\,dx,\end{aligned}$$

and we see that we can find for any $\varepsilon^* > 0$, an

$$\varepsilon = \frac{\varepsilon^*}{[(b-a)^2/2] + [2(b-a)/\sqrt{2}] + 1}$$

and, in turn, a $\delta_\varepsilon > 0$ such that

$$\left| \int_a^b (\varepsilon_1 h^2 + 2\varepsilon_2 hh' + \varepsilon_3 h'^2)\, dx \right| < \varepsilon^* \int_a^b h'^2\, dx, \tag{7.5.5}$$

provided that $||h|| < \delta_\varepsilon$.

Let us now assume that $R(x) > 0$ for all $x \in [a,b]$ and that $(a,b]$ does not contain a conjugate point to a. Then, by the corollary to Lemma 7.4.2, we have that

$$\delta^2 I_\lambda[h] = \int_a^b [Ph^2 + 2Qhh' + (R - \lambda^2)h'^2]\, dx$$

is positive definite for all λ that are sufficiently small, that is,

$$\delta^2 I[h] = \int_a^b (Ph^2 + 2Qhh' + Rh'^2)\, dx > \lambda^2 \int_a^b h'^2\, dx$$

for all $h \in \mathcal{H}$, $||h|| \neq 0$.

Hence, by (7.5.1) and (7.5.5),

$$\Delta I > \tfrac{1}{2}\left[\delta^2 I[h] - \varepsilon^* \int_a^b h'^2\, dx\right] > \tfrac{1}{2}[\lambda^2 - \varepsilon^*] \int_a^b h'^2\, dx > 0$$

if $\lambda^2 > \varepsilon^*$ and $||h|| \neq 0$. Since ε^* can be chosen arbitrarily small, we see that ΔI remains positive so long as $||h|| \neq 0$ is sufficiently small, and we have:

Theorem 7.5: Sufficient condition for a weak minimum *The regular extremal $y = y_o(x) \in C^1[a,b]$ yields a weak relative minimum (maximum) if $(a,b]$ does not contain a conjugate point to a and if $f_{y'y'}(x,y_o(x),y_o'(x)) > 0$ (< 0) for all $x \in [a,b]$.*

(This is essentially Theorem 3.8.1, where we called an extremal that satisfies the strengthened Legendre condition a *weakly regular extremal*—see Definition 3.8.)

PROBLEMS 7.5

*1. Show that $\lim_{\|h\|\to 0} \varepsilon_i(x) = 0$ uniformly in x, where $\varepsilon_i(x)$ are defined on page 405.

2. Are the conditions of Theorem 7.5 satisfied for the following variational problems?

(a) $\displaystyle\int_0^{\pi/2} [y'^2(x) - y^2(x)]\,dx \to \text{minimum},\ y(0) = 0,\ y(\pi/2) = 1$

(b) $\displaystyle\int_0^{\pi} [y'^2(x) - y^2(x)]\,dx \to \text{minimum},\ y(0) = 0,\ y(\pi) = 0$

(c) $\displaystyle\int_0^{1} y'^3(x)\,dx \to \text{minimum},\ y(0) = 0,\ y(1) = 0$

(d) $\displaystyle\int_0^{b} y(x)\sqrt{1 + y'^2(x)}\,dx \to \text{minimum},\ y(0) = 1,\ y(b) = \cosh b$

(e) $\displaystyle\int_0^{1} (y'(x) - 1)^2\,dx \to \text{minimum},\ y(0) = 0,\ y(1) = 0$

7.6 SCHEMATIC REVIEW OF THE SIMPLEST VARIATIONAL PROBLEM

We find it appropriate at this time to let all the necessary and sufficient conditions which we have found for a solution of the simplest variational problem pass in review.

To wit, we consider the problem

$$I[y] = \int_a^b f(x,y(x),y'(x))\,dx \to \text{minimum (maximum)}$$

$$y(a) = y_a$$

$$y(b) = y_b.$$

NECESSARY CONDITIONS

For the *extremal* $y = y_o(x) \in C^1[a,b]$ to yield a *strong* relative minimum (maximum) for $I[y]$, it is necessary that y_o satisfy the conditions N1 *and* N2 *and* N4.

For the *regular extremal* $y = y_o(x) \in C^1[a,b]$ to yield a *strong* relative minimum (maximum) for $I[y]$, it is necessary that y_o satisfy the conditions N1 *and* N3 *and* N4 *and* N5.

For the *extremal* $y = y_o(x) \in C^1[a,b]$ to yield a *weak* relative minimum (maximum) for $I[y]$, it is necessary that y_o satisfy the conditions N1 *and* N2.

For the *regular extremal* $y = y_o(x) \in C^1[a,b]$ to yield a *weak* relative minimum (maximum) for $I[y]$, it is necessary that y_o satisfy the conditions N1 *and* N3 *and* N5.

N1 Euler-Lagrange equation (Corollary 2 to Theorem 2.3)

$$f_y(x,y_o(x),y_o'(x)) - \frac{d}{dx} f_{y'}(x,y_o(x),y_o'(x)) = 0.$$

N2 Legendre condition (Theorem 7.2)

$$f_{y'y'}(x,y_o(x),y_o'(x)) \begin{Bmatrix} \geq 0 \text{ for minimum} \\ \leq 0 \text{ for maximum} \end{Bmatrix} \qquad \text{for all } x \in [a,b].$$

N3 Strengthened Legendre condition (7.3.1)

$$f_{y'y'}(x,y_o(x),y_o'(x)) \begin{Bmatrix} > 0 \text{ for minimum} \\ < 0 \text{ for maximum} \end{Bmatrix} \qquad \text{for all } x \in [a,b].$$

N4 Weierstrass' necessary condition (Theorem 3.9.1)

$$\mathcal{E}(x,y_o(x),y_o'(x),\bar{y}') \begin{Bmatrix} \geq 0 \text{ for minimum} \\ \leq 0 \text{ for maximum} \end{Bmatrix} \qquad \text{for all } x \in [a,b]$$

and all $-\infty < \bar{y}' < \infty$.

N5 Jacobi condition (Theorem 7.3)

(a,b) contains *no* conjugate point to a.

SUFFICIENT CONDITIONS

The *extremal* $y = y_o(x) \in C^1[a,b]$ yields a *strong* relative minimum (maximum) for $I[y]$ if the following conditions are satisfied: N1 *and* S1 *and* S2, *or* N1 *and* S3 *and* S2.

The *extremal* $y = y_o(x) \in C^1[a,b]$ yields a *weak* relative minimum (maximum) for $I[y]$ if the following conditions are satisfied: N1 *and* S3 *and* N3.

S1 y_o is embeddable in a field (Definition 3.1.2)

S2 Weierstrass condition (Theorem 3.3.2)

$$\mathcal{E}(x,y,y',\bar{y}') \begin{Bmatrix} \geq 0 \text{ for minimum} \\ \leq 0 \text{ for maximum} \end{Bmatrix} \qquad \text{for all } (x,y) \in N_w^\delta(y_o)$$

and all $-\infty < y' < \infty$ and $-\infty < \bar{y}' < \infty$.

S3 Strengthened Jacobi condition (Theorem 3.7.2)

$(a,b]$ does *not* contain a conjugate point to a.

For the special example

$$\int_0^b (y'^2(x) - y^2(x))\,dx \to \text{minimum}, \qquad y(0) = 0, \qquad y(b) = 1, \qquad b \neq k\pi,$$

we found the *Euler-Lagrange equation*

$$y'' + y = 0$$

with the *extremal*

$$y = y_o(x) \equiv \frac{\sin x}{\sin b}.$$

We obtained the *Jacobi equation*

$$\eta'' + \eta = 0$$

with a solution for $\eta(0) = 0$,

$$\eta = \sin x.$$

We have $f_{y'y'} = 2$ and the *excess function*

$$\mathcal{E}(x,y,y',\bar{y}') = (y' - \bar{y}')^2.$$

For $0 < b < \pi$, a field is defined by $(1,\phi)$, where

$$\phi(x,y) = \frac{y \sin b + \tan\Delta \cos x}{\tan\Delta \cos x + \sin x}\left(\frac{\cos x - \tan\Delta \sin x}{\sin b}\right) + \frac{\sin x}{\sin b}\tan\Delta,$$

whereby $0 < \Delta < \pi - b$, $b < \pi$.

If $0 < b < \pi$, then $y = y_o(x) \equiv (\sin x)/(\sin b)$ satisfies the *Euler-Lagrange equation* (N1), the *Legendre condition* $f_{y'y'} = 2 \geq 0$ (N2), the *strengthened Legendre condition* $f_{y'y'} = 2 > 0$ (N3), the *Weierstrass necessary condition* $\mathcal{E}(x,y_o(x),y_o'(x),\bar{y}') = (y_o'(x) - \bar{y}')^2 \geq 0$ for all $-\infty < \bar{y}' < \infty$ (N4), and the *Jacobi condition* $0^* = \pi \geq b$ (N5).

On the basis of these observations, we *cannot* state that $y = y_o(x) \equiv (\sin x)/(\sin b)$ is the solution to our problem. *Nor* can we state that it is not the solution to our problem, because all the necessary conditions are satisfied.

But we know further that $y = y_o(x) \equiv (\sin x)/(\sin b)$ is *embeddable in the field* given above (S1), satisfies the *Weierstrass condition* $\mathcal{E}(x,y,y',\bar{y}') = (y' - \bar{y}')^2 \geq 0$ for all $-\infty < y' < \infty$, $-\infty < \bar{y}' < \infty$ (S2) [this, incidentally, suffices for the conclusion that $y = y_o(x)$ yields a strong relative minimum], and satisfies the *strengthened Jacobi condition* $0^* = \pi > b$ (S3). Also, since $f_{y'y'} = 2 > 0$, we know that y_o is a *regular extremal.* So we may conclude *on several counts* that $y = y_o(x) \equiv (\sin x)/(\sin b)$ yields a *strong relative minimum* for $I[y] = \int_0^b (y'^2(x) - y^2(x))\,dx$ for $0 < b < \pi$.

If $\pi < b < 2\pi$, then $y = y_o(x) \equiv (\sin x)/(\sin b)$ still satisfies the necessary conditions N1 to N4 but does *not* satisfy the Jacobi condition N5 because now $(0,b)$ contains

$0^* = \pi$. On this basis, we may already conclude that $y = y_o(x) \equiv (\sin x)/(\sin b)$ does *not* yield a *weak relative minimum* for $I[y]$ for the case $\pi < b < 2\pi$ and, by the same token, does not yield a *strong relative minimum* either.

7.7 THE SECOND VARIATION OF FUNCTIONALS OF n VARIABLES

In order to derive a formula for the second variation of the functional

$$I[\hat{y}] = \int_a^b f(x,y_1(x), \ldots ,y_n(x),y_1'(x), \ldots ,y_n'(x))\, dx,$$

$$y_i(a) = y_i^a, \qquad y_i(b) = y_i^b, \tag{7.7.1}$$

at $\hat{y} = \hat{y}_o(x) \equiv (y_1(x), \ldots ,y_n(x))$, we shall assume that $f \in C^2(\mathcal{R})$ and that $\hat{y}_o \in C^1[a,b]^n$, where $\mathcal{R}$ denotes a domain in $(x,\hat{y},\hat{y}')$ space such that

$$(x,\hat{y}_o(x),\hat{y}_o'(x)) \in \mathcal{R} \qquad \text{for all } x \in [a,b].$$

By Definition 1.8, we have

$$\delta^2 I[\hat{h}] = \frac{d^2}{dt^2} I[\hat{y}_o + t\hat{h}]_{t=0}$$

$$= \frac{d^2}{dt^2} \int_a^b f(x,\, \hat{y}_o(x) + t\hat{h}(x),\, \hat{y}_o'(x) + t\hat{h}'(x))\, dt\, |_{t=0},$$

where $\hat{h} = (h_1, \ldots ,h_n) \in \mathcal{H}$, $\mathcal{H} = \{\hat{h} \mid \hat{h} \in C^1[a,b]^n,\ \hat{h}(a) = \hat{h}(b) = 0\}$. We find

$$\frac{d^2}{dt^2} \int_a^b f(x,\, \hat{y}_o(x) + t\hat{h}(x),\, \hat{y}_o'(x) + t\hat{h}'(x))\, dx\, |_{t=0}$$

$$= \int_a^b \sum_{i=1}^n \sum_{k=1}^n [f_{y_iy_k}h_ih_k + f_{y_iy'_k}h_ih_k' + f_{y'_iy_k}h_i'h_k + f_{y'_iy'_k}h_i'h_k']\, dx,$$

where $(x,\hat{y}_o(x),\hat{y}_o'(x))$ is the argument of $f_{y_iy_k}$, $f_{y_iy'_k}$, $f_{y'_iy_k}$, $f_{y'_iy'_k}$ and where $h_j = h_j(x)$.

Since $f \in C^2(\mathcal{R})$, we have $f_{y_iy_k} = f_{y_ky_i}$, $f_{y'_iy'_k} = f_{y'_ky'_i}$. Hence,

$$P = (f_{y_iy_k})_{i=1,\ldots,n,\ k=1,\ldots,n}, \qquad R = (f_{y'_iy'_k})_{i=1,\ldots,n,\ k=1,\ldots,n}$$

are real symmetric matrices: $P^T = P$, $R^T = R$. (The superscript T denotes the transpose of a matrix.)

If $\hat{h}$ denotes the row vector $(h_1, \ldots ,h_n)$ and, accordingly,

$$\hat{h}^T = \begin{pmatrix} h_1 \\ \vdots \\ h_n \end{pmatrix}$$

and $\hat{h}' = (h'_1, \ldots, h'_n)$, we have

$$\sum_{i=1}^{n}\sum_{k=1}^{n} f_{y_iy_k}h_ih_k = \hat{h}P\hat{h}^T, \qquad \sum_{i=1}^{n}\sum_{k=1}^{n} f_{y_iy'_k}h_ih'_k = \hat{h}Q\hat{h}'^T$$

$$\sum_{i=1}^{n}\sum_{k=1}^{n} f_{y'_iy_k}h'_ih_k = \hat{h}'Q^T\hat{h}^T, \qquad \sum_{i=1}^{n}\sum_{k=1}^{n} f_{y'_iy'_k}h'_ih'_k = \hat{h}'R\hat{h}'^T,$$

where $Q = (f_{y_iy'_k})_{i=1,\ldots,n,\ k=1,\ldots,n}$, $\quad Q^T = (f_{y'_iy_k})_{i=1,\ldots,n,\ k=1,\ldots,n}$

Since $\hat{h}Q\hat{h}'^T$ is a scalar, we have $(\hat{h}Q\hat{h}'^T)^T = \hat{h}Q\hat{h}'^T$ and, on the other hand, $(\hat{h}Q\hat{h}'^T)^T = \hat{h}'Q^T\hat{h}^T$. Hence

$$\hat{h}Q\hat{h}'^T + \hat{h}'Q^T\hat{h}^T = 2\hat{h}Q\hat{h}'^T,$$

and we can write the second variation in the form

$$\delta^2 I[\hat{h}] = \int_a^b (\hat{h}(x)P(x)\hat{h}^T(x) + 2\hat{h}(x)Q(x)\hat{h}'^T(x) + \hat{h}'(x)R(x)\hat{h}'^T(x))\,dx. \tag{7.7.2}$$

For a relative minimum, it is necessary that

$$\delta^2 I[\hat{h}] \geq 0 \qquad \text{for all } \hat{h} \in \mathcal{H} \tag{7.7.3}$$

(see Theorem 1.8.2). This condition can again be expressed as a condition on the $f_{y'_iy'_k}$ in a manner similar to that of Theorem 7.2.

However, the integration-by-parts process that we carried out in Sec. 7.2 in order to find a simpler expression for the second variation cannot be generalized unless Q is a symmetric matrix, which, in general, it is not. Therefore, we shall have to work with (7.7.2) as it stands.

In order to express the condition (7.7.3) in a practical manner, we need the concept of positive semidefiniteness of a matrix:

Definition 7.7 *The real symmetric matrix* $A = (a_{ik})$ *is called* positive (negative) semidefinite *if*

$$\hat{x}A\hat{x}^T \geq 0 \qquad (\leq 0)$$

for all real row vectors $\hat{x} = (x_1, \ldots, x_n)$. *A is* positive (negative) definite *if the equals sign holds only for* $\hat{x} = 0$.

Now we can state:

Theorem 7.7: Legendre condition *For* $\hat{y} = \hat{y}_o(x) \in C^1[a,b]^n$ *to yield a relative minimum (maximum) for* $I[\hat{y}]$ *as given in* (7.7.1), *it is necessary* *that*

$$R(x) = \left(f_{y'_i y'_k}(x, y_o(x), y_o'(x)) \right)_{i=1,\ldots,n,\; k=1,\ldots,n}$$

be positive (negative) semidefinite for all $x \in [a,b]$.

Proof: Suppose that for some $x_o \in (a,b)$, $R(x_o)$ is not positive semidefinite, i.e., that there exists a constant vector $\hat{h}_o \neq 0$ such that

$$\hat{h}_o R(x_o) \hat{h}_o^T < 0.$$

We take

$$\hat{h}_m = \begin{cases} \left[\dfrac{1}{m} \sin^2 m(x - x_o)\right] \hat{h}_o & \text{for } |x - x_o| \leq \dfrac{\pi}{m} \\ 0 & \text{for } |x - x_o| > \dfrac{\pi}{m}. \end{cases}$$

Since $P(x)$ and $Q(x)$ are continuous matrix functions in $[a,b]$, we have

$$|\hat{h}_o P(x) \hat{h}_o^T| \leq M_1, \qquad |\hat{h}_o Q(x) \hat{h}_o^T| \leq M_2, \qquad x \in [a,b],$$

for some real numbers M_1, M_2.

Hence,

$$\left| \int_a^b \hat{h}_m(x) P(x) \hat{h}_m^T(x)\, dx \right|$$

$$= \left| \frac{1}{m^2} \int_{x_o - \pi/m}^{x_o + \pi/m} \sin^4 m(x - x_o)\, \hat{h}_o P(x) \hat{h}_o^T\, dx \right| \leq \frac{2\pi M_1}{m^3},$$

$$\left| 2 \int_a^b \hat{h}_m(x) Q(x) \hat{h}_m'^T(x)\, dx \right|$$

$$= 4 \left| \int_{x_o - \pi/m}^{x_o + \pi/m} \frac{1}{m} \sin^3 m(x - x_o) \cos m(x - x_o)\, \hat{h}_o Q(x) \hat{h}_o^T\, dx \right| \leq \frac{8\pi M_2}{m^2}.$$

Finally,

$$\int_a^b \hat{h}_m'(x) R(x) \hat{h}_m'^T(x)\, dx$$

$$= 4 \int_{x_o - \pi/m}^{x_o + \pi/m} \sin^2 m(x - x_o) \cos^2 m(x - x_o)\, \hat{h}_o R(x) \hat{h}_o^T\, dx$$

$$= \hat{h}_o R\left(x_o + \Theta \frac{\pi}{m}\right) \hat{h}_o^T \frac{\pi}{m}, \qquad \text{where } |\Theta| \leq 1.$$

If m_o is sufficiently large, we have

$$\hat{h}_o R\left(x_o + \Theta\frac{\pi}{m}\right)\hat{h}_o^T < -\mu^2$$

for some μ and all $m > m_o$ and hence

$$\delta^2 I[\hat{h}] \leq \frac{\pi}{m}\left(\frac{2M_1}{m^2} + \frac{8M_2}{m} - \mu^2\right) < 0$$

for all $m > m_1$ for some $m_1 \geq m_o$. This contradicts (7.7.3).

As an example, consider

$$\int_a^b (y'^2(x) - z'^2(x))\,dx \to \text{minimum},$$

$y(a) = y_a$, $z(a) = z_a$, $y(b) = y_b$, $z(b) = z_b$. We have

$$R = \begin{pmatrix} f_{y'y'} & f_{y'z'} \\ f_{y'z'} & f_{z'z'} \end{pmatrix} = \begin{pmatrix} 2 & 0 \\ 0 & -2 \end{pmatrix}.$$

This matrix is *not* positive semidefinite since one of its eigenvalues is negative. Hence no matter what functions $y = y(x)$, $z = z(x)$ one may choose, the Legendre condition will not be satisfied, and it follows that the problem has no solution. Neither can there be a maximum because the matrix is not *negative semidefinite* either since one of its eigenvalues is positive.

By contrast, we obtain for

$$\int_a^b (y'^2(x) + z'^2(x) + 2y'(x)z'(x))\,dx \to \text{minimum},$$

with the same boundary conditions as above,

$$R = \begin{pmatrix} 2 & 2 \\ 2 & 2 \end{pmatrix},$$

and we see that

$$(h_1', h_2')\begin{pmatrix} 2 & 2 \\ 2 & 2 \end{pmatrix}\begin{pmatrix} h_1' \\ h_2' \end{pmatrix} = 2(h_1' + h_2')^2 \geq 0.$$

PROBLEMS 7.7

1. Show that if $Q = Q^T$, then

$$2\int_a^b \hat{h}(x)Q(x)\hat{h}'^T(x)\,dx = -\int_a^b \hat{h}(x)\,\frac{dQ(x)}{dx}\,\hat{h}^T(x)\,dx$$

for all $\hat{h} \in C^1[a,b]^n$, $\hat{h}(a) = \hat{h}(b) = 0$.

2. Consider the variational problem

$$\int_a^b \sqrt{1 + y'^2(x) + z'^2(x)}\,dx \to \text{minimum}$$

with any boundary conditions. Is the Legendre condition (Theorem 7.7) satisfied?

7.8 THE STRENGTHENED LEGENDRE CONDITION

$\hat{y} = \hat{y}_o(x)$ is called a *regular extremal* of $I[\hat{y}]$, as given by (7.7.1), if it is a solution of the Euler-Lagrange equations of $I[\hat{y}]$ and if

$$\left.\frac{\partial(f_{y'_1}, \ldots, f_{y'_n})}{\partial(y'_1, \ldots, y'_n)}\right|_{\hat{y}=\hat{y}_o(x)} = \det |f_{y'_i y'_k}|_{\hat{y}=\hat{y}_o(x)} \neq 0 \qquad \text{for all } x \in [a,b].$$

(See also Prob. 2.11.9, Theorem 2.12, and Definition 6.4.)

The following lemmas on real symmetric matrices will enable us to strengthen the Legendre condition of Theorem 7.7 in the case that the extremal is regular.

Lemma 7.8.1 *A real symmetric (constant) matrix A is positive semidefinite (positive definite) if and only if all its eigenvalues are nonnegative (positive).*

Proof: For every real symmetric matrix A, there exists an orthogonal matrix U with $U^T U = I$, $\det U = \pm 1$, such that

$$UAU^T = D(\lambda_1, \ldots, \lambda_n),$$

where
$$D(\lambda_1, \ldots, \lambda_n) = \begin{pmatrix} \lambda_1 & 0 & \cdots & 0 \\ 0 & \lambda_2 & \cdots & 0 \\ \cdots & \cdots & \cdots & \cdots \\ 0 & 0 & \cdots & \lambda_n \end{pmatrix}$$

and where $\lambda_1, \ldots, \lambda_n$ are the (real) eigenvalues of A.†

By hypothesis,

$$\hat{x} A \hat{x}^T \geq 0 \qquad (> 0)$$

for all $\hat{x} \neq 0$ and, in particular, for $\hat{x} = \hat{y}U$ for all $\hat{y} \neq 0$. Then

$$\hat{x} A \hat{x}^T = \hat{y} U A U^T \hat{y}^T = \hat{y} D(\lambda_1, \ldots, \lambda_n) \hat{y}^T = \lambda_1 y_1^2 + \cdots + \lambda_n y_n^2 \geq 0 \qquad (> 0)$$

for all $\hat{y} \neq 0$, and it follows that $\lambda_i \geq 0$ (>0), $i = 1, 2, \ldots, n$.

On the other hand, if $\lambda_i \geq 0$ (>0), $i = 1, 2, \ldots, n$, then, because of $\hat{y} = \hat{x} U^T$,

$$\hat{x} A \hat{x}^T = \lambda_1 (\hat{x} U^T)_1^2 + \cdots + \lambda_n (\hat{x} U^T)_n^2 \geq 0 \qquad (>0)$$

for all $\hat{x} \neq 0$. Here, $(\hat{x}U^T)_i$ denotes the ith component of the vector $\hat{x}U^T$.

Lemma 7.8.2 *If the real symmetric (constant) matrix A is positive semidefinite (positive definite), then* $\det A \geq 0$ (>0).

† F. E. Hohn, "Elementary Matrix Algebra," 2d ed., p. 296, The Macmillan Company, New York, 1965.

Proof: With $A = U^T D(\lambda_1, \ldots, \lambda_n) U$, we have

$$\det A = \det U^T \det D(\lambda_1, \ldots, \lambda_n) \det U = \lambda_1 \lambda_2 \cdots \lambda_n \geq 0 \qquad (>0)$$

since $\det U^T = \det U = \pm 1$ and since $\lambda_i \geq 0$ (>0) by Lemma 7.8.1.

Lemma 7.8.3 *If A is positive semidefinite and* $\det A \neq 0$, *then A is positive definite.*

Proof: Since A is positive semidefinite, by Lemma 7.8.1, $\lambda_i \geq 0$. Since $\det A = \lambda_1 \lambda_2 \cdots \lambda_n \neq 0$, we have to have $\lambda_i > 0$, and consequently,

$$\hat{x} A \hat{x}^T = \lambda_1 (\hat{x} U^T)_1^2 + \cdots + \lambda_n (\hat{x} U^T)_n^2 > 0$$

for all $\hat{x} \neq 0$.

Note that if $\det A > 0$, then A is not necessarily positive definite or positive semidefinite. Take $A = \begin{pmatrix} -1 & 0 \\ 0 & -1 \end{pmatrix}$. Then $\det A = 1 > 0$, but $\hat{x} A \hat{x}^T = -x_1^2 - x_2^2 \leq 0$.

If

$$A = \begin{pmatrix} -1 & 0 & 0 \\ 0 & -1 & 0 \\ 0 & 0 & 1 \end{pmatrix},$$

then $\det A = 1$ again, but $\hat{x} A \hat{x}^T = -x_1^2 - x_2^2 + x_3^2 \gtrless 0$, depending on the values of x_1, x_2, x_3.

We saw in Theorem 7.7 that if $\hat{y} = \hat{y}_o(x)$ is to yield a relative minimum, then, by necessity, $R(x) = (f_{y'_i y'_k})_{\hat{y} = \hat{y}_o(x)}$ is positive semidefinite for all $x \in [a,b]$. Then, from Lemma 7.8.3 and in view of the definition of a regular extremal, we have:

Theorem 7.8 Strengthened Legendre condition *If the regular extremal $\hat{y} = \hat{y}_o(x)$ is to yield a relative minimum for $I[\hat{y}]$, then it is necessary that*

$$R(x) = (f_{y'_i y'_k})_{\hat{y} = \hat{y}_o(x)}$$

be positive definite *for all $x \in [a,b]$.*

PROBLEMS 7.8

1. Consider the variational problem

$$\int_a^b \sqrt{y'^2(x) + z'^2(x)}\, dx \to \text{minimum}.$$

Find

$$R = \begin{pmatrix} f_{y'y'} & f_{y'z'} \\ f_{y'z'} & f_{z'z'} \end{pmatrix}$$

for the extremals $y = \alpha x$, $z = \beta x$. (See also Secs. A7.10 and A7.11.)

2. Same as in problem 1 for the variational problem

$$\int_a^b \sqrt{y'^2(x) + (\sin^2 y(x))z'^2(x)}\, dx \to \text{minimum}$$

for the extremal
(a) $y = \pi/2,\ z = x$ (b) $y = x,\ z = 0$
(See also Secs. A7.10 and A7.11.)

7.9 CONJUGATE POINTS AND JACOBI'S NECESSARY CONDITION

We shall assume in this section that $f \in C^3(\mathcal{R})$ and that the *regular* extremal $\hat{y} = \hat{y}_o(x) \in C^2[a,b]^n$ yields a relative minimum for $I[\hat{y}]$. Then $\det |R(x)| \neq 0$, and by Theorem 7.8, $R(x) = (f_{y'_i y'_k})_{y=\hat{y}_o(x)}$ is *positive definite.*

It is our ultimate goal in this section to arrive at a natural generalization of the concept of a conjugate point via a secondary variational problem, as in Sec. 7.3, and to state a necessary condition for a relative minimum in terms of the location of the conjugate point.

Toward this end, we shall first subject the second variation to a transformation. We have

$$\delta^2 I[\hat{h}] = \int_a^b [\hat{h}(x)P(x)\hat{h}^T(x) + 2\hat{h}(x)Q(x)\hat{h}'^T(x) + \hat{h}'(x)R(x)\hat{h}'^T(x)]\, dx \tag{7.9.1}$$

and we observe that the integrand

$$\Omega(\hat{h},\hat{h}') = \hat{h}P\hat{h}^T + 2\hat{h}Q\hat{h}'^T + \hat{h}'R\hat{h}'^T$$

is homogeneous of the second degree in $(h_1, \ldots, h_n, h'_1, \ldots, h'_n)$. Hence we have, from Euler's identity,

$$\sum_{i=1}^n (\Omega_{h_i} h_i + \Omega_{h'_i} h'_i) = 2\Omega,$$

which we may write in vector form as

$$\hat{h}\Omega_{\hat{h}}^T + \hat{h}'\Omega_{\hat{h}'}^T = 2\Omega.$$

Since $$\int_a^b \hat{h}'\Omega_{\hat{h}'}^T\, dx = \hat{h}\Omega_{\hat{h}'}^T \Big|_a^b - \int_a^b \hat{h}\frac{d}{dx}\Omega_{\hat{h}'}^T\, dx$$

and since $\hat{h}(a) = \hat{h}(b) = 0$, we obtain

$$\delta^2 I[\hat{h}] = \int_a^b \Omega\, dx = \frac{1}{2}\int_a^b \hat{h}\left(\Omega_{\hat{h}}^T - \frac{d}{dx}\Omega_{\hat{h}'}^T\right) dx.$$

Since $$\begin{aligned} \Omega_{\hat{h}}^T &= 2P\hat{h}^T + 2Q\hat{h}'^T \\ \Omega_{\hat{h}'}^T &= 2Q^T\hat{h}^T + 2R\hat{h}'^T \end{aligned} \tag{7.9.2}$$

(see Prob. 7.9.1 and note that $P^T = P$, $R^T = R$), we have, finally,

$$\delta^2 I[\hat{h}] = \int_a^b \hat{h}\left[P\hat{h}^T + Q\hat{h}'^T - \frac{d}{dx}(Q^T\hat{h}^T + R\hat{h}'^T)\right] dx. \qquad (7.9.3)$$

We shall now consider the *secondary variational problem*

$$\delta^2 I[\hat{h}] = \int_a^b \Omega(\hat{h}(x),\hat{h}'(x))\, dx \to \text{minimum}, \qquad \hat{h}(a) = \hat{h}(b) = 0,$$

which has

$$\Omega_{\hat{h}} - \frac{d}{dx}\Omega_{\hat{h}'} = 0, \text{ or, equivalently, } \Omega_{\hat{h}}{}^T - \frac{d}{dx}\Omega_{\hat{h}'}{}^T = 0$$

as Euler-Lagrange equations. But these are exactly the equations

$$P\hat{h}^T + Q\hat{h}'^T - \frac{d}{dx}(Q^T\hat{h}^T + R\hat{h}'^T) = 0 \qquad (7.9.4)$$

which we shall call the *Jacobi equations.* (See also Prob. 7.9.2.)

From (7.9.3), we see that any solution $\hat{h} = \hat{h}(x)$ of the Jacobi equations (7.9.4) will make the second variation $\delta^2 I[\hat{h}]$ vanish.

The Jacobi equations (7.9.4) represent a system of n linear differential equations of second order with continuous coefficients. Since det $|R(x)| \neq 0$ in $[a,b]$, we can solve these equations for $\hat{h}''$, apply the general existence and uniqueness theorems for linear differential equations,† and obtain n solutions $\hat{h} = \hat{h}_i(x)$, $i = 1, 2, \ldots, n$, for the initial conditions

$$\hat{h}_i(a) = 0, \qquad \hat{h}_i'(a) = \hat{e}_i,$$

where $\hat{e}_i = (0, \ldots, 0,1,0, \ldots, 0)$ is the ith unit vector.

If

$$\hat{h} = \hat{h}_i(x) \equiv (h_1{}^i(x), \ldots, h_n{}^i(x)),$$

then we have, in view of the initial conditions, that

$$\begin{vmatrix} h_1{}^1(a) & \cdots & h_n{}^1(a) \\ \vdots & & \\ h_1{}^n(a) & \cdots & h_n{}^n(a) \end{vmatrix} = 0.$$

Definition 7.9 *a^* is called the conjugate point to a if $a^* > a$ is the smallest value for which*

$$\begin{vmatrix} h_1{}^1(a^*) & \cdots & h_n{}^1(a^*) \\ \vdots & & \\ h_1{}^n(a^*) & \cdots & h_n{}^n(a^*) \end{vmatrix} = 0.$$

† G. Birkhoff and G. C. Rota, "Ordinary Differential Equations," p. 116, Ginn and Company, Boston, 1962.

Let us assume that $a < a^* < b$. Then there exist n constants $(c_1, \ldots, c_n) \neq (0, \ldots, 0)$ such that

$$c_1\hat{h}_1(a^*) + \cdots + c_n\hat{h}_n(a^*) = 0.$$

Hence
$$\hat{h} = c_1\hat{h}_1(x) + \cdots + c_n\hat{h}_n(x)$$

is a solution of the Jacobi equations (7.9.4) for which

$$\hat{h}(a) = 0, \qquad \hat{h}(a^*) = 0,$$

but $\hat{h}(x) \neq 0$ for $x \in (a,a^*)$.

Then, because of (7.9.3), we obtain for

$$\hat{\hat{h}} = \hat{\hat{h}}(x) \equiv \begin{cases} \hat{h}(x) & \text{for } a \leq x \leq a^* \\ 0 & \text{for } a^* < x \leq b \end{cases}$$

that
$$\delta^2 I[\hat{\hat{h}}] = 0.$$

If 0 were the minimum of $\delta^2 I[\hat{h}]$, then $\hat{h} = \hat{\hat{h}}(x)$ would have to satisfy the first corner condition (2.11.7), whereby

$$\Omega_{\hat{h}'}\big|_{a^*-0} = \Omega_{\hat{h}'}\big|_{a^*+0}$$

has to hold. From (7.9.2), we have

$$\Omega_{\hat{h}'^T}\big|_{a^*-0} = 2Q^T(a^*)\hat{h}^T(a^*) + 2R(a^*)\hat{h}'^T(a^*) = 2R(a^*)\hat{h}'^T(a^*) \neq 0$$

since $\hat{h}(a^*) = 0$ and since $R(a^*)$ is positive definite by hypothesis, and $\hat{h}'(a^*) \neq 0$. [Otherwise, $\hat{h} = \hat{h}(x) \equiv 0$ would be uniquely determined by the terminal conditions $\hat{h}(a^*) = \hat{h}'(a^*) = 0$.]

On the other hand,

$$\Omega_{\hat{h}'^T}\big|_{a^*+0} = 0,$$

and we see that the corner conditions are not satisfied. Hence, 0 is *not* the minimum of $\delta^2 I[\hat{h}]$ and there exists a sectionally smooth function $\hat{g} = \hat{g}(x)$, $\hat{g}(a) = \hat{g}(b) = 0$, such that $\delta^2 I[\hat{g}] < 0$. Then, by an obvious generalization of the fairing theorem (Corollary 1 to Theorem 2.10), there exists a smooth function $\hat{g} = \hat{g}_o(x)$, $\hat{g}_o(a) = \hat{g}_o(b) = 0$, such that $\delta^2 I[\hat{g}_o] < 0$, and we have a contradiction to the necessary condition $\delta^2 I[\hat{h}] \geq 0$ for $\hat{h} \in C^1[a,b]$, $\hat{h}(a) = \hat{h}(b) = 0$.

Hence we may state the following theorem:

Theorem 7.9: Jacobi's necessary condition *If the regular extremal $\hat{y} = \hat{y}_o(x) \in C^2[a,b]$ is to yield a relative minimum for $I[\hat{y}]$ with $f \in C^3(\mathcal{R})$, then it is necessary that (a,b)* not *contain a conjugate point to a, or as we may put it, that $a^* \geq b$.*

PROBLEMS 7.9

*1. Prove

$$[(\hat{h}A\hat{h}^T)_{\hat{h}}]^T = (A + A^T)\hat{h}^T, \qquad [(\hat{h}B\hat{h}'^T)_{\hat{h}}]^T = B\hat{h}'^T,$$

$$[(\hat{h}B\hat{h}'^T)_{\hat{h}'}]^T = B^T\hat{h}^T,$$

where A, B are square matrices and where $\hat{h}$ is a row vector.

2. Show that if $Q = Q^T$, then the Jacobi equations (7.9.4) can be written in the form

$$(Q' - P)\hat{h}^T + \frac{d}{dx}(R\hat{h}'^T) = 0.$$

[Compare with (7.3.3).]

3. Prove: If A is positive definite and $\hat{x} \neq 0$, then $A\hat{x}^T \neq 0$.

4. Prove: If $\hat{h} = \hat{h}(x)$ is a solution of the Jacobi equations and if $\hat{h}(a^*) = \hat{h}'(a^*) = 0$, then $\hat{h}(x) \equiv 0$.

5. Given the n-parameter family of solutions $y_i = Y_i(x,c_1, \ldots, c_n)$, $i = 1, 2, \ldots, n$, of the Euler-Lagrange equations $f_{y_k} - (d/dx) f_{y'_k} = 0$, $k = 1, 2, \ldots, n$. Impose suitable conditions on $Y_i(x,\hat{c})$ and on f and show that

$$\hat{\eta} = \sum_{k=1}^{n} \lambda_k \left(\frac{\partial \hat{Y}}{\partial c_k}\right)_{\hat{c}_o}$$

for arbitrary constants $\lambda_1, \ldots, \lambda_n$ is a solution of the Jacobi equations

$$P\hat{\eta}^T + Q\hat{\eta}'^T - \frac{d}{dx}(Q^T\hat{\eta}^T + R\hat{\eta}'^T) = 0,$$

where P, Q, R are to be taken at $\hat{y} = \hat{Y}(x,\hat{c}_o)$.

6. Suppose there is no conjugate point to a in $(a,b]$. Show that $y_i = Y_i(x,\hat{c})$, $i = 1, 2, \ldots, n$, $Y_i(a,\hat{c}) = y_i^a$, can be solved uniquely for $c_k = C_k(x,y_1, \ldots, y_n)$ in a neighborhood of $\hat{c}_o$ and for all $a < x \leq b$.

7. Under the conditions of problem 6, there is a slope $\hat{y}' = \hat{Y}'(x,\hat{C}(x,\hat{y})) = \hat{\phi}(x,\hat{y})$ associated with every point $(x,\hat{y})$ in a weak neighborhood $N_w^\delta(\hat{y}_o)$. Show that

$$U[\hat{y}] = \int_{P_a}^{P_b} [f(x,\hat{y},\hat{\phi}(x,\hat{y})) + (\hat{y}' - \hat{\phi}(x,\hat{y}))f_{\hat{y}'}{}^T(x,\hat{y},\hat{\phi}(x,\hat{y}))]\, dx$$

is independent of the path $\hat{y} = \hat{y}(x)$ so long as $\hat{y} = \hat{y}(x)$ joins P_a to P_b where $\hat{y}_b = \hat{y}_o(b)$ and so long as $\hat{y} = \hat{y}(x)$ remains in $N_w^\delta(\hat{y}_o)$.

8. Prove that it is sufficient for a weak relative minimum of $I[\hat{y}]$ that the extremal $\hat{y} = \hat{y}_o(x)$ can be embedded in a Mayer field (see Definition 3.10.2) and that $R(x) = (f_{y'_i y'_k})_{\hat{y}=\hat{y}_o(x)}$ is positive definite.

9. Prove: If there is no conjugate point to a in $(a,b]$ and if $R(x)$ is positive definite, then $\hat{y} = \hat{y}_o(x)$ yields a weak relative minimum for $I[\hat{y}]$.

BRIEF SUMMARY

The second variation of the functional

$$I[y] = \int_a^b f(x,y(x),y'(x))\, dx, \qquad y(a) = y_a, \qquad y(b) = y_b,$$

at $y = y_o(x) \in C^1[a,b]$ is given by

$$\delta^2 I[h] = \int_a^b [(f_{yy})_o h^2(x) + 2(f_{yy'})_o h(x)h'(x) + (f_{y'y'})_o h'^2(x)]\, dx,$$

where $(\quad)_o$ indicates that the quantities so designated are to be taken at $(x,y_o(x),y_o'(x))$.

If $y = y_o(x)$ is to yield a weak relative minimum for $I[y]$, then it is necessary that

$$\delta^2 I[h] \geq 0$$

for all $h \in \mathcal{H} = \{h \mid h \in \mathcal{C}^1[a,b], h(a) = h(b) = 0\}$.

This condition leads, in turn, to the *Legendre condition*

$$f_{y'y'}(x,y_o(x),y_o'(x)) \geq 0 \qquad \text{for all } x \in [a,b]$$

and, in case of a regular extremal, to the *strengthened Legendre condition*

$$f_{y'y'}(x,y_o(x),y_o'(x)) > 0 \qquad \text{for all } x \in [a,b]$$

(Sec. 7.2).

For a regular extremal, one obtains an additional necessary condition, namely, the *Jacobi condition*, which states that (a,b) must not contain a conjugate point to a (Sec. 7.3).

For a regular extremal $y = y_o(x)$ to yield a weak relative minimum for $I[y]$ it is sufficient that $(a,b]$ does not contain a conjugate point to a *and* that

$$f_{y'y'}(x,y_o(x),y_o'(x)) > 0 \qquad \text{for all } x \in [a,b]$$

(Sec. 7.5).

For the functional

$$I[\hat{y}] = \int_a^b f(x,y_1(x), \ldots, y_n(x), y_1'(x), \ldots, y_n'(x))\, dx,$$

$$y_i(a) = y_i^a, \qquad y_i(b) = y_i^b,$$

one obtains as a generalization of the Legendre condition that the matrix $R(x) = (f_{y'_i y'_k})_{\hat{y}=\hat{y}_o(x)}$ has to be positive semidefinite.

In case of a regular extremal $\hat{y} = \hat{y}_o(x)$, that is, $\det |f_{y'_i y'_k}|_{\hat{y}=\hat{y}_o(x)} \neq 0$, $R(x)$ has to be positive definite (Secs. 7.7 and 7.8).

For a regular extremal, one obtains again the Jacobi condition as a necessary condition for a weak relative minimum, namely, that (a,b) must not contain a conjugate point to a (Sec. 7.9).

APPENDIX

A7.10 THE LEGENDRE CONDITION FOR THE HOMOGENEOUS PROBLEM

We shall now sketch a method by which the homogeneous problem can be transformed into an x problem so that the results that have been obtained in our study of the second variation of the x problem may then be translated into the formalism of the homogeneous problem.

As in Chap. 4, we consider the problem of finding a simple, smooth arc

$$\gamma_o \begin{cases} x = x_o(t) \\ y = y_o(t) \end{cases} t \in [t_a,t_b]$$

(see Definition 4.1.1) such that

$$I[\gamma] = \int_{P_a}^{P_b} F(x,y,\dot{x},\dot{y})\, dt \to \text{minimum},$$

where P_a and P_b have the prescribed coordinates (a,y_a) and (b,y_b), respectively. Hereby it is assumed that F has continuous partial derivatives of second order with respect to all variables and that F is positive homogeneous of the first degree in $\dot{x}$, $\dot{y}$. (See Sec. 4.1.)

In order to obtain a lead, let us first consider the case where it is possible to represent the extremal arc γ_o as $y = y(x) \in C^1[a,b]$ by means of the transformation $x = x_o(t)$, $\dot{x}_o > 0$. Then

$$\int_{P_a}^{P_b} F(x_o(t),y_o(t),\dot{x}_o(t),\dot{y}_o(t))\, dt = \int_{P_a}^{P_b} F\left(x_o(t),\, y_o(t),\, 1,\, \frac{\dot{y}_o(t)}{\dot{x}_o(t)}\right) \dot{x}_o(t)\, dt$$

$$= \int_a^b F\left(x,\, y,\, 1,\, \frac{dy}{dx}\right) dx,$$

and we see that the Legendre condition $f_{y'y'} \geq 0$ becomes

$$F_{\dot{y}\dot{y}}\left(x,y,1,\frac{dy}{dx}\right) \geq 0.$$

By (4.2.7c), we have $F_{\dot{y}\dot{y}}(x,y,1,(dy/dx)) = F_1(x,y,1,(dy/dx))$, and we see that the Legendre condition can be expressed by $F_1(x,y,1,(dy/dx)) \geq 0$ or, in view of $\dot{x}_o > 0$,

$$F_1(x_o(t),y_o(t),\dot{x}_o(t),\dot{y}_o(t)) \geq 0 \qquad \text{for all } t \in [t_a,t_b].$$

In general, such a transformation as we have just used is, of course, not possible. (Otherwise, there would not be much point in considering the homogeneous problem to begin with. See also the remarks in Sec. 4.1,

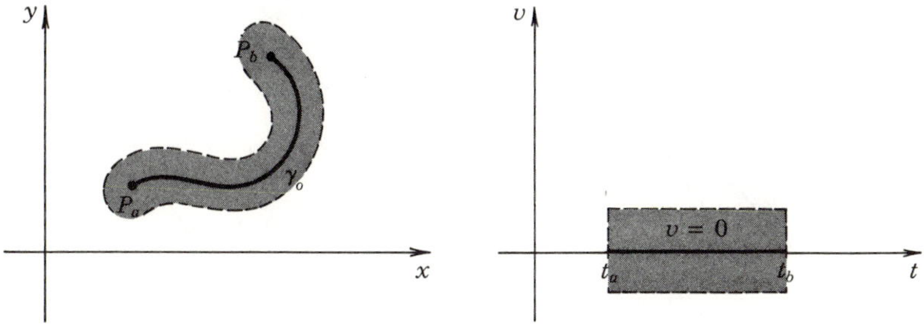

Figure 7.3

page 202.) However, if we also assume that $x_o(t), y_o(t) \in C^2[t_a,t_b]$, then a more general transformation will lead to the same goal.

The transformation that we shall consider is the following one:

$$\begin{aligned} x &= x_o(t) - \dot{y}_o(t)v \\ y &= y_o(t) + \dot{x}_o(t)v. \end{aligned} \tag{A7.10.1}$$

This transformation maps the t,v plane into the x,y plane and, in particular, maps the smooth curve $v = 0$, $t \in [t_a,t_b]$ into the extremal arc γ_o. It also maps a weak δ neighborhood of $v = 0$, $t \in [t_a,t_b]$ into a weak δM neighborhood of the extremal arc γ_o, where $m^2 \leq \dot{x}_o^2(t) + \dot{y}_o^2(t) \leq M^2$, $t \in [t_a,t_b]$. (See Fig. 7.3.) This can be seen as follows:

From (A7.10.1),

$$(x - x_o(t))^2 + (y - y_o(t))^2 = v^2(\dot{x}_o^2(t) + \dot{y}_o^2(t)),$$

and hence, if $|v| < \delta$, then

$$(x - x_o(t))^2 + (x - y_o(t))^2 < \delta^2 M^2.$$

(See also Definition 4.3.1.)

Further, a *strong* δ neighborhood of $v = 0$ in the t,v plane is mapped into a strong δ_1 neighborhood of γ_o in the x,y plane with $\delta_1 \to 0$ as $\delta \to 0$. (See Prob. A7.10.3.)

The transformation (A7.10.1) also has a unique inverse, provided that $|v| < (m^2/\mu^2)$, where $|\ddot{x}_o\dot{y}_o - \dot{x}_o\ddot{y}_o| \leq \mu^2$, because

$$\frac{\partial(x,y)}{\partial(t,v)} = \begin{vmatrix} \dot{x}_o - \ddot{y}_o v & \dot{y}_o + \ddot{x}_o v \\ -\dot{y}_o & \dot{x}_o \end{vmatrix} = \dot{x}_o^2 + \dot{y}_o^2 + v(\ddot{x}_o\dot{y}_o - \dot{x}_o\ddot{y}_o)$$

$$\geq m^2 - |v|\,|\ddot{x}_o\dot{y}_o - \dot{x}_o\ddot{y}_o| > m^2 - \frac{m^2}{\mu^2}\mu^2 = 0.$$

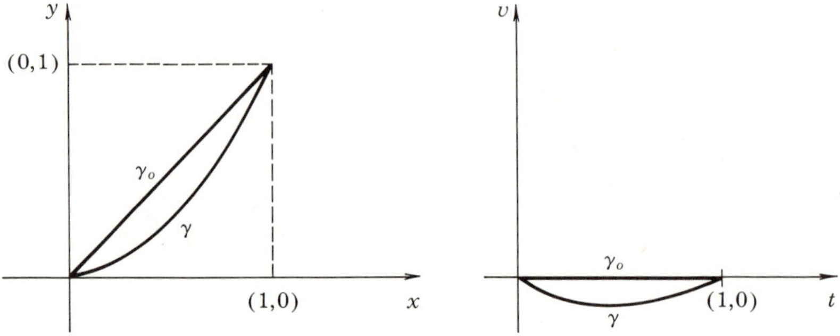

Figure 7.4

Hence the inverse transformation

$$\begin{aligned} v &= V(x,y) \\ t &= T(x,y) \end{aligned} \tag{A7.10.2}$$

is uniquely determined in a weak (m^2/μ^2) neighborhood of $v = 0$.

As an example, we consider γ_o: $x_o = t$, $y_o = t$, $t \in [0,1]$. Then the transformation (A7.10.1) will assume the form

$$\begin{aligned} x &= t - v \\ y &= t + v \end{aligned}$$

with the unique inverse

$$v = \frac{y - x}{2}, \qquad t = \frac{x + y}{2}.$$

If we wish to map γ: $x = \tau$, $y = \tau^2$, $\tau \in [0,1]$ into the t,v plane, we obtain from

$$v = \frac{\tau^2 - \tau}{2}, \qquad t = \frac{\tau^2 + \tau}{2}$$

that
$$\tau = \frac{-1 + \sqrt{1 + 8t}}{2}, \qquad \frac{d\tau}{dt} = \frac{2}{\sqrt{1 + 8t}} > 0, \qquad t \in [0,1],$$

and hence $v = \frac{1}{2} - \frac{1}{2}\sqrt{1 + 8t} + t$·
(see Fig. 7.4).

In general, if we wish to map the simple smooth arc γ which joins P_a to P_b into the t,v plane by means of the transformation (A7.10.2), we first introduce a parameter τ such that $\tau = t_a$ yields P_a and $\tau = t_b$ yields P_b. (This is always possible by a translation and a scale change.) Then we have $x = x(\tau)$, $y = y(\tau)$, $\tau \in [t_a,t_b]$ as the representation of γ, and we obtain from (A7.10.2)

$$\begin{aligned} v &= V(x(\tau),y(\tau)). \\ t &= T(x(\tau),y(\tau)), \end{aligned}$$

We express $\tau = \tau(t)$ from the second equation and obtain

$$v = V(x(\tau(t)),y(\tau(t))) \equiv v(t).$$

(See Problem A7.10.1.) If $x_o,y_o \in C^2[t_a,t_b]$ and $x,y \in C^1[t_a,t_b]$, then $v \in C^1[t_a,t_b]$. (See Problem A7.10.2.)

We shall now consider $\int_{P_a}^{P_b} F(x,y,\dot{x},\dot{y})\,dt$ for all arcs γ which join P_a to P_b and lie in a strong δ neighborhood of γ_o. Then, if τ is appropriately chosen, we have

$$\int_{P_a}^{P_b} F(x(\tau),y(\tau),\dot{x}(\tau),\dot{y}(\tau))\,d\tau$$

$$= \int_{t_a}^{t_b} F\Bigg(x_o(t) - \dot{y}_o(t)v,\ y_o(t) + \dot{x}_o(t)v,\ \dot{x}_o(t)\frac{dt}{d\tau} - \ddot{y}_o(t)\frac{dt}{d\tau}v - \dot{y}_o(t)\frac{dv}{dt}\frac{dt}{d\tau},$$

$$\dot{y}_o(t)\frac{dt}{d\tau} + \ddot{x}_o(t)\frac{dt}{d\tau}v + \dot{x}_o(t)\frac{dv}{dt}\frac{dt}{d\tau}\Bigg)\Bigg|_{t=t(\tau)} d\tau$$

$$= \int_{t_a}^{t_b} F\Bigg(x_o(t) - \dot{y}_o(t)v,\ y_o(t) + \dot{x}_o(t)v,\ \dot{x}_o(t) - \ddot{y}_o(t)v - \dot{y}_o(t)\frac{dv}{dt},$$

$$\dot{y}_o(t) + \ddot{x}_o(t)v + \dot{x}_o(t)\frac{dv}{dt}\Bigg)\frac{dt}{d\tau}\,d\tau.$$

The latter step was possible because F is supposed to be positive homogeneous of the first degree in $\dot{x}$, $\dot{y}$ (third and fourth positions of argument).

We introduce the new integration variable $t = t(\tau)$, $dt = (dt/d\tau)\,d\tau$ and obtain, finally,

$$\int_{P_a}^{P_b} F(x(\tau),y(\tau),\dot{x}(\tau),\dot{y}(\tau))\,d\tau = \int_{t_a}^{t_b} G\left(t,v,\frac{dv}{dt}\right) dt.$$

Suppose now that $v = 0$ (the image of γ_o) does *not* yield a weak relative minimum for

$$\bar{I}[v] = \int_{t_a}^{t_b} G\left(t,v,\frac{dv}{dt}\right) dt \to \text{minimum},$$

$$v(t_a) = 0, \qquad v(t_b) = 0.$$

Since a strong δ neighborhood of $v = 0$ in the t,v plane is mapped into a strong δ_1 neighborhood of γ_o in the x,y plane, with $\delta_1 \to 0$ as $\delta \to 0$, it follows that γ_o does not yield a weak relative minimum for $I[\gamma]$ either, and we see that $v = 0$ yields, by necessity, a weak relative minimum for

$\bar{I}[v]$. Hence, by Theorem 7.2, the Legendre condition

$$G_{\dot{v}\dot{v}}(t,0,0) \geq 0 \qquad \text{for all } t \in [t_a,t_b]$$

has to be satisfied.

Since

$$G\left(t, v, \frac{dv}{dt}\right) = F(x_o - \dot{y}_o v,\ y_o + \dot{x}_o v,\ \dot{x}_o - \ddot{y}_o v - \dot{y}_o \dot{v},\ \dot{y}_o + \ddot{x}_o v + \dot{x}_o \dot{v}), \tag{A7.10.3}$$

we obtain

$$G_{\dot{v}} = -F_{\dot{x}}\dot{y}_o + F_{\dot{y}}\dot{x}_o$$
$$G_{\dot{v}\dot{v}} = F_{\dot{x}\dot{x}}\dot{y}_o^2 - 2F_{\dot{x}\dot{y}}\dot{x}_o\dot{y}_o + F_{\dot{y}\dot{y}}\dot{x}_o^2.$$

In view of (4.2.7), we can write this as

$$G_{\dot{v}\dot{v}} = (\dot{x}_o^2 + \dot{y}_o^2)^2 F_1, \tag{A7.10.4}$$

and hence

$$G_{\dot{v}\dot{v}}(t,0,0) = (\dot{x}_o^2 + \dot{y}_o^2)^2 F_1(x_o(t),y_o(t),\dot{x}_o(t),\dot{y}_o(t)).$$

Thus we obtain as *Legendre condition for the homogeneous problem*

$$F_1(x_o(t),y_o(t),\dot{x}_o(t),\dot{y}_o(t)) \geq 0 \qquad \text{for all } t \in [t_a,t_b]. \tag{A7.10.5}$$

For a *regular extremal arc* (see Definition 4.5), we obtain the *strengthened Legendre condition*

$$F_1(x_o(t),y_o(t),\dot{x}_o(t),\dot{y}_o(t)) > 0 \qquad \text{for all } t \in [t_a,t_b]. \tag{A7.10.6}$$

As a first example, let us discuss the geodesics on a right circular cylinder of radius 1 with the z axis as axis.

We represent this cylinder in cylindrical coordinates

$$x = \cos\varphi$$
$$y = \sin\varphi$$
$$z = z,$$

and we obtain for any curve $\varphi = \varphi(t)$, $z = z(t)$ on the cylinder,

$$ds^2 = (d\varphi)^2 + (dz)^2.$$

Hence the problem of finding the geodesics on the cylinder amounts to solving the variational problem

$$\int_{P_a}^{P_b} \sqrt{\varphi'^2 + z'^2}\, dt \to \text{minimum}. \tag{A7.10.7}$$

We have

$$F_{\varphi'\varphi'} = \frac{z'^2}{(\varphi'^2 + z'^2)^{3/2}}, \qquad F_{z'z'} = \frac{\varphi'^2}{(\varphi'^2 + z'^2)^{3/2}},$$

and hence

$$F_1 = \frac{F_{\varphi'\varphi'} + F_{z'z'}}{\varphi'^2 + z'^2} = \frac{1}{(\varphi'^2 + z'^2)^{3/2}} > 0$$

for all lineal elements of any conceivable simple, smooth curve on the cylindrical surface. Hence all extremals on the cylindrical surface are regular extremals and the Legendre condition is satisfied. (More about this problem in the next section.)

As a second example, we consider the geodesics on the unit sphere with center at the origin. We represent the unit sphere in spherical coordinates

$$x = \sin\theta\cos\varphi$$
$$y = \sin\theta\sin\varphi$$
$$z = \cos\theta,$$

where θ denotes the geographic latitude measured from the north pole ($\theta = 0$) and where φ denotes the geographic longitude.

If $\varphi = \varphi(t)$, $\theta = \theta(t)$ represents a curve on the sphere, then

$$ds^2 = (d\theta)^2 + \sin^2\theta\,(d\varphi)^2,$$

and we obtain the variational problem for the geodesics

$$\int_{P_a}^{P_b} \sqrt{\theta'^2 + \sin^2\theta\varphi'^2}\,dt \to \text{minimum}. \tag{A7.10.8}$$

Since

$$F_{\theta'\theta'} = \frac{\varphi'^2\sin^2\theta}{(\theta'^2 + \varphi'^2\sin^2\theta)^{3/2}}, \qquad F_{\varphi'\varphi'} = \frac{\theta'^2\sin^2\theta}{(\theta'^2 + \varphi'^2\sin^2\theta)^{3/2}},$$

we have

$$F_1 = \frac{\sin^2\theta}{(\theta'^2 + \varphi'^2\sin^2\theta)^{3/2}} \geq 0.$$

We see again that all extremals (geodesics) are regular and that the only singular points occur at the north pole ($\theta = 0$) and the south pole ($\theta = \pi$). (More about this problem in the next section.)

PROBLEMS A7.10

*1. Under the conditions stated in the text, show that it is possible to solve $t = T(x(\tau),y(\tau))$ for τ in terms of t so that $d\tau/dt > 0$ provided that $|v|$ is sufficiently small.

*2. Show that $v = v(t) \in C^1[t_a,t_b]$, where $v(t) = V(x(\tau(t)),y(\tau(t)))$ and where $x_o,y_o \in C^2[t_a,t_b]$, $x(\tau),y(\tau) \in C^1[t_a,t_b]$, provided that $|v|$ is sufficiently small.

*3. Show that (A7.10.1) maps a strong δ neighborhood of $v = 0$ into a strong δ_1 neighborhood of γ_o so that $\delta_1 \to 0$ as $\delta \to 0$.

4. Apply the Legendre condition to the variational problem

$$\int_{(0,0)}^{(1,1)} \frac{\dot{y}^2}{\dot{x}}\,dt \to \text{minimum}$$

with the extremal $x_o = t$, $y_o = t$, $t \in [0,1]$. (See also the example in Sec. 4.3, where it was demonstrated that $x_o = t$, $y_o = t$ does *not* yield a strong relative minimum.)

5. Apply the Legendre condition to the variational problem

$$\int_{(0,0)}^{P_b} \frac{\dot{x}^2 + \dot{y}^2}{\sqrt{2(\dot{x}^2 + \dot{y}^2)} + \dot{x}}\, dt \rightarrow \text{minimum}$$

for:

(*a*) $P_b = (-1,0)$ (*b*) $P_b = (-1,\frac{1}{2})$

(For extremals of this problem, see Sec. 4.7, page 234.)

6. Find $G(t,v,\dot{v})$ for the variational problems in problems 4 and 5*a*.

A7.11 THE JACOBI CONDITION FOR THE HOMOGENEOUS PROBLEM

The Jacobi equation (7.3.3) for the variational problem

$$\bar{I}[v] = \int_{t_a}^{t_b} G\left(t, v, \frac{dv}{dt}\right) dt \rightarrow \text{minimum}$$

$$v(t_a) = 0, \qquad v(t_b) = 0,$$

which was obtained by means of the transformation (A7.10.1) from the homogeneous problem

$$I[\gamma] = \int_{P_a}^{P_b} F(x,y,\dot{x},\dot{y})\, dt \rightarrow \text{minimum}$$

with the extremal γ_o: $x = x_o(t)$, $y = y_o(t)$ and which has the solution $v = 0$, assumes the form

$$\frac{d}{dt}\left(G_{\dot{v}\dot{v}}(t,0,0)h'\right) + \left(\frac{d}{dt} G_{v\dot{v}}(t,0,0) - G_{vv}(t,0,0)\right) h = 0. \qquad \text{(A7.11.1)}$$

In view of (A7.10.4), we have

$$G_{\dot{v}\dot{v}} = (\dot{x}_o^2 + \dot{y}_o^2)^2 F_1,$$

and in view of (A7.10.3),

$$\begin{aligned} G_{v\dot{v}} = &-\dot{y}_o(-F_{x\dot{x}}\dot{y}_o + F_{x\dot{y}}\dot{x}_o) + \dot{x}_o(-F_{y\dot{x}}\dot{y}_o + F_{y\dot{y}}\dot{x}_o) \\ &- \ddot{y}_o(-F_{\dot{x}\dot{x}}\dot{y}_o + F_{\dot{x}\dot{y}}\dot{x}_o) + \ddot{x}_o(-F_{\dot{y}\dot{x}}\dot{y}_o + F_{\dot{y}\dot{y}}\dot{x}_o) \end{aligned} \qquad \text{(A7.11.2)}$$

and

$$\begin{aligned} G_{vv} = &-\dot{y}_o(-F_{xx}\dot{y}_o + F_{xy}\dot{x}_o - F_{x\dot{x}}\ddot{y}_o + F_{x\dot{y}}\ddot{x}_o) \\ &+ \dot{x}_o(-F_{yx}\dot{y}_o + F_{yy}\dot{x}_o - F_{y\dot{x}}\ddot{y}_o + F_{y\dot{y}}\ddot{x}_o) \\ &- \ddot{y}_o(-F_{\dot{x}x}\dot{y}_o + F_{\dot{x}y}\dot{x}_o - F_{\dot{x}\dot{x}}\ddot{y}_o + F_{\dot{x}\dot{y}}\ddot{x}_o) \\ &+ \ddot{x}_o(-F_{\dot{y}x}\dot{y}_o + F_{\dot{y}y}\dot{x}_o - F_{\dot{y}\dot{x}}\ddot{y}_o + F_{\dot{y}\dot{y}}\ddot{x}_o), \end{aligned} \qquad \text{(A7.11.3)}$$

where $(t,0,0)$ is to be taken as the argument of G and its partial derivatives and where $(x_o(t),y_o(t),\dot{x}_o(t),\dot{y}_o(t))$ is to be taken as the argument of F and its partial derivatives.

Since we need $(d/dt)\, G_{v\dot{v}}$ for the Jacobi equation (A7.11.1), we see that we have to assume now that $x_o,y_o \in C^3[t_a,t_b]$.

By Theorem 7.3, $v = 0$ ceases to yield a weak relative minimum for $\bar{I}[v]$ if (t_a,t_b) contains a conjugate point to t_a. Hence, by the same argument as in the preceding section, γ_o will cease to yield a weak relative minimum for $I[\gamma]$ in this case, and we obtain as the *necessary condition for a weak relative minimum of $I[\gamma]$, that any solution $h = h(t)$ of (A7.11.1) that vanishes at t_a shall not vanish again in (t_a,t_b).*

As a first example, we consider again the geodesics on a right circular cylinder of radius 1 with the z axis as axis. By (A7.10.7), these geodesics are the solutions of the variational problem

$$\int_{P_a}^{P_b} \sqrt{\varphi'^2 + z'^2}\, dt \to \text{minimum}, \qquad P_a(a,z_a), \qquad P_b(b,z_b).$$

We saw in the last section that

$$F_1 = \frac{1}{(\varphi'^2 + z'^2)^{3/2}},$$

and hence, by (A7.10.4),

$$G_{\dot{v}\dot{v}} = \sqrt{\varphi'^2 + z'^2}.$$

In order to find G_{vv} and $G_{v\dot{v}}$ by (A7.11.2) and (A7.11.3), we need

$$F_{zz} = 0, \qquad F_{z\varphi} = 0, \qquad F_{zz'} = 0, \qquad F_{z\varphi'} = 0,$$

$$F_{\varphi\varphi} = 0, \qquad F_{\varphi z} = 0, \qquad F_{\varphi z'} = 0, \qquad F_{\varphi\varphi'} = 0,$$

$$F_{\varphi' z'} = -\frac{z'\varphi'}{(\varphi'^2 + z'^2)^{3/2}}.$$

Since the cylinder is a developable surface, the extremals through P_a (geodesics) are given by $\varphi = \lambda_1(t - t_a) + a$, $z = \lambda_2(t - t_a) + z_a$, (see Fig. 7.5), and we obtain

$$G_{\dot{v}\dot{v}}(t,0,0) = \sqrt{\lambda_1^2 + \lambda_2^2},$$

$$G_{v\dot{v}}(t,0,0) = 0,$$

$$G_{vv}(t,0,0) = 0.$$

Thus, we have as Jacobi equation

$$\sqrt{\lambda_1^2 + \lambda_2^2}\, h'' = 0$$

with the solution $h = \alpha t + \beta$, and we see that any solution of the Jacobi equation vanishes at most once and that there is no conjugate point to $t = t_a$ on any of the helices $\varphi = \lambda_1(t - t_a) + a$, $z = \lambda_2(t - t_a) + z_a$.

As a second example, we consider the geodesics on the sphere. These are, by (A7.10.8), the solutions of the variational problem

$$\int_{P_a}^{P_b} \sqrt{\theta'^2 + \sin^2\theta\varphi'^2}\, dt \to \text{minimum}, \qquad P_a(a,\theta_a), \qquad P_b(b,\theta_b).$$

We saw in the preceding section that

$$F_1 = \frac{\sin^2\theta}{(\theta'^2 + \varphi'^2 \sin^2\theta)^{3/2}},$$

and hence, by (A7.10.4),

$$G_{\dot v\dot v} = \frac{\sin^2\theta(\theta'^2 + \varphi'^2)^2}{(\theta'^2 + \varphi'^2 \sin^2\theta)^{3/2}}.$$

It is well known† that arcs of great circles are the geodesics on the sphere. We shall investigate, in particular, the *equator* $\varphi = t - t_a + a$, $\theta = \pi/2$, and a *meridian* $\varphi = 0$, $\theta = t - t_a + \theta_a$. Then we have

$$G_{\dot v\dot v}\,|_{\text{equator}} = 1, \qquad G_{\dot v\dot v}\,|_{\text{meridian}} = \sin^2(t - t_a + \theta_a).$$

In order to find G_{vv} and $G_{v\dot v}$, we need

$$F_{\varphi\theta} = 0, \qquad F_{\varphi\theta'} = 0, \qquad F_{\varphi\varphi} = 0, \qquad F_{\varphi\varphi'} = 0,$$

$$F_{\theta\varphi'} = \frac{\varphi'\sin\theta\cos\theta(2\theta'^2 + (\sin^2\theta)\varphi'^2)}{(\theta'^2 + (\sin^2\theta)\varphi'^2)^{3/2}},$$

$$F_{\theta\theta'} = -\frac{\sin\theta\cos\theta\varphi'^2\theta'}{(\theta'^2 + \sin^2\theta\varphi'^2)^{3/2}},$$

$$F_{\theta\theta} = \frac{4\cos 2\theta\varphi'^2(\theta'^2 + \sin^2\theta\varphi'^2) - \sin^2 2\theta\varphi'^4}{4(\theta'^2 + \varphi'^2\sin^2\theta)^{3/2}}.$$

From Sec. A7.10, we have

$$F_{\theta'\theta'} = \frac{\varphi'^2\sin^2\theta}{(\theta'^2 + \varphi'^2\sin^2\theta)^{3/2}}, \qquad F_{\varphi'\varphi'} = \frac{\theta'^2\sin^2\theta}{(\theta'^2 + \varphi'^2\sin^2\theta)^{3/2}}$$

and, finally,

$$F_{\theta'\varphi'} = -\frac{\theta'\varphi'\sin^2\theta}{(\theta'^2 + \varphi'^2\sin^2\theta)^{3/2}}.$$

Hence, we have on the *equator* $\varphi = t - t_a + a$, $\theta = \pi/2$,

$$F_{\varphi\theta} = 0, \quad F_{\varphi\theta'} = 0, \quad F_{\varphi\varphi} = 0, \quad F_{\varphi\varphi'} = 0, \quad F_{\theta\varphi'} = 0, \quad F_{\theta\theta'} = 0,$$
$$F_{\theta\theta} = -1, \quad F_{\theta'\theta'} = 1, \quad F_{\varphi'\varphi'} = 0, \quad F_{\theta'\varphi'} = 0$$

and on the *meridian* $\varphi = 0$, $\theta = t - t_a + \theta_a$,

$$F_{\varphi\theta} = 0, \quad F_{\varphi\theta'} = 0, \quad F_{\varphi\varphi} = 0, \quad F_{\varphi\varphi'} = 0, \quad F_{\theta\varphi} = 0, \quad F_{\theta\theta'} = 0,$$
$$F_{\theta\theta} = 0, \quad F_{\theta\varphi'} = 0, \quad F_{\theta'\theta'} = 0, \quad F_{\varphi'\varphi'} = \sin^2(t - t_a + \theta_a), \quad F_{\theta'\varphi'} = 0.$$

Consequently, we have from (A7.11.2) and (A7.11.3),

$$G_{v\dot v}\,|_{\text{equator}} = 0, \qquad G_{vv}\,|_{\text{equator}} = -1$$
$$G_{v\dot v}\,|_{\text{meridian}} = 0, \qquad G_{vv}\,|_{\text{meridian}} = 0.$$

† If not, see A. R. Forsyth, "Calculus of Variations," p. 105, Dover Publications, Inc., New York, 1960.

Figure 7.5

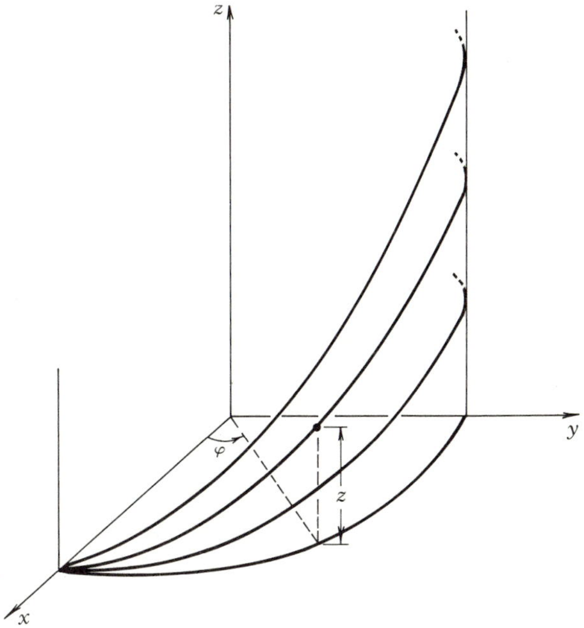

Hence, we obtain for the Jacobi equation (A7.11.1) on the *equator* $\varphi = t - t_a + a$, $\theta = \pi/2$,

$$h'' + h = 0$$

with the solution $h = \sin(t - t_a)$. We see that to every point P_a on the equator, there is a conjugate point halfway around the equator, and any arc of a great circle that is longer than half the great circle does not yield a relative minimum anymore.

Since $F_1 = 0$ for $t - t_a + \theta_a = 0, \pi$ on the meridian, we cannot carry out the Jacobi transformation in the entire interval $[0,\pi]$ but only in $0 < t - t_a + \theta_a < \pi$, that is, north pole and south pole have to be excluded.

We obtain for the Jacobi equation on the *meridian* in $(0,\pi)$

$$\frac{d}{dt}(\sin^2(t - t_a + \theta_a)h') = 0$$

and hence

$$h = \alpha \cot(t - t_a + \theta_a),$$

and we see that there is no conjugate point to any point $\varphi = 0$, $\theta = \theta_a$ on the meridian between north pole and south pole.

A shortcut to the computation of conjugate points is frequently achieved by the following procedure:

Suppose that a two-parameter family of solutions $v = v(t,a,b)$ of the

Euler-Lagrange equation of the problem $\bar{I}[v] \to$ minimum is known and contains the extremal $v = 0$ for the parameter values (a_o,b_o). Then $(\partial v/\partial a)_{a_o,b_o}$, $(\partial v/\partial b)_{a_o,b_o}$ are solutions of the Jacobi equation (A7.11.1). (See Prob. 3.5.4.)

The Euler-Lagrange equation of $\bar{I}[v] \to$ minimum is given by

$$G_v - \frac{d}{dt} G_{\dot{v}} = 0,$$

where

$$G_v = -\dot{y}_o F_x + \dot{x}_o F_y - \ddot{y}_o F_{\dot{x}} + \ddot{x}_o F_{\dot{y}}$$

$$G_{\dot{v}} = -\dot{y}_o F_{\dot{x}} + \dot{x}_o F_{\dot{y}}.$$

The geodesics on the right circular cylinder are given by $\varphi = \lambda_1(t - t_a) + a$, $z = \lambda_2(t - t_a) + z_a$. Hence the transformation (A7.10.1) will assume the form

$$\varphi = \lambda_1(t - t_a) + a - \lambda_2 v$$

$$z = \lambda_2(t - t_a) + z_a + \lambda_1 v,$$

and we obtain

$$G(t,v,\dot{v}) = \sqrt{\left(\lambda_1 - \lambda_2 \frac{dv}{dt}\right)^2 + \left(\lambda_2 + \lambda_1 \frac{dv}{dt}\right)^2}$$

and hence

$$G_v(t,v,\dot{v}) = 0,$$

$$G_{\dot{v}}(t,v,\dot{v}) = \frac{(\lambda_1^2 + \lambda_2^2)\dfrac{dv}{dt}}{\sqrt{\left(\lambda_1 - \lambda_2 \dfrac{dv}{dt}\right)^2 + \left(\lambda_2 + \lambda_1 \dfrac{dv}{dt}\right)^2}}.$$

A first integral of the Euler-Lagrange equation is given by

$$G_{\dot{v}} = \text{constant},$$

from which, in turn, we find $\dot{v} = \alpha$ and hence $v = \alpha t + \beta$.

Therefore, $h_1 = \partial v/\partial\alpha \equiv t$, $h_2 = \partial v/\partial\beta \equiv 1$ are two linearly independent solutions of the Jacobi equation, and we obtain $h = \alpha t + \beta$ as the general solution of the Jacobi equation. This is, of course, the same solution that was obtained before, but this time we found it with much less effort.

PROBLEMS A7.11

1. Show that the geodesics on the sphere, other than the meridians, can be written in the form

$$\theta = \text{arc cot } (\alpha \cos \varphi + \beta \sin \varphi).$$

2. In the problem of finding the geodesics on the right circular cylinder, one would naively expect a conjugate point to $t = 0$ on $\varphi = t$, $z = 0$ at $t = \pi$. We have seen that there are no conjugate points on any of the geodesics on the cylinder. Explain.
3. Transform the variational problem of the geodesics on the sphere into a problem of the form $\bar{I}[v] \to$ minimum by means of the transformation (A7.10.1) with

$\varphi_o = 0$, $\theta_o = t$. Solve the Euler-Lagrange equation of the transformed problem, and find two solutions of the Jacobi equation (A7.11.1) from a two-parameter family of solutions of the Euler-Lagrange equation.

4. Formulate the problem of finding the geodesics on a surface $x = x(u,v)$, $y = y(u,v)$, $z = z(u,v)$.
5. Specialize the result of the preceding problem for the case where the surface is a surface of revolution with the z axis as axis and find F_1.
6. Let $v = v(t,c)$ represent a one-parameter family of solutions of the Euler-Lagrange equation $G_v - (d/dt)\, G_{\dot v} = 0$ of the variational problem $\bar{I}[v] \to$ minimum. Assume that $v = 0$ is contained in this family for $c = c_o$. Find the image $x = x(\tau,c)$, $y = y(\tau,c)$ of this family in the x,y plane, where $\tau = \tau(t,c)$, $\partial\tau/\partial t > 0$, $\tau(t,c_o) = t$, and show that

$$\frac{\partial(x,y)}{\partial(\tau,c)} = \frac{\partial v}{\partial c} \cdot \frac{\partial t}{\partial \tau} \cdot \frac{\partial(x,y)}{\partial(t,v)},$$

where $\partial(x,y)/\partial(t,v)$ is the Jacobian of the transformation (A7.10.1).

7. $h = (\partial v/\partial c)_{c_o}$, as defined in the preceding problem, is a solution of the Jacobi equation (A7.11.1). Show that if

$$\left(\frac{\partial v}{\partial c}\right)_{c_o,t_a} = 0 \qquad \text{but} \qquad \left(\frac{\partial v}{\partial c}\right)_{c_o} \neq 0 \qquad \text{for all } t \in (t_a,t_b],$$

then, under suitable continuity assumptions, it is possible to embed the extremal arc γ_0: $x = x_o(t)$, $y = y_o(t)$, $t \in [t_a,t_b]$, in a field that satisfies the requirements of Definition 4.9. (*Hint*: Utilize the result of problem 6 and apply Theorem 3.6.2.)

BIBLIOGRAPHY

TEXT AND REFERENCE BOOKS ON CALCULUS OF VARIATIONS, OPTIMAL CONTROL THEORY, AND DYNAMIC PROGRAMMING

N. I. Akhiezer: The Calculus of Variations
Translated from the Russian by Aline H. Frink
Blaisdell Publishing Company, a division of Ginn and Company
Boston, 1962

M. Athans and P. L. Falb: Optimal Control
McGraw-Hill Book Company
New York, 1966

R. Pallu De La Barrière: Optimal Control Theory
Translated from the French
W. B. Saunders Company
Philadelphia, 1967

R. E. Bellman: Dynamic Programming
Princeton University Press,
Princeton, N.J., 1957

R. E. Bellman: Adaptive Control Processes: A Guided Tour
Princeton University Press,
Princeton, N.J., 1961

R. E. Bellman: Introduction to the Mathematical Theory of Control Processes—Linear Equations and Quadratic Criteria
Academic Press Inc.
New York, 1967

R. E. Bellman and S. E. Dreyfus: Applied Dynamic Programming
Princeton University Press
Princeton, N.J., 1962

G. A. Bliss: Calculus of Variations
The Open Court Publishing Company
La Salle, Ill., 1935

G. A. Bliss: Lectures on the Calculus of Variations
University of Chicago Press
Chicago, 1946

O. Bolza: Lectures on the Calculus of Variations
Dover Publications Inc.
New York, 1961

O. Bolza: Vorlesungen über Variationsrechnung
B. G. Teubner, Verlagsgesellschaft, mbH
Stuttgart, 1909

C. Carathéodory: Calculus of Variations and Partial Differential Equations of the First Order
Volumes 1 and 2
Translated from the German by R. B. Dean and J. J. Brandstatter
Holden-Day, Inc.
San Francisco, 1965, 1967

P. Cicala: An Engineering Approach to the Calculus of Variations
Libreria Editrice Universitaria Levrotto & Bella
Torino, Italy, 1964

S. E. Dreyfus: Dynamic Programming and the Calculus of Variations
Academic Press Inc.
New York, 1965

L. E. El'sgol'c: Calculus of Variations
Translated from the Russian
Addison-Wesley Publishing Company, Inc.
Reading, Mass., 1962

A. A. Fel'dbaum: Optimal Control Systems
Translated from the Russian by A. Kraiman
Academic Press Inc.
New York, 1965

A. R. Forsyth: Calculus of Variations
Dover Publications, Inc.
New York, 1960

C. Fox: An Introduction to the Calculus of Variations
Oxford University Press
Fair Lawn, N.J., 1950

P. Funk: Variationsrechnung und ihre Anwendung in Physik und Technik
Springer-Verlag, OHG
Berlin, 1962

I. M. Gelfand and S. V. Fomin: Calculus of Variations
Translated from the Russian by R. A. Silverman
Prentice-Hall, Inc.
Englewood Cliffs, N.J., 1963

S. H. Gould: Variational Methods for Eigenvalue Problems
University of Toronto Press
Toronto, Canada, 1957

J. S. Hadamard: Leçons sur le calcul des variations
A. Herman and Fils
Paris, 1910

M. R. Hestenes: Calculus of Variations and Optimal Control Theory
John Wiley & Sons, Inc.
New York, 1966

A. Kneser: Lehrbuch der Variationsrechnung
Friedr. Vieweg & Sohn
Braunschweig, Germany, 1900, 1925

C. Lanczos: The Variational Principles of Mechanics
University of Toronto Press
Toronto, Canada, 1949

E. B. Lee and L. Markus: Foundations of Optimal Control Theory
John Wiley & Sons, Inc.
New York, 1967

C. B. Morrey, Jr.: Multiple Integral Problems in the Calculus of Variations and Related Topics
University of California Press,
Berkeley, Cal., 1943

M. Morse: The Calculus of Variations in the Large
American Mathematical Society
Providence, R.I., 1934

F. D. Murnaghan: The Calculus of Variations
Spartan Books
Washington, D.C., 1962

L. A. Pars: An Introduction to the Calculus of Variations
John Wiley & Sons, Inc.
New York, 1963

Iu. P. Petrov: Variational Methods in Optimum Control Theory
Translated from the Russian by M. D. Friedman
Academic Press Inc.
New York, 1968

L. S. Pontryagin, V. G. Boltyanskii, R. V. Gamkrelidze, E. F. Mishchenko: The Mathematical Theory of Optimal Processes
Translated from the Russian by K. N. Trirogoff
Interscience Publishers
New York, 1962

H. Rund: The Hamilton-Jacobi Theory in the Calculus of Variations
D. Van Nostrand Company, Inc.
Princeton, N.J., 1966

L. Tonelli: Foudamenti di Calcolo delle Variazioni
N. Zanichelli
Bologna, Italy, 1921, 1923

M. M. Vainberg: Variational Methods for the Study of Nonlinear Operators
Translated from the Russian by Amiel Feinstein
Holden-Day, Inc.
San Francisco, 1964

R. Weinstock: Calculus of Variations, with Applications to Physics and Engineering
McGraw-Hill Book Company
New York, 1952

L. C. Young: Calculus of Variations and Optimal Control Theory
W. B. Saunders Company
Philadelphia, 1969

INDEX

INDEX

A CATALOG OF SELECTED

DOVER BOOKS

IN SCIENCE AND MATHEMATICS

RELATIVITY, THERMODYNAMICS AND COSMOLOGY, Richard C. Tolman. Landmark study extends thermodynamics to special, general relativity; also applications of relativistic mechanics, thermodynamics to cosmological models. 501pp. 5⅜ × 8½. 65383-8 Pa. $12.95

APPLIED ANALYSIS, Cornelius Lanczos. Classic work on analysis and design of finite processes for approximating solution of analytical problems. Algebraic equations, matrices, harmonic analysis, quadrature methods, much more. 559pp. 5⅜ × 8½. 65656-X Pa. $12.95

SPECIAL RELATIVITY FOR PHYSICISTS, G. Stephenson and C.W. Kilmister. Concise elegant account for nonspecialists. Lorentz transformation, optical and dynamical applications, more. Bibliography. 108pp. 5⅜ × 8½. 65519-9 Pa. $4.95

INTRODUCTION TO ANALYSIS, Maxwell Rosenlicht. Unusually clear, accessible coverage of set theory, real number system, metric spaces, continuous functions, Riemann integration, multiple integrals, more. Wide range of problems. Undergraduate level. Bibliography. 254pp. 5⅜ × 8½. 65038-3 Pa. $7.95

INTRODUCTION TO QUANTUM MECHANICS With Applications to Chemistry, Linus Pauling & E. Bright Wilson, Jr. Classic undergraduate text by Nobel Prize winner applies quantum mechanics to chemical and physical problems. Numerous tables and figures enhance the text. Chapter bibliographies. Appendices. Index. 468pp. 5⅜ × 8½. 64871-0 Pa. $11.95

ASYMPTOTIC EXPANSIONS OF INTEGRALS, Norman Bleistein & Richard A. Handelsman. Best introduction to important field with applications in a variety of scientific disciplines. New preface. Problems. Diagrams. Tables. Bibliography. Index. 448pp. 5⅜ × 8½. 65082-0 Pa. $11.95

MATHEMATICS APPLIED TO CONTINUUM MECHANICS, Lee A. Segel. Analyzes models of fluid flow and solid deformation. For upper-level math, science and engineering students. 608pp. 5⅜ × 8½. 65369-2 Pa. $13.95

ELEMENTS OF REAL ANALYSIS, David A. Sprecher. Classic text covers fundamental concepts, real number system, point sets, functions of a real variable, Fourier series, much more. Over 500 exercises. 352pp. 5⅜ × 8½. 65385-4 Pa. $9.95

PHYSICAL PRINCIPLES OF THE QUANTUM THEORY, Werner Heisenberg. Nobel Laureate discusses quantum theory, uncertainty, wave mechanics, work of Dirac, Schroedinger, Compton, Wilson, Einstein, etc. 184pp. 5⅜ × 8½. 60113-7 Pa. $4.95

INTRODUCTORY REAL ANALYSIS, A.N. Kolmogorov, S.V. Fomin. Translated by Richard A. Silverman. Self-contained, evenly paced introduction to real and functional analysis. Some 350 problems. 403pp. 5⅜ × 8½. 61226-0 Pa. $9.95

PROBLEMS AND SOLUTIONS IN QUANTUM CHEMISTRY AND PHYSICS, Charles S. Johnson, Jr. and Lee G. Pedersen. Unusually varied problems, detailed solutions in coverage of quantum mechanics, wave mechanics, angular momentum, molecular spectroscopy, scattering theory, more. 280 problems plus 139 supplementary exercises. 430pp. 6½ × 9¼. 65236-X Pa. $11.95

NUMERICAL METHODS FOR SCIENTISTS AND ENGINEERS, Richard Hamming. Classic text stresses frequency approach in coverage of algorithms, polynomial approximation, Fourier approximation, exponential approximation, other topics. Revised and enlarged 2nd edition. 721pp. 5⅜ × 8½.
65241-6 Pa. $14.95

THEORETICAL SOLID STATE PHYSICS, Vol. I: Perfect Lattices in Equilibrium; Vol. II: Non-Equilibrium and Disorder, William Jones and Norman H. March. Monumental reference work covers fundamental theory of equilibrium properties of perfect crystalline solids, non-equilibrium properties, defects and disordered systems. Appendices. Problems. Preface. Diagrams. Index. Bibliography. Total of 1,301pp. 5⅜ × 8½. Two volumes. Vol. I 65015-4 Pa. $12.95
Vol. II 65016-2 Pa. $12.95

OPTIMIZATION THEORY WITH APPLICATIONS, Donald A. Pierre. Broad-spectrum approach to important topic. Classical theory of minima and maxima, calculus of variations, simplex technique and linear programming, more. Many problems, examples. 640pp. 5⅜ × 8½. 65205-X Pa. $13.95

THE MODERN THEORY OF SOLIDS, Frederick Seitz. First inexpensive edition of classic work on theory of ionic crystals, free-electron theory of metals and semiconductors, molecular binding, much more. 736pp. 5⅜ × 8½.
65482-6 Pa. $15.95

ESSAYS ON THE THEORY OF NUMBERS, Richard Dedekind. Two classic essays by great German mathematician: on the theory of irrational numbers; and on transfinite numbers and properties of natural numbers. 115pp. 5⅜ × 8½.
21010-3 Pa. $4.95

THE FUNCTIONS OF MATHEMATICAL PHYSICS, Harry Hochstadt. Comprehensive treatment of orthogonal polynomials, hypergeometric functions, Hill's equation, much more. Bibliography. Index. 322pp. 5⅜ × 8½. 65214-9 Pa. $9.95

NUMBER THEORY AND ITS HISTORY, Oystein Ore. Unusually clear, accessible introduction covers counting, properties of numbers, prime numbers, much more. Bibliography. 380pp. 5⅜ × 8½. 65620-9 Pa. $8.95

THE VARIATIONAL PRINCIPLES OF MECHANICS, Cornelius Lanczos. Graduate level coverage of calculus of variations, equations of motion, relativistic mechanics, more. First inexpensive paperbound edition of classic treatise. Index. Bibliography. 418pp. 5⅜ × 8½. 65067-7 Pa. $10.95

MATHEMATICAL TABLES AND FORMULAS, Robert D. Carmichael and Edwin R. Smith. Logarithms, sines, tangents, trig functions, powers, roots, reciprocals, exponential and hyperbolic functions, formulas and theorems. 269pp. 5⅜ × 8½. 60111-0 Pa. $5.95

THEORETICAL PHYSICS, Georg Joos, with Ira M. Freeman. Classic overview covers essential math, mechanics, electromagnetic theory, thermodynamics, quantum mechanics, nuclear physics, other topics. First paperback edition. xxiii + 885pp. 5⅜ × 8½. 65227-0 Pa. $18.95

ORDINARY DIFFERENTIAL EQUATIONS, Morris Tenenbaum and Harry Pollard. Exhaustive survey of ordinary differential equations for undergraduates in mathematics, engineering, science. Thorough analysis of theorems. Diagrams. Bibliography. Index. 818pp. 5⅜ × 8½. 64940-7 Pa. $16.95

STATISTICAL MECHANICS: Principles and Applications, Terrell L. Hill. Standard text covers fundamentals of statistical mechanics, applications to fluctuation theory, imperfect gases, distribution functions, more. 448pp. 5⅜ × 8½. 65390-0 Pa. $9.95

ORDINARY DIFFERENTIAL EQUATIONS AND STABILITY THEORY: An Introduction, David A. Sánchez. Brief, modern treatment. Linear equation, stability theory for autonomous and nonautonomous systems, etc. 164pp. 5⅜ × 8¼. 63828-6 Pa. $5.95

THIRTY YEARS THAT SHOOK PHYSICS: The Story of Quantum Theory, George Gamow. Lucid, accessible introduction to influential theory of energy and matter. Careful explanations of Dirac's anti-particles, Bohr's model of the atom, much more. 12 plates. Numerous drawings. 240pp. 5⅜ × 8½. 24895-X Pa. $5.95

THEORY OF MATRICES, Sam Perlis. Outstanding text covering rank, nonsingularity and inverses in connection with the development of canonical matrices under the relation of equivalence, and without the intervention of determinants. Includes exercises. 237pp. 5⅜ × 8½. 66810-X Pa. $7.95

GREAT EXPERIMENTS IN PHYSICS: Firsthand Accounts from Galileo to Einstein, edited by Morris H. Shamos. 25 crucial discoveries: Newton's laws of motion, Chadwick's study of the neutron, Hertz on electromagnetic waves, more. Original accounts clearly annotated. 370pp. 5⅜ × 8½. 25346-5 Pa. $9.95

INTRODUCTION TO PARTIAL DIFFERENTIAL EQUATIONS WITH APPLICATIONS, E.C. Zachmanoglou and Dale W. Thoe. Essentials of partial differential equations applied to common problems in engineering and the physical sciences. Problems and answers. 416pp. 5⅜ × 8½. 65251-3 Pa. $10.95

BURNHAM'S CELESTIAL HANDBOOK, Robert Burnham, Jr. Thorough guide to the stars beyond our solar system. Exhaustive treatment. Alphabetical by constellation: Andromeda to Cetus in Vol. 1; Chamaeleon to Orion in Vol. 2; and Pavo to Vulpecula in Vol. 3. Hundreds of illustrations. Index in Vol. 3. 2,000pp. 6⅛ × 9¼. 23567-X, 23568-8, 23673-0 Pa., Three-vol. set $41.85

ASYMPTOTIC EXPANSIONS FOR ORDINARY DIFFERENTIAL EQUATIONS, Wolfgang Wasow. Outstanding text covers asymptotic power series, Jordan's canonical form, turning point problems, singular perturbations, much more. Problems. 384pp. 5⅜ × 8½. 65456-7 Pa. $9.95

AMATEUR ASTRONOMER'S HANDBOOK, J.B. Sidgwick. Timeless, comprehensive coverage of telescopes, mirrors, lenses, mountings, telescope drives, micrometers, spectroscopes, more. 189 illustrations. 576pp. 5⅜ × 8¼. (USO) 24034-7 Pa. $9.95